2nd International Conference on Food Science and Engineering (ICFSE 2018)

Current Perspectives and Future Challenges in Food Science and Engineering for Sustainable Development

IOP Conference Series: Materials Science and Engineering
Volume 633

Surakarta, Indonesia
25 - 26 September 2018

Editors:

Koen Dewettinck
Jinap Selamat
Jirawat Yongsawatdigul
Sastia Prama Putri
Danar Praseptiangga
Venty Suryanti
Zukhrufuz Zaman Muhammad
Cahyadi
Lilik Retna Kartikasari Ahmad
Marzuki
Gusti Fauza
Elisa Herawati

ISBN: 978-1-5108-9835-6
ISSN: 1757-8981

Printed from e-media with permission by:

Curran Associates, Inc.
57 Morehouse Lane
Red Hook, NY 12571

proceedings.com

Some format issues inherent in the e-media version may also appear in this print version.

Institute of Physics
Dirac House, Temple Back
Bristol BS1 6BE UK

Phone: 44 1 17 929 7481
Fax: 44 1 17 920 0979

techtracking@iop.org

Additional copies of this publication are available from:

Curran Associates, Inc.
57 Morehouse Lane
Red Hook, NY 12571 USA
Phone: 845-758-0400
Fax: 845-758-2633
Email: curran@proceedings.com
Web: www.proceedings.com

Table of contents

Volume 633

International Conference on Food Science and Engineering

25–26 September 2018, Surakarta, Indonesia

Accepted papers received: 29 August 2019
Published online: 8 October 2019

Preface

Papers

Food Microbiology and Biotechnology

Physical Food Analysis and Chemistry

Miscellaneous Topics

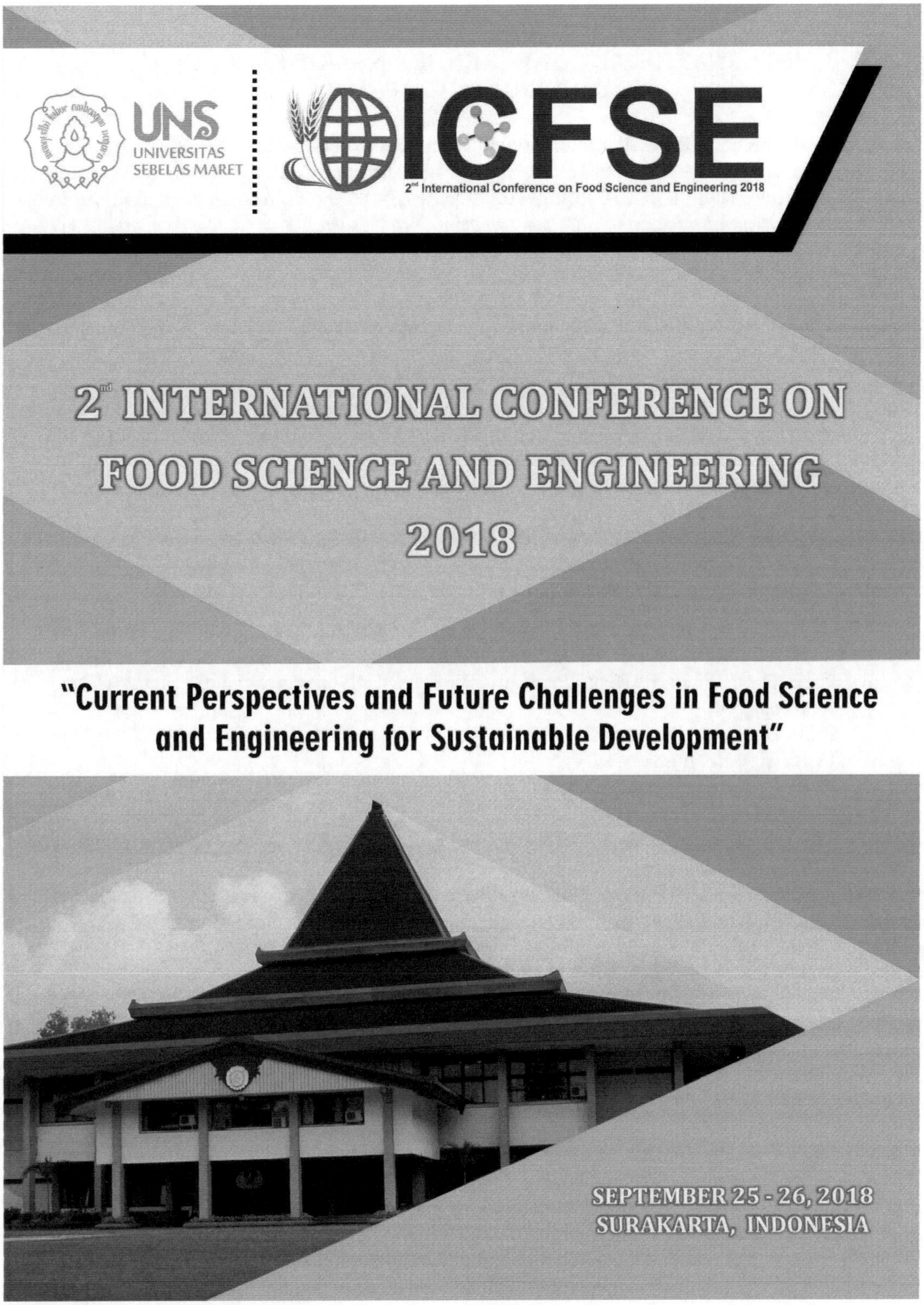
UNS
UNIVERSITAS
SEBELAS MARET
ICFSE
2nd International Conference on Food Science and Engineering 2018
2nd INTERNATIONAL CONFERENCE ON
FOOD SCIENCE AND ENGINEERING
2018
"Current Perspectives and Future Challenges in Food Science
and Engineering for Sustainable Development"
SEPTEMBER 25 - 26, 2018
SURAKARTA, INDONESIA

IOP Conf. Series: Materials Science and Engineering **633** (2019) 011001 doi:10.1088/1757-899X/633/1/011001

PREFACE
2nd INTERNATIONAL CONFERENCE ON FOOD SCIENCE AND ENGINEERING (ICFSE) 2018

Following the success of the first International Conference on Food Science and Engineering (ICFSE) that has been held on 2016, the second International Conference on Food Science and Engineering (ICFSE) 2018 brings the theme of "*Current perspectives and future challenges in food science and engineering for sustainable development*". The transformational vision of the 2030 Agenda for Sustainable Development calls on all countries and stakeholders to work together to end hunger and prevent all forms of malnutrition by 2030. This goal can only be fulfilled if agriculture and food systems become sustainable, so that food supplies are stable and all people have access to adequate nutrition and health. Food nutrition, safety, and security are important factors that required a serious attention. Decreasing amount of arable farmland and the rapid growth of human population forced the experts to find new concepts to feed the world through innovative research in the field of food science and engineering. The effort must be supported by the understanding and cooperation between scholars and practitioners all over the world in facing and solving the problems that might appears. Hence, the 2nd ICFSE 2018 is designed to facilitate the needs of academicians, researchers, policy makers, governmental, and private agencies in disclosing and sharing their knowledge and experiences in current topics and future issues related to food science and engineering, as well as to strengthen the international networking between Universitas Sebelas Maret (UNS) and other universities or institutions worldwide.

The 2nd ICFSE 2018 was attended by leading scientists and researchers working in the field of food science and engineering. The outstanding invited speakers were:

1. Prof. Dr. Sutarno (Department of Biology, Universitas Sebelas Maret (UNS), Indonesia)
2. Prof. Dr. Koen Dewettinck (Department of Food Safety and Food Quality, Ghent University, Belgium)
3. Prof. Dr. Jinap Selamat (Department of Food Science, Universiti Putra Malaysia, Malaysia)
4. Prof. Dr. Senay Simsek (Department of Plant Science, North Dakota State University, USA)
5. Assoc. Prof. Dr. Jirawat Yongsawatdigul (School of Food Technology, Suranaree University of Technology, Thailand)
6. Assist. Prof. Dr. Sastia Prama Putri (Department of Biotechnology, Osaka University, Japan)
7. Assoc. Prof. Dr. Jung Samooel (College of Agriculture and Life Science, Chungnam National University, Korea)

The expertise and experiences of these distinguished speakers in their field of work made this conference excel, and we are very grateful that they are able to participate and to share their best knowledge in the 2nd ICFSE 2018.

In this proceeding, the authors of our accepted papers come from 4 different countries, such as Indonesia, Belgium, Japan, and Thailand. We do realize that the success of this conference lies in the participation of the delegate and the quality of accepted papers. However, we also know this will not possible without helps and efforts put in by the reviewers, sponsors, volunteers as well as the members of the Organizing Committee. Their dedicated contributions to this conference are very much acknowledge and appreciated.

IOP Conf. Series: Materials Science and Engineering **633** (2019) 011001 doi:10.1088/1757-899X/633/1/011001

Date and Venue

Date: September 25-26, 2018
Venue: Solo Paragon Hotel and Residences, Jl. Dr. Soetomo, Solo 57125, Central Java, Indonesia

Danar Praseptiangga, S.T.P., M.Sc., Ph.D.
Chairman of the 2nd ICFSE 2018
Universitas Sebelas Maret (UNS), Indonesia

IOP Conf. Series: Materials Science and Engineering **633** (2019) 011002 doi:10.1088/1757-899X/633/1/011002

ORGANIZING COMMITTEE

Chairman	: Danar Praseptiangga, S.T.P., M.Sc., Ph.D.
Co-chairman	: Prof. Venty Suryanti, S.Si., M.Phil., PhD.
Secretary	: M. Zukhrufuz Zaman, S.P., M.P., Ph.D.
Treasury	: Bayu Setya Hertanto, S.Pt., M.Sc.
	Dr. Sayekti Wahyuningsih, S.Si., M.Si.
Secretariat Team	: Dr. Ahmad Pramono, S.Pt., M.P. Dr. Tetri Widiyani, S.Si., M.Si.

STEERING COMMITTEE

1. Prof. Dr. Ir. Bambang Pujiasmanto, M.S.
2. Prof. Ir. Ari Handono Ramelan, M.Sc. (Hons), Ph.D.
3. Prof. Dr. Ir. Samanhudi, S.P., M.P.
4. Dr. Desi Suci Handayani, S.Si., M.Si.

EDITORIAL BOARDS

1. Prof. Dr. Ir. Koen Dewettinck – Ghent University, Belgium
2. Prof. Dr. Jinap Selamat – Universiti Putra Malaysia, Malaysia
3. Assoc. Prof. Dr. Jirawat Yongsawatdigul – Suranaree University of Technology, Thailand
4. Assist. Prof. Dr. Sastia Prama Putri – Osaka University, Japan
5. Danar Praseptiangga, S.T.P., M.Sc., Ph.D. – Sebelas Maret University, Indonesia
6. Prof. Venty Suryanti, S.Si., M.Phil., PhD – Sebelas Maret University, Indonesia
7. M. Zukhrufuz Zaman, S.P., M.P., Ph.D. – Sebelas Maret University, Indonesia
8. Dr.agr. Muhammad Cahyadi, S.Pt., M. Biotech. – Sebelas Maret University, Indonesia
9. Ir. Lilik Retna Kartikasari, M.P., M.Agr.Sc., Ph.D. – Sebelas Maret University, Indonesia
10. Ahmad Marzuki, S.Si., Ph.D. – Sebelas Maret University, Indonesia
11. Gusti Fauza, S.T., M.T., Ph.D. – Sebelas Maret University, Indonesia
12. Elisa Herawati S.Si., M.Eng., Ph.D. – Sebelas Maret University, Indonesia

Peer review statement

All papers published in this volume of *IOP Conference Series: Materials Science and Engineering* have been peer reviewed through processes administered by the proceedings Editors. Reviews were conducted by expert referees to the professional and scientific standards expected of a proceedings journal published by IOP Publishing.

IOP Conf. Series: Materials Science and Engineering **633** (2019) 012001 doi:10.1088/1757-899X/633/1/012001

Current trends in bioactive peptides from muscle foods and their potential application

J Yongsawatdigul and A Hamzeh

School of Food Technology, Institute of Agricultural Technology
Suranaree University of Technology, Nakhon Ratchasima 30000, Thailand

E-mail: jirawat@sut.ac.th

Abstract. Muscle protein is a valuable source of energy and essential amino acids, which are needed for growth and maintenance of physiological functions. It also provides peptides known as bioactive peptides exerting some health benefits, including antihypertension, antioxidant activity, immunomodulatory activity, and improving brain function, among others. Bioactive peptides can be generated from digestion or enzymatic reaction. Cooking condition is found to critically affect digestibility of chicken breast and release of bioactive peptides. Extreme thermal treatment at 121°C/15 min reduces simulated gastrointestinal digestion (GI) of chicken breast. In addition, release of peptides inhibiting angiotensin converting enzyme (ACE) that regulates blood pressure via renin–angiotensin system decreases. This implies that extreme thermal process of muscle food lower nutritional values and available ACE inhibitor peptides. Enzymatic hydrolysis is also an effective means to produce bioactive peptides, providing appropriate protease(s) is applied under the optimal condition. Hydrolysates or bioactive peptides would likely be modified through GI digestion. Thus, changes of bioactivities upon GI digestion should be taken into consideration for optimization of protein hydrolysate production. Structural changes of bioactive peptides further take place during transepithelial transport. This would definitely affect bioactivities of peptides reaching the target organ in either positive or negative manner. To further develop functional food or nutraceutical products, efficacy of bioactive peptides should be tested in vivo to assure their health benefits.

1. Introduction

Bioactive peptides are protein fragments composed of 2-20 amino acids which can exert a physiological effect, such as antioxidant, antihypertension, immunomodulatory activity [1-3]. These peptides generated from various sources including egg, milk, bean, meat and fish, after being exposed to hydrolysis process. The hydrolysis can be carried out using either enzymes or chemicals, in which chemical hydrolysis of proteins via acid or alkali is less effective due to harsh condition destroying some amino acids [4]. Enzymatic process can be conducted in two categories including addition of exogenous proteases or fermentation by protease-producing microorganisms (Table 1).

Addition of exogenous proteases is more preferable because many factors affecting hydrolysis can be effectively controlled. Production of peptides with high ACE inhibition from Pacific hake can be achieved using Protamex at pH 6.5 for 125 min at 3% enzyme [5]. The optimum conditions to produce antioxidant peptides from duck meat were at 50 °C, pH 7 for 8 h by Protamex [6]. In addition, crude proteases from hepatopancreas of Pacific white shrimp was used to generate antioxidant peptides from splendid squid [7].

Table 1. Bioactive peptides derived from muscle foods by exogenous proteases and fermentation

Origin	Bioactivity	Peptide generation	Reference
Dry-cured beef	ACE-inhibitory Antioxidant	Aging	[8]
Beef	ACE-inhibitory Antioxidant	Aging	[9]
Duck meat	Antioxidant	Aging	[10]
Dry sausage (pork)	ACE-inhibitory Antioxidant	Fermentation	[11]
Shrimp	ACE-inhibitory Antioxidant	Fermentation	[12]
Fish	Antioxidant	Fermentation	[13]
Pork	ACE-inhibitory	Pepsin	[14]
Fish	DPP-IV and ACE inhibitory	Alkalase, Flavorzyme, Corolase PP and Promod 144	[15]
Chicken	Antioxidant	Papain	[16]

Bioactive peptides can also be released by gastrointestinal digestion of muscle foods. In this regard, food processing and preparation play an important role in generation of peptides with certain bioactivities [17]. Sangsawad, Roytrakul and Yongsawatdigul [18] reported that peptides with higher ACE inhibition ability could be generated from chicken breast as exposed to a mild heat at 70 °C for 30 min, while an extreme heat at 121 °C resulted in peptides with lower activity.

2. Purification and identification of bioactive peptides

Hydrolysates prepared from a proteinous substrates contain a mixture of active and inactive peptides with different properties of size, charge, hydrophobicity, and others [19]. To separate and purify bioactive peptides, chromatographic techniques are used, in which a consecutive chromatography including size exclusion chromatography (SEC), ion exchange chromatography (IEC), hydrophobic interaction chromatography (HIC) and reversed phase chromatography (RPC) can be applied to obtain the most active peptides. SEC, IEC were applied to separate peptides based on their size and charge characteristics, respectively. Hydrophobicity separation of peptides can be conducted by HIC and RPC. Liquid chromatography tandem-mass spectrometry (LC-MS/MS) is a common method to identify peptide sequence. Some purification and identification methods used in bioactive peptides derived from muscle foods are shown in Table 2.

Table 2. Purification schemes and identification applied for bioactive peptides derived from various muscle foods.

Source	Bioactivity	P[a]. and I[b]. Methods	Identified peptide	Reference
Tuna dark muscle	Antiproliferation	SEC (Sephadex G-25), RPC[c] (C-18)/ LC-MS/MS	LPHVLTPEAGAT PTAEGVYMVT	[20]
Croceine croaker	Antioxidant	UF[d], AEC (DEAE-52 cellulose), SEC (Sephadex G-15), RPC (C-18)/ LC-MS/MS	YLMSR VLYEE MILMS	[21]
Beef	ACE-inhibitory	UF, SEC (Sephadex G-25), RPC (C-18)/ Peptide sequencer	VLAQYK	[22]
Silver carp	DPP-IV inhibitory	UF, TLC[e](silica gel), RPC (C-18)/ LC-MS/MS	LPIIDI APGPAGP	[23]
Chicken	ACE-inhibitory	IEC[f], SEC, RPC/ LC-MS/MS	KPLLCS ELFTT KPLL	[18]
Alaska pollock	Immunomodulatory	SEC, RPC/ LC-MS/MS	PTGADY	[24]

[a]Purification. [b]Identification. [c]Reverse phase chromatography. [d]Ultrafiltration. [e]Thin layer chromatography. [f]Ion exchange chromatography.

3. Gastrointestinal (GI) digestion of bioactive peptides

GI digestion and intestinal epithelial absorption in human tract is the great challenges for bioavailability of peptides [25]. After oral ingestion, gastrointestinal enzymes may break down peroteins/peptides, thereby increasing or decreasing their activity. It has been reported that the GI digestion improved bioactivities of pork and chicken meat [26,18]. Peptides prepared from Tilapia showed a reduction in ACE inhibitory after GI digestion [3]. Some researchers reported an increase in bioactivities of peptides after in vitro GI digestion such as antioxidant activity of hydrolysates prepared from splendid squid [7] and poultry proteins [27]. Peptides can be hydrolyzed in stomach by pepsin, but the main hydrolsis takes place in intestine by pancreatin. You et al. [28] reported an increase in DH of fish protein hydrolysates from 37.4 to 38% at stomach digestion in 2 h, while the DH raised up to 46.6% after pancreatin digestion at the same incubation time.

4. Peptide transport

Transport of peptides across intestine into blood is another challenge for bioactive peptides to reach to target organ and exert a physiological function. Peptides can pass through intestinal cells via four mechanisms including para-cellular, passive diffusion, endocytosis and carrier mediated transport. Smaller peptides (di or/and tri) can pass easily, however there is some differences between adults and infants due to incomplete development of GI tract in infant [3,5,29]. Toopcham et al. [3] reported that tri and tetra-peptides of tilapia hydrolysate could pass through Caco-2 cell line. Sangsawad et al. [30] identified 9 transepithelial transported tri-peptides from chicken breast, in which APP had the greatest ability to inhibit ACE. Peptides structure can also be transformed during transport by brush border enzymes. Sangsawad et al. [30] reported that KPLL was degraded to KP and LL by brush border proteases. Peptides derived from dry-cured ham, AAATP, AAPLAP and KPVAAP have been affected by brush border proteases, resulting in generation of mostly di and tri-peptides [31]. Escudero, Toldrá, Sentandreu, Nishimura and Arihara [32] reported that KAPVA, PTPVP, and RPR from pork meat could decrease blood pressure in rats. Three peptides, ALTA, SLTA and VT, from pork actomyosin showed antioxidant effects in vitro and in vivo in rats [33].

5. Conclusions

Bioactive peptides can be applied in food formula as nutraceutical compounds, exerting various health benefits. Among various substrates, muscle food can be a good candidate to use for bioactive peptides generation because of their high protein content. The challenge with these peptides is that their structure is likely to be affected by digestive proteases and epithelial transportation. Study of transport would shed lights on how peptides are modified. Although peptides might be modified in GI tract, bioactivities of peptides have been shown in vivo, suggesting that permeated peptides in blood in spite of possible structural change could still exhibit functions.

References

[1] Piyadhammaviboon P, Wongngam W, Benjakul S and Yongsawatdigul J 2012 *J. Aquat. Food Prod. T.* **21** 265–78
[2] Lafarga T, O'Connor P and Hayes M 2014 *Peptides* **59** 53–62
[3] Toopcham T, Mes JJ, Wichers HJ and Yongsawatdigul J 2017 *Food Chem.* **224** 320–8
[4] Kristinsson HG and Rasco BA 2000 *Crit. Rev. Food Sci. Nutr.* **40** 43–81
[5] Cinq-Mars CD and Li-Chan ECY 2007 *J. Agric. Food Chem.* **55** 9380–8
[6] Lee SJ, Kim EK, Hwang JW, Oh HJ, Cheong SH, Moon SH, Jeon BT, Lee SM and Park PJ 2010 *Food Chem.* **123** 216–20
[7] Hamzeh A, Benjakul S and Senphan T 2016 *J. Food Sci. Technol.* **53** 3615–23
[8] Deniz E, Mora L, Aristoy MC, Candoğan K and Toldrá F 2016 *Food Res. Int.* **89** 194–201
[9] Fu Y, Young JF and Therkildsen M 2017 *Meat Sci.* **123** 134–42
[10] Liu D, Chen X, Huang J, Zhou X, Huang M and Zhou G 2017 *Int. J. Food Sci. Technol.* **52** 2513–21
[11] Gallego M, Mora L, Escudero E and Toldrá F 2018 *Int. J. Food Microbiol.* **276** 71–8
[12] Kleekayai T, Harnedy PA, O'Keeffe MB, Poyarkov AA, Cunhaneves A, Suntornsuk W and Fitzgerald RJ 2015 *Food Chem.* **176** 441–7
[13] Giri A, Nasu M and Ohshima T 2012 *Int. J. Food Sci. Nutr.* **1** 13–22
[14] Katayama K, Anggraeni HE, Mori T, Ahhmed AM, Kawahara S, Sugiyama M., Nakayama T, Maruyama M and Muguruma M 2008 *J. Agric. Food Chem.* **56** 355–60
[15] Neves AC, Harnedy PA, O'Keeffe MB, Alashi MA, Aluko RE and FitzGerald RJ 2017 *Food Res. Int.* **100** 112–20
[16] Sun Y, Pan D, Guo, Y and Li J 2012 Purification of chicken breast protein hydrolysate and analysis of its antioxidant activity. *Food Chem. Toxicol.* **50** 3397–404
[17] Dallas DC, Guerrero A, Parker EA, Robinson RC, Gan J, German JB, Barile D and Lebrilla CB 2015 *Proteomics* **15** 1026-38
[18] Sangsawad P, Roytrakul S and Yongsawatdigul J 2017 *J. Funct. Foods* **29** 77–83
[19] Sila A and Bougatef A 2016 *J. Funct. Foods* **21** 10–26
[20] Hsu KC, Li-Chan ECY and Jao CL 2011 *Food Chem.* **126** 617–22
[21] Chi C F, Hu F Y, Wang B, Ren X J, Deng S G and Wu C W 2015 *Food Chem.* **168** 662–7
[22] Jang A and Lee M 2005 *Meat Sci.* **69** 653–61

[23] Zhang Y, Chen R, Chen X, Zeng Z, Ma H and Chen S 2016 *J. Agric. Food Chem.* **64** 831–9
[24] Hou H, Fan Y, Wang S, Si L and Li B 2016 *J. Funct. Foods* **24** 37–47
[25] Ao J and Li B 2013 *Food Res. Int.* **52** 334–41
[26] Escudero E, Sentandreu M A, Arihara K and Toldrá F 2010 *J Agric Food Chem.* **58** 2895–901
[27] Anna T, Alcxcy K, Anna B, Mikhail T and Ulia M 2016 *Curr. Res. Nutr. Food Sci.* **4** 77–86.
[28] You L, Zhao M, Regenstein J M and Ren J 2010 *Food Chem.* **120** 810–6
[29] Vermeirssen V, Camp J Van and Verstraete W 2004 *Br. J. Nutr.* **92** 357–66
[30] Sangsawad P, Roytrakul S, Choowongkomon K, Kitts DD, Chen X M, Meng G, Li-Chan ECY and Yongsawatdigul J 2018 *Food Chem.* **251** 77–85
[31] Gallego M, Mora L, Escudero E and Toldrá F 2018 *Int. J. Food Microbiol.* **276** 71–8
[32] Escudero E, Toldrá F, Sentandreu M A, Nishimura H and Arihara K 2012 *Meat Sci.* **91** 382–4
[33] Arihara K and Ohata M 2006 Advanced technologies for meat processing ed. Toldrá F (New York: Springer) pp. 245-74

IOP Conf. Series: Materials Science and Engineering **633** (2019) 012002 doi:10.1088/1757-899X/633/1/012002

Effect of fermentation time by *Lactobacillus plantarum* FNCC 0027 on chemical, physical and physico-chemical properties of modified breadfruit flour

T Wayantika, D Ishartani, A Nursiwi, M Z Zaman

Department of Food Science and Technology, Sebelas Maret University (UNS), Jl. Ir. Sutami 36 A Kentingan 57126, Surakarta, Indonesia

E-mail : dwiishartani@staff.uns.ac.id

Abstract. Breadfruit contains a high amount of carbohydrate, but it has short shelf life. Processing it into flour has the advantage of extending its shelf life and making the next process easier. The aim of this work was to study the effect of fermentation time by *Lactobacillus plantarum* FNCC 0027 on the chemical, physical and physico-chemical properties of modified breadfruit flour. The experimental design of this research was Completely Randomized Design (CRD) with variation of fermentation time (0, 12, 24, 36, 48, and 60 hours) as the factor. Reducing sugar values had a fluctuative trend. Longer fermentation time produced breadfruit flour with lower bulk density but higher whiteness index. No significant differences were found in the solubility contents. Water absorption, oil absorption, swelling power, gelatinization temperature, gelatinization time and setback viscosity of modified breadfruit flour were higher than unfermented breadfruit flour. Twenty-four hours was chosen as the optimum fermentation time to produce modified breadfruit flour.

1. Introduction

Breadfruit is an agricultural commodity that has the potential to be used as a raw material of carbohydrate source. In addition, breadfruit (*Artocarpus communis*) can be used as an alternative food because it contains 25% carbohydrate, 1.5% protein and 0.3% fat from its weight [13]. Nevertheless, breadfruit has a short shelf live. One alternative method to extend its shelf life is making breadfruit into breadfruit flour. Flour production has advantages such as extending the shelf life of flour and making the next processing step easier. Every 100 grams of breadfruit flour can give three times more calories than ripe breadfruit itself [10]. In addition, making breadfruit flour can increase the diversification of food products from local materials.

Breadfruit flour has been used in processed products such as breadfruit round cake, breadfruit porridge, breadfruit cake, breadfruit cookies, breadfruit cake, breadfruit pastel and fresh role cake [10]. But based on literature research, breadfruit flour as a raw material and substitution for some food products still has shortcomings. The lack of sensory value in some processed breadfruit products is thought to be due to some physical and physico-chemical characteristics of the flour that affect the sensory properties. Amilographic profile, swelling power, water absorption and oil absorption of flour have an effect on texture and general acceptance of the product. Oil absorption of flour also affects the taste of food products, while the whiteness index affects their appearance [18, 5, 20, 22].

IOP Conf. Series: Materials Science and Engineering **633** (2019) 012002 doi:10.1088/1757-899X/633/1/012002

According to Medikasari [15], breadfruit flour still has weaknesses in its functional properties, which are low volume improvement and viscosity breakdown when it is heated at 95°C and stirred mechanically. The lack of functionality of breadfruit flour has also discussed by Budiyati [7], who noted that although it has good water solubility (55.27 g /100 g), breadfruit flour has low swelling power (1.55 g/g). In addition, breadfruit flour has a susceptibility to heat, too high viscosity, low resistance to shearing and high tendency to retrogradation and sineresys.

Lactic acid bacteria can produce enzymes that hydrolyze starch into sugar and then convert it to organic acids, especially lactic acid. It will increase viscosity, gelation ability and rehydration power of produced flour. In addition, the taste will become neutral by being able to cover the taste of the original raw material up to 70%. During the fermentation process, there will be degradation of the color-forming component that can cause brown color when heating, so the brighter flour can be produced.

Fermentation time is one factor to be considered in the fermentation process. According to Nurani et al. [17], modified starch is optimally generated at the stationary phase in which secondary metabolite products are produced, such as enzymes that can hydrolyze complex compounds of flour into simpler compounds. Therefore, this research used 0, 12, 24, 36, 48 and 60 hours of fermentation time, considering that microbial growth has exceeded the stationary phase. The aim of this work was to study the effect of fermentation time by *Lactobacillus plantarum* FNCC 0027 on the chemical, physical and physico-chemical properties of modified breadfruit flour.

2. Materials and Methods

2.1. Making Modified Flour

The breadfruit flesh is reduced in size by slicing vertically with ± 5 mm thickness. The 300 g of breadfruit were soaked in 400 ml sterile aquades with 6% (v/v) starter Lactobacillus plantarum FNCC 0027 and then fermented for 0 hours (control), 12 hours, 24 hours, 48 hours and 60 hours at 37°C. Then it was dried in cabinet dryer (70°C for 4 hours), powdered and sieved (80 mesh).

2.2. Chemical, Physical and Physico-chemical Analysis

Several analyses were conducted on modified breadfruit flour. A chemical properties test was conducted on reducing sugar analysis using Nelson Somogyi's method [21]. Physical properties tests were conducted on bulk density [6] and whiteness index by total colour different method [4]. Physico-chemical properties tests were conducted on gelatinization profile by Rapid Visco Analyzer method, water absorption by centrifugation method [1], solubility by centrifugation method [9] and swelling power by centrifugation method [9].

3. Results and Discussion

3.1. Chemical Characteristics

Table 1. Chemical Properties of Modified Breadfruit Flour

Fermentation Length (hour)	Reducing Sugar (%)
0	$4.86^{bc} \pm 0.63$
12	$5.78^{c} \pm 0.61$
24	$3.45^{a} \pm 0.65$
36	$4.90^{bc} \pm 0.89$
48	$4.46^{b} \pm 0.19$
60	$5.08^{bc} \pm 0.59$

Value in the same line followed by a different letter was significantly different (p=0.05).

Table 1 shows that the lowest reducing sugar content was found in breadfruit flour with fermentation treatment for 24 hours, while the modified breadfruit flour with the highest reducing sugar content was found in the fermentation treatment for 12 hours at 5.78%. The yield trend of reducing sugar obtained tended to fluctuate.

3.2. Physical Analysis

3.2.1. Bulk Density, Whiteness Index, Water Absorption, Oil Absorption, Swelling Power and Solubility

Table 2. Chemical, Physical and Physico-Chemical Properties of Modified Breadfruit Flour

Fermentation Length (hour)	Bulk Density (g/ml)	Whiteness Index	Water Absorption (g/g)	Oil Absorption (g/g)	Swelling Power (g/g)	Solubility (%)
0	$0.62^{c} \pm 0.02$	$84.59^{a} \pm 0.04$	$3.09^{a} \pm 0.33$	$1.13^{a} \pm 0.06$	$1.99^{a} \pm 0.29$	$9.93^{a} \pm 0.09$
12	$0.55^{b} \pm 0.03$	$84.68^{a} \pm 0.02$	$4.94^{b} \pm 0.11$	$1.18^{a} \pm 0.12$	$5.50^{b} \pm 0.53$	$9.92^{a} \pm 0.06$
24	$0.51^{a} \pm 0.03$	$85.15^{b} \pm 0.33$	$4.92^{b} \pm 0.34$	$1.58^{c} \pm 0.17$	$5.49^{b} \pm 0.52$	$9.89^{a} \pm 0.06$
36	$0.50^{a} \pm 0.00$	$85.32^{bc} \pm 0.08$	$4.97^{b} \pm 0.12$	$1.64^{c} \pm 0.13$	$5.96^{b} \pm 0.10$	$9.90^{a} \pm 0.03$
48	$0.51^{a} \pm 0.01$	$85.47^{c} \pm 0.06$	$4.91^{b} \pm 0.17$	$1.39^{b} \pm 0.13$	$5.98^{b} \pm 0.24$	$9.88^{a} \pm 0.01$
60	$0.50^{a} \pm 0.02$	$85.34^{bc} \pm 0.09$	$4.59^{b} \pm 0.26$	$1.29^{ab} \pm 0.06$	$5.39^{b} \pm 0.39$	$9.86^{a} \pm 0.05$

Value in the same line followed by a different letter was significantly different (p=0.05).

Table 2 shows that the longer fermentation, bulk density of breadfruit flour has a trend that tends to decrease. The decrease in bulk density during fermentation can be attributed to the increase in porosity of the flour with the longer fermentation time. According to [2], the longer the fermentation time, the greater the degradation of macromolecules into simpler molecules. The compact macromolecules become porous because they were broken into simpler molecules with small mass. Because of the stretching of the structure of the flour due to its increased porosity, the volume of flour increases, which can cause decreased of bulk density.

The result of the lowest white degree was found in the flour of breadfruit without fermentation (control) that is equal to 84.59, while the highest yield of breadfruit flour with 48 hours fermentation treatment is 85.47. According to [11], breadfruit has riboflavin (yellow pigment), so the degree of whiteness tends to increase during fermentation because the riboflavin is decreased.

The lowest water absorption result was obtained by native breadfruit flour and the highest is shown by 36 hours fermented breadfuit flour. The similar trend was shown on oil absorption result. The lowest oil absorption result was obtained by controlled breadfruit flour (without fermentation) with value of 1.13 g/g. The results obtained were not significantly different from the breadfruit flour with 12 hours of fermentation or breadfruit flour with 60 hours of fermentation. The highest absorption of oil is shown by breadfruit flour with a 36-hour-long fermentation treatment, which is equal to 1.64 g/g. According to [3], with longer fermentation in the flour, the amylose content will increase as the amylopectin content is broken down into amylose. The increased amylose content will lead to an increase in oil absorption.

Table 2 shows that fermented breadfruit flour has a higher swelling power compared to controlled breadfruit flour (without fermentation). The results are relevant to the results of [23], who found that the longer the process of corn fermentation, the greater the swelling power of modified corn flour. This is allegedly because the longer the fermentation process, the more starch is broken down. Starch is able to increase swelling power because broken starch granules bind more easily to water when heated.

Table 2 shows that the fermentation time has no significant effect ($p > 0.05$) on the solubility of modified breadfruit flour. Although statistically not significantly different, this decreasing trend of solubility was due to hydrolysis process during fermentation, which causes a decrease in micellar bond strength in starch granules [15].

3.2.2. *Amilographic Profile*

Table 3. Pasting Properties of Modified Breadfruit Flour in Various Fermentation Time

Fermentation Length (hours)	Gelatinization		Peak Viscosity (cP)			Breakdown Viscosity (cP)
	Temperature (°C)	Time (minutes)	Temperature (°C)	Time (minutes)	Viscosity (cP)	
0	84.20	15.00	-	-	-	4815.00
12	83.10	15.00	-	-	-	3915.00
24	84.90	15.00	-	-	-	3630.00
36	84.20	15.00	-	-	-	3995.00
48	88.60	17.00	-	-	-	3240.00
60	84.40	15.00	-	-	-	4935.00

As Table 3 shows, the gelatinization temperature of the breadfruit control flask was 84.20°C. After 12 hours of fermentation, it decreased to 83.10°C. The other fermentation times did not show a significant decrease or increase (ranging from 84.20°C to 84.90°C) compared to the control breadfruit flour (84.20°C). But after 48 hours of fermentation treatment, the gelatinization temperature of modified starch flour increased to 88.6°C. Because the time of gelatinization can be known, the control flour is different from breadfruit flour after fermentation for 48 hours but not different from other fermentation treatments. This suggests that the fermentation treatment has little effect on the increase or decrease in gelatinization time of modified breadfruit flour. Table 3 shows that the unfermented breadfruit flour (control) has higher breakdown viscosity than the breadfruit flour with fermentation treatment of 12 hour to 48 hours.

4. Conclusion

Reducing sugar value has a fluctuating trend with the highest value in fermentation for 12 hours. The fermentation treatment decreases the bulk density but increases whiteness index compared to native breadfruit flour. The longer the fermentation, the more water absorption, oil absorption and swelling power increase compared with native breadfruit flour. Fermentation did not have a real effect on the solubility value. Based on the gelatinization profile obtained because of the fermentation treatment, gelatinization temperature, gelatinization time and breakdown viscosity tended to decrease compared with unfermented breadfruit flour. Breadfruit modification by fermentation method can increase water absorption, oil absorption, swelling power and whiteness index. This result shows that flour modification by fermentation method has more affect of increasing the physico-chemical properties that the previous studies about breadfruit modification [12, 16, 19]

References

[1] Adebowale Y A, Adeyemi I A , Oshodi, A A 2005 *African Journal of Biotechnology* **4** 1461-1468
[2] Aini N, Wijonarko G, Sustriawan B 2016 *AGRITECH.* **36** 160-168
[3] Akbar M R, Yunianta 2014 *Jurnal Pangan dan Agroindustri* **2** 91-102
[4] Angulo-Bejarano, Paola I, Nadia M, Verdugo-Montoya, Edith O, Rodrı´guez C, Mila´n-Carrillo J, Mora-Escobedo R, Jose A, Valenzuela L, Jose A, Garzo´n-Tiznado, Cuauhte´moc, Reyes-Moreno 2008 *Food Chemistry* **106** 106-112
[5] Borja J C S, Sedano D S, Noel M G 2013 *Functional Properties of Flours Prepared from Glucossinolate – Rich Vegetables; Alugbati (Basellarubra)* Research Congress De La Salle University Manilla Filiphina
[6] Bryant R J, Kadan R S, Champagne E T, Vinyard B T, Boykin D 2001 *Cereal Chemistry* **78** 131-137
[7] Budiyati C S, Kumoro A C, Ratmawati, Retnowati D S 2016 *Jurnal Teknik* **37** 32-40
[8] Collado L S, Mabesa L B, Oates C G, Corke H 2001 *Journal Food Science* **66** 604-609

[9] Darmawan M R, Andreas P, Jos B, Sumardiono S 2013 *Jurnal Teknologi Kimia dan Industri* **2** 137-145
[10] Gardjito M, Djuwardi A, Harmayani E 2013 *Pangan Nusantara* (Jakarta: Penerbit Kencana)
[11] Graham H D, Bravo E N 1981 *Journal of Food Science* **4** 538
[12] Hartanti F D, Amanto B S, Rahardian D 2013 *Jurnal Teknosains Pangan* **4** 54-61
[13] Hendri, Marlina L, Liferdi 2010 *Diversifikasi Pangan dan Gizi dengan Alpukat, Pisang dan Sukun* Seminar Nasional Program dan Strategi Pengembangan Buah Nusantara
[14] Hidayat B, Kalsum N, Surfiana 2009 *Jurnal Teknologi Industri dan Hasil Pertanian* **14** 148-159
[15] Medikasari S N, Yuliana N, Lintang N 2009 *Jurnal Teknologi Industri dan Hasil Pertanian* **4** 173-177
[16] Mutmainah F, Rahardian D, Amanto B S 2013 *Jurnal Teknosains Pangan* **2** 46-53
[17] Numfor F A, Walter Jr W M, Schwartz S J 1995 *Starch/Starke* **47** 86-91
[18] Nurani D, Sukotjo S, Nurmalasari I 2013 *Jurnal IPTEK* **8** 65-71
[19] Pudjihastuti I 2010 *Pengembangan Proses Inovatif Kombinasi Reaksi Hidrolisa Asam dan Reaksi Photokimia UV untuk Produksi Pati Termodifikasi dari Tapioka* Tesis Magister Teknik Kimia Universitas Diponegoro Semarang
[20] Putri W D R, Zubaidah E 2015 *Prosiding Seminar Agroindustri dan Lokakarya nasional FKPT-TPI*
[21] Sahan Y, Gocmen D, Cansev A, Celik G, Aydin E, Ayse N, Dundar, Duger, Kaplan, H P Kilci A Gucer S2015 *Journal of Applied Botany and Food Quality* **88** 34-41
[22] Sudarmadji S, Haryono B, Suhardi 2003 (Yogyakarta: Liberty)
[23] Wang S, Li C, Copeland L, Niu Q, Wang S 2015 *Comprehensive Reviews in Food Science and Food Safety* **14** 568-585
[24] Yuan N E, Taylor J, Patrick R 2008 *Journal of Food Engineering* **85** 12-17

IOP Conf. Series: Materials Science and Engineering **633** (2019) 012003 doi:10.1088/1757-899X/633/1/012003

Rapid identification of *Salmonella enterica* serovars Typhi and *Salmonella enterica* serovars Paratyphi A from chicken meat

D Raharjo[1,2], R Yulistiani[3,4], W Setyarini[2], R Y Arizandy[2], W Prayoga[5], and T Shirakawa[6]

[1]Veterinary Public Health Laboratory, Faculty of Veterinary Medicine, Airlangga University, Surabaya, Indonesia. 60115.
[2]Gastroenteritis and Salmonellosis Laboratory, Institute of Tropical Disease, Airlangga University, Surabaya, Indonesia. 60115
[3]Department of Food Technology, Faculty of Engineering, Universitas Pembangunan Nasional Veteran Jawa Timur,Jalan Rungkut Madya Gunung Anyar, Surabaya, Indonesia. 60294
[4]Graduate School Program of Sebelas Maret University (UNS), JalanIr.Sutami36A, Kentingan, Surakarta,Indonesia, 57126.
[5]Salmonella Research Centre, Animal Biosafety Level 3, Airlangga University, Surabaya, Indonesia. 60115
[6]Graduate School of Science, Technology and Innovation, Kobe University, Kobe Japan

E-mail: Dadik_tdc@yahoo.co.id

Abstract. Rapid diagnosis of pathogenic bacteria in food is closely related to safety. We modified a Culture-based methods for *Salmonella* sp. which rapidly identified *Salmonella enterica* serovars Typhi and *Salmonella enterica* serovars Paratyphi A from chicken meat samples. The World Health Organization method consists of six steps of the test, sequentially pre-enrichment, selective enrichment, selective diagnostic isolation, pick presumptive Salmonella colonies, biochemical and serological confirmation. Modifications made at three stages of pick presumptive Salmonella colonies, biochemical confirmation and serological confirmation into a single stage of molecular testing to detect genes *rfbE* and, using *tyv* and *prt* primer. In summary, from 120 samples, the modified technique successfully identified 39 (32.5%) *Salmonella enterica* serovars Typhi and one *Salmonella enterica* serovars Paratyphi A (0,83%) compare with 13 (10.8%) and negative result with conventional protocol. The modified were faster and more sensitive than conventional techniques.

1. Introduction

Salmonella sp is one the major source to morbidity and mortality in the world, Infections by *Salmonella* manifest as Typhoid fever and Nontyphoidal salmonellae. Typhoid fever, a life-threatening infection caused by *Salmonella enterica* serotype Typhi, every year there are 17 million cases of typhoid fever and 600.000 die. Clinical symptoms include fever, fatigue, headache, nausea, abdominal pain, constipation or diarrhea [1,2]. Nontyphoidal salmonellae estimated to cause 153 million cases of gastroenteritis and 57,000 deaths globally each year. Gastroenteritis is the most common clinical presentation, commonly manifested as acute diarrhea, abdominal pain, fever, and sometimes vomiting. Clinical symptoms of salmonellosis are generally mild and patients will recover without treatment, except in pediatric and elderly patients, severe dehydration and death can occur.

IOP Conf. Series: Materials Science and Engineering **633** (2019) 012003 doi:10.1088/1757-899X/633/1/012003

Salmonella is a member of the *Enterobacteriaceae*, there are two species, *Salmonella bongori* and *Samonella enterica* with 2,500 different serotypes [3], some serotypes have limited host such as *Salmonella enterica* serotype Typhi in human, *Salmonella enterica* serotype Gallinarum in poultry, *Salmonella enterica* serotype Dublin in cattle, *Salmonella enterica* serotype Choleraesuis in pigs but all serotypes of *Salmonella* can cause disease in humans [4].

Consumer protection in Indonesia regulated by Indonesian National Standard No 3924-2009 which states that carcass and chicken meat should free from *Salmonella* sp., the total plate count less than 10^6 cfu/g, the total coliform less than 10^2 cfu/g, the total *Staphylococcus aureus* less than 10^2 cfu/g [5]. *Salmonella* sp. characterized using the Kauffmann- White scheme basis of surface antigen identification using antiserum to detect the O (somatic) and H (flagellar) antigen [6] unfortunately, there are difficulties in getting antiserum in Indonesia so that genoserotyping can be used as an alternative for *Salmonella sp.*detection [7,8]. Availability of reliable and rapid tests for food-borne pathogen detection play an important role in the food industry and regulators in food security.

2. Experimental

One hundred and twenty samples chicken leg collected_from the wet market in Surabaya, Indonesia. Samples were packed in sterile plastic bags, transferred directly to the laboratory in the ice box for an immediate bacteriological testing.

Isolation and identification of bacteria. Twenty five grams of each chicken leg incised using a sterile scalpel and forceps, then transferred to sterile Erlenmeyer containing 225 ml of buffered peptone water (Oxoid CM0509, UK), the mixture incubated aerobically at 37 °C for 24 h. From each sample, 1 ml added to 9 ml of Selenite Cystine Broth base in 0,4% sodium biselenite (Oxoid LP0121A, UK) and incubated aerobically at 37 °C for 24 h. one hundred micro liter transferred and streaked on xylose lysine deoxycholate (XLD) agar (Oxoid CM 0469, UK). The inoculated plates incubated aerobically at 37 °C for 24 h. Five of presumptive *salmonella* colonies from XLD agar plates identified biochemically using IMViC test. Biochemically confirmed *Salmonella* isolates by genoserotyping for identification of *Salmonella* serovars [9, 10]. The developed of rapid test for identification of *SalmonellaSalmonella enterica* serovars Typhi and *Salmonella enterica* serovars Paratyphi A from chicken meat done by directly scrub colony on XLD agar for identification of *Salmonella* serovars by genoserotyping. This work carried out at the Gastroenteritis and Salmonellosis Laboratory, Institute of Tropical Disease Airlangga University, Surabaya, Indonesia.

The DNA samples for conventional method obtained from isolate that tested with IMViC by plating on nutrient agar (Merck, Darmstadt, Germany). The DNA samples for rapid method obtained from scraping of the bacterial colonies on XLD agar (Oxoid CM 0469, Basingstoke, UK). The bacterial colonies suspended in 1000 µl of distilled water in 1.5 ml microcentrifuge tube. Tubes boiled for 15 min, and cooled to room temp. The cell debris was pelleted by centrifugation at 10,000 RPM for 10 min. 5 µl of supernatant used as the template for PCR [11].

The specific primer sets and the cycling of the PCR for *Salmonella enterica* serovars Typhi and *Salmonella enterica* serovars Paratyphi A presented in Table 1. PCRs carried out in a Bio-rad PCR system T100™ Thermal Cycler (BioRad). A total of 25 µl PCR mixture contained 12.5 µl PCR mix (MytaqTM HS Redmix 2X, Bioline, USA), 10 µM of each primer, 5 µl PCR template and 5.5 µl ddH2O. The PCR condition was 95°C for 4 min, followed by 35 cycles of 95°C for 30 s, 55°C for 30 s, 72°C for 30 s, and last stage of 72°C for 5 min (6). PCR products loaded on 2% (wt/Vol) agarose gels, DNA marker used sizerTM-100 (Intron Biotechnology Inc.), stained with ethidium bromide and visualized on a UV Transilluminator.

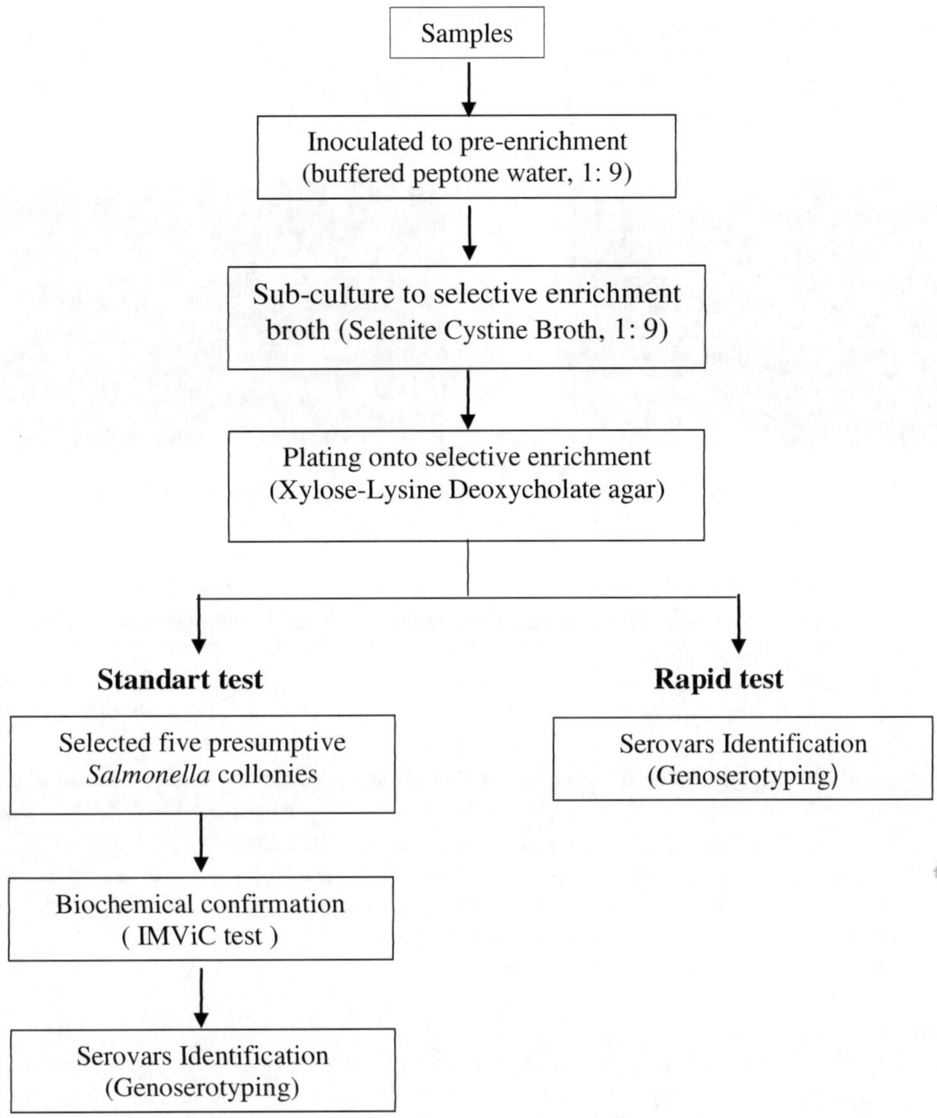

Figure 1. Conventional and Rapid Method for the identification of *Salmonellaenterica* serovars Typhi and *Salmonella enterica* serovars Paratyphi [9, 10]

Table 1. Primers used in PCR assays and the Size of PCR product (bp)

Primer	Primer sequence (5' to 3')	PCR product (bp)	Reference
tyv-for	GAGGAAGGGAAATGAAGCTTTT	614	11, 12
tyv-rev	TAGCAAACTGTCTCCCACCATAC		
prt-for	CTTGCTATGGAAGACATAACGAACC	256	11, 12
prt-rev	CGTCTCCATCAAAAGCTCCATAGA		

3. Results and Discussion

The results showed that modified method was more sensitive and rapid, from 120 samples, the modified technique successfully identified 39 (32.5%) *Salmonella enterica* serovars Typhi and one

Salmonella enterica serovars Paratyphi A (0,83%) compare with 13 (10.8%) and negative result with conventional protocol. *Prt* gene is found in both *Salmonella enterica serovars Typhi* and *Salmonella enterica serovars Paratyphi* while the *tyv* gene is only obtained in *Salmonella enterica serovars Typhi*. PCR used *prt* and *tyv* primers was proven abilities to detect *Salmonella enterica serovars Typhi* and *Salmonella enterica serovars Paratyphi* from bacterial mixtures obtained from XLD agar as well as from purified isolates [12]. Result of PCR using *prt* primer showed in figure 2, and result of PCR using *tyv* primer showed in figure 3.

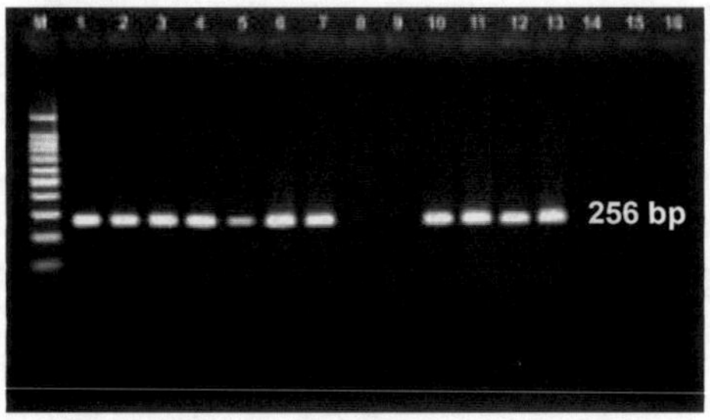

Figure 2. PCR products of *prt* primers and DNA marker Sizer TM-100 (Intron Biotech, Inc).

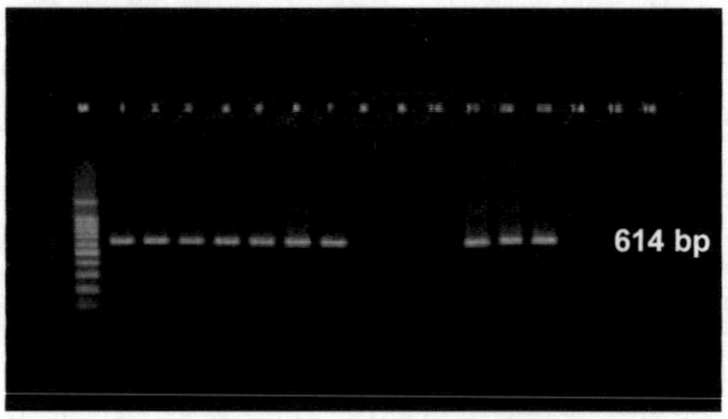

Figure 3. PCR products of tyv primers and DNA marker Sizer TM-100 (Intron Biotech, Inc).

The bacteria that can grow on XLD agar and have the same colony properties are *Salmonella, Edwardsiella, Shigella, Providencia,* H_2S-negative *Salmonella a*nd bacteria with different colonies properties are *Escherichia, Enterobacter, Klebsiella, Citrobacter, Proteus, Serratia* [13].Conventional method for the isolation and identification of *Salmonella sp* carried out by picking five presumptive *Salmonella* colonies from selective agar[12]. Selection of colonies, IMViC test and Serotyping requires special technical expertise and laborious. Rapid diagnostic allowing for direct testing of various types of colonies in the XLD agar and avoided the colony's misrepresentation and worked it out more simply. PCR method has advantages related to sensitivity, specificity, speed, simplicity, as well as in a quantitative [14]. Based on Indonesia National Standard No 3924-2009, chicken meat required free of *salmonella* sp. and the amount of total plate count did not exceed 10^6cfu/g, *Staphylococcus aureus* did not exceed 10^2cfu/g [5]. Identify and count of bacteria on chicken carcasses will be very difficult to carry out conventionally, but make it possible to carry out used molecular method [14]

4. Conclusion

These results showed that the rapid test using the molecular technique which applied directly to the colonies on selective media has a higher success rate than conventional methods, the modified technique has identified 39 from 120 samples (32.5%) *Salmonella enterica* serovars Typhi comparative conventional method which only identified 13 from 120 samples (10,8%). The modified technique successfully identified one *Salmonella enterica* serovars Paratyphi A from 120 samples (0,83%) where the conventional technique has failed. Modified method has advantages on rapidity, sensitivity, and simplicity where detection of *Salmonella enterica* serovars Typhi and *Salmonella enterica* serovars Paratyphi A from chicken meat only needed four steps compared six steps on conventional method.

References

[1] Typhoid World Health Organization http://www.who.int/news-room/fact-sheets/detail/typhoid Accessed July 10, 2018
[2] Background document 2003 The diagnosis, treatment and prevention of typhoid fever Communicable Disease Surveillance and Response Vaccines and Biologicals (Switzerland: World Health Organization) chapter 1 pp 1-3

[3] Salmonella (non-typhoidal). World Health Organization. http://www.who.int/news-room/fact-sheets/detail/salmonella-(non-typhoidal). Accessed July 17, 2018

[4] Foley SL, TJ Johnson, SC Ricke, R Nayak, J Danzeisen 2013 *Salmonella* Pathogenicity and Host Adaptation in Chicken-Associated Serovars Microbiology and Molecular Biology Reviews **77** 582–607

[5] Badan Standardisasi Nasional. Mutu karkas dan daging ayam. SNI 3924:2009. ICS 67.120.20.

[6] Forbes BA, DF Sahm, AS Weissfeld 2007 Bailey & Scott's Diagnostic Microbiology 12th Ed. (Missouri : Mosby Elsevier) chapter 28 pp 362-384

[7] Malorny B, J Hoorfar, C Bunge, and R Helmuth 2003 Multicenter Validation of the Analytical Accuracy of *Salmonella* PCR:towards an International Standard Applied and Environmental Microbiology **69** 290–296

[8] Berghaus RD, SG Thayer, BF Law, RM Mild, CL Hofacre, RS Singer 2013 Enumeration of *Salmonella* and *Campylobacter* spp. in Environmental Farm Samples and Processing Plant Carcass Rinses from Commercial Broiler Chicken Applied and Environmental Microbiology. **79** 4106–4114

[9] Perilla PJ 2003 Manual for the Laboratory Identification and Antimicrobial Susceptibility Testing of Bacterial Pathogens of Public Health Importance in the Developing World. (Switzerland: World Health Organization) chapter 7 pp 103-162

[10] Hendriksen RS 2003 Global Salm-Surv A global *Salmonella* surveillance and laboratory support project of the World Health Organization Laboratory Protocols Level 1 Training Course Isolation of Salmonella (Switzerland: World Health Organization) chapter 1 pp 1-7

[11] Levy H, S Diallo, SM Tennant, S Livio, SO Sow, M Tapia, P Fields, M Mikoleit, B Tamboura, KL Kotloff, R Lagos, JP Nataro, JE Galen, and MM Levine 2008 PCR Method to Identify *Salmonella enterica* Serovars Typhi, Paratyphi A, and Paratyphi B among *Salmonella* Isolates from the Blood of Patients with Clinical Enteric Fever. J. Clin. Microbiol **46** 1861–1866

[12] Hirose K, K Itoh, H Nakajima, T Kurazono, M Yamaguchi, K Moriya, T Ezaki, Y Kawamura, K Tamura, and H Watanabe 2002 Selective amplification of *tyv* (*rfbE*), *prt* (*rfbS*), *viaB*, and *fliC* genes by multiplex PCR for identification of *Salmonella enterica* serovars Typhi and Paratyphi A. J. Clin. Microbiol. **40** 633–636

[13] XLD Agar. Thermo scientific. Oxoid Microbiology Product http://www.oxoid.com/UK/blue/prod_detail/prod_detail.asp?pr=CM0469&c=UK&lang=EN. Accessed July 17, 2018.

[14] Dennis Lo YM, W Rossa WK Chiu, KC Allen Chan 2006 Clinical Applications of PCR. 2nd Ed. (New Jersey : Humana) chapter 4 pp 33-45

IOP Conf. Series: Materials Science and Engineering **633** (2019) 012004 doi:10.1088/1757-899X/633/1/012004

The addition of sorbitol and glycerol to improve the physico-chemical and sensory characteristics of chocolate dodol

R Nurhayati[1], E R N Herawati[1], A S Putri[2] and G P Adyana[2]

[1]Research Institute for Natural Product Technology, Indonesian Institute of Sciences Jl. Yogya-Wonosari km 31,5 Gading, Playen, Gunungkidul, Yogyakarta 55861, Indonesia.
[2]Study Program of Agricultural Technology, Faculty of Agriculture, Sebelas Maret University, Jl. Ir. Sutami 36A Kentingan, Surakarta 57126 Indonesia

E-mail: rifa.lipi@gmail.com

Abstract. Dodol is a traditional food from Indonesia which made from glutinous rice flour, coconut milk, and sugar. Chocolate dodol is one variant of dodol that use cocoa powder on its formulation. Dodol has a sweet taste and chewy texture, but its shelf life is very short, i.e., 4-5 days only. Therefore it needs an additional humectant, such as glycerol and sorbitol, to decrease the water activity (a_w) of the product and increase its shelf life. This research aimed to determine the effect of glycerol and sorbitol addition to the physical, chemical and sensory characteristics of chocolate dodol. The addition of glycerol and sorbitol to the chocolate dodol formulation were 0%, 6%, 8%, and 10%. The addition of sorbitol and glycerol to chocolate dodol product decreased water activity, increased texture characteristic, and have no significant difference in sensory preferences. In the same concentration, the addition of glycerol has a lower a_w than sorbitol. Panelists most preferred chocolate dodol with 6% sorbitol addition. It have a_w 0,74; L*-a*-b* value 25,47; 3,66 and 0,73.

1. Introduction

Indonesia is the largest producer of cocoa beans in the world after Cote d'Ivoire and Ghana [1]. More than 90% of cocoa plantations are managed by smallholder with the land ownership under 5 hectares [2]. Cocoa and chocolate become a concern because of the content of its active compounds. Polyphenol, catechin, and epicatechin are the bioactive compounds inside the cocoa seed. Its polyphenol being as an antioxidant which has a protective role against oxidative stress that is involved in cardiovascular and cerebrovascular disease [3].

The way to improve the value added of cocoa beans is product diversification[4]. Cocoa can proceed into several products, one of them is chocolate dodol. Dodol is a traditional food from West Jawa, Indonesia that formulated from glutinous rice flour, coconut milk, and sugar. Dodol is an intermediate moisture food which the moisture 10-40 %; a_w 0,70-0,85; sweet taste, soft texture and elastic. Chocolate dodol is proceeded by mixing and heating coconut milk, glutinous rice flour, rice flour, chocolate powder, and sugar [5, 6]. The shelf life of chocolate dodol is fairly short, 4-5 days.

As an intermediate moisture food, dodol is easily damaged. The deterioration of dodol is caused by its a_w 0,7-0,85. These water activity condition will cause an activity of foodborne microorganism and chemical reaction that damage food quality, such as rancidity. This deterioration can be inhibited by adding a humectans [7]. Humectants are a food additive that binds water of the product and control a_w.

The hydroxyl group of humectants has an affinity to form hydrogen bonds with water; this will decrease the free water ability [8]. The decreasing of free water ability will inhibit microorganism activity and chemical reactions that accelerate food deterioration. The humectant used in a food product is sucrose, fructose, lactose, mannitol, honey, sorbitol, and glycerol [9].

Among the various type of humectants, sorbitol and glycerol are often used for a food product. Some previous studies stated that the addition of glycerol and sorbitol can improve product quality. Glycerol reduced the product humidity until 60%. The addition of 4% glycerol in *khoa* caused decreasing the a_w without significantly affect its textural parameter [10]. The best *jerky* formulation was produced by adding 15% glycerol compare to sorbitol and control sample [11]. Badola [8] reported that the addition of 20% sorbitol to *khoa* product has the maximum water activity reduction compared with corn syrup, maltodextrin, polydextrose, and mannitol. While Syafutri [12] reported that panelists most preferred the jelly candies with the addition of 20% sorbitol. The addition of humectant to reduce the a_w of chocolate dodol has not been studied yet. So, this study was conducted to determine the effect of adding sorbitol and glycerol to the physico-chemical and sensory characteristics of chocolate dodol.

2. Materials and Methods

2.1. Materials

Material for making chocolate dodol are glutinous rice flour *Rosebrand*, rice flour *Rosebrand*, coconut milk, chocolate powder from Griya Cokelat Nglanggeran, sugar *Gulaku*. Glycerol and Sorbitol food grade (70%) were obtained from Bratachem.

2.2. Chocolate Dodol Processing

Chocolate dodol processing followed the method of Nurhayati [13]. Thick coconut milk and sugar were mixed and heated (70°C, 15 minutes). Glutinous rice flour, rice flour, chocolate powder, and runny coconut milk were mixed, then added to sugar dough. Sorbitol and glycerol (6%, 8% and 10%) were added then the dough was mixed until perfectly gelatinized. Chocolate dodol was cooled to a temperature 28°C.

2.3. Physicochemical Analysis

Each sample of chocolate dodol was tested for water activity with $A_w$meter Pawkit Decagon, color using *Chromameter* CR-400 Konika Minolta and texture using Llyod Texture Analyzer.

2.4. Sensory Analysis

Sensory evaluation using hedonic test method involving 25 panelists for 7 samples (original choclate dodol, 3 samples chocolate dodol+glycerol and 3 samples chocolate dodol+sorbitol). Taste, color, aroma, adhesiveness, elasticity and overall were evaluated by providing score 1-5 (5=really like, 4=like, 3=rather like, 2=rather dislike, 1=dislike).

3. Result and Discussion

3.1. Water Activity of Chocolate Dodol

Dodol is an intermediate moisture food that has a water activity (a_w) 0,65-0,90 [14]. Based data on Table 1 the a_w of chocolate dodol is 0,54-0,79. The highest a_w was chocolate dodol original, while the lowest was chocolate dodol with the addition of 10% glycerol. At the same concentration, glycerol could reduce the water activity lower than sorbitol. The higher concentration of sorbitol and glycerol, the lower water activity of chocolate dodol. Glycerol and sorbitol have hydroxyl group. The hydroxyl group functions to bind free water from product and form the hydrogen bonds with water. So, the more humectants added, the lower a_w of product [15].

Jariyah [14] reported that the addition of 8% sorbitol in Mangulu, traditional food from Sumba Timur-Nusa Tenggara Timur, decreased its water activity into 0,57. While this research data was the sae as Syamsir and Sitanggang [7] did. They reported that the ability of glycerol to reduced the water activity of product higher than sorbitol. The addition of 10% glycerol to dodol products produced a_w 0,83; while 10% sorbitol produced a_w 0,86.

Table 1. Water Activity of Chocolate Dodol.

Chocolate Dodol	**Water activity (a_w)**
Original	0,79^{e}
Glyserol 6%	0,69^{c}
Glyserol 8%	0,59^{b}
Glyserol 10 %	0,54^{a}
Sorbitol 6%	0,73^{d}
Sorbitol 8%	0,73^{d}
Sorbitol 10%	0,72^{d}

The amount of free water in food is known as water activity (a_w). Water activity plays an important role in damaged microbiological food. Water needs to add or remove in order to modify the a_w of food for preserving and control food quality. Each microorganism requires a different amount of water for growth. Bacteria grow well in a high a_w 0,91. Yeast requires medium a_w 0,87-0,91 while fungi require a_w 0,80-0,87, but they still grow slowly at a_w 0,75 such as *Aspergillus* [16]. Xerophilic fungi showed growth in the a_w 0,6-0,75 [17, 18]. The addition of glycerol and sorbitol reduced the a_w of chocolate dodol under 0,8. This condition prevented the growth of microorganism in the chocolate dodol.

3.2. Sensory Characteristic

The hedonic tests were conducted to determine the level of preference of the panelists for the product. Chocolate dodol original and chocolate dodol with the addition of sorbitol and glycerol were evaluated by 25 panelists. Six parameters (taste, color, aroma, adhesiveness, elasticity and overall) of each chocolate dodol were tested by giving score 1-5.

Based on sensory's data (Table 2), chocolate dodol with the addition of 6% sorbitol was most preferred by panelists. Its preference score for all parameters has no significant difference with chocolate dodol original. The level of panelist preference to the chocolate dodol product was in the range of rather like to like. The preference result showed that there was no significant different between chocolate dodol original and chocolate dodol with the addition of 10% glycerol, 6% sorbitol and 8% sorbitol. The result of this study was the same as those conducted by Syamsir and Sitanggang [7] which reported that the addition of 5% and 10% glycerol to dodol have no significant different for all the sensory parameter compared with the original.

Table 2. Sensory Characteristics of Chocolate Dodol.

Chocolate Dodol	Sensory Parameters					
	Taste	Colour	Aroma	Adhesiveness	Elasticity	Overall
Original	3,81^{a}	3,89^{a}	3,69^{a}	3,47^{a}	3,56^{ab}	3,86^{a}
Glyserol 6%	3,28^{b}	3,42^{b}	3,47^{a}	3,17^{ab}	3,22^{b}	3,28^{b}
Glyserol 8%	3,5^{ab}	3,83^{a}	3,67^{a}	2,78^{b}	3,17^{b}	3,56^{ab}
Glyserol 10%	3,64^{ab}	3,69^{ab}	3,67^{a}	3,33^{a}	3,58^{ab}	3,72^{a}
Sorbitol 6%	3,67^{ab}	3,94^{a}	3,81^{a}	3,64^{a}	3,86^{a}	3,89^{a}
Sorbitol 8%	3,64^{ab}	3,69^{ab}	3,67^{a}	3,25^{a}	3,58^{ab}	3,58^{ab}
Sorbitol 10%	3,53^{ab}	3,89^{a}	3,58^{a}	3,53^{a}	3,36^{b}	3,67^{a}

3.3. Physical Characteristics

Colour test used Chromameter, while texture used Llyod texture analyzer. L* value with the scale 0-100 showed the brightness level of the product, from dark into white [19]. L* value of chocolate dodol was 23,93-26,57. This indicated that the color of chocolate dodol approach to black. Sugar caramelization or Maillard reaction occurs during chocolate dodol processing. Maillard is a complex reactions between reducing sugar and free amine groups from protein or amino acid [9]. Beside that, chocolate powder became natural brown coloring for the product. These two things that caused the color of chocolate dodol is brown-black.

Table 3. The Colour of Chocolate Dodol.

Chocolate dodol	L*	a*	b*
Original	$26{,}24^{cd}$	$4{,}68^{d}$	$1{,}73^{c}$
Glyserol 6%	$23{,}93^{a}$	$4{,}26^{bc}$	$1{,}82^{c}$
Glyserol 8%	$25{,}93^{cd}$	$4{,}41^{cd}$	$0{,}44^{a}$
Glyserol 10%	$25{,}28^{bc}$	$4{,}01^{b}$	$0{,}90^{b}$
Sorbitol 6%	$25{,}47^{bcd}$	$3{,}66^{a}$	$0{,}73^{ab}$
Sorbitol 8%	$24{,}83^{ab}$	$4{,}04^{b}$	$0{,}94^{b}$
Sorbitol 10%	$25{,}52^{bcd}$	$4{,}05^{b}$	$1{,}04^{b}$

Positive a* value 0-60 showed redness level, while negative a* value 0-(-60) showed the greenish level. a* value of chocolate dodol range between 3,53-4,88; this showed that the color of chocolate dodol was a little red. Positive b* value 0-60 showed yellowish level, while negative b* value 0-(-60) showed a bluish level. b* value of chocolate dodol range between 0,73-1,82; this showed that the color of chocolate dodol was a little yellow.

The texture of chocolate dodol measured by texture analyzer. There were five parameters; hardness, cohesiveness, adhesiveness, gumminess, chewiness (Table 4). The addition of sorbitol and glycerol has significant effect to the texture of chocolate dodol. The addition of sorbitol and glycerol made the chocolate dodol softer, more cohesive, and less adhesive.

Table 4. The Texture of Chocolate Dodol.

Chocolate dodol	Hardness (N)	Cohesiveness (N)	Adhesiveness (Nmm)	Gumminess (N)	Chewiness (N)
Original	$1545{,}17^{a}$	$0{,}46^{c}$	$278{,}55^{a}$	$710{,}54^{ab}$	$5424{,}24^{ab}$
Glyserol 6%	$872{,}83^{b}$	$0{,}66^{b}$	$34{,}09^{b}$	$576{,}74^{abc}$	$4856{,}59^{ab}$
Glyserol 8%	$715{,}17^{bc}$	$0{,}73^{a}$	$6{,}32^{b}$	$512{,}49^{bc}$	$4426{,}82^{b}$
Glyserol 10%	$552{,}00^{c}$	$0{,}79^{a}$	$9{,}21^{b}$	$433{,}25^{c}$	$3722{,}71^{b}$
Sorbitol 6%	$860{,}83^{b}$	$0{,}74^{a}$	$15{,}80^{b}$	$632{,}87^{abc}$	$5501{,}99^{ab}$
Sorbitol 8%	$990{,}33^{b}$	$0{,}77^{a}$	$10{,}55^{b}$	$769{,}65^{a}$	$6771{,}10^{a}$
Sorbitol 10%	$737{,}50^{bc}$	$0{,}76^{a}$	$6{,}63^{b}$	$558{,}98^{abc}$	$4817{,}93^{b}$

Hardness is the resistance of the material to rupture due to the pressure applied. The addition of humectants reduced the value of hardness. There was a significant difference in the value of hardness between chocolate dodol original and chocolate dodol with the addition of humectants. Cohesiveness is the product's ability to resist deformation. All chocolate dodol with the addition of humectants have higher cohesiveness value than the original. The adhesiveness is the work needed to pull away from the product from the surface. The addition of humectants reduced the adhesiveness of the product. Gumminess is the durability of the material to release or break due to compressive force. The addition of humectants decreased the gumminess of the product compared with the original. Chewiness is the pressure needed to chew samples to become the right consistency to swallow. The addition of humectants decreased the chewiness of the product compared to the original chocolate dodol.

Generally, the addition of humectants improved the texture quality of chocolate dodol. The same result also reported by Badula [18] Syamsir and Sitanggang [7], Sorapukdee [11] and Karimi [20].

4. Conclusion

Based on research data, we concluded that the addition of sorbitol and glycerol to chocolate dodol product, decreased water activity, increased texture characteristic, and have no significant difference in sensory preferences. Panelists most preferred chocolate dodol with the addition of 6% sorbitol. It have characteristics a_w 0,74; L*-a*-b* value 25,47; 3,66 and 0,73. The hardness 860,83 N; cohesiveness 0,74 N; adhesiveness 15,80 Nmm, gumminess 632,87 N and the chewiness 5501,99 N.

Acknowledgments
We gratefully acknowledge to IPTEKDA LIPI for funding this research activities; Mrs Sri Endartini and Mrs Wiwin Widiastuti for all support.

References

[1] Nielson J, Hafid H, Hiswaty H, Mount T and Mckenzie F 2012 *Sustainability Impact Assessment of a Certification Scheme in the Indonesian Cocoa Industry: 2012 Pilot Survey Results* University of Sydney, Australia. p 1-77
[2] Sub Directorate of Estate Crop Statistics 2017 Statistics Indonesia, Jakarta
[3] Alia M, Horcajo C, Bravo L and Goya L 2003 *Nutrition Research* **23**(9) p 1251-1267
[4] Nurhayati R, Pratiwi R, Anandito B K, Herawati E R N and Angwar M. 2018 *Reaktor* **18**(2) p:63-70
[5] Haryadi, Sai'in, and Suhardi 1998 *Agritech* **18**(1) p 29-33
[6] Chuah T G, Nisah H H, Thomas Choong S Y, Chin N L and Nazimah Sheikh AH 2007 *Journal of Food Engineering* **80** p 423-430
[7] Syamsir E and Sitanggang P 2011 *Jurnal Ilmu dan Teknologi Pangan* **9**(1) p 65-76
[8] Badola R., Singh R R B, Panjagari N R., Singh A K and Hussain S A 2017 *Indian Journal Dairy Sci* **70**(2) p 145-154
[9] Martins F C O L, Sentanin M A and de Souza D 2018 *Food Chemistry* **272** p 732-750
[10] Sawhney I K, Patil G R., Kumar B and Rao J 1994 *J Food sci Tech* **31**(3) p 252-254
[11] Sorapukdee S, Uesakulrungrueng C and Pilasombut K 2016 *Korean J. Food Sci.* **36(**3), p 326-334
[12] Syafutri, Indriyani M, Lidiasari E and Indrawan H 2010 *Jurnal Gizi dan Pangan* **5(**2) p 78-86
[13] Nurhayati R, Herawati E R N, and Angwar M 2015 *Karakteristik Sifat Sensoris dan Kimiawi Dodol Cokelat dengan Penggunaan Variasi Sumber Lemak*. Prosiding Seminar Nasional PATPI 2015 p 883-889
[14] Jariyah, Rosida and Inda R H. 2015 *J. Rekapangan*, **9(**2) p 1-6.
[15] Fontana AJ *Water Activity for Confectionery Quality and Shelf Life* https://courses.ecolechocolat.com/lobjects/pdf/water_activity_confectionery.pdf accessed 3rd February 2019
[16] Buckle K A, Edwards R A, Fleet G H and Wooton M 1985 *Food Science* Translation H Purnomo and Adiono Universitas Indonesia Press Jakarta
[17] Adam M R and Moss M O 2008 *Food Microbiology* 3rd edition RSC Publishing, Cambridge, UK p 40
[18] Beuchat L R 1981 *Microbial stability as affected by water activity* Cereal Foods World
[19] Misnawi and Selamat 2005 *Cita Rasa, Tekstur dan Warna Cokelat* Penebar Swadaya Jakarta
[20] Karimi M, Sahraiyan B, Naghipour F, Sheikholeslami Z and Davoodi M G 2013 *Intl J Agri Crop Sci.* **5**(11) 1209-1213

IOP Conf. Series: Materials Science and Engineering **633** (2019) 012005 doi:10.1088/1757-899X/633/1/012005

Antimicrobial effect of *Zingiber officinale* var. officinale essential oil and nisin against pathogenic and spoilage microorganisms

D A Ashari, A Nissa, A Nursiwi, A M Sari and R Utami

Department of Food Science and Technology, Universitas Sebelas Maret, Jl. Ir Sutami No. 36A, Surakarta, Central Java, Indonesia

E-mail: rohulautami@staff.uns.ac.id

Abstract. This study was conducted to investigate the antimicrobial effect of *Zingiber officinale* var. officinale essential oil (EO), nisin, and their combination against some pathogenic and spoilage microorganisms. Minimum Inhibitory Concentration (MIC), Minimum Bactericidal Concentration (MBC), Minimum Fungicidal Concentration (MFC) and the synergism effect were assayed. The MIC values of nisin on *Aspergillus niger*; *Eschericia coli* and *Pseudomonas fluorescens*; *Bacillus cereus;* and *Salmonella typhimurium* and *Staphylococcus aureus* were 250; 500; 2000; and >2000 IU, respectively. On the other hand, MIC values of EO on *E. coli*; *B. cereus* and *S. aureus*; *Pseudomonas fluorescens* and *A. niger;* and *Salmonella typhimurium* were 0,125; 0.25; 0.5; and 4%, respectively. 62.5 IU of nisin combined with 2% of EO could inhibit the growth of *Salmonella typhimurium, S. aureus,* and *E. coli.* In addition, the combination of nisin and EO had synergistic effect against *B. cereus, A. niger,* and *Salmonella typhimurium.* The combination of nisin and EO had no bactericidal effect against all five bacterias but it had fungicidal effect against *A. niger* at concentration 62.5 IU of nisin and 1% of EO. Sabinene (16.88%), Z-citral (11.25%), and camphene (10.05%) were the major components in *Zingiber officinale* var. officinale EO which contributed on its antimicrobial activity.

1. Introduction

Microbial contamination was one of the biggest cases found on food, this kind of contamination spoils the food and/or causing foodborne disease. In 2014, suspected foodborne outbreaks caused by microbial contamination in Indonesia reached 51.06%, that number was higher than the outbreaks caused by chemical and physical contamination [1]. The data decreased in 2015 and reached 42.62%, eventhough it was still the highest amongst others [2]. Data from *Center for Disease Control and Prevention* also showed the same trend. From 2011 until 2014, microbial contamination, caused by for example *Salmonella, Eschericia coli, Staphylococcus aureus,* and *Bacillus cereus*, was the biggest outbreaks in America [3,4,5,6]. Natural antimicrobial and preservative agents are needed to be investigated.

Nisin, is a bacteriocin which ribosomally synthesised by *Lactococcus lactis* subs. *Lactis.* It has had classified by FDA as a safe food preservative agent/*Generally Recognizes as Safe* (GRAS) [7]. On several previous studies, nisin was identified as an effective agent on inhibiting the growth of microorganisms. Some microorganisms were inhibited by nisin, such as *Micrococcus luteus* ATCC

10240 [8] *Listeria monocytogenes* [9], *Bacillus cereus*, *Lactobacillus sakei,* and *Listeria monocytogenes* HPB 2812 [10]. 100 IU/ml and 500 IU/ml of nisin could protect the serro cheese from the development of *Staphylococcus aureus* [11]. Nisin also produced low impact on sensory value, only 1/40 of nisin was detected by saliva. In addition, WHO recommends nisin as an additive with Acceptable Daily Intake/ADI is 33000 IU [12], it makes the use of nisin becomes more beneficial than the use of other antimicrobial agents. Unfortunately, nisin is less effective against Gram negative and fungi and it is also expensive [13,14].

On the other hand, the essential oil from *Zingiber officinale var. officinale* or usually known as ginger, is one of the most potential antimicrobial agent to be developed in Indonesia. Data from Central Bureau of Statistics/Badan Pusat Statistik (BPS) Indonesia showed that in 2014 the production of ginger (as the raw material of ginger EO) reached 226,114.8 ton [15]. In addition, ginger EO also had effectivity against several microorganisms. 2.0 mg/ml inhibited the growth of *E. aurogenes* and *S. marcescens*, 4.0 and 8.0 mg/ml inhibited the growth of *E. coli* and *S. typhi* respectively[16]. But, higher concentration of essential oil may disturb the sensory value of food when it was added [17,18].

The combination of nisin and ginger EO may reduce the concentration of nisin and/or essential oil, increase the effectivity of their antimicrobial activity, and it may also protect the sensory value of food. On the previous research, the combination of nisin and *Zingiber officinale* var. rubrum essential oil showed an effective activity on *Bacillus cereus,* it also rose up the effectivity of nisin on *Salmonella typhimurium, Pseudomonas fluorescens,* and *Aspergillus niger* [19]. The aim of this research was to investigate the antimicrobial effect of the combination between nisin and ginger (*Zingiber officinale* var. officinale) essential oil on *Bacillus cereus* FNCC 0057, *Salmonella typhimurium* FNCC 0050, *Staphylococcus aureus* FNCC 0047, *Eschericia coli* FNCC 0091, *Pseudomonas fluorescens* FNCC 0070 and *Aspergillus niger* FNCC 6080.

2. Material and Method

2.1 Essential oil preparation

Fresh ginger was obtained from Pasar Legi, Surakarta, Indonesia. Then, it was sorted and washed to remove the soil. After that, it was sliced into 2-3 mm thickness and dried under the air-drying condition for 4-6 days. After the simplicia reached 15-16% of moisture content, it was distillated. The distillation process was using the Stahl apparatus to obtain the essential oil [20].

2.2 Gas Chromatography-Mass Spectrometry (GC-MS) *analysis*

GC-MS Shimadzu QP 2010 (Tokyo, Japan) was used to analyse the chemical components of the ginger EO. 1 μl of EO was injected into the GC-MS. The oven temperature was setted on 50°C and it increased 5°C/minute until it reached its final temperature on 240°C. The injection temperature was 300°C. Helium as the carrier gas, flowed on 0.55 ml/minute flow rate and the pressure was 14.0 kPa [21].

2.3 Inoculum preparation

Bacillus cereus FNCC 0057, *Salmonella typhimurium* FNCC 0050, *Staphylococcus aureus FNCC 0047, Eschericia coli* FNCC 0091, *Pseudomonas fluorescens* FNCC 0070, and *Aspergillus niger* FNCC 6080 were obtained from Pusat Studi Pangan dan Gizi, Gadjah Mada University, Indonesia. All of the inoculum stocks were refreshed and incubated at 35±2°C. The bacteria cultures were growth on Nutrient Agar (NA) for 24 hours while the fungi culture was growth on Potato Dextrose Agar (PDA) for 3 days. The inoculum was suspended into saline water and its turbidity was adjusted with 0.5 McFarland [22].

2.4 Minimum Inhibitory Concentration (MIC) test

MIC test was conducted using microdilution method [21]. 1000 IU, 500 IU, 250 IU, 125 IU and 62.5 IU of nisin were made from nisin (10^6 IU) that was diluted in 10 ml of media (Mueller Hinton Broth or RPMI 1640 2% of glucose without bicarbonate). The suspension of 4%, 2%, 1%, 0.5% and 0.25%

(v/v) of EO were made by using the same method as making nisin dilution. 0.5% (v/v) of Tween 20 was added into the suspension to make the dilution becomes more stable.

The 96-wells microplate was filled with 50 µl of nisin solution and ginger EO solution, then 10 µl of bacteria suspension or 100 µl of fungi suspension were added. For bacteria inhibition test, the microplates should be incubated for 20 hours while for fungi inhibition test was 48 hours. The MIC were determined from the minimum concentration that could inhibit the growth of the microorganisms.

2.5 Minimum Bactericidal Concetration (MBC) and Minimum Fungicidal Concentration (MFC) test

The suspension that showed MIC value were regrowth on Mueller Hinton Agar (MHA) at 35±2°C for 24 hours to determine the MBC while 20 µl from fungi MIC test were used and regrowth on Saboraud Dextrose Agar (SDA) at 35±2°C for 48 hours to determine the MFC [23,24]. MBC and MFC were identified as the minimum concentration that could eliminate the growth of the microorganisms.

2.6 Synergism Effect Analysis

FIC index were calculated to analyse the synergism effect of the combination. Fractional inhibitory concentration of the first and the second antimicrobial agents were added (FICA and FICB). A is identified as the MIC of the first agent that inhibited the microorganism when it is combined with the second agent. While MICA is described as the MIC of the first agent without combination. Whereas B and MICB have the same description as A and MICA but on the second agent.

$$\text{FIC Index} = \text{FICA} + \text{FICB}$$

$$\text{FIC A} = \frac{\text{A}}{\text{MICA}}$$

$$\text{FIC B} = \frac{\text{B}}{\text{MICB}}$$

Synergistic effect of the combination is determined with FIC < 0.5, partial synergistic is determined when 0.5<FIC< 0.76. Additive effect is determined with 0.76< FIC <1, indifference effect is identified with $1 < \text{FIC} \leq 4$, and antagonistic effect is identified when FIC > 4 [25].

3. Result and Discussion

3.1 Chemical composition

The ginger EO was analysed by using GC-MS. Sabinene (16.88%), Z-citral (11.25%), camphene (10.05%), citral (8.57%), and zingiberene (8,49%) were the major components of ginger EO.

Table 1. Chemical Composition of Ginger EO.

Chemical Components	Ginger EO (%)
Sabinene	16.88
Z-citral	11.25
Camphene	10.05
Citral	8.57
Zingiberene	8.49
Beta-sesquiphellandrene	4.21
Alpha-pinene	4.03
Alpha-farnesene	3.41
Benzene	2.61
Beta-bisabolene	2.20
Myrcene	2.11

On the other hand, beta-sesquiphellandrene (4.21%), alpha-pinene (4.03%) and alpha-farnesene (3.41%) and other minor components also helped to determine the antimicrobial activity of ginger EO (**Table 1**). Environmental condition and genetic of the plant may contribute on the composition of EO [26].

3.2 MIC, Synergism effect, MBC, and MFC

The MIC value are shown on **Table 2**. MIC is described as the minimum concentration of the antimicrobial agent that inhibits the growth of microorganism [27,28]. Most of the target microorganisms needed no more than 2000 IU to inhibit their growth (**Table 2**). 250 IU inhibited the growth of *Aspergillus niger* FNCC 6080, 500 IU was needed on the inhibition of *Eschericia coli* FNCC 0091 and *Pseudomonas fluorescens* FNCC 0070, while the inhibition of *Bacillus cereus* FNCC 0057 needed 2000 IU of nisin. No activity was observed on *Salmonella typhimurium* FNCC 0050 and *Staphylococcus aureus* FNCC 0047 at 2000 IU of nisin. The sensitivity of *E. coli* and *Staphylococcus aureus* were different from the previous studies, which stated that Gram positive was more sensitive than Gram negative and fungi [10,13,14]. But, the growth of *Staphylococcus aureus* may also be affected by the capability of that microorganism to adapt with new habitat [29], and it becomes more resistant with nisin.

Ginger EO showed an effective activity. At the lowest concentration, 0.125% of ginger EO could inhibit the growth of *Eschericia coli* FNCC 0091. 0.25% inhibited *Bacillus cereus* FNCC 0057 and *Staphylococcus aureus* FNCC 0047 to grow normally. *Pseudomonas fluorescens* FNCC 0070 and *Aspergillus niger* FNCC 6080 were affected by 0.5% ginger EO, while 4% of EO was active against *Salmonella typhimurium* FNCC 0050. The EO activity had the same result with the previous study, where it showed inhibition activity on Gram positive, Gram negative and fungi [16].

Combination results are reported on **Table 2**. 62.5 IU of nisin and 0.125% ginger EO inhibited *Bacillus cereus* FNCC 0057 and *Aspergillus niger* FNCC 6080 to grow. 62.5 IU with 2% ginger EO was active against *Salmonella typhimurium* FNCC 0050. *Pseudomonas fluorescens* FNCC 0070 was inactive against 62.5 IU of nisin with 0.5% of ginger EO. To inhibit the growth of *Staphylococcus aureus* FNCC 0047 and *Eschericia coli* FNCC 0091 62.5 IU of nisin should be combined with 2% of ginger EO.

Table 2. The MIC of Nisin, Essential Oil, Their Combination, and Their Synergism Effect.

Microorganism	Minimum Inhibitory Concentration			FIC	Synergism effect	MBC or MFC
	Nisin (IU)	Essential oil (%)	Nisin (IU) + Essential Oil (%)			
Bacillus cereus FNCC 0057	2000	0,25	62,5 IU + 0,125%	0,5312	P	-
Salmonella typhimurium FNCC 0050	> 2000	4	62,5 IU + 2%	0,5312	P	-
Staphylococcus aureus FNCC 0047	> 2000	0,25	62,5 IU + 2%	8,0312	A	-
Eschericia coli FNCC 0091	500	0,125	62,5 IU + 2%	16,125	A	-
Pseudomonas fluorescens FNCC 0070	500	0,5	62,5 IU + 0,5%	1,125	I	-
Aspergillus niger FNCC 6080	250	0,5	62,5 IU + 0,125%	0,5	S	62,5 IU + 1 %

S : Synergistic effect
P : Partial synergistic effect
I : Indifference effect
A : Antagonistic effect
- : had no bactericidal or fungicidal effect

Combination could rise the effectivity of both EO and nisin. From the previous research [10], to inhibit the growth of *B. cereus* LPSQ 2872, *Salmonella typhimurium* SL 1344, and *Staphylococcus aureus* ATCC 2913 172 ppm of nisin were needed, while on *E. coli* O157:H7 more than 172 ppm was needed, these results were higher than when nisin was combined with ginger EO on this assay. Those activity caused by the activity of both agents. The hydrophobic cation in nisin will bind the phospate of Gram positive bacteria. It causes the disturbance of cell membrane transglycosylation, it also creates pores that causes lysis [30, 31, 12, 32]. The addition of ginger EO may increase that activity. EO which is also hydrophobic creates pores on the cell, so the antimicrobial activity increases [33, 34]. While the activity on the fungi, ginger EO may disrupts the ergosterol metabolism, that ergosterol, has function on keeping the strenght of the cell wall [35]. If the metabolism is disrupted than the cell is also deteriorated. Nisin helps the inhibition process after the cell wall was degradated [36].

The synergism effects between two antimicrobial agents were showed by their FIC index value. The synergism effect was greater when the FIC index value was lower. The combination of ginger EO and nisin showed a total synergistic effect against *Aspergillus niger* FNCC 6080. The combination effects against *Bacillus cereus* FNCC 0057 and *Salmonella typhimurium* FNCC 0050 showed partial synergistic effect with FIC index value was 0.5312. The combination indicated indifference effect when it was used for against the *Pseudomonas fluorescens* FNCC 0070. Whereas, when the combination was used on *Staphylococcus aureus* FNCC 0047 and *Eschericia coli* FNCC 0090, it showed antagonistic effect. Nisin may inhibit the EO penetration, it also makes the outer membrane more stable [28, 29].

Furthermore, the bactericidal and fungicidal effect of the combination were tested (**Table 2**). The combination did not show bactericidal effect on all targeted bacterias. Contrastly, it killed the *Aspergillus niger* FNCC 6080 and showed fungicidal effect at 62.5 IU of nisin and 1% of ginger EO. Optimum condition, including temperature, pH, water activity, redox potential, and nutrient, may affect on MBC test [14].

4. Conclusion

In conclusion, the combination of ginger EO and nisin increased the activity of both agents and showed total and partial synergistic effect against *Bacillus cereus* FNCC 0057, *Salmonella typhimurium* FNCC 0050 and *Aspergillus niger* FNCC 6080. 62.5 IU plus 2% of ginger EO were effective on *Salmonella typhimurium* FNCC 0050, *Eschericia coli* FNCC 0091, and *Staphylococcus aureus* FNCC 0047. While *Pseudomonas fluorescens* FNCC 0070; *Bacillus cereus* FNCC 0057 and *Aspergillus niger* FNCC 6080 needed lower concentration at 62.5 IU plus 0.5% and 62.5 IU plus 0.125%.

Acknowledgement

This work was financially supported by research project of PNBP UNS 543/UN27.21/LT/2018. The authors fully acknowledged Sebelas Maret University for the approved fund which makes this important research viable and effective.

References

[1] Badan Pengawas Obat dan Makanan 2015 *Laporan Tahunan Tahun 2014 Badan Pengawas Obat dan Makanan* (Jakarta: BPOM) p 135-136

[2] Badan Pengawas Obat dan Makanan 2016 *Laporan Tahunan Tahun 2015 Badan Pengawas Obat dan Makanan* (Jakarta: BPOM) p 117-118

[3] Center for Disease Control and Prevention 2011 *Surveillance for Foodborne Disease Outbreaks United States, 2011: Annual Report* (USA: CDC) p 2-6

[4] Center for Disease Control and Prevention 2012 *Surveillance for Foodborne Disease Outbreaks United States, 2012: Annual Report* (USA: CDC) p 2-6

[5] Center for Disease Control and Prevention 2013 *Surveillance for Foodborne Disease Outbreaks United States, 2013: Annual Report* (USA: CDC) p 2-6

[6] Center for Disease Control and Prevention 2014 *Surveillance for Foodborne Disease Outbreaks United States, 2014: Annual Report* (USA: CDC) p 2-5

[7] Ghrairi T, Chaftar N and Hani K 2012 *Progress in Food Preservation* (United Kingdom: Willey Blackwell)
[8] Mauriello G, de Luca E, Storia L A, Villani F and Ercolini D 2005 *Letters in Applied Microbiology* **41** 464-9
[9] Solomakos N, Govaris A, Koidis P and Botsoglou N 2008 *Elsevier Food Microbiology* **25** 120-7
[10] Turgis M, Vu K D, Dupont C and Lacroix M 2012 *Elsevier Food Research International* **48** 696-702
[11] Pinto M S, de Carvalho A F, Pires A C S, Souza A A C, da Silva P H F, Sobral D, de Paula J C J and Santos A L 2011 *Elsevier International Dairy Journal* **21** 90-6
[12] Jozala A F, Novaes, L C L and Junior, A P 2015. *Nisin, Concepts, Compounds and the Alternatives of Antibacterials* (Croatia: Intech) chapter 5 p 103-119
[13] Nattress F M and Baker L P 2003 *International Journal of Food Microbiology* **85** 259-267
[14] Delves B and Weber G 2011 *Nisin, Natamycin and Other Commercial Fermentates Used in Food Biopreservation* (England: Woodhead Publishing Limited)
[15] Badan Pusat Statistik 2015 *Statistik 70 Tahun Indonesia Merdeka* (Jakarta:Badan Pusat Statistik)
[16] Lopez E I C, Balcazar M F H, Mendoza J M R, Ortiz A D R, Melo R S P and Delgado T H 2017 *American Journal of Plant Sciences* **8** 1511-24
[17] Negi P S 2012 *Elsevier International Journal of Food Microbiology* **156** 7-17
[18] Dwijatmoko M I, Praseptiangga D and Muhammad D R A 2016 *Nusantara Bioscience* **8** 301-5
[19] Nissa A, Utami R, Sari A M and Nursiwi A 2018 *AIP Conference Proceedings* **2014** 1-6
[20] Sadgrove N and Jones G 2015 *Agriculture* **2015** 48-102
[21] Jayanudin 2011 *Jurnal Teknik Kimia Indonesia* **10** 37-42
[22] CLSI 2012 *Methods for Dilution Antimicrobial Susceptibility Tests for Bacteria that Grow Aerobically; Approved Standard – Ninth Edition* (USA: Clinical and Laboratory Standards Institute)
[23] Yu J, Lei J, Yu H, Cai X and Zou G 2004 *Elsevier Phytochemistry* **65** 881-4
[24] Ingroff A E, Fothergill A, Peter J, Rinaldi M G and Walsh T J 2002 *Journal of Clinical Microbiology* **40** 3204-8.
[25] Sasidharan N K, Sreekala S R, Jacob J and Nambisan B 2014 *Hindawi Publishing Corporation BioMed Research International* **2014** 1-8
[26] Singh G, Kapoor I P S, Singh P, de Heluani C S, de Lampasona M P and Catalan C A N 2008 *Food and Chemical Technology* **46** 3295-302
[27] Coyle M B 2005 *Manual of Antimicrobial Susceptibility Testing* (USA: American Society for Microbiology)
[28] EUCAST 2015 *EUCAST Definitive Document E.DEF 9.3: Method for the Determination of Broth Dilution Minimum Inhibitory Concentrations of Antifungal Agents for Conidia Forming Moulds* (Denmark: European Comittee of Antimicrobial Susceptibility Testing) p 1-23
[29] Afifurrahman, Samadin K H and Aziz S 2014 *MKS* **46** 266-70
[30] Gulluce M, Karaday M and Bars O 2013 *Formatex* **2013** 1016-27
[31] Yusuf M A and Hamid T H A T A 2013 *IOSR Journal of Pharmacy* **3** 44-50
[32] Garg N, Oman T J, Wang T S A, de Gonzalo C V G, Walker S and van der Donk W A 2014 *The Journal of Antibiotics* **67** 133-6
[33] Burt S 2004 *Elsevier International Journal of Food Microbiology* **94** 223-53
[34] Bouayed J 2012 *Nutrition Well Being and Health* (Croatia: InTech)
[35] Hu Y, Zhang J, Kong W, Zhao G and Yang M 2017 *Elsevier Food Chemistry* **220** 1-8
[36] Steeg P F T, Hellemons J C and Kok A E 1999 *Applied and Environmental Microbiology* **65** 4148-54

IOP Conf. Series: Materials Science and Engineering **633** (2019) 012006 doi:10.1088/1757-899X/633/1/012006

Isolation oligosaccharides from gembili (*Dioscorea esculenta* Lour. Burkill) as prebiotics

Y Khasanah[1], R Nurhayati[1], Miftakhussholihah[1], S Btari[2], E Ratnaningrum[2]

[1]Research Unit for Natural Product Technology, Indonesian Institute of Sciences
Jln. Jogjakarta – Wonosari km, 31.5 Yogyakarta, Indonesia
[2]Biology Faculty, Universitas Gadjah Mada

E-mail: yuniar.khasanah@gmail.com

Abstract. Oligosaccharides is a potential prebiotic and used as an alternative diet therapy for diabetes mellitus. Gembili (*Dioscorea esculenta* (Lour.) Burkill) is known as functional food that contains oligosaccharides. This study was done to isolate oligosaccharides from gembili and evaluate its prebiotics potential. Oligosaccharides was extracted and analysed by High Performance Liquid Chromatography (HPLC). Potential of prebiotics was analysed using faecal samples from healthy volunteers in anaerobic condition at certain time interval (0, 24, and 48 hours) and bacterial growth (Bifidobacteria and Lactobacillus as probiotic bacteria, Bacteroides and Clostridium as patogen bacteria), prebiotic index were analysed. Short Chain Fatty Acid (SCFA) production (acetic acid, propionic acid, and butyric acid) analyzed using HPLC. The results showed that oligosaccharides from gembili contains lactulose (0,231 %); inulin (2,541 %) and rafinose (1,485 %). Prebiotic index increased from 0,86 ± 0,20 (24 hours incubation) to 1,12 ± 0,05 (48 hours). This oligosaccharides increased probiotics and decreased patogen bacteria. It could be concluded that oligosaccharides from gembili were potential as prebiotics.

1. Introduction

Diabetes mellitus (DM), as an international public health concern and it is one of the five leading causes of death in the world and about six deaths per minute are attributable to diabetes complications [1]. In 2015, about 415 million adults had diabetes, a 4-fold increase from 108 million in the 1980s. If there are no precautions, this number will continue to increase and in 2040 estimated it will increase to 642 million sufferers [2].

Prebiotic may play important role in prevention rather than in the treatment of human type 2 diabetes [3]. Disaccharides (lactulose and lactylol), oligosaccharides (oligosaccharides include rafinose, oligofructose, palatimose, isomaltose, lactosucrose) and polysaccharides (inulin and resistant starch) are prebiotics [4] [5].

Gembili *(Dioscorea esculenta* (Lour.) *Burkill)* is a functional food and contains oligosaccharide inulin [6]. Gembili also have several bioactive compounds such as dioscorin, diosgenin that can improve the body's defense mechanism (immunomodulator), prevention of metabolic disease (hypercholesterolemia, dyslipidemia, diabetes and obesity), inflammation and cancer [7]. Because of its functional content especially on oligosaccarides, this study was conducted to isolation olligosaccharides from gembili and evaluate its prebiotic potential.

2. Methods

2.1. Materials

Gembili flour, glucose standard (inulin, lactulose, glucose, galactose and fructose), chemical (for analysis), agar media (Plate count agar, columbia agar, rogossa agar, reinforced clostridial agar and tryptone soya agar), antibiotic (kanamycin, vancoycin and collistin) and feses from healthy people.

2.2. Oligosaccarides extraction

Oligosaccharides extraction Using ethanol (70%) by stirring for 15-20 hours at room temperature, filtered and washed with ethanol (70%), evaporated and drying at 60°C [8]. Total Dissolve Solids (TDS) from extract was analysed using vacuum oven method [9] and oligosaccharides analysed using HPLC with Metacharb 87°C with RID detector, eluent flow rate H_2O, 0.6 ml/min and temperature 85°C [10].

2.3. The potential of prebiotic

Faecal samples were obtained from a healthy human volunteer who had not been prescribed antibiotics for at least 6 months prior to the study and had no history of any gastrointestinal diseases from a healthy volunteer (35 – 37 years old). Fresh faecal dissolved in 0.1 M phosphate buffer saline (PBS) sterile, pH 7 with ratio 1: 10 (w / v), and homogenised with magnetic stirrer (200 rpm, 2 min). The liquid was taken and used as an inoculum for subsequent testing [11].

For preparation microbial culture, faecal slurry (10 ml) and oligosaccharide extract (1 % v/v) with the basal nutrient medium (90 ml, peptone water 2 g/l), yeast extract 2 g/l, NaC l0.1 g/l, K_2HPO4 0.04 g/l, KH_2PO4 0.04 g/l, $MgSO_4.7H_2O$ 0.01 g/l, $CaCl_2.6H_2O$ 0.01, $NaHCO_3$ 2 g/l, Tween 80 2 ml, haemin 0.02 g/l, vitamin K 110 µL, cysteine.HCl 0.5 g/l, and bile salts (sodium glycocholate andsodium taurocholate) 0.5 g/l, pH 7.0). The culture temperature was set at 37^0 C and with CO_2 gas for 10 minutes (anaerobic). The sample is taken at certain time interval (hours 0, 24, and 48) [11], and parameters are bacterial growth (*Bifidobacteria* and *Lactobacillus*) and microbial pathogens (*Clostridium* and *Bacteroides*) [12][13], prebiotic index [14], pH [10], and SCFA production (acetic acid, propionic acid, and butyric acid) by HPLC [13].

3. Result and Discussions

3.1. Oligosaccharides extraction

Extraction gembili flour (200 g) in ethanol 70% (2000 ml) yielded 165 ml of concentrated extract. It was analysed for total disolve solids (TDS) to determine amount of dissolve sugar which represented the content of oligosaccharides. The TDS content were 19.5%. Oligosaccharide test using HPLC result the parameter of oligosaccharides were lactulose, inulin and rafinose (table 1). The presence of rafinose show the monosaccharides content (glucose, galactose and fructose), so needs to be known (table 1).

IOP Conf. Series: Materials Science and Engineering **633** (2019) 012006 doi:10.1088/1757-899X/633/1/012006

Table 1. HPLC test results for analysis of oligosaccharides and monosaccharides from gembili

Parameter of oligosaccharides	Results (%)	Parameter of monosaccharides	Results (%)
Lactulose	0,231	Glucose	15,411
Inulin	2,541	Galactose	1,122
Rafinose	1,485	Fructose	9,042

The highest oligosaccharide in gembili was inulin (2,541 %), but it's lower than the other studies (14,629 %) [15]. The differences is due to extraction methode, using single extraction using ethanol (70 %) [8] and double extraction using thanol (30 %) [16]. The highest monosaccharide content is glucose (15,411 %), affect the prebiotic potential of gembili. It'is related to microorganisms both probiotic and pathogens bacteria using glucose for metabolism. The fructose content (9,042 %) proof of high inulin, because inulin is a polymer of fructose units which is associated with β- (2-1) glycosidic bonds with glucose terminal group [17].

3.2. The potential of prebiotic

3.2.1. Bacterial growth

The diversity of colon bacteria in faeces at the beginning showed (figure 1). *Bifidobacteria*, *Lactobacillus* and *Clostridium* have same population of 4 logs, while Bacteroides are 2 logs. This is influenced by many factors, especially diet, type of food consumed, illness, antibiotic exposure, etc. Colon microbiota in infant dominated by Actinobactera and Proteobacteria. However, getting older, more diverse compositions of colonic microbiota, and 65 years old there is an increase in Bacteroides and Clostridium custer IV [18][19].

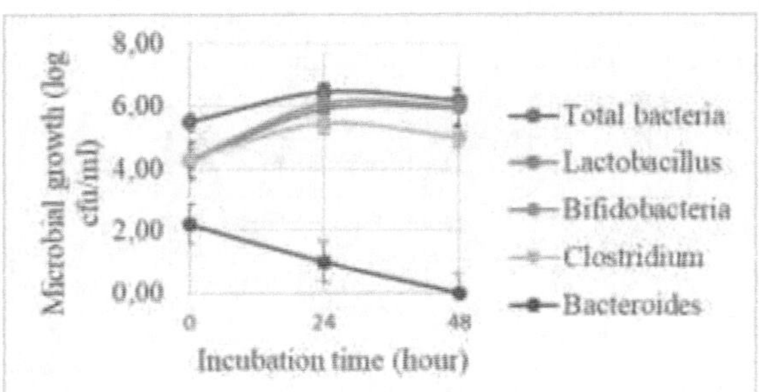

Figure 1. Colon bacterial growth in basal media containing oligosaccharides extract of gembili for 48 hours of incubation time

Incubation 24 hours increase Total bacteria, *Bifidobacteria and Lactobacillus* 2 logs and *Clostridium* 1 logs. The high growth of *Bifidobacteria* and *Lactobacillus* is supported by pH decrease (acidity) of the media (figure 2.). *Bifidobacteria* and *Lactobacillus* will produce lactic acid as the primary metabolite, and reduce the pH media. *Bacteroides* decrease (48 hours), because of acidic media environment and the inability of *Bacteroides* to ferment oligosaccharides for their growth.

Total bacteria and *Clostridium* decrease at 48 hours, because substrat in medium begins to diminish and pressure of the acidic media environment. The oligosaccharides can support the growth of probiotic bacteria and inhibit growth of pathogenic bacteria in the colon. Inulin, rafinosa and lactulosa in gembili extract can be a prebiotic source and support probiotics growth in human digestive tract [20][21].

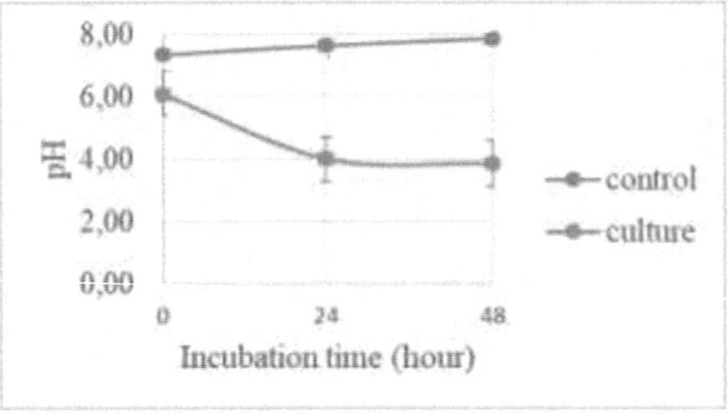

Figure 2. pH on the fermentation oligosaccharides extract of gembili for 48 hours of incubation time

3.2.2. Prebiotic index

The prebiotic index oligosaccharide extract of gembili increase from 0.86 ± 0.20 (24 hours) to 1.12 ± 0.05 (48 hours) (Table 2). Substrate from oligosaccharide extracts of gembili supports the growth of probiotic bacteria and decrease the pathogenic bacteria. The prebiotic index optimal at 24 hours and decreases at 48 hours, in accordance with the availability of substrate and environmental conditions [11]. Oligosaccharides extract of gembili used in this study is crude extract, where the purity of oligosaccharides is low and there is high glucose. The presence of glucose still be used by pathogenic to grow, especially in the first 24 hours.

Table 2. Prebiotic index on the fermentation oligosaccharides extract of gembili for 48 hours of incubation time

Incubation time	Prebiotic index
24	0,86 ± 0,20
48	1,12 ± 0,05

3.2.3. Production of short chain fatty acid (SCFA)

There was an increase on SCFA at the end of fermentation. It show the activity of prebiotic consumption and the production of metabolites in the form of SCFA. The highest concentration of acetic acid compared to propionic acid and butyric acid. The highest concentration of acetic acid was followed by the concentration of propionic acid and butyric acid [22]. This can be caused by other probiotic bacteria that are able to produce SCFA, but pathogenic bacteria such as members of the *Bacteroides* also have the potential to produce health by producing SCFA, example, *Bacteroides vulgatus* which is capable of producing acetic acid and formic acid.

Table 3. SCFA compotition on the fermentation oligosaccharides extract of gembili

Fatty acid	0 hours	48 hours
Acetic acid	21,07	32,34
Pronionic acid	4,75	8,18
Butyric acid	0	0,23

4. Conclusions

Olligosaccarides from gembili contains lactulosa (0,231 %); inulin (2,541 %) and rafinosa (1,485 %) and prebiotic index increased from 0,86 ± 0,20 (24 hours) to 1,12 ± 0,05 (48 hours). This olligosacarides increased probiotics and decreased patogen bacteria. So, olligosaccarides from gembili were potential as prebiotics.

References

[1] Li HT, Wu XD, Davey AK and Wang J. 2011 Antihyperglycemic effects of baicalin on streptozotocin – nicotinamide induced diabetic rats *Phytother. Res.* 25, pp 189–194

[2] Wild S, Roglic G, Green A, Sicree R and King H. 2004 Global Prevalence of Diabetes: Estimates for the year 2000 and Projections for 2030 *Diabetes Care*, vol 27 (5), pp 1047–1053

[3] Pandersen C, Gallagher E, Horton F, Ellis RJ, Ijaz UZ, Wu H, Jaiyeola E, Diribe O, Duparc T, Cani PD, Gibson GR, Hinton P, Wright J, La Ragione R and Robertson MD 2016 Host-microbiome interactions in human type 2 diabetes following prebiotic fibre (galactooligosaccharide) intake. *British Journal of Nutrition*, **116** (11). pp. 1869-1877. ISSN 0007-1145 doi: https://doi.org/10.1017/S0007114516004086

[4] Fotiadis CI, Stoidis CN, Spyropuolos BG and Zografos ED 2008 Role of probiotics, prebiotics and synbiotics in chemopreention for colorectal cancer. *World Journal of Gastroenterology*: **14** (42), pp 6453-6457

[5] Gibson GR., Scott KP, Rastall RA., Tuohy KM., Hotchkiss A, Dubert FA, Gareau M, Murphy EF, Saulnier D, Loh G, Macfarlane S, Delzeme N, Ringle Y, Kozianowski G, Dickmann R, Lenoir WI, Walker C and Buddington R 2010 Dietary prebiotics: current status and new definition. *Food Science and Technology Bulletin: Functional Foods* **7**, pp 1-9

[6] Serafica BCK, Sembiring L. Retnaningrum E, Khasanah Y 2015 *Isolasi oligosakarida dari tepung umbi gembili (Dioscorea esculanta (Lour.) Burkill) dan potensinya sebagai prebiotik*, Skripsi, Fakultas Biologi, Universitas Gadjah Mada , Yogyakarta

[7] Prabowo AY, Estiasih, T, Purwantiningrum I 2014 Umbi Gembili (Dioscorea Esculenta L.) Sebagai Bahan Pangan Mengandung Senyawa Bioaktif : Kajian Pustaka. *Jurnal Pangan dan Agroindustri* **2** (3), p 129 –135

[8] Muchtadi D 1989 *Petunjuk Laboratorium Evaluasi Nilai Gizi Pangan*. Pusat Antar Universitas Pangan dan Gizi. Institut Pertanian Bogor. Bogor

[9] AOAC (Association of Official Analytical Chemists) 1995 Official Methods of Analysis of AOAC International 2 **vols. 16a** edition. Arlington, VA, USA, Asociacion de Comunidades Analiticas.

[10] Apriyantono A, Fardiaz D, Puspitasari NL, Yasni S and Budiyanto S 1989 *Petunjuk Praktikum Analisis Pangan*. IPB Press. Bogor

[11] Vulevic J, Rastall RA and Gibson GR 2004 Developing a quantitative approach for determining the in vitro prebiotic potential of dietary oligosaccharides. *FEMS Microbiology Lett*, **236**, pp 153–159

[12] Vardakou M, Palop CN, Christakopoulos P, Faulds CB, Gasson MA and Narbad A 2008 Evaluation of the prebiotic properties of wheat arabinoxylan fractions and induction of hydrolase activity in gut microflora. *International Journal Food Microbiology*, **123,** pp 166–170

[13] Polnaya FJ 2013 *Karakteristik Fisiko-Kimia dan Potensi Prebiotik Pati Sagu (Metroxylon rumphii) Fosfat*. Disertasi. Program Pascasarjana Fakultas Teknologi Pertanian, Universitas Gadjah Mada, Yogyakarta

[14] Palframan R., Gibson GR and Rastall RA 2003 Development of a quantitativetool for the comparison of the prebiotic effect of dietary oligosaccharides. *Lett Appl Microbiol* **37**, pp 281–284

[15] Winarti S, Harmayani E and Nurismanto R 2011 Karakteristik dan profil inulin beberapa jenis uwi (Dioscorea spp.) *Agritech*, vol **31**, no 4, November, pp 378 – 383

[16] Widowati TC, Sunarti and Zaharani A 2005 Esktraksi, karakterisasi dan kajian potensi prebiotik inulin umbi dahlia (Dahlia pinnata L. Seminar rutin puslitbang tanaman pangan, Bogor

[17] Yang Z, Hu J and Zhao M. 2011. Isolation and quantitative determination of inulin type oligosaccharides in roots of Morinda officinalis. *Carbohydrate Polymers* **83**, pp 1997-2004.

[18] Rodriguez JM et.al 2015 The Composition of the Gut Microbiota Throughout Life, with an Emphasis on Early Life. *Microbial Ecology in Health and Disease* **26**, 26050.

[19] Thursby E and Juge N 2017 Introduction to the Human Gut Microbiota. Review Article. *Biochemical Journal* **474**, pp 1823-1836

[20] Tuohy KM, Finlay RK, Wynne AG, and Gibson GR 2001 A Human Volunteer Study on the Prebiotic Effects of HP-Inulin Faecal Bacteria Enumerated using Fluorescent in Situ Hybridization (FISH). *Anaerobe*, vol **7**, pp 113-118.

[21] Pacifici S, Song J, Zhang CK, and Tako E 2016 Evaluating the effect of plant origin prebiotics (Raffinose and Stachyose) on iron status, intestinal functionality and intestinal bacterial populations in vivo. *The FASEP Journal* 692.17

[22] Mandalari G, Palop CN, Tuohy, Gibson GR, Bennet RN, Waldron KW, Bisignano G, Narbad A and Faulds CB 2007 In vitro evaluation of the prebiotic activity of a pectic oligosaccharide exctract enzymatically derived from bergamot peel. *Appl Microbio Biotechnol* vol **73**, p 1173-1179

Acknowledgments

We gratefully acknowledge for all support from colleagues, special work by technical staff and financial support from Research funding project LIPI authors would like to express their sincere gratitude to Kegiatan Unggulan LIPI 2017.

IOP Conf. Series: Materials Science and Engineering **633** (2019) 012007 doi:10.1088/1757-899X/633/1/012007

Occurrences of *Salmonella* spp. and *Escherichia coli* in chicken meat, intestinal contents and rinse water at slaughtering place from traditional market in Surabaya, Indonesia

R Yulistiani[1,2], D Praseptiangga[3], Supyani[4], Sudibya[5]

[1] Doctoral Program of Agricultural Science, Graduate School Program of Sebelas Maret University (UNS), Jalan Ir. Sutami 36 A, Kentingan, 57126, Surakarta, Indonesia
[2] Department of Food Technology, Faculty of Engineering, Universitas Pembangunan Nasional Veteran Jawa Timur, Jalan Raya Rungkut Madya Gunung Anyar 60294, Surabaya, Indonesia
[3] Department of Food Science and Technology, Faculty of Agriculture, Sebelas Maret University (UNS), Jalan Ir. Sutami 36 A, Kentingan 57126, Surakarta, Indonesia
[4] Department of Agrotechnology, Faculty of Agriculture,Sebelas Maret University (UNS), Jalan Ir. Sutami 36 A, Kentingan 57126, Surakarta, Indonesia
[5] Department of Animal Husbandry, Faculty of Agriculture,Sebelas Maret University (UNS), Jalan Ir. Sutami 36 A, Kentingan 57126, Surakarta, Indonesia

Email: ratnayulistiani@yahoo.co.id

Abstract. The purpose of this study was to compare the presence of *Salmonella* spp. and *Escherichia coli* in chicken meat, intestinal contents and rinse water at slaughtering place from traditional markets in Surabaya, Indonesia. A total of 120 samples (chicken meat, intestinal contents and rinse water, each of the 40 samples) were collected from the slaughtering place from traditional markets in Surabaya. The samples were analyzed the presence of *Salmonella* spp. and *Escherichia coli* in chicken meat, intestinal contents and rinse water. Isolation and identification *Salmonella* spp.and *Escherichia coli* were carried out according to the conventional culture-based method (ISO 6579:2002). This study showed *Salmonella* spp. was isolated by 85.00%, 57.50%, 52.50% of chicken meat, intestinal contents and rinse water, respectively. While *E.coli* was isolated by 77.50%, 65.00%, 75.00 % of chicken meat, intestinal contents and rinse water, respectively. Prevalence of *Salmonella* spp. and *E. coli* in chicken meat is higher than intestinal contents and rinse water. Thus, it is indicated that the contamination of *Salmonella* spp. and *Escherichia coli* in chicken meat not only comes from intestinal contents and rinse water.

1. Introduction

The process of poultry slaughtering includes the following phases : stunning and bleeding, scalding, defeathering, evisceration, rinsing/washing and chilling [1]. Chicken carcasses have higher pathogenic and spoilage bacteria counts than most other foods, where carcass can be contaminated at several points of processing operations such as during scalding, defeathering and evisceration as well as cross contamination from other birds and processing equipment. During chicken slaughtering, carcass can be contaminated with fecal matters from the chicken intestine.

Contamination of poultry meat with foodborne pathogens is an important public health problem, where many food poisoning bacteria contaminate chicken meat [2]. *Salmonella* spp. and *Escherichia coli* (*E. coli*) causes health problems in the world because they are the two most important foodborne pathogens of *Enterobacteriaceae* family as a cause of foodborne illness which transmitted through poultry meat. *Salmonella* is the second most common cause of foodborne illness, it is responsible for millions of cases of foodborne illness every year [3]. *Salmonella* could infect and caused food poisoning in humans through the handling of raw carcasses and their products, or through the consumption of undercooked poultry meat [4]. *E. coli*, a natural inhabitant of the human intestinal tract and warm-blooded animals, is used as an indicator bacterium because these bacteria acquires antimicrobial resistance faster than other conventional bacteria [5]. According to [6], diarrheal diseases caused by *E. coli* account for more than 4% of the total daily global disease burden every day and about 1.8 million deaths occur every year, of which 90% are children .

Food safety is an issue of growing public health concern [7], especially foodborne diseases associated with the consumption of poultry meat and its processed products [8]. During and after slaughtering, the bacteria from animal microbiota, the slaughterhouse environment, and the equipment used contaminate carcasses, their subsequent cuts, and processed meat products. Carcasses and cuts after animal killing could be contaminated by animal microbiota and the slaughterhouse environment [9]. Some of these bacterial contaminants can grow or survive during food processing and storage. Previous research [10], indicated that *Salmonella* spp. and *E. coli* found in chicken meat at traditional markets in Surabaya were resistant to some antibiotics. Thus, it is important to evaluate the source of *Salmonella* spp. and *E. coli* contamination at slaughtering place from traditional market. The purpose of this study was to compare the presence of *Salmonella* spp. and *Escherichia coli* in chicken meat, intestinal contents and rinse water at slaughtering place from traditional market in Surabaya, Indonesia.

2. Experimental

A total of 120 samples (chicken meat, intestinal contents and rinse water, each of 40 samples) was collected from chicken slaughtering places at traditional market locations in Surabaya, Indonesia.The collected samples were packed in a sterile plastic bag, transferred directly to the laboratory in the ice box for an immediate bacteriological testing.

The chicken meat, intestinal contents and rinse water samples were processed immediately for isolation and identification of *Salmonella* and *E. coli*. Isolation and identification *Salmonella* spp. from all samples were carried out according to the horizontal method [11] with modification. From each sample (25 gram of chicken meat/intestinal content samples or 25 ml rinse water samples) were pre-enriched in 225 ml buffered peptone water (BPW, Oxoid), and incubated at 37°C for 24 h. One ml of BPW pre-enrichment step transferred to 10 ml Selenite Cystine Broth (SCB, Oxoid) for selective enrichment, and incubated at 37°C for 24 h. A loopful of inoculum from selective enrichment broth was streaked onto Xylose-Lysine Deoxycholate (XLD) agar plate and incubated at 37°C for 24 h. For confirmation, single colonies that growing on the XLD agar surface with specific criteria for *Salmonella* spp. (pink colonies with black centers) was evaluated to gram staining and biochemical tests (TSI and IMViC test) were carried out [12].

Isolation and identification *E. coli* from all samples were carried out according to ISO 6579 : 2002. From each sample (25 gram of chicken meat/intestinal content samples or 25 ml rinse water samples) were chopped aseptically and homogenized with 225 ml of 0.1% Buffered Pepton Water (BPW, Oxoid) ; and then incubated at 37°C for 24 h. A loop full pre-enrichment broth was streaked on MacConkey agar plates (MCA, Oxoid) and incubated at 37°C for 24 h. For confirmation, single colonies with pink colour colonies on the MacConkey agar surface (specific criteria for *E. coli* spp.) was evaluated to gram staining and biochemical tests (TSI and IMViC test) were carried out [12]. The flow diagram in this research as shown in Figure 1.

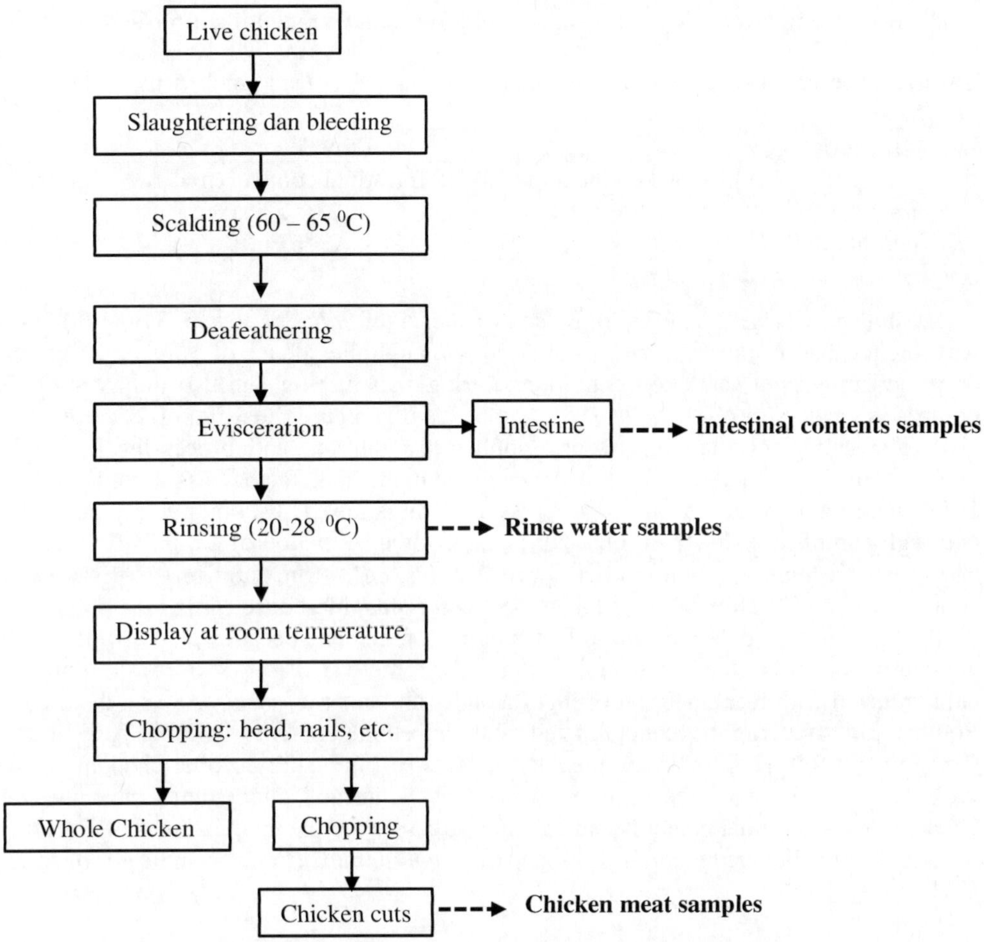

Figure 1. The flow diagram of chicken slaughtering in traditional markets with detailed description of *Salmonella* spp. dan *E. coli* sampling point

3. Results and Discussion

Table 1. Biochemical characterization of *Salmonella* spp. and *E. coli* isolates

Bacteria	Biochemical test					
	Triple Sugar Iron agar test	Motility	Indole production test	Methyl-red	Voges-Proskauer test	Simmon's citarate test
Salmonella spp.	A/A gas +, H_2S+ or K/A, or K/A gas + or K/A , H_2S or K/A gas, H_2S	+	-	+	-	±
Escherichia coli	A/A, or A/A gas + or K/A, or K/A gas +	+	+	+	-	-

Notes : K, Alkali (red) ; A, Acid (yellow)
+ , positive; - negative ; ± , positive/negative

Salmonella spp and *E. coli* isolates were obtained met the criteria for microscopic examination (short rod form, gram negative) and positive biochemical test results (TSI and IMViC) for *Salmonella* spp and *E. coli* (Table 1). TSI (Triple Sugar Iron) tests were carried out to check their ability to ferment glucose, lactose and sucrose sugars, gas and H_2S production. The IMViC test includes Indole production, Methyl Red, Voges-Proskauer, and Simmon Citrate utilization [13] [14].

Table 2. The prevalence (%) of *Salmonella* spp. and *E. coli* in chicken meat, intestinal contents and rinse water

Bacteria	Prevalence (%)		
	Chicken meat (n=40)	Intestinal contens (n=40)	Rinse water(n=40)
Salmonella spp.	34(85.00)	23(57.50)	21(52.50)
Escherichia coli	31(77.50)	25(65.00)	30(75.00)

Notes: n = sample numbers collected

As shown in Table 2 and Figure 2, the prevalence of *Salmonella* spp. 34 (85.00 %) in chicken meat was higher than *Escherichia coli* 31 (77.50 %). High prevalence of *Salmonella* in chicken meat, not only indicating poor sanitation conditions during slaughtering but also indicates the health status of poultry as carriers of *Salmonella* [15], where poultry is the most important reservoir of *Salmonella* [16]. The unhygienic handling during poultry slaughtering and processing of chicken meat using unclean equipment and contaminated water were the risk factors associated with the presence of *Salmonella* with chicken meat due to cross contamination [17]. Poor sanitary and hygiene conditions and high humidity at slaughterhouses are ideal for the formation of *Salmonella* biofilms. This biofilm can last for a long time and tends to protect *Salmonella* from sanitizers, so it is very risky to cross contamination of *Salmonella* in chicken meat at the slaughterhouse [18].

Poultry food products are important sources of *E. coli*, because at the time of slaughter, faecal contamination from the intestines contaminated the carcass. As a result, poultry meat can be contaminated with fecal material or ingesta and with bacteria associated with these contaminants [19]. Poultry carcasses can be contaminated with intestinal contents which *E. coli* can spill out as a contaminant. Ref. [20] found the prevalence of *E. coli* in healthy broiler chickens was 88%, 38% and 25%, respectively, from the samples of caecum, ileum and duodenum (intestinal contents). Broiler chicken meat, intestinal contents and chicken carcasses are the main sources of *E. coli* contaminants in meat and warm-blooded animals. Presence of *E. coli* in meat is a good indicator of fecal contamination [21].

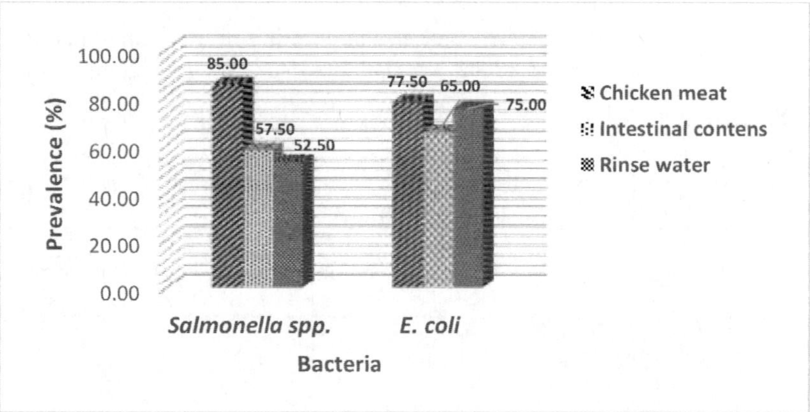

Figure 2. The comparison of *Salmonella* spp. and *E. coli* in chicken meat, intestinal contents and rinse water.

Our study (Figure 2), demonstrated *Salmonella* spp. was isolated by 85.00%, 57.50%, 52.50% of chicken meat, intestinal contents and rinse water, respectively. While *E.coli* was isolated by 77.50%, 65.00%, 75.00 % of chicken meat, intestinal contents and rinse water, respectively. Evisceration is the first stage of the clean part of the slaughter process. It is consisting of several stages, evisceration starts with head removal followed by the opening of the body cavity, removal of intestines, and ends with the cleaning of the carcass [22]. From the hygienic point of view, attention is paid to the removal of the intestines and the prevention of cross-contamination with faecal material. Prevalence of *Salmonella* spp. and *E. coli* in chicken meat are higher than intestinal contents and rinse water. Thus, it indicates that contamination of *Salmonella* spp. and *Escherichia coli* in chicken besides come from the intestinal contents and rinse water, but also can come from others, such as equipment surfaces, animal microbiota and the slaughtering place environment. During slaughtering, bacterial contamination may occur from equipment surfaces, water, and animal microbiota. Chicken carcasses can be contaminated by the slaughterhouse environment [9].

4. Conclusion

The presence of *Salmonella* spp. and *Escherichia coli* in chicken meat are higher than intestinal contents and rinse water. Thus, it is indicated that the contamination of *Salmonella* spp. and *Escherichia coli* in chicken meat not only comes from intestinal contents and rinse water. The need of implementing a stricter hygiene and sanitation standard in slaughtering place to reduce the incidence of *Salmonella* and *E. coli*.

References

[1] Escudero-Gilete ML, Gonzalez-Miret ML and Heredia FJ 2005. Multivariate study of the decontamination process as function of time, pressure and quantity of water used in washing stage after evisceration in poultry meat production. *J. Food Eng.* **69** 245-51.
[2] Mbata TI 2005. Poultry meat pathogens and its control. *Internet J Food Safety* **7** 20-28.
[3] FSIS 2008. FSIS Issues Public Health Alert for Frozen, Stuuffed Raw Chicken products.
[4] Panisello PJ, Rooney R, Quantick PC and Stanwell-Smith R 2000. Application of foodborne disease outbreak data in the development and maintenance of HACCP systems. *Int. J. Food Microbiol.* **59** 221-234.
[5] Miranda, JM, Vázquez BI, Fente CA, Barros-Velázquez J, Cepeda A and Franco CM 2008. Evolution of resistance in poultry intestinal *Escherichia coli* during three commonly used antimicrobial therapeutic treatments in poultry. *Poult. Sci.* **87** 1643-1648.
[6] WHO 2004. The World Medicines Situation. Geneva, Switzerland:World Health Organization
[7] Nata Menabde. 2015. Message from WHO representative to India on World Health Day 2015. http://www.searo. who.int/india/about/who_representative
[8] Bhaisare DB, Thyagarajan D, Churchil RR and Punniamurthy N 2014. Bacterial pathogens in chicken meat: Review. *Int J Life Sci Res* **2** 1-7.
[9] Firildak G, Asan A and Goren E 2015. Chicken carcasses bacterial concentration at poultry slaughtering facilities. *Asian J. Biol. Sci* **8** 16-29
[10] Yulistiani R, Praseptiangga D, Raharjo D, Supyani, Sudibya and Shirakawa, T 2017. Prevalence of antibiotic-resistance *Enterobacteriaceae* strains isolated from chicken meat at traditional markets in Surabaya, Indonesia. IOP Conference Series: Materials Science and Engineering 193 (2017) 012007. IOP Publishing. doi:10.1088/1757-899X/193/1/ 012007
[11] International Organization for Standardization (ISO) 2002. Microbiology of Food and Animal Feeding Stuffs : Horizontal Method for the Detection of *Salmonella* spp. ISO 6579 : 2002. Geneva, Switzerland.
[12] Roy SR , Rahman MB, Hassan J and Nazir KNH 2012. Isolation and identification of bacterial flora from internal organs of broiler and their antibiogram studies. *Microbes and Health* **1** 72-75

[13] Morello JA, Granato PA and Mizer HE 2002. Laboratory manual and workbook in microbiology. Applications to patient care. 7th Ed. ISBN :0-07-246354-0. WCB/McGraw-Hill.
[14] Paiao FG. Arisitides LGA, Murate LS,Vilas-Bôas GT, Vilas-Boas LA and Shimokomaki M 2013. Detection of *Salmonella spp, Salmonella enteritidis* and *typhimurium* in naturally infected broiler chickens by a multiplex PCR-based assay. *Braz. J. Microbiol.* **44** 37-42
[15] Shah AH and Korejo NA 2012.Antimicrobial resistance profile of *Salmonella* serovars isolated from chicken meat. *J. Vet. Anim. Sci.* **2** 40-46
[16] Balakrishnan S, Sangeetha A and Dhanalakshmi M 2018. Prevalence of *Salmonella* in chicken meat and its slaughtering place from local markets in Orathanadu, Thanjavur district, Tamil Nadu. *J. Entomol Zool. Stud* **6** 2468-2471
[17] Nidaullah, H., Abirami N, Shamila-Syuhada AK, Chuah LO, Nurul H, Tan TP, Abidin FWZ and Rusul G 2017. Prevalence of *Salmonella* in poultry processing environments in wet markets in Penang and Perlis, Malaysia. *Vet. World* **10** 286 – 292
[18] Smith DP, Northcutt JK, Cason JA, A. Hinton JR, Buhr RJ and Ingram KD 2007. Effect of external or internal faecal contamination on numbers of bacteria on prechilled broiler carcasses. *Poult. Sci.* **86** 1241–1244
[19] Seidavi A, Mirhosseini SZ, Shivazad M, Chamani M, Sadeghi AA and Pourseify R 2010. Detection and investigation of *Escherichia coli* in contents of duodenum, jejunum, ileum and cecum of broilers at different ages by PCR. *Asia Pac. J. Mol. Biol. Biotechnol* **18** 321-6
[20] Mead GC 2000. Fresh and Further-Processed Poultry. In: The Microbiological Safety and Quality of Food, Lund BM, Baird-Parker TC and Gould GW(Eds.). Vol. 1, Chapter 20, Aspen Publication, Gaithersburg, MD., USA., ISBN-13: 9780834213234, pp: 445-471
[21] Cox JM and Pavic A 2010: Advances in Enteropathogen control in poultry production. *J. Appl.Microbiol.* **108** 745-755
[22] Rouger A, Tresse O and Zagorec M 2017. Bacterial contaminants of poultry meat: sources, species, and dynamics. *Microorganisms* **5** 50

IOP Conf. Series: Materials Science and Engineering **633** (2019) 012008 doi:10.1088/1757-899X/633/1/012008

Characteristics stability of cheese whey-tomato juice fermented beverages with pectin and CMC addition

N Saidah, E Nurhartadi, A Nursiwi, A M Sari and R Utami
Departement of Food Science and Technology, Faculty of Agriculture, Sebelas Maret University, Indonesia

Email: edhi.nr@gmail.com

Abstract. The aim of this study was to investigate the chemical and microbiological stability of 12 days of storage (±4°C) of fermented beverages from cheese whey and tomato juice with addition of pectin:carboxy methyl cellulose combination (0%:0%; 0.15%:0.25%; 0.25%:0.15%; 0.2%:0.2%; 0.1%:0.3%) and to determine the best formula to find out it's hypocholesterolemic effects. Fermented cheese whey beverages was prepared from cheese whey:tomato juice (95%:5%), sucrose (10% w/v), and 5 variation of pectin:CMC combination. This research used a Completely Random Design (CRD) with one factor, combination of pectin and CMC. The result was analyzed by applying one-way analysis of variance (ANOVA) followed by Duncan test with 5% significant level. The result showed that during 12 days of storage, the addition of pectin:CMC combination in fermented beverages from cheese whey reduced the rate of pH, antioxidant activity, total lactic acid bacteria, and the rate of lactic acid rise. The best formula of fermented beverage from cheese whey and tomato juice is fermented beverage with 0.1%:0.3% (w/v) of pectin:CMC addition which has 34.636% of hypocholesterolemic effect.

1. Introduction

Whey is a liquid separated from milk curd after coagulation of milk, cream, skim milk or butter of milk on cheese, casein or similar products especially with rennet type enzymes [1]. Whey consists of 96% water, a total milk solid of 50-55%, 70% lactose, 20% protein, 70-90% minerals, and almost all water-soluble vitamins such as vitamin B complex and vitamin C [2]. Conversion of whey into a fermented beverage is one of the best ways of utilizing whey for human consumption [3]. The addition of fruit juice can improve the physicochemical properties of fermented beverages [4].

Fermented cheese whey beverages using 1% *Lactobacillus acidophilus* and *Lactobacillus plantarum* with a ratio 1:1 gives a good taste. *Lactobacillus acidophilus* produced more lactic acid than other LAB, *Lactobacillus plantarum* could improved sensory properties and inhibited pathogenic microorganisms, so the combination of both starter improved physicochemical and sensory properties of fermented beverages [5]. *Lactobacillus acidophilus* and *Lactobacillus plantarum* are lactic acid bacteria that can lower cholesterol [6, 7, 8, 9]. Tomato juice addition on fermented beverage can increase nutrical values of fermented beverages. Tomato contains phytosterols, B vitamins, and potassium which can lower cholesterol, and contains high antioxidants [10, 11]. Tomato fruit is also a source of fructooligosaccharide (FOS). FOS serves as a prebiotic that can stimulate the growth of lactic acid bacteria [12].

Separation of fermented cheese whey beverage might take place during storage. This separation occurs because high methoxyl pectin can form gel in acidic environments [13]. So to apply pectin in an acidic fermented beverage, pectin needs to be combined with a more acid-resistant stabilizer i.e. carboxy methyl cellulose.

Pectin is a polymer of D-galacturonic acid linked to α- (1,4) -glucose bonds having gelling, thickening, and stabilizing properties [14]. According to a study by Schonbrun (2002), the addition of pectin in yoghurt showed high viscosity and overall acceptability. Pectin has a large water holding capacity [15]. However, high pectin metoxyl can form gel in acidic environments. Carboxy methyl cellulose (CMC) is a polysaccharide hydrocolloid containing a α- (1,4) -D-glucose- polymer derived with carboxy methyl groups. Carboxy methyl cellulose can be used as a stabilizer in dairy products, because it can react with casein near the isoelectric point, producing a soluble complex [13].

This study was to examine the effect of pectin and carboxy methyl cellulose on whey fermented beverage characteristics with the addition of tomato juice (*Solanum lycopersicum*) using starter *Lactobacillus acidophilus* FNCC 0051 and *Lactobacillus plantarum* FNCC 0027, to find the stability of formula chemically and microbiologically during storage of low temperatures, as well as to find the cholesterol-lowering activity of selected formulas in-vitro.

2. Materials and Methods

2.1. Materials

The ingredients used for the manufacture of fermented beverages in this study are cheese whey mozarella wastes collected from the cheese industry in Boyolali, fresh ripe tomatoes (tomato varieties of permata) bought from Pasar Gede Surakarta, sucrose (Gulaku), *Lactobacillus acidophillus* FNCC 0051 and *Lactobacillus plantarum* FNCC 0027 taken from Food and Nutrition Culture Collection (FNCC), Food and Nutrition Laboratory, Gadjah Mada University, Yogyakarta.

2.2. Methods

2.2.1.Mother Starter

Each pure culture of *Lactobacillus acidophilus* FNCC 0051 and *Lactobacillus plantarum* FNCC 0027 propagated by aseptically taking 1 ose bacterial culture then put in test tubes containing 5 ml sterile MRS broth and then incubated at 37°C for 24 hours. The results of bacterial culture in MRS broth was then centrifuged at 1000 rpm for 10 minutes. Natan which was obtained would then be used as many as 3 ose and inoculated in 5 ml of 10% skim milk solution which has passed the pasteurization process at a temperature of 85°C for 30 minutes. Furthermore, they were then to be incubated at 37°C for 24 hours [16, 17].

2.2.2. Culture Starter

Manufacture of culture starter was with 100 ml of 10% skim milk pasteurized at 85°C for 30 minutes and then cooled to 37°C. After that, inoculated with 2% mother starter, then incubated at 37°C for 24 hours [17].

2.2.3. Tomato Juice

Fresh red tomato fruit washed and cut into small pieces then put into juicer to separate the pulp and its juice. The tomato juice that has been obtained then filtered using a filter cloth.

2.2.4. Fermented Whey Beverage

The preparation of fermented beverages was done by mixing homogenous cheese whey (95%) with sucrose (10% w/v of total cheese whey and tomato juice), and pectin: CMC (0%: 0%; 0.15%: 0, 25%, 0.25%: 0.15%, 0.2%: 0.2% and 0.1%: 0.3% w/v of the total volume of cheese whey and tomato juice) which have been mixed dry. Tomato juice (5%) was then added inlto the cheese whey mixture, sucrose, and stabilizer (pectin: CMC). The ingredients are mixed, then homogenized and continued pasteurization at 75°C for 30 minutes. The heat was cooled and inoculated with 1% starter *Lactobacillus acidophilus* FNCC 0051 and *Lactobacillus plantarum* FNCC 0027 with 1: 1 ratio (v/v of total cheese whey and tomato juice) with 10^8 CFU / ml starter. Furthermore, it was incubated at 37°C for 18 hours [18, 19].

2.2.5. Sample Analysis
The characteristic stability of the five varieties of fermented cheese whey beverage samples were analyze (pH, lactic acid content, antioxidant activity, and total lactic acid bacteria) during 12 days of low temperature storage to obtain the preferred formula. The selected formula was then tested for in-vitro cholesterol-lowering activity. The packaging is done during storage using a sealed glass bottle.

2.2.6. Cholesterol Lowering Activity Analysis
10 μl of standard cholesterol was added to 50 ml of tomato juice inoculated 1% of probiotic bacteria *Lactobacillus acidophilus* FNCC 0051: *Lactobacillus plantarum* FNCC 0027 (1: 1) and not inoculated bacteria (control sample) which then incubated in incubator shaker 37°C for 24 hours. The samples were incubated for 24 hours, then taken 40 μl and added with 4000 μl of Cholesterol FS Diasys reagent. The mixture of sample and reagent then homogenized continued incubation at room temperature for 20 minutes. The sample absorbance was subjected to a spectrophotometer with λ 500nm [20]. Cholesterol levels are then calculated using standard curve equations.

3. Results and Discussion

3.1. Antioxidant Activity
During 12 days of storage, all samples of fermented drinks decreased antioxidant activity. The table for decreasing antioxidant activity during storage can be seen in **Figure 1**. The activity of antioxidants during storage due to degradation of antioxidant compounds contained in fermented beverages [21]. Antioxidant compounds such as ascorbic acid would eliminate oxygen, control the activity of polyphenol oxidase enzymes, and increase the stability of lycopene, so that during storage the amount would decrease. During storage, the phenolic component in the fermented beverage would be combined with ascorbic acid to inhibit the occurrence of oxidation reactions, so the amount also decreased. An increase in antioxidant activity on day 3 might be caused by antioxidant activity produced by *Lactobacillus acidophilus* and *Lactobacillus plantarum*. 0.1% pectin and 0.3% CMC might be the best prebiotic combination that could stimulated bacterial growth to produced higher antioxidant compound. Probiotic lactic acid bacteria demonstrated antioxidant ability both in-vivo and in-vitro. Lactic acid bacteria might resist reactive oxygen species (ROS), peroxide radicals, superoxide anions, and hydroxyl radicals.

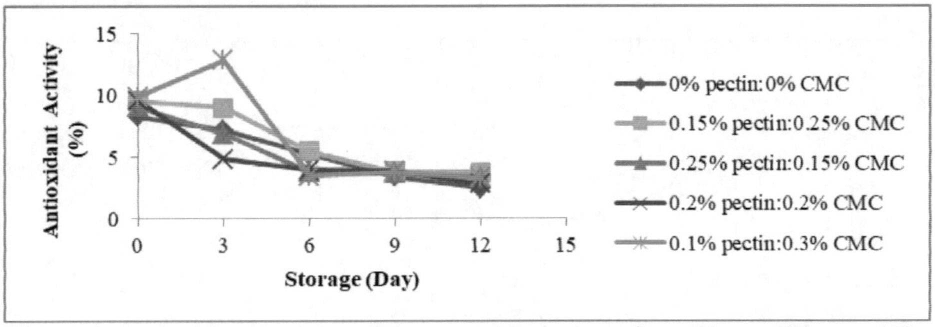

Figure 1. Antioxidant Activity of Fermented Beverages from Cheese Whey and Tomato Juice with Variation of Pectin:CMC Combination During Cold Storage ± 4°C

3.2. Total Lactic Acid Bacteria, Lactic Acid Level, and pH
The results in **Figure 2** showed that the amount of lactic acid bacteria in the cheese whey fermented beverage sample decreased for 12 days of storage. The reducing viability of lactic acid bacteria during storage can be caused by the decreasing amount of nutrients available in fermented beverages from cheese whey and tomato juice. During storage, lactose and protein as a source of nutrients lactic

acid bacteria would reduce because it has been converted into lactic acid and other organic acids. This caused lactic acid bacteria could not survive due to lack of nutrients. During 12 days of storage, all samples of fermented beverages increased lactic acid levels. This was because during the storage fermented beverages still underwent a fermentation process that made lactic acid bacteria had more time to utilize nutrients especially lactose in its medium [22, 23]. The amount of *Lactobacillus acidophilus* and *Lactobacillus plantarum* during this reasonably high storage temperature of ± 4°C (**Figure 2**), caused lactic acid as *Lactobacillus acidophilus* and *Lactobacillus plantarum* metabolited to increase during storage. This increasing of lactic acid levels, caused the pH of fermented drinks to decrease during the storage of low temperatures.

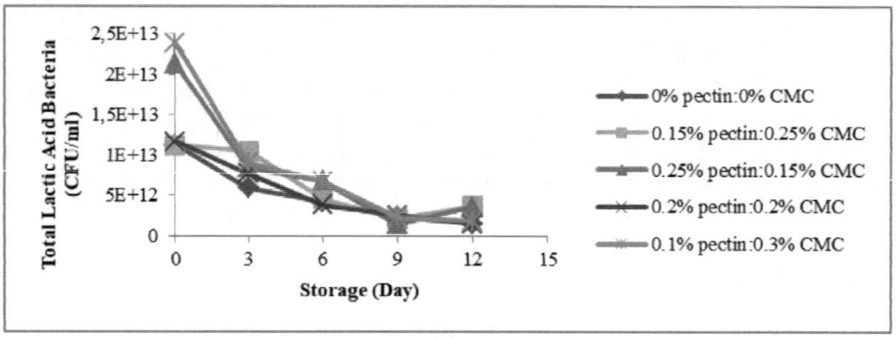

(a)

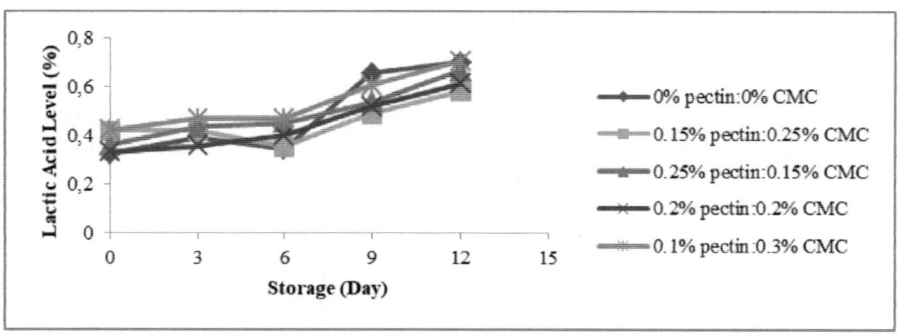

(b)

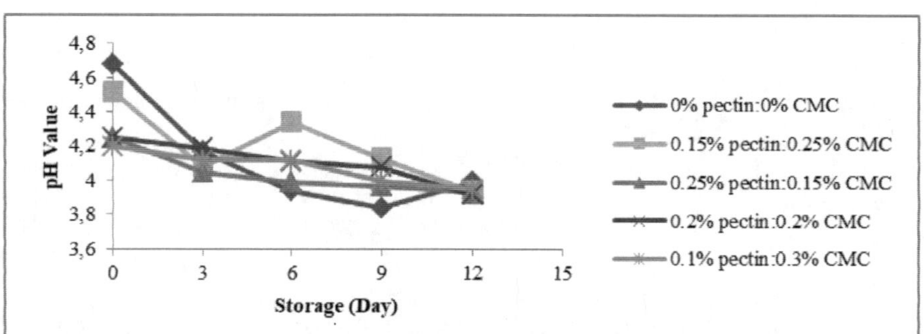

(c)

Figure 2. (a) Total Lactic Acid Bacteria (b) Lactic Acid Level (c) pH Value Fermented Beverages from Cheese Whey and Tomato Juice with Variation of Pectin:CMC Combination During Cold Storage Storage ± 4°C

IOP Conf. Series: Materials Science and Engineering **633** (2019) 012008 doi:10.1088/1757-899X/633/1/012008

3.3. Cholesterol Lowering Activity of Fermented Beverage from Cheese Whey and Tomato Juice Selected Formula

Formulas of fermented beverage from cheese whey and tomato juice with the addition of pectin:CMC stabilizers was selected using an effectiveness index test [24]. Based on the results of effectiveness index test, fermented cheese whey with 0.1% pectin: 0.3% CMC selected for continued cholesterol-lowering activity test in vitro. The results of cholesterol measurements of the sample can be seen in **Table 1**. Beverages of cheese whey fermentation with the addition of tomato juice and stabilizer can lower cholesterol levels through several mechanisms. There are three possible mechanisms of probiotics in lowering cholesterol, the first mechanism of fermented products by LAB inhibits synthetic cholesterol thereby lowering cholesterol production. The second mechanism is through the removal of bile salts through the feces, where the conjugated bile salts are not absorbed by the intestine, and are more easily removed from the gastrointestinal tract than the conjugated bile salts. This results in more cholesterol needed to synthesize bile salts again so that it will lower cholesterol levels. The third mechanism is the ability of LAB to bind cholesterol so as to lower cholesterol levels in the media [25].

Table 1. Results Measurement of Cholesterol Levels In-vitro

Sample	Absorbance	Cholesterol Level (mg/dl)	Cholesterol Reduction (mg/dl)	Cholesterol Reduction (%)
Control	0.206	1.941	0.059	2.957
Sample	0.136	1.307	0.693	34.636

4. Conclusion

The result showed that during 12 days of storage, the addition of pectin:CMC combination in fermented beverages from cheese whey improved the rate of viscosity rise, reduced the rate of dispersion index decline, pH, antioxidant activity, total lactic acid bacteria, and the rate of lactic acid rise. The best formula of fermented beverage from cheese whey and tomato juice was fermented beverage with 0.1%:0.3% (w/v) of pectin:CMC addition which has 34.636% of hypocholesterolemic effect.

References

[1] Badan Standarisasi Nasional 1996 *SNI 01-4220:1996 Tentang Tajian Susu (Whey) Bubuk* (Jakarta: BSN)
[2] Archer RH 2013 III-Dairy-G-Whey Products *Protein and Powder Technology Section* (New Zealand: Dairy Research Institute Press)
[3] Kumar PA and Bangaraiah P 2014 *Int J Pharm Bio Sci* **5(4)** 1101-11
[4] Widagdha S and Nisa FC 2015 *Jurnal Pangan dan Agroindustri* **3(1)** 248-258
[5] Nursiwi A, Nurhartadi E, Utami R, Sari AM, Laksono PW and Aprilia EN 2016 *IOP Conf. Series: Materials Science and Engineering* ***193*** *012009*
[6] Gilliland SE, Nelson CR and Maxwell C 1985 *Applied and Environmental Microbiology* **49(2)** 377-381
[7] Guo LD, Yang LJ and Huo GC 2011 *Czech J. Food Sci* **29(3)** 219–225
[8] Vani M, Prakash MS and Devi PY 2012 *Indian Journal of Clinical Practice* **23(4)** 224-230
[9] Fuentes MC, Lajo T, Carrio JM and Cune J 2013 *British Journal of Nutrition* **109** 1866–72
[10] Bhowmik D, Kumar KPS, Paswan S and Srivastava S 2012 *Journal of Pharmacognosy and Phytochemistry* **1(1)** 33-43
[11] Nasir MU, Hussain S and Jabbar S 2015 *Science* **3(1)** 1-5
[12] Maltos DAF, Mussatto SI, Esquivel, Herrera RR, Teixeira JA and Aguilar CN 2014 *Critical Reviews in Biotechnology* **36(2)** 259-267
[13] Schonbrun R 2002 *Thesis* University of Florida
[14] Winarno FG 2004 *Kimia Pangan dan Gizi* (Jakarta: PT. Gramedia Pustaka Utama)

[15] Arioui F, Saada DA and Cheriguene A 2016 *Food Science and Nutrition* (Algeria: Wiley Periodical, Inc)
[16] Setioningsih E, Setyaningsih R and Susilowati A 2004 *Bioteknologi* **1(1)** 1-6
[17] Utami R, Andriani MAM and Putri ZA 2010 *Caraka Tani* **25(1)** 50-55
[18] Maryana D 2014 *Thesis* Hasanuddin University Makassar
[19] Krasaekoopt W and Cabraal TL 2011 *AU J.T.* **14(4)** 253-258
[20] Fadhilah AN, Hafsan and Nur F 2015 *Prosiding Seminar Nasional Mikrobiologi Kesehatan dan Lingkungan*
[21] Porto MRA, Okina VS, Pimentel TC and Prudencio SH 2017 *Beverages* **3(36)** 1-12
[22] Gianti I and Evanuarini H 2011 *Jurnal Ilmu dan Teknologi Hasil Ternak* **6(1)** 28-33
[23] Temesgen M 2015 *Food Science and Quality Management* **37** 90-102
[24] De Garmo EP, Sullivan WG and Canada JR 1984 *Engineering Economy (7th ed)* (New York: Macmillan Publishing Company)
[25] Lee DK, Jang S, Baek EH, Kim MJ, Lee KS, Shin HS, Chung MJ, Kim JE, Lee KO and Ha NJ 2009 *Lipids in Health and Disease* **8(21)** 1-8

Aknowledgement
This work was financially supported by research project by PNBP UNS No. 1073/UN27.21/PP/2017 from Sebelas Maret University, Indonesia.

IOP Conf. Series: Materials Science and Engineering **633** (2019) 012009 doi:10.1088/1757-899X/633/1/012009

Nata de cincau: Pacitan's black *cincau* (*Mesona palustris* BL) product innovation with respect to its physical, chemical and sensory characteristics

H L Naja, A M Sari and M Z Zaman

Department of Food Science and Technology, Faculty of Agriculture, Universitas Sebelas Maret, 57126 Indonesia.

Email: ardheamustikasari@staff.uns.ac.id

Abstract. Black *cincau* (*Mesona palustris* BL) is a particularly abundant commodity in Pacitan, Indonesia. It has a higher fibre and antioxidant content compared to other varieties. Black *cincau* is potentially processed into *nata de cincau* in order to diversify food products in Pacitan. The aim of this study is to define the best formulation of *nata de cincau*, considering its physical (yield, thickness and texture) and chemical (water and crude fibre) characteristics. The best three formulations, based on physical and chemical characteristics, were chosen for sensory evaluation (colour, flavour, taste, texture and overall). *Nata de cincau* processing used various concentration of substrates (extract black *cincau*) (2.5%, 7.5% and 12.5%) and a variety of starter concentrations (5%, 10% and 15%) were added. *Nata de cincau* with 12.5% substrate concentration and 5%, 10% and 15% starter concentration were the best three formulations in terms of all physical and chemical characteristics. Sensory evaluation was performed on the best three formulations; 12.5% substrate concentration and 15% starter concentration was found to be the best formulation.

1. Introduction

Black *cincau* (*Mesona palustris* BL) is a plant that grows wild on hillsides and forests in Pacitan, Wonogiri and Ponorogo, which is known locally as Janggelan. Black *cincau* is a shrub with a height of 30–60 cm and grows at an altitude of 150–1,800 above sea level [1]. Black *cincau* is highly abundant in Pacitan. The success of cultivating black *cincau* in this region has not been followed by a good post-harvest process. The most common post-harvest problem is the lack of processed black *cincau* diversification. Normally, people only sell black *cincau* in the form of dried black *cincau* and gel [2]. Black *cincau* has a nutritional content of 122 kcal energy, 6 g protein, 1 g fat, 26 g carbohydrates, 17 mg vitamin C and 40 % digestible ingredients [3]. Further, the antioxidant content of black *cincau* is higher compared to green *cincau* and the shrub *cincau* [4].

Nata is a processed food product which has the form of a gel, chewy texture and white-coloured [5]. According to some studies, nata can be made from coconut water [6], seaweed [7], pineapples [8], cassava [9] and melon with papaya [10]. Nata is an organic product that contains high fibre [11]. The ideal liquid substrate for *Acetobacter xylinum* in making nata has to contain at least carbohydrates, proteins and vitamins to provide a significant effect on growth and production of nata. Sodium, potassium, calcium, magnesium and iron have been found to be important nutrients in the formation of polysaccharides [12]. This study analyzes food products based on

Pacitan's local wisdom: nata made from black *cincau* (*nata de cincau*), This might be the first report on the utilization of black *cincau* as raw material of nata production. The influence of substrate and starter culture concentration on physical and chemical characteristic as well as sensory attributes of nata de cincau was studied in this study.

2. Experimental

2.1. Sample preparation

Black *cincau* (*Mesona palustris* BL) was weighed to 2.5%, 7.5% and 12.5% (w/v). Black *cincau* was then boiled with one litre of water for five minutes at 100^{0}C. The next process was the disposal of black *cincau* dregs, followed by obtaining the extract water. The water was boiled again, then 12.5% sugar, 2.5% ammonium sulphate and 1.5% citric acid were added to the boiling water. The extract water was poured into a tray, which was tightly closed. The extract water was left to stand for 24 hours at room temperature. *Acetobacter xylinum* obtained from local producer of nata de coco at Sragen City, Central Java was then inoculated into the extract water at 5%, 10% and 15% (v/v) concentration and then fermented for seven days at room temperature.

2.2. Analytical methods

The analysis of this research included physical parameters such as yield (using a gravimetric method) [13], thickness (with a Kenmaster Vernier caliper), texture (texture analyzer CT3) and chemical (water content: using a gravimetric method; crude fibre content: employing a gravimetric method) [14]. The best three formulations were chosen from physical and chemical characteristics and were examined in terms of their organoleptic properties (colour, flavour, taste, texture and overall) with a 5-point scoring scale 1) Dislike, 2) Dislike moderately, 3) Neither like nor dislike, 4) Like moderately, 5) Like [15]. This sensory evaluation for nata de *cincau* involved a group of panellists, comprising 60 random untrained and semi-untrained panellist members.

2.3. Statistical analyses

Data physical and chemical tests were analyzed using two-way ANOVA to determine whether there were any differences in each treatment. If differences were noted, the test was then continued by using a Duncan Multiple Range Test (DMRT) at a significance level of $\alpha = 0.05$.

3. Results and Discussion

3.1. Physical characteristics

3.1.1. Yield

Table 1. Physical characteristics of *Nata de cincau*

Substrate (%)	Starter (%)	Yield (%)	Thickness (cm)	Texture (gram)
2.5	5	42.04 ± 1.41^{Aa}	0.60 ± 0.02^{Aa}	521.33 ± 0.02^{Aa}
	10	50.62 ± 1.46^{Ab}	0.70 ± 0.02^{Ab}	549.42 ± 0.08^{Ab}
	15	59.26 ± 1.63^{Ac}	0.79 ± 0.02^{Ac}	560.58 ± 0.04^{Ac}
7.5	5	52.96 ± 1.34^{Ba}	0.78 ± 0.03^{Ba}	571.33 ± 0.03^{Ba}
	10	60.33 ± 1.17^{Bb}	0.89 ± 0.02^{Bb}	581.50 ± 0.01^{Bb}
	15	68.48 ± 1.16^{Bc}	1.00 ± 0.02^{Bc}	588.58 ± 0.04^{Bc}
12.5	5	61.77 ± 1.47^{Ca}	0.91 ± 0.01^{Ca}	600.33 ± 0.02^{Ca}
	10	69.27 ± 1.22^{Cb}	1.10 ± 0.02^{Cb}	615.67 ± 0.02^{Cb}
	15	76.86 ± 1.20^{Cc}	1.30 ± 0.02^{Cc}	634.17 ± 0.03^{Cc}

Values within a column with different superscripts (A–C) denote significant differences from each other as affected by different substrate concentrations, $p \leq 0.05$

Values within a row with different superscripts (a–c) denote significant differences from each other, as affected by different starter concentrations, $p \leq 0.05$

Yield is the percentage ratio between the final weight product and initial weight of the material in the form of substrate, water and nutrients (sugar, ammonium sulphate, pH) multiplied by 100% [16,17]. Based on Table 1, substrate and starter concentrations were significantly different between samples. Results also showed that the percentage of yield increased with increasing substrate and starter concentration. The sample with the highest yield (76.86%) was nata de *cincau* with a substrate concentration of 12.5% and starter concentration of 15%. During fermentation, *Acetobacter xylinum* hydrolyzes sugar into polysaccharides or extracellular cellulose. A high yield indicates that *Acetobacter xylinum* was able to break down sugar to form nata. If the nutrients (carbon, nitrogen and minerals) and pH were not suitable for the fermentation process, the growth of *Acetobacter xylinum* would be inhibited [18]. Appearance of nata de cincau samples are showed at Figure 1 and Figure 2.

Figure 1. Nata de *cincau* after fermentation in nine sample formulations

Figure 2. nata de *cincau* prior to the organoleptic test.

3.1.2. Thickness

Based on Table 1, the thickness of nata is significantly influenced by substrate and starter culture. A strong correlation between factors and thickness were exhibited by Pearson correlation test, with the result of 0.918. Results showed that the thickness level increased with increasing substrate and starter concentration. The thickest sample (1.3 cm) was nata de *cincau* with a substrate concentration of 12.5% and starter concentration of 15%. Some factors that lead to increased thickness include substrate and starter concentration, nitrogen and carbon source, size of tray and duration of fermentation [19, 20].

3.1.3. Texture

Texture of the samples was tested using a CT3 Texture Analyzer with Texture Profile Analysis (TPA) mode [21]. Based on Table 1, substrate and starter concentration were significantly different between samples and there were interactions between factors. The interactions between factors were tested using a Pearson correlation test; the result of 0.704 is indicative of a strong correlation between them. Results also demonstrated that the texture level increased with increasing substrate and starter concentration. The most textured sample (634.17 g) was nata de *cincau* with a substrate concentration of 12.5% and starter concentration of 15%. The texture of nata is related to its crude fibre and water content. High crude fibre lead to nata's texture becoming chewy, while high water content causes excessive amounts of water to enter the cellulose tissue and subsequently the texture becomes harder [22].

3.2. Chemical Characteristics

3.2.1. Water Content

Water content is very important in determining food shelf life because it affects physical and chemical properties, microbiological changes and enzymatic reactions. Based on Table 2, substrate and starter concentrations were found to be significantly different between samples. Results showed that the water content percentage increased with increasing substrate and starter concentration. The sample with the highest percentage of water content (93.34%) was nata de *cincau* with a substrate concentration of 12.5% and starter concentration of 15%. The main factor that affected water content is sugar concentration. Water content percentage increased with increasing sugar addition. Sugar leads to several cavities in nata, where water will most likely be trapped [17].

Table 2. Chemical characteristics data

Substrate (%)	Starter (%)	Water content (%)	Crude fiber (%)
2.5	5	83.19 ± 0.62^{Aa}	1.53 ± 0.07^{Aa}
	10	85.23 ± 0.52^{Ab}	1.76 ± 0.07^{Ab}
	15	87.27 ± 0.67^{Ac}	1.98 ± 0.09^{Ac}
7.5	5	86.20 ± 0.72^{Ba}	1.86 ± 0.07^{Ba}
	10	88.33 ± 0.56^{Bb}	2.07 ± 0.08^{Bb}
	15	90.38 ± 0.59^{Bc}	2.26 ± 0.06^{Bc}
12.5	5	89.31 ± 0.66^{Ca}	2.14 ± 0.04^{Ca}
	10	91.36 ± 0.57^{Cb}	2.37 ± 0.08^{Cb}
	15	93.34 ± 0.66^{Cc}	2.57 ± 0.07^{Cc}

Values within a column with different superscripts (A–C) denote significant differences from each other as affected by different substrate concentrations, $p \leq 0.05$
Values within a row with different superscripts (a–c) denote significant differences from each other, as affected by different starter concentrations, $p \leq 0.05$

3.2.2. Crude Fibre Content

Crude fibre is a part of food that cannot be hydrolyzed by acids and bases. The acid used to test for crude fibre is sulfuric acid (H_2SO_4) and the base is sodium hydroxide (NaOH). Based on Table, substrate and starter concentrations were significantly different between samples. Results showed that the crude fibre percentage increased with increasing substrate and starter concentration. The sample with the highest percentage of crude fibre content (2.54%) was nata de *cincau* with a substrate concentration of 12.5% and starter concentration of 15%. High fibre in nata is affected by bacterial activity, while such activity is affected by sugar as the source of carbon in the fermentation process. *Acetobacter xylinum* converts sugar into cellulose to produce fibre [23].

The best three formulations based on physical and chemical test results were then evaluated using sensory tests. These best three formulations were nata de *cincau* with a substrate concentration of 12.5% and starter concentrations of 5%, 10% and 15%.

3.3. Sensory Characteristics

An organoleptic test was carried out by using a food-preference test, with in this study, the higher score implying better quality. Data regarding the sensory characteristics test (colour, flavour, taste, texture, overall) are presented in Table 3.

Table 3. Sensory characteristics data

Substrate (%)	Starter (%)	Colour	Flavour	Taste	Texture	Overall
2.5	5	2.80 ± 0.8^{a}	2.55 ± 0.8^{a}	2.20 ± 0.7^{a}	2.35 ± 0.6^{a}	2.48 ± 0.4^{a}
	10	3.06 ± 0.9^{a}	3.11 ± 0.8^{b}	2.64 ± 0.7^{b}	2.64 ± 0.6^{a}	2.86 ± 0.3^{b}
	15	3.61 ± 0.9^{b}	3.69 ± 0.7^{c}	3.35 ± 0.7^{c}	3.30 ± 0.8^{b}	3.49 ± 0.4^{c}

Values within a row with different superscripts (a–c) denote significant differences from each other, as affected by different starter concentrations, $p \leq 0.05$

3.3.1. Colour

The colour of a product is an important factor that shows the first impression of sensory characteristics seen by panellists. Determining food's initial quality often begins with its colour [24]. Results showed that the colour preference score increased with increasing bacterial concentration. The sample with the highest score in colour preference (3.61 or like moderately) was nata de *cincau* with a substrate concentration of 12.5% and starter concentration of 15%. Common nata of good quality should be transparent white, slippery and rather shiny, but nata de *cincau* is blackish brown due to the addition of black *cincau* [25]. Black *cincau* has flavonoid compounds, which cause a distinctive dark colour that will change nata or any other products' colour to be rather dark [26, 27].

3.3.2. Flavour

Flavour is one of the quality parameters in food products that determines whether or not a product is good, using the sense of smell [25]. Based on Table 3, the sample with a starter concentration of 15% was significantly different from the others. Results showed that the flavour preference score increased with increasing bacterial concentration. The sample with the highest score in flavour preference (3.69 or like moderately) was nata de *cincau* with a substrate concentration of 12.5% and starter concentration of 15%. Human senses can only detect at least four smells: sour, sweet, rancid, and charred/burnt, although the flavor of nata de *cincau* is neutral with a little sour [28]. The sour aroma can be removed by soaking nata in fresh water and boiling it for ten minutes at 100 ^{0}C during the harvesting process after fermentation. The boiling needs to be repeated until the smell has completely vanished [29, 25].

3.3.3. Taste

Results demonstrated that the taste preferences score increased with increasing bacterial concentration. The sample with the highest taste preference (3.35 or neither like nor dislike) was nata de *cincau* with a substrate concentration of 12.5% and starter concentration of 15%. Nata de *cincau* was tasteless with a rather sour and slightly sweet. The sour taste of nata is produced by glucose, which is hydrolyzed into cellulose in the fermentation media. While nata is being soaked and boiled, the osmosis process releases the sour taste and nata becomes tasteless [22].

3.3.4. Texture

Texture is one of the organoleptic parameters that determines product quality. The human senses used in observing the texture are usually fingers, teeth and palate. A good nata appears in the form of a gel, with its texture tending to be chewy, tender and transparently white in colour [30, 29, 31]. Results illustrated that the texture preferences score increased with increasing bacterial concentration. The sample with the highest texture preference (3.3 or neither like nor dislike) was nata de *cincau* with a substrate concentration of 12.5% and starter concentration of 15%. This result indicated that nata de cincau had an acceptable texture according to SNI 01-4317-1996, which states that the texture of a nata must be normal. The texture parameter is affected by several factors such as water content, duration of fermentation and crude fibre content [30, 29, 19].

3.3.5. Overall

Results demonstrated that the overall preferences score increased with increasing bacterial concentration. The sample with the highest overall preference (3.49 or neither like nor dislike) was nata de *cincau* with a substrate concentration of 12.5% and starter concentration of 15%.

4. Conclusion

In sum, Pacitan's black *cincau* (*Mesona palustris* BL) can be processed into nata de *cincau*. The best three samples in terms of physical and chemical characteristics were nata de *cincau* with a substrate concentration of 12.5% and starter concentrations of 5%, 10% and 15%. These samples were then subjected to a sensory evaluation using an organoleptic test, considering their performance with respect to colour, aroma, taste, texture and overall. The sample with a substrate concentration of 12.5% and starter concentration of 15% was found to be the best formulation of nata de *cincau*.

5. References

[1] Heyne K 1950 *De Nuttige Planten Van Indonesie* (Bandung: I.N.V. Vitgeverjn Van Hoevers Gravenhage)
[2] Kartikaningrum A A, Haryadi, and Marseno D W 2001 *Agritech. J.* **21** 2
[3] Hung C Y and Yen C C 2002 *J. Agric. Food Chem.* **50** 10
[4] Sasmita A O and Ling A P K 2017 *JESR* 1 (2): 47-53.
[5] Santosa B, Ahmadi K, and Taeque D 2012 *IJSTE* **1** 1
[6] Jagannath A, Kalaiselvan A, Manjunatha S S, Raju P S and Bawa A S 2008 *World J. Microbio. Biotech.* **24**
[7] Wenno M R, Wattimena M L, Rumakey Z A and Thenu J L 2015 *1st Int. Symp. Aquatic Product.* **1**
[8] Sutanto A 2012 *Makara. J. Tek.* **16** 1
[9] Setyawaty R, Hirayama K K, Kaneko H and Hirayama K 2011 *World. Appl. Sci. J.* **14** 5
[10 Afreen S S and Lokeshappa B 2014 *The IJST* **2** 8
[11] Mesomya W, Pakpeankitvatana V, Komindr S, Leelahakul P, Cuptapun Y, Hengsawadi D, Tammarate P and Tangkanakul P 2006 *Songklanakarin J. Sci. Tech.* **28** 1
[12] Almeida D M, Prestes R A, da Fonseca A F, Woiciechowsk A L and Wosiacki G 2013 *Brazilian J. Microbiol* **44** 1
[13,14] AOAC 2006 *Official Methods of Analysis 16th Ed.* (USA: Association of Official Analytical Chemists)
[15] Lawless H T and Heymann H 1998 *Sensory Evaluation of Food: Principles and Practices* (New York: Chapman & Hall)
[16] Indrianingsih AW, Rosyida V T, Jatmiko T H, Prasetyo D J, Poeloengasih C D, Apriyana W, Nisa K, Nurhayati S, Hernawan, Darsih C, Pratiwi D, Suwanto A and Ratih D 2017 *IOP Conf. Ser.: Earth Environ. Sci.* **101**
[17] Syukroni I, Yuliati K and Baehaki A 2013 *Fishtech J.* **2** 1
[18] Brock T D and Madigan M T 1998 *Biology of Microorganisms* (New Jersey: Prentice Hall)
[19] Arifiani N, Sani T A and Utami A S 2015 *Biotek.* **12** 2
[20] Rachmawati N A, Haryati S and Munandar A 2017 *JPK* **7** 2
[21] Nikoofar E, Hojjatoleslami M and Shariaty M A 2013 *Int. J. Farming Allied Sci.* **2** 20
[22] Al Awwaly K U, Puspadewi A and Radiati L E 2011 *JITEK* **6** 2
[23] Rizal H M, Pandiangan D M and Saleh A 2013 *JTK* **1** 19
[24] Negara J K, Sio A K, Rifkhan, Arifin M, Oktaviana A Y, Wihansah R R S and Yusuf M 2016 *J. IPTHP* **4** 2
[25] Tamimi A, Sumardi H S and Hendrawan Y 2015 *JBKT* **3** 1
[26] Yulianto R R and Widyaningsih T D 2013 *JPA* **1** 1
[27] Fauzziyah I N, Widyaningsih T D and Widyastuti E 2016 *JPA* **4** 2
[28] Pakaya R, Mandey L C and Lumoindong F 2015 *J. ITP* **3** 2
[29] Putriana I and Aminah S 2013 *JPDG* **4** 7
[30] Herawaty N and Moulina M A 2015 Agritepa **2** 1
[31] Saputra F and Hidaiyanti R 2015 *Agritepa* **1** 2

IOP Conf. Series: Materials Science and Engineering **633** (2019) 012010 doi:10.1088/1757-899X/633/1/012010

Microbiological profile of fresh goat milk: Impact of goat farmer practices in "Taruna Mukti" goat farmer group, Sragen, Central Java

A D M Nuhriawangsa, L R Kartikasari, R F Hadi, A K Wati and B S Hertanto
Laboratory of Industry and Animal Product Processing, Department of Animal Science, Faculty of Agriculture, Universitas Sebelas Maret

E-mail: bayusetya@staff.uns.ac.id

Abstract. Goat milk is susceptible to be contaminated by many microorganisms including microbial pathogens responsible for causing diseases. Various contaminations come from every stage of goat milk produce including feeding, milking, and milk treatments. The aim of this study was to assess the practice of dairy goat farmer and evaluate the microbiological profile of raw goat milk in dairy goat farmer group. The study showed that row milk had high total plate count ($7{,}8x10^6$ CFU/ml). Thus, *Coliform* and *E. coli* were detected in the samples of fresh goat milk. Overall, it can be seen that fresh goat milk was relatively poor hygienic quality and was not conformed to the Indonesian Nasional Standard.

1. Introduction

Dairy goats are important groups of milk producing animals throughout the world and 80% majority of the milk goat is produced in developing countries [1]. Dairy small ruminants especially goats approximately 52.1% of total goats in the world are spread in Asia, and producing 8.04 Mt or 52.7% of world's goat milk [2]. Goats, mainly located in subtropical-temperate areas of Asia, Europe, and Africa, are concentrated in low-income, food deficit countries [3]. The demand of dairy goats' milk is increasing because of the growing population of people, the increasing awareness of medicinal and nutritional status associated with goat milk, and also the special interest in goat milk products, especially cheeses and yoghurt, in many developed countries which has led to increasing levels of disposable incomes [4,5]. As the demand of row milk and dairy products continuely grows with an increasing population, there is a great pressure to maintain the quality of the products especially in microbe quality because total viable bacterial count is considered as one of the acceptance criteria for categorizing milk for human consumption and processing for dairy products [6].

The contamination of *Campylobacter*, *Salmonella*, *Escherichia coli* and *Staphylococcus aureus* in milk can occur at different stages in the milking procedure and from different sources, mainly from the external surface of the udder and teats, from the surface of the milking utensils, storage equipment, and transport [7–9]. Milk contaminated with high level of spoilage bacteria often becomes unsuitable for further processing [10]. Thus, ensuring supply of safe food has been one of the major challenges and concerns for producers, consumers and public health officials in the world. This is because foods excessively contaminated with pathogenic and spoilage microorganism are undesirable and can cause food borne illnesses [11].

Microbiological quality of goat milk from dairy goat farm has received a big attention to date. Potential health hazards transmissible through milk and milk products were started from the farm. Therefore, the study concentrated on dairy goat farmer group must be carried out, because it is facing hygiene and safety problems in all stages of food production. This study focused on assessment of practice of dairy goat farmer and evalution of the microbiological profile of raw goat milk in dairy goat farmer group.

2. Materials and Methods

The study was conducted in Dairy Goat Farmer "Taruna Mukti" located in village of Purworejo, Gemolong, Sragen, Central Java. The materials of this study consisted of lactating dairy goats, questionnaires, milking and storage equipments, and milk samples. The hygienic practices of dairy goat farmers were evaluated by observation included milking process and milking handling. Researchers conducted an assessment using Yes or No to express the suitability of hygienic practices in dairy goat farmer group by interviewing farmers using questionnaires consisted of the farm characteristics. Milk Samples were taken from the dairy goat in the existing condition with 2 replications. Milk was collected aseptically in bottle for analysis of microbiological profile included total plate count, *E. coli*, and *Coliform* in Laboratory of Veterinary and Animal Health Center of Boyolali District.

3. Result and discusion

3.1 The existing condition of dairy goat farmer group

The supply chain of fresh goat milk in sragen district was generally produced by local farmer group with small scale dairy goat who have not considered to aspects of food safety. One of goat milk producer in Sragen comes from "Taruna Mukti" dairy goat farmer group. The group has has 10 members with 15 lactating dairy goats consisted of Etawah crossbred and Sanen with milk production between 0.5 - 1.5 per head / liter / day. Dairy goat is a type of goat that produces milk at the rate exceeding the needs of their kids. The typical dairy goats raised by farmers in Indonesia predominantly was Etawah crossbred because it has long period adaption to tropical condition and capable to produce milk [12]. Milk production of Etawah crossbred goat ranged between 0.5–1.2 L/head/day. The developing Saanen goat in Indonesia leads to generate crossbred goats, such as Sapera, a result of crossing between female Etawah crossbred and Saanen buck [13]. Purebred exotic dairy breeds and their crosses were mostly found in commercial dairy farms. Milk production of Sapera goat was observed to range between 0.8-1.2 L/day [14]. The existing condition in Sragen was less resources during the dry season especially water. This condition causes a lack of water supply for drinking and sanitation. The supply of forage was lacking so farmers use dry forage feed. Likewise, feeding activity in farmer group still use traditional animal feeding. Only small amount of concentrates was given to the dairy goat without considering feed requirements.

3.2 Practices of milking process

The existing condition of milking practices was counted to be poor because farmers did not apply hygienic practices in milking process such as cleaning goats udder with warm water, dipping the teats, using a washcloth to dry the udder, discarde first milk flow, and using good milking equipments (Table 1.). Good hygienic quality of milk for consumers requires good hygienic practices, such as clean milking utensils, washing milker's hands, cleaning udder, and use of individual towels during milking and handling, before delivery to consumers or processors [15]. The practical milking preparation caused undetectable of milk abnormality [16]. The prevalence of mastitis is higher in dairy cattle with poor hygiene of the milking process than in dairy cattles with good hygiene of milking process [17]. Before milking starts, the cow should be as clean as possible. It is not advisable to rinse the cow since this will only dissolve the dirt and make it pour down along the skin and accumulate on the udder [18]. A statistically significant association between the use of udder-towel and the occurrence of subclinical mastitis. It is known that failure in hand washing and in cleaning of the udder exposes the udder and milk to microbiological contamination [19,20].

Table 1. Practices of milking process in existing dairy goat farm

Activities	Responses of milking process
Milking frecuency per day	2
Streamed floor with clean water before and after milking	Yes
Washing a dairy goat before milking	No
Cleaning goats udder with warm water	No
Dipping teats into the disinfectant before and after milking	No
Provide a washcloth to dry the udder before and after milking	No
Washing milk milking equipment using hot water	No
Using milking equipments made of stainless steel	No
Discarded first milk flow	No
Hand milking until the milk in the udder has run out	Yes

According to [7], teat dipping after milking can reduce the numbers of bacteria on the teat skin, help to protect the stressed teat orifice after milking and also prevent bacteria from entering the teat canal. Teat dipping should be practiced according to national recommendations and regulations [21]. Higher bacterial counts when the same towel was used for drying multiple cows after washing the udder compared to using a single towel for each cow [18,22]. The equipment that is used must be suitable for effective cleaning and sanitization [9]. Buckets should be non-corrosive and easy to clean and disinfect [23]. Thus materials that are non-absorbent and non-corrosive with smooth surfaces, minimal joints and free from dents are recommended [9,23]. Plastic equipment will become scratched on the surface after some time, be nearly impossible to clean and is therefore not advisable [9]. A clean brush with good bristles, hot water and a detergent should be used during washing. The equipment should be rinsed with clean water and dried immediately, preferable upside down and in the sun. [24] reported that, besides udder infection and water quality, hygienic behavior with respect to hand washing, containers cleaning and disinfection are the key areas that need hygiene intervention.

3.3 Practices of milking handling

Milk is one of the most common food sources in the human diet and is also a product that is directly available for consumption [25]. Being a nutritious food, it is an excellent growth medium for bacteria, Being a nutritious food, it is an excellent growth medium for bacteria, originating from either mastitis or from contamination of the milk with environmental spoilage as well as pathogenic microorganisms during milking or milk handling process [26]. Due to the highly perishable nature of milk and mishandling, the amount produced is subjected to high post-harvest losses. Post harvest losses up to 40% of milk and its derivatives have been reported from milking to consumption [27]. Further losses incurred are quality losses by storing in unclean storage utensil, which is prone to high microbial contamination. The equipment used for milking, transportation and storage determine the quality of milk and milk products. Types of milk containers determine the qualities of milk, especially during transportations of milk to the selling point. Milk storage and transportation are aimed at having good quality milk available where and when needed for processing [28]. Therefore, producers need to pay particular attention for the type as well as cleanliness of milk equipments. The use of plastic containers was associated with high coliform counts in raw milk [29]. Plastic containers are not recommended for

IOP Conf. Series: Materials Science and Engineering **633** (2019) 012010 doi:10.1088/1757-899X/633/1/012010

handling milk as they are known to be vulnerable to bacterial contamination. Milk handling problems coupled with lack of quality assurance of milk delivered to most of the retailers and household consumers pose potential sources of public health risks to consumers [30].

Table 2. Practices of milking handling in existing dairy goat farm

Activities	Responses of milking handling
Filtering milk harvest	Yes
Using milking equipments made of stainless steel	No
Using personal hygienic tool for milking handling	No
Milk was directly pasteurized	No
milk was directly stored in the freezer	Yes
Need more than 30 minutes to store the milk	Yes

Milk products are generally more stable than fresh milk because they are more acidic and/or contain less moisture [20], so milk should be harvested and stored under hygienic conditions. Equipment used to harvest and store milk should be suitable and well maintained [23]. Milk cooling and refrigerated storage are necessary after milking to reduce bacterial growth rates [31]. Pasteurized milk stored in cold temperatures can inhibit the growth of microorganisms. Pasteurisation significantly decreases the amount of available nutrients and the second is that this decrease negatively impacts an individual's nutrient intake [32]. Coliform bacteria usually cannot survive at the pasteurization temperature [33].

3.4 Microbiological profile of goat milk

Microbiological quality of goat milk in dairy goat farmer group must be improve. It can be showed by result of microbes analysis including total plate count, *E. Coli*, and *Coliform*. The importance of udder cleanliness for psychrotolerant spore contamination of raw milk is logical, as spore forming bacteria have been found throughout the dairy farm environment, including in bedding, feed, manure, soil, and water [34,35]. Cleaning the udder of cows before milking is one of the most important hygienic practices required to ensure clean milk production [20]. A study found that poor teat end cleanliness was associated with higher bacteria counts in bulk tank milk [36]. Further, adopting the hygienic practice of cleaning teats with moist paper towels reduced the number of Clostridia tyrobutyricum spores isolated from raw milk [37].

Table 3. Microbiological profile of goat milk in existing dairy goat farm

Microbiological analysis	Number of microbes (CFU/ml)
Total plate count	7.8×10^6
E. coli	7
Coliform	93

Besides udder infection and water quality, hygienic behavior with respect to hand washing, containers cleaning and disinfection are the key areas that need hygiene intervention [24]. *Coliform* and *E. Coli* can find in milk, water and on milking equipment [21]. Degree of cleanliness of milking equipments depends on the procedure which is adopted for cleaning and sanitizing. The microbiological quality of the water supply at the farm could be improved by addition of chlorine or by boiling the water [21,38].

4. Conclusion

Overall, fresh goat milk was relatively poor hygienic quality and was not conformed to the Indonesian Nasional Standard due to unhygienic conditions at dairy goat farm. Improving milking and milk handling practices by providing information, education and guidance of hygienic practices to the farmers will be able to increase microbiological quality of milk.

References

[1] Boyazoglu J, Hatziminaoglou I and Morand-Fehr P 2005 *Small Ruminant Research* **60** 13–23
[2] FAOSTAT (Food and Agriculture Organization of the United Nations 2018 Statistics database
[3] Pulina G, Milán M J, Lavín M P, Theodoridis A, Morin E, Capote J, Thomas D L, Francesconi A H D and Caja G 2018 *J. Dairy Sci.* **101** 6715–29
[4] Morand-Fehr P, Boutonnet J P, Devendra C, Dubeuf J P, Haenlein G F W, Holst P, Mowlem L and Capote J 2004 *Small Ruminant Research* **51** 175–83
[5] Taufik E, Hildebrandt G, Kleer J N, Wirjantoro T I, Kreausukon K, Zessin K H, Baumann M P O and Pasaribu F H 2011 *MediaPeternakan* **34** 105–11
[6] PMO (revised) 2015 *Grade "A" Pasteurized Milk Ordinance* (Washington, DC)
[7] Murphy S C and Boor K J 2000 *Dairy Food Environ. Sanit.* **20** 606–11
[8] Oliver S P, Jayarao B M and Almeida R A 2005 *Foodborne Pathog. Dis.* **2** 115–29
[9] Pandey G S and Voskuil G C S 2011 *Manual on Milk Safety, Quality and Hygiene for dairy extension workers and dairy farmers* (Lusaka, Zambia)
[10] Mhone T A, Matope G and Saidi P T 2011 *Int. J. Food Microbiol.* **151** 223–8
[11] Su H P, Chiu S I, Tsai J L, Lee C L and Pan T M 2005 *J. Infect. Chemother.* **11** 146–51
[12] Yuni Suran Y, Haidar Ali D, Firdaus N and . R 2018 *Int. J. Dairy Sci.* **13** 1–6
[13] Astuti D A and Sudarman A 2012 Dairy goats in Indonesia: Potential, opportunities and challenges *Proc. the 1st Asia Dairy Goat Conference* (Kuala Lumpur, Malaysia) pp 47–51
[14] Sutama I 2014 Dairy goat production on smallholder agriculture in Indonesia *Proc. the 2nd Asian-Australasian Dairy Goat Conference* (Bogor, Indonesia) pp 8–17
[15] Lencho G K and Seblewongel A M 2018 *J. Vet. Med. Anim. Heal.* **10** 45–54
[16] Suranindyah Y, Wahyuni E, Bintara S and Purbaya G 2015 *Procedia Food Sci.* **3** 150–5
[17] Lakew M, Tolosa T and Tigre W 2009 *Trop. Anim. Health Prod.* **41** 1525–30
[18] Olofsson L 2013 *Milking routines and hygiene in small-scale dairy farms in Mapepe, Choma and Batoka districts in Zambia* 1–57
[19] Kivaria F M, Noordhuizen J P T M and Kapaga A M 2004 *Trop. Anim. Health Prod.* **36** 581–92
[20] Zalalem Yilma 2010 *Epidemiol. Insights* 297–18
[21] Gran H M, Mutukumira A N, Wetlesen A and Narvhus J A 2002 Smallholder dairy processing in Zimbabwe: Hygienic practices during milking and the microbiological quality of the milk at the farm and on delivery *Food Control* **13** 41–7
[22] Elmoslemany A M, Keefe G P, Dohoo I R, Wichtel J J, Stryhn H and Dingwell R T 2010 *Prev. Vet. Med.* **95** 32–40
[23] FAO-Food and Agriculture Organization 2011 *Guide to good dairy farming practice*
[24] Bonfoh B, Roth C, Traoré A N, Fané A, Simbé C F, Alfaroukh I O, Nicolet J, Farah Z and Zinsstag J 2006 *Food Control* **17** 153–61
[25] Grimaud P, Sserunjogi M, Wesuta M, Grillet N, Kato M and Faye B 2009 *Trop. Anim. Health Prod.* **41** 883–90

[26] A P and E A 2009 *Lucr. ştiinţifice Zooteh. şi Biotechnol.* **42** 501–6
[27] Getachew F 2003 *FAO Prevention of Food Losses Program*me 1–60
[28] Walstra P, Geurts T J, Noomen A, Jellema A and van Boekel M A J S 2005 *Dairy Technology: Principles of Milk Properties and Processes*
[29] Omore., Lore T, Staal S, Kutwa J, Ouma R, Arimi S and Kang'ethe E 2005 *Addressing the public health and quality concerns towards marketed milk in Kenya*
[30] Gratian K K and Kurwijila L R 2011 *Food Safety in Milk Markets of Smallholder Farmers in Tanzania: a Case of Peri Urban Wards in Temeke Municipality* 01-71
[31] Paludetti L F, Kelly A L, O'Brien B, Jordan K and Gleeson D 2018 *J. Dairy Sci.* **101** 1921–9
[32] Nandy S K and Venkatesh K V 2010 *African J. Microbiol. Res.* **4** 61–70
[33] Dey S and Karim M H 2013 *Int. J. Sci. Invent. Today* **2** 150–7
[34] Scheldeman P, Pil A, Herman L, De Vos P and Heyndrickx M 2005 *Appl. Environ. Microbiol.* **71** 1480–94
[35] Huck J R, Sonnen M and Boor K J 2008 *J. Dairy Sci.* **91** 1218–28
[36] Elmoslemany A M, Keefe G P, Dohoo I R and Jayarao B M 2009 *J. Dairy Sci.* **92** 2634–43
[37] Magnusson M, Christiansson A, Svensson B and Kolstrup C 2006 *J. Dairy Sci.* **89** 3866–75
[38] Debela G T 2015 *Glob. J. Food Sci. Technol.* **3** 153–8

Acknowledgment
The authors gratefully acknowledge to Indonesian Higher Education for funding the research "program kemitraan masyarakat" (PKM) with project number: 475/UN27.21/PM/2018.

IOP Conf. Series: Materials Science and Engineering **633** (2019) 012011 doi:10.1088/1757-899X/633/1/012011

Effects of drying time on essential oil production of kaffir lime (*Citrus hystrix* DC) leaves at ambient temperature

Y W Wulandari[1,2], C Anwar[3] and S Supriyadi[4]

[1]Post Graduate School of Food Science Doctor Program Gadjah Mada University (UGM), Jl. Folra no.1, Bulaksumur, Yogyakarta 552812, Central Java, Indonesia
[2]Department of Agricultural Products Technology, Faculty of Industrial and Food Technology, Slamet Riyadi University, Surakarta 57136, Central Java, Indonesia
[3]Department of Chemistry, Faculty of Mathematics and Natural Science, Gadjah Mada University, Yogyakarta, Indonesia
[4]Department of Food and Agricultural Products Technology, Faculty of Agricultural Technology, Gadjah Mada University, Jl. Flora no.1, Bulaksumur, Yogyakarta 552812, Indonesia

Email: suprif248@ugm.ac.id

Abstract. Drying was one of the primary processes in essential oil production. Time factor in drying affects both the content and chemical compositions of essential oils. This study aims to evaluate the influence of the drying time of kaffir lime leaves on the yield at room temperature and the major volatile compounds in the essential oils by observing day-to-day treatments. As a key odorants in essential oils, the quantity of citronellal was determined by Gas Chromatography (GC) using the standard compound. Based on the daily observation results, a long-time drying gave a significant effect on the yield and the major volatile compounds. The highest yield 1.26±0.07 % (w/w) was obtained on the fourth day, while the highest content of citronellal compounds, 87.58±0.33% was achieved on the fifth day of drying,. A prolonged drying time resulted in a negative effect on not only the quantity, but also the quality of essential oils. Accordingly, this study suggests that the drying technique at room temperature best lasts for four until five days to obtain the most favorable quality and quantity of essential oils.

1. Introduction

Essential oils from kaffir lime (*Citrus hystrix.* DC) leaves are natural flavoring materials of commercial importance. Eessential oil is the secondary metabolite from this plant. Kaffir lime leaves oil possesses some important bioactivities; such as antioxidant [1], antileukemic [2], antitussive, antihemorrhagic, antioxidative stress properties [3], and antibacterial properties [4]. The enantiomer of citronellal in kaffir lime leaves oil is responsible for odor characteristic of these leaves. Twenty-nine compounds are found in the essential oils of kaffir lime leaves [5]. Furthermore, ß-citronellal is the major compound, amounting to 66.85% of the total yield.

Essential oils are natural products obtained from different parts of plants by extraction and distillation process. Many studies had examined the effects of different drying methods on the content and chemical compositions of essential oils. For instance, extraction on dried leaves both reduced the moisture content and affected the volatile compounds of the produced essential oils. Although the concentration of the

IOP Conf. Series: Materials Science and Engineering **633** (2019) 012011 doi:10.1088/1757-899X/633/1/012011

volatile compounds tends to change during the process, its magnitude depends on the type of the product, drying methods and conditions [6]. Meanwhile, drying methods also can affect both oil content and the composition of aromatic plants [7],[8],[9]. [10] investigated the effect of different temperature on the aromatic profile of fresh ginger in microwave-drying process. [11] studied the effect of freeze-drying time on the levels of the two main components of fennel essential oil. Higher drying temperature decreased the content of *Artemisia annua* essential oil [12]. Meanwhile, air drying at ambient temperature and infrared-drying at 45°C significantly increase the content of the essential oil of *Laurus nobilis* L. leaves [13]. Interaction between different temperatures and various air flow rates in drying condition considerably affected both content and chemical composition of the essential oil of lemon verbena (*Lippia citriodora*) [14].

Citronellal is the major aromatic profile in kaffir lime leaves oils, and the highest value of essential oil depends on it. The drying method is proven to significantly influence the quality of the content and composition of aromatic plants. Therefore, this study aims to investigate the major volatile compounds of kaffir lime leaves that had been subjected to different drying processes.

2. Experimental

2.1. Materials
Samples of kaffir lime leaves were collected from a local orchard in Klaten, Central Java. After the samples arrived in the laboratory, the mature leaves were sorted out and divided into nine groups, each of that received different treatments during the drying process. All chemicals and solvents used in this research were of analytical grade and procured from Merck. The citronellal standards ($C_{10}H_{18}O$) were procured from Santa Cruz Biotechnology.

2.2. Drying treatment
Kaffir lime leaves were dried at a room or ambient temperature (i.e., averagely 25±1°C). About 300g of the sample were placed on a stainless steel mesh with known weight and volume. In all experiments, the samples had the same bed height and were spread as a thin layer with a bulk density of 300 g/11.25 cm^3. A similar drying temperature is also preferred in the drying of lemon Myrtle leaves [7]. During the drying process, the samples were regularly weighed at a fixed time interval for eight days. The weighing process used a digital balance with 0.01 accuracy and 1,500 g capacity.

2.3. Determination of moisture content
Moisture content of the leaves was determined by a gravimetric method using a laboratory-scale drying chamber by specifying the initial wet basis and final dry matter content in a sample that had been dried at a temperature of 105°C for three hours. The results were presented in mean values [9].

2.4. Essential oil isolation
The essential oil was isolated by using hydrodistillation method with the ratio of leaves to water 1:5. The distillation was carried out until the maximum possible quantity of oil was obtained (i.e., two hours). The yield was determined based on dry weight. The essential oil was collected in numbered and colored vials, dehydrated with anhydrous sodium sulphate, capped under nitrogen, and kept at 4°C for preservation until the next analysis.

2.5. Citronellal determination
Gas Chromatography (SHIMADZU-QP2010S) method was used to determine the citronellal content. GC analyses were performed using a Hewlett-Packard 5890A apparatus equipped with a fused silica capillary column (SH-Rxi-5Sil MS 30 m x 0,25 mm ID x 0.25 μm film thickness). These analyses used the following temperature program, i.e., oven isotherm at 40°C for 10 minutes followed by a temperature increase from 40°C to 280 °C at a rate 4°C/min and isotherm at 280 °C for 10 minutes. Injector and detector (FID) temperatures were both set at 290 °C. Helium, with a linier velocity of 32 cm/s was used

IOP Conf. Series: Materials Science and Engineering **633** (2019) 012011 doi:10.1088/1757-899X/633/1/012011

as the carrier gas. Percentages of the compounds were calculated by the area normalization method without considering the response factors. A total of 2.0 μL diluted samples was injected in the split mode (1:60). The volatile compounds were identified by comparing their mass spectra with those of the mass spectral database [14].

Components of essential oil were identified by calculating their retention indices under a temperature-programmed condition for $C_{10}H_{18}O$ and comparing them with the bibliographic data, whereas individual compounds were identified by comparing their mass spectra with those of the authentic compounds in the mass spectral reference library, and the results were confirmed by comparing their retention indices with the authentic compounds.

2.6. Statistical analysis

The data were the average values of the three samples presented in mean ± standard deviation. The statistical analysis was performed in STATISTICA. Differences were tested for their significance by using analysis of variance (ANOVA) at a significance level of $P< 0.05$.

3. Results and Discussion

3.1. Drying time and moisture content

Fresh kaffir lime leaves had moisture content of 72,61% (wb). Influence on drying process on moisture content in eight days treatment are presented in Table 1. Drying process was simultaneous heat and mass transfer. The rate of moisture loss in drying was assumed to be proportional to the moisture remaining to be lost [15]. Based on this research, after four-day drying dehydration process was slower than that on the days before. The phenomenon was influenced by the movement of moisture occurring by diffusion and was analogous to that of heat conduction in material. The drying constant of the process achieved on equilibrium moisture content, increasing drying time did not decrease the moisture content.

Table 1. Moisture content of fresh and dried kaffir lime leaves during the eight-day drying

Time (day)	Moisture content (%)
0	$72.61{\pm}0.02^{e}$
1	$71.14{\pm}0.03^{cd}$
2	$70.19{\pm}0.01^{c}$
3	$68.79{\pm}0.72^{bc}$
4	$67.01{\pm}0.01^{bc}$
5	$66.80{\pm}0.01^{b}$
6	$65.99{\pm}0.04^{a}$
7	$65.55{\pm}0.02^{a}$
8	$64.83{\pm}0.03^{a}$

[a]Means±SD; n=3. Values in same row marked by the different letters are significantly different $p<0.05$

In general, there were three important stages in a drying process: The first phase or initial period happens when sensible heat was transferred to the product from the inlet to the process condition. The rate of evaporation would increase dramatically during this period with mostly free moisture content being removed. During the second phase or constant period, free moisture persists on the surfaces and the rate of evaporation altered very little as the moisture content reduced. There was a gradual and relatively small increase in the product temperature during this period. Finally, the third phase or falling rate period was the phase when migration of moisture from the inner interstices of each particle to the outer surface become the limiting factor that reduced the drying rate.

IOP Conf. Series: Materials Science and Engineering **633** (2019) 012011 doi:10.1088/1757-899X/633/1/012011

3.2. Volatile oil yield

The yield of kaffir lime essential oil produced in this experiment was monitored by taking a daily reading on oil yield for eighth days. It was then calculated on a wet basis and the experiment was repeated three times for every different treatments drying. After accumulated in percentage during the extraction process at different treatments drying, it was tabulated in Table 2. The yield of essential oils of fresh leaves was 0.56±0.06 %. Although the fresh leaves and leaves with the drying treatment on first day was not significantly different, the yield tended to increase until the fourth day of drying time. Furthermore, increasing drying time would reduce the yield value. Hence, the yield of the essential oils of dried leaves were not significantly different on either the second or eight day.

Table 2. Effect of different drying time on the yield of kaffir lime essential oil

Time (day)	Essential oil yield (%)
0	$0.56{\pm}0.06^{a}$
1	$0.62{\pm}0.02^{a}$
2	$0.99{\pm}0.14^{bc}$
3	$1.05{\pm}0.07^{bcd}$
4	$1.26{\pm}0.07^{d}$
5	$1.18{\pm}0.06^{cd}$
6	$1.05{\pm}0.04^{bcd}$
7	$0.97{\pm}0.01^{bc}$
8	$0.88{\pm}0.03^{b}$

[a]Means±SD; n=3. Values in same row marked by the different letters are significantly different $p<0.05$

Table 2 shows that different treatment at long-day drying gave a significant effect on the essential oil content of kaffir lime leaves ($p < 0.05$). These results were in agreement with those numerous studies confirming that the yield essential oil was affected by different drying conditions [7, 8, 12, 16]. Our findings showed that increased drying time (from zero to four days) resulted in a significant increase in the essential oil content. Meanwhile, the decreased amount of yield due to increased long time drying might be caused by oil accumulated in storage cells of essential oil at different developmental stages [17]. Drying was simultaneous process of heat and mass transfer that affected schizogenous oil glands in kaffir lime leaves. Essential oil is a volatile compound, therefore the damage of oil gland cells due to drying time process of kaffir lime leaves could reduce essential oil yield by hydrodistillation methods. Similar observations were made in the previous work with peppermint [18,19] demonstating that the accumulation of leaf essential oil correlated with the developmental distribution of grandular trichomes.

3.3. Characterization of citronellal compounds in kaffir lime essential oil

Volatile oil obtained from kaffir lime leaves that had been subjected to treatment with different durations of drying was analyzed with GC. This experiment identified that citronellal was the major compounds of essential oil in *Citrus hystrix* leaves. Effects of drying methods on the citronellal content in essential oil are shown in Table 3. Findings clarified that drying the leaves influenced the chemical composition of the essential oil. The level of citronellal increased while the drying time increased. The highest retention level was obtained on the fifth day, i.e., 87.58%, and the level of citronellal decreased afterward.

The essential oil of kaffir lime leaves significantly decreased after five days of drying. It was importantly noted that the effects of drying time on volatile compounds and oil content were similar. In another research, cavity filling in *Citrus* peel appeared to be a relatively slow process when compared with the accumulation of essential oil in the glandular trichomes of peppermint. Genes encoding all previously characterized terpene syntheses of *Citrus* peel are expressed at high levels in the resultant essential oil [17]. Drying process before essential oil extraction aimed to reduce the moisture content.

During this process, the volatile compounds were altered. Previous research had successfully proven that the concentration of volatile compounds were always affected in this process, the magnitude of that depended on the type of the product, drying methods, and conditions [6].

Table 3. Effect of different drying time on the citronellal compounds of kaffir lime essential oil

Time (day)	Citronellal compounds (%)
0	44.79±0.02[a]
1	47.87±0.08[b]
2	57.14±0.02[c]
3	63.18±0.54[d]
4	72.12±0.19[g]
5	87.58±0.33[i]
6	85.97±0.21[h]
7	69.88±0.70[f]
8	67.44±0.39[e]

[a]Means±SD; n=3. Values in same row marked by the different letters are significantly different $p<0.05$

4. Conclusion

Essential oil yields tended to increase in kaffir lime leaves that have been dried for four-five days at an ambient temperature. Longer drying period had been proven to decrease the yield. The presence of citronellal as the volatile and major compound of essential oils signified the quality of the yield. Citronellal gave a specific flavor to kaffir lime leaves. Finally, this research concluded that the drying period of kaffir lime leaves at an ambient temperature lasted no longer than five days.

Acknowledgement

This work was funded by HIBAH DOKTOR-Kemenristek DIKTI Project by The Ministry of Science and Research and Technology to the Republic of Indonesia in 2017.

References

[1] Hutadilok-Towanata N, Chaiyamutti P, Pathong K, Mahabusarakam W, and Rukachaisirikul V 2006 *Pharm. Biol.* **44** 221-228
[2] Ampasavate C, Okonogi S, and Anuchapreeda S 2010 *Afr. J. Pharm. Pharmacol.* **4** 13-21
[3] Laohavechnich P, Muangnoi C, Butryee C, and Kriengsinyos W 2010 *Science Asia* **36** 112-117
[4] Siripongvutikorn S, Thummaratwasik P, and Huang Y W 2005 *Tom-Yum. LWT* **38** 347
[5] Loh F S, Awang R M, Omar D, and Rahman M 2011 *J. Med. Plant. Res.* **5** 3739-3744
[6] Barbieri S, Elustondo M, and Urbicain M 2004 Journal of Food Engineering. **65** 109-115
[7] Buchaillot A, Caffin N, and Bhandari B 2009 *Drying Technology* **27** 445-450
[8] Huang B, Wang G, Chu Z, and Qin L 2012 *Drying Technology* **30(3)** 248-255
[9] Antal T, Figiel A, Kerekes B, and Sikolya I 2011 *Drying Technology* **29(15)**1836-1844
[10] Hussain A, Li Z, Ramanah DR, Niamnuy C,and Raghavan G S V 2019 *Drying Technology* **28(1)** 42-48
[11] Gardeli C, Evageliou V, Poulus C, Yanniotis S, and Komaitis M 2010 *Drying Technology* **28** 542-549
[12] Khangholi S H and Rezaeinodehi A 2008 *Pakistan Journal of Biological Science* **11(6)** 934-937
[13] Sellami I H, Wannes W A, Bettaieb I, Berrima S, Chahed T, Marzouk B, and Limam F 2011 *Food Chemistry* **126** 691-697
[14] Shahhoseini R, Ghorbani H, Karimi S R, Estaji A, and Moghaddam M 2013 *Drying Technology* **31** 1020-1028
[15] Phoungchandang S, Srinukroh W, and Leenanon B 2008 *Drying Technology* **26** 1602-1609
[16] Asekum O T, Grierson D S, and Afolayan A J 2007 *Journal of Applied Sciences* **7(7)** 1005-1008

[17] Voo S S, Howard D, Grimes, and Lange B M 2012 *Plant Physiol.* **159** 81-94
[18] Turner G W, Gershenzon J, and Croteau 2000 *Plant Physiol.* **124** 665-680
[19] Rios-Estepa R, Lange I, Lee J M,and Lange B M 2010 *Plant Physiol.* **152** 2015-2119

The utilizing of zeolite and manures for increasing potassium availability, uptake, and yield on soybean in Alfisols

I L Haniati[1], S Minardi[2], A H L Nastiti[2], S Harieni[3], J Syamsiyah[2], and S Hartati[2]

[1]Postgraduate of Soil Science, Sebelas Maret University, Surakarta, Indonesia
[2]Department of Soil Science, Faculty of Agriculture, Sebelas Maret University, Surakarta, Indonesia
[3]Department of Agrotechnolgy, Faculty of Agriculture, Tunas Pembangunan University, Surakarta, Indonesia

Email: sminardi@staff.uns.ac.id

Abstract. Enhancement soybean production is needed to fulfill its high demand in Indonesia by utilizing the widespread potential land, such as Alfisols. Zeolite and manure addition used to overcome the nutrient availability in Alfisols and decrease the use of inorganic fertilizer. This study was conducted in field condition in a factorial randomized complete block design with three replicates in an Alfisols aimed to assess the effect of applying three rates of zeolites (Z0: 0 $t.ha^{-1}$, Z1: 2.5 $t.ha^{-1}$ and Z2: 5 $t.ha^{-1}$) and three types of 5 $t.ha^{-1}$ manures (P0: 0 No Manure, P1: quail manure (5 $t.ha^{-1}$), P2: cattle manure (5 $t.ha^{-1}$)) on K availability, its uptake, number of pods and seed weight/plot on soybean [*Glycine max. (L) Merrill*]. Zeolites showed a significant result of potassium availability at the rate 5 $t.ha^{-1}$. The similar result is given by cattle manure for potassium uptake in soybean. An interaction between zeolites at the rate 5 $t.ha^{-1}$ and cattle manure showed the significant result on a number of $pods.plant^{-1}$, 100-seeds weight and seed yield.

1. Introduction

Soybean has a great role as a staple food in Indonesia, important for cropping system moreover as a feeder. According to statistics Indonesia merely able to fulfill one-third of their needs, which is about 7.8 tons and imported about 1.8 tons that year [1]. An enhancement is needed to fulfill high demand of soybean but still eco-friendly. Utilizing widespread potential land such as Alfisols is one effort that can be done.

Alfisols with an area of 7 hectares scattered in Java and West Nusa Tenggara. Alfisols are known to have undergone moderate leaching, are susceptible to surface compaction, erosion and acidify rapidly under continuous cultivation. They have low water retention capacity and are tend to be deficient in macronutrient [2],[3]. Alfisols formed by crust and seal process on the soil surface, resulting very low infiltration even under dry soil condition [4]. Alfisols inert mineral composition and its lack of interparticle bonding leads to low organic matter content. Unstable soil aggregation together with a conventional system of farming increase the potential for soil surface sealing and crusting [5]. Nitrogen (N), Phosphorus (P), and potassium (K) have great effects on plant growth and development. Deficiencies or excesses of N, P, K resulting marked effects on the growth and yield of crops. Potassium as a macronutrient required for regulation of metabolic system in plants, therefore its availability needs to be improved to increase grain yield. Therefore, productivity is significantly affecteded nutrient deficiency [6],[7].

Potassium availability can be improved by zeolite addition which prevents nutrient leaching with its ability to absorb K cation. Zeolites are crystalline hydrated aluminosilicates of alkali and alkaline earth cations [8]. Zeolites when dehydrated develop a porous structure made up of Si-O and Al-O tetrahedrons, which bond together sharing vertices and forming a square and hexagonal structures. Zeolites have an important role in improving the soil fertility and in the prevention of nutrition losses [9],[10]. Cattle and quail manure selected as fertilizer to provide nutrient for soils. Manure application improves physical and chemical properties of soil, in addition, increases microorganism activity and water retention capacity. Nutrient availability and soil structure improved by application of manure and usually increased crop yield [11]. Zeolite and manure combination expected to prevent leaching and provide nutrient continuously to overcome problems in Alfisols.

The present study aimed to assess the effects of the use of two zeolite rates (2.5 $t.ha^{-1}$ and 5 $t.ha^{-1}$) and two types of manure (quail manure and cattle manure) on the Potassium availability and uptake on soybean its yield adapted to Alfisols.

2. Materials and Methods

This study was conducted in field condition at Sebelas Maret University experimental field, located in Jumantono district, Karanganyar Regency, Central Java, Indonesia at the geographical coordinates of 7^0 37’ 49,9” S and 110^0 56’54,4” W. The soil from this study area is Alfisols according to the USDA (United States Department of Agriculture) soil classification.

The soybean variety used in this study is “Gema”, crossing result of Shirome and Wilis variety. Three blocks prepared with 30 cm spacing between block, each block contains 9 plot (2x1 m) with 30 cm spacing between plot. Plant spacing in each plot was 20 x 25 cm, 120 seeds sown in each plot and thinned to 40 plants per plot. The watering was estimated visually as the soil surface dried.

This study was arranged in a factorial randomized complete block design with 3 x 3 factorial three rates of zeolites (Z0: 0 $t.ha^{-1}$, Z1: 2.5 $t.ha^{-1}$and Z2: 5 $t.ha^{-1}$) and three types of manures (P0:0 No Manure, P1: quail manure (5 $t.ha^{-1}$), P2: cattle manure (5 $t.ha^{-1}$) with three replicates. The zeolite used in this study was collected from Klaten Regency in Central Java. Particle size in this zeolite varied between 2-4 mm. Zeolite and manure in each treatment applied to the plot 5 days before seeds were planted. Before the treatment applied, initial soil samples were collected from each plots and mixed together to get a composite sample. Chemical characteristics of the initial soil, zeolite, and manure are shown in Table 1.

Table 1. Chemical characteristics of experimental plot soil, manure, and zeolite

Chemical Characteristics	Soil	Manure		Zeolite
		Quail	Cattle	
pH H_20	5.4	-	-	-
C-organic ($g.kg^{-1}$)	1.15	17.56	19.94	-
Soil Organic Matter ($g.kg^{-1}$)	1.93	24.9	28.3	-
CEC ($cmol.kg^{-1}$)	18.93	-	-	78.20
Base Saturation ($g.kg^{-1}$)	14.32	-	-	-
C/N Ratio	4.42	13.31	14.30	-
N Total ($g.kg^{-1}$)	0.26	1.32	1.40	-
P Availability ($mg.kg^{-1}$)	4.45	4.54	4.41	-
K Availability ($cmol.kg^{-1}$)	0.22	1.51	1.40	-
Ca ($cmol.kg^{-1}$)	1.72	-	-	-
Mg ($cmol.kg^{-1}$)	0.56	-	-	-
Na ($cmol.kg^{-1}$)	0.20	-	-	-

Plant sampling for potassium uptake was taken when the flower starts to bloom. Each plant dried in forced circulation oven at 70°C and subsequently ground for K uptake determination. Whole plants of each plot were harvested for yield determination. After harvesting the soybean, soil samples were collected from each plot.

The post-harvest soil samples were air dried at room temperature before analysis. The soil pH determined by glass electrode pH meter [12]. Soil organic matter quantified by Walkley-Back method [13]. Base saturation (exchangeable K) estimated with ammonium acetate extraction method [12]. The dried plant tissues ground and digested with concentrated HNO_3 [14], for determination of K using a flame photometer. The results were subjected to analysis of variance (ANOVA), and means compared using Duncan's Multi Range Test DMRT with SPSS ver 23.0 (SPSS, Inc., Chicago, IL, USA).

3. Results and Discussion

Soil pH showed no significant difference between treatment but a slightly enhancement in every treatment compared with control. The highest pH given by zeolite 5 $t.ha^{-1}$ and cattle manure while the lowest pH given by control with 5.7. Polat et al. [15] stated that zeolite is not acidic but marginally alkaline and its use with fertilizers can help buffer soil pH levels, thus reducing the need for lime application.

Soil base saturation describes how completely the soil particle surface is filled with the basic cations such as K, Ca, Mg, Na [15]. The base saturation showed no significant difference between treatment. Highest result given by zeolite 5$t.ha^{-1}$ and cattle manure, increasing 59% compared with control.

The potassium availability in soil varied due to the different rates of zeolite and manure types application. The findings responded differently to zeolite, with increase 11% and 22% in potassium availability with zeolite rates 0 $t.ha^{-1}$ to 2 $t.ha^{-1}$ and 5 $t.ha^{-1}$ respectively (Table 2). Regarding manure types, cattle manure showed the higher result compared with quail manure in each zeolite rate. Potassium often limits production and needs to be included in a soil fertility program; potassium should be included as a corrective nutrient [17]. Application of zeolite in combination with organic fertilizer had also shown impact in terms of potassium availability. This result comparable to that reported by Basri et al. [18].

Table 2. Soil Chemical characteristic and Potassium uptake on soybean

Treatments	pH	Base Saturation (%)	Availability K ($cmol.kg^{-1}$)	K Uptake (mg/plant)
Z0P0	5.70	42.41	0.32	138.12
Z0P1	5.83	57.32	0.37	223.58
Z0P2	5.77	58.62	0.38	286.51
Z1P0	5.73	46.17	0.33	175.15
Z1P1	5.80	57.97	0.41	279.73
Z1P2	5.80	61.14	0.45	309.20
Z2P0	5.90	62.22	0.39	170.82
Z2P1	6.00	62.92	0.44	345.71
Z2P2	6.10	67.33	0.49	373.57

Description: Z0P0: control, Z0P1: quail manure 5 $t.ha^{-1}$, Z0P2: cattle manure 5 $t.ha^{-1}$, Z1P0: zeolites 2.5 $t.ha^{-1}$, Z1P1: zeolites 2.5 $t.ha^{-1}$+quail manure 5 $t.ha^{-1}$, Z1P2: zeolites 2.5 $t.ha^{-1}$+cattle manure 5 $t.ha^{-1}$, Z2P0: zeolites 5 $t.ha^{-1}$, Z2P1: zeolites 5 $t.ha^{-1}$a+quail manure 5 $t.ha^{-1}$, Z2P2: zeolites 5 $t.ha^{-1}$+cattle manure 5 $t.ha^{-1}$.

The present study showed that cattle manure give the higher result for potassium uptake on soybean in each zeolite rate (Table 2). Quail manure and cattle manure increased potassium uptake 17% and 27% compares to without manure treatment respectively (Table 2). Moreira et al. [6] also reported that potassium concentration in the plants increased significantly with increasing potassium rates. The fertility treatment significantly increased accumulation of potassium uptake. These findings suggest that supplemental fertilization in this study was used as a strategy to maintain greater nutrient availability, leading to increased nutrient uptake, and ultimately grain yield Increased nutrient accumulation did not always lead to greater nutrient removal [19].

Table 3. Effect of treatment to height, dry weight, number of pods, 100 seeds weight and seed yield of soybean

Treatments	Height (cm)			Dry Weight (g)	No. of Pods ($pods.plant^{-1}$)	100-seeds Weight (g)	Seed Yield ($t.ha^{-1}$)
	2nd week	3rd week	4th week				
Z0P0	7.5	12.1	23.1	15.68	56 a	11.36 a	1.86 a
Z0P1	10.3	15.2	24.5	16.07	61 c	12.84 b	2.02 c
Z0P2	10.4	15.6	27.2	17.66	74 d	13.45 c	2.45 d
Z1P0	9.7	15.6	27.6	18.33	58 b	12.85 b	1.95 b
Z1P1	10.7	15.6	27.7	18.74	61 c	13.42 c	2.63 e
Z1P2	10.0	16	26.7	18.53	74 d	14.35 d	2.65 e
Z2P0	9.3	15.5	27.1	17.77	60 c	13.08 bc	1.97 b
Z2P1	10.3	16.3	29.3	19.63	74 d	13.45 c	2.70 f
Z2P2	10.2	16.9	29.7	20.26	101 e	15.29 e	2.90 g

The number followed by the same letters on the same row are not significantly different according to Duncan Multiple Range Test (DMRT) at 5% level.

Plant height showed slightly enhancement but no significant difference between treatments in every week (Table 3). The shortest plant height at 4th week was 23.1 cm given by control; the highest one was 29.7 cm, given by zeolite 5 $t.ha^{-1}$ and cattle manure [17].

The dry weight result showed no significant difference but slightly enhancement in every treatment. Zeolite 5 $t.ha^{-1}$ combined with cattle manure increase 29% of dry weight compared to control. Soybean 100-seeds weight showed significant differences between each manure type treatment. The highest result is given by zeolite 5 $t.ha^{-1}$ and cattle manure, increasing 35% compared with control.

The number of $pods.plant^{-1}$ was the highest with 5 t.ha-1 of zeolites and cattle manure, which was significantly highest than other treatments. The maximum number of $pods.plant^{-1}$ was recorded from treatments with highest potassium availability (Table 2). Parvej et al. [20] showed the same result. The result is in agreement with the findings of Xiang *et al.* [21] who reported that the number of pods.plant-1 was influenced significantly by potassium rates.

The highest soybean yield (2.90 $t.ha^{-1}$) given by zeolite 5 t.ha-1 and cattle manure (Table 3). The maximum result was recorded from treatments with highest potassium availability (Table 2). The result was in agreement who also reported a similar finding on seed yield of soybean under the condition of the proper potassium rate [22]. The lowest seed yield was recorded from 0 $t.ha^{-1}$ zeolite and without manure that might due to the lowest performance yield attributes under the same treatment. Increases in soybean yield were obtained in response to K fertilization [17].

Table 4. Correlation between soil characteristic, soybean growth, and yield

	pH	No. of Pods	Seed Yield	K Uptake	Base Saturation	Dry Weight	100-seeds weight	K Availability
pH	1	.582**	.475*	.342	.561**	.251	.515**	.425*
No. of Pods	.582**	1	.800**	.646**	.526**	.144	.848**	.643**
Seed Yield	.475*	.800**	1	.754**	.540**	.210	.824**	.718**
K Uptake	.342	.646**	.754**	1	.592**	-.004	.614**	.562**
Base Saturation	.561**	.526**	.540**	.592**	1	-.058	.590**	.385*
Dry Weight	.251	.144	.210	-.004	-.058	1	.226	.195
100-seeds weight	.515**	.848**	.824**	.614**	.590**	.226	1	.699**
K Availability	.425*	.643**	.718**	.562**	.385*	.195	.699**	1

*. Correlation is significant at the 0.05 level (2-tailed).
**. Correlation is significant at the 0.01 level (2-tailed).

Dry weight showed a negative correlation between potassium uptake and soil base saturation. The result indicates that a potassium uptake not related to dry weight. Positive and strong correlation of potassium availability, its uptake on soybean and number of pods with grain yield revealed the importance of these characters in determining grain yield. A lot of potassium needed for pod forming and seed replenishment [23]. Seed will be formed inside the pod and continues until it's fully formed. The role of potassium in agriculture is intimately connected with photosynthesis. The number of pods $plant^{-1}$ and seed yield increased indicated a relation with potassium role in plant photosynthesis [24]. Higher potassium rates increase plant photosynthesis causing a number of $pods.plant^{-1}$and seed yield increased.

4. Conclusion

It may be concluded from the result of this study that the application of Zeolite and manure influenced soil fertility, soybean growth, and yield. A number of $pods.plant^{-1}$, 100-seeds weight, and seed yield were significantly affected by zeolite and manure combination. The application of 5 $t.ha^{-1}$ zeolite and cattle manure was found to be the most appropriate combination for maximum productivity of soybean.

Acknowledgements

The authors would like to express their gratitude for PNBP research grant 2018 0f Universitas Sebelas Maret (UNS), Indonesia.

References

[1] FAO (Food and Agriculture Organization) of the United Nations Database 2017 (http://faostat3.fao.org/home/E)

[2] Kang B T and Juo A S R 1983 Management of low activity clay soils in tropical Africa for foodcrop production. In: Beinroth FH, Neel H, Eswaran H. (Eds.). *Proc. of the Fourth Int. Soil Classification Workshop* (Kigali: Rwanda). 1983 pp 450-470.

[3] Achieng J, Ouma G, and Odhiamb G 2013 Use of organic inputs in management of Alfisols and ultisols for sustainable maize production in Western Kenya *American J. of Experimental Agric.* **3**(4) pp 884-895

[4] Hussein M H, Awad M M, Abdul-Jabbar A S 2010 Effect of surface crust on rainfall infiltration in an aridisol in northern Iraq. *European water* **32** pp 25-34

[5] Wakindiki I I C, Ben-Hur M 2002 Soil mineralogy and texture effects on crust micromorphology, infiltration and erosion *Soil Sci. Society of American J.* **66** pp 897-905

[6] Pettigrew W T 2008 Potassium influences on yield and quality production for maize, wheat, soybean, and cotton. *Physiologia Plantarum* **133** pp 670–681

[7] Moreira A, Moraes L A C, Fageria N K 2015 Variability on yield, nutritional status, soil fertility and potassium-use efficiency by soybean cultivar in acidic soil. *Communication in Soil Sci. and Plant Analysis* **46** pp 2490-2508

[8] Eroglu N, Emekci M, Athanassiou C G 2017 Applications of natural zeolites on agriculture and food production *J. Food Agric* **97** (11) pp 3487-3499

[9] Daković A M, Tomašević-Čanović E G, Rottinghaus S, Matijašević Ž, Sekulić 2007 Fumonisin b1 adsorption to octadecyldimetylbenzyl ammonium-modified clinoptilolite-rich zeoliteic tuff. *J. Microporous and Mesoporous Materials* **105** pp 285-290

[10] Khodaeijoghan A, Ghavaland A, Aghaalikhani M, Gholamhoseni M, Dolatabadian 2010 Comparison among different integrated nutrition management for soil micro and macro elements after winter wheat harvesting and yield. *Sci. Biol.* **2** (2) pp 107-111

[11] Nozari R, Tohidi M, Zahedi H 2013 Effect of cattle manure and zeolite application on physiological and biochemical changes in soybean [*Glycine max* (L) merr.] grown under water deficit stress. *Agricola* **13** (1) pp 76-8

[12] Jackson M L 1973 *Soil Chemical Analysis* (New Delhi: Prentice Hall)

[13] Nelson D W, Sommers L E 1982 Total carbon, organic carbon, and organic matter 539-579. In: A L Page, R H Miller, D R Keeney (eds). American Society of Agronomy, Madison, WI, USA.
[14] Page A L, Miller R H, Keeney D R 1982. Methods of soils analysis: Part 2, chemical and microbiological properties (eds). Agronomy series No 9, American Society of Agronomy, Madison, WI.
[15] Polat E, Karaca M, Demir H, Onus A N 2004. Use of natural zeolite (clinoptilolite) in agriculture. *J. Fruit and ornamental plant research* **12** pp 183-189.
[16] Johnston, A. and R. Karamanos 2005 Base saturation and basic cation saturation ratios–how do they fit in northern great plains soil analysis? In: News & views. Potash and Phosphate Institut (PPI) Nacros pp 1-4.
[17] Yagoub S O, Ahmed W M A, Mariod A A 2012 Effect of urea NPK and compost on growth and yield of soybean (*Glycine max* L.) in semi-arid region of Sudan *Int. Scholarly Res. Network* pp 1-6
[18] Basri M H A, Arifin A, Shamshuddin J, Sumanu H A, Hazandy A H, Mohd A K, Baharm Z, Abdul L S, Nasima J, 2013. Effects of mixed organic and inorganic fertilizers application in soil properties and the growth of kenaf (*Hibiscus cannabinnus* L.) cultivated on bris soils. *American J. of Applied Sci.* **10** (12) pp 1586-159
[19] Bender R R, Haegele J W, Below F E 2015 Nutrient uptake, portioning, and remobilization in modern soybean varieties *Agron. J.* **107** (2) pp 563-573
[20] Parvej M R, Nathan A S, Larry C P, Trenton L R 2016 Soybean yield components and seed potassium concentration response among nodes to potassium fertility. *Agron. J.* **108** (2) pp 854-863
[21] Xiang D B, Yong T W, Yang W Y, Wan Y, Gong W Z, Cui L, Lei T 2012. Effect of phosphorus and potassium nutrition on growth and yield of soybean in relay strip intercropping system *Sci. Res. Essays* **7** (3) pp 342-351
[22] Devi K N, Singh L N K, Devi T S, Devi H N, Singh T B, Singh K K, Singh W M 2012. Response of soybean [*Glycine max* (l.) merrill] to sources and levels of phosphorus. *J. Agril. Sci* **4** (6) pp 1916-9752
[23] Mascarenhas 2004 Potassium for Soybeans *Better Crops* **88** (3) pp 26-27
[24] Jin S H, Huan J Q, Li X Q, Zheng B S, Wu J S, Wang Z J, Liu G H, Chen M 2011 Effects of potassium supply on limitations of photosynthesis by mesophyll diffusion conductance in caryacathayensis *Tree Physiol.* **31** pp 1142-1151

IOP Conf. Series: Materials Science and Engineering **633** (2019) 012013 doi:10.1088/1757-899X/633/1/012013

Effect of dietary supplementation of frutescens seeds (*Perilla Frutescens* L.) on growth performance, slaughter weight, carcass weight, and meat quality of male local duck

R F Hadi and Sudiyono

Department of Animal Science, Faculty of Agricultural, Universitas Sebelas Maret, Jl. Ir. Sutami 36A, Surakarta, 57126

E-mail: rendi_fathoni@staff.uns.ac.id

Abstract. This experiment was aimed to observe the effects of Frutescens seeds (*Perilla frutescens* L.) levels in the diet on the growth performance, slaughter weight, carcass weight, and meat quality of male local duck. A total of 90 ducks were divided into 3 treatment groups with five replicates. The treatments were: basal diet without supplementation frutescens seeds meal (P0), experimental diets with 2.5% frutescens seeds meal (P1); and 5% frutescens seeds meal (P2). The experiment was conducted for 60 days. Average daily gain (ADG), slaughter weight, carcass weight, crude protein and fat contents of duck meat were observed. The data were analyzed with one way analysis of variance (ANOVA). If there was significant differences among treatments, Duncan's multiple range test (DMRT) were applied. The body weight, feed intake, slaughter weight, carcass weight, dry matter, and meat protein content were not affected by the dietary treatments. However, the effect of the Frutescens supplementation in the diets significantly affected ADG, non-carcass weight, and fat content in the duck meat.The highest fat content of meat has been found in the duck raised with 5% of Frutescens seeds supplementation in the diet. These results suggested that the use of 5% Frutescens seeds supplementation in the diet effectively increased ADG, non carcass weight, and fat content of meat. Whereas other traits observed in this study were not affected.

1. Introduction

The performance, production and quality of production of local duck genotypes, are still relatively small, to date more than chickens, both local breeds in the laying and broiler type [1]. Nutritional composition of duck meat (*Anas plathyrynchos*) water content 74.13-74.93%; 19.28-20.50% protein; 2.80-3.78% fat [2]. Perilla frutescens is an annual plant and the Lamiaceae family, which is native to Asian countries (India, China, Japan, Korea, Thailand, and several Asian countries). It is edible in medicinal plant and in cooking [3,4].

Frutescens seeds (*Perilla frutescens*) are a nice source of the α-linolenic acid (ALNA) and other aspects of their dietary value have been studied [3], perilla seed and meal are suited as animal feed [5]. Livestock meat can be enriched with n-3 polyunsaturated fatty acid (n-3 PUFA) by adding oilseed rich in α-linolenic acid [6]. *Perilla frutescens* oil, which constitutes approximately 40% of the seed weight, is primarily composed of fatty acids, such as palmitic acid (C16:0), stearic acid (C18:0), oleic acid (C18:1n-9), linoleic acid (C18:2n-6), and α-linolenic acid (ALNA)[4,7,8] .

The fatty acids contained in perilla seeds are unsaturated fatty acids such as oleic acid 14-23%, linoleic acid 11-16%, linolenic acid 54-64%. Perilla seed oils also contains 6.7-7.6% saturated fatty acids [9]. In addition, researchers reported the plant perilla mint (*Perilla frutescens*) as another potential source of this fatty acid. Particularly, the seeds of this plant have been reported with high omega-3 contents. When used as feed in swine and chicken, perilla seed and meal were found to increase omega-3 fatty acid contents in meat and other animal products such as eggs [5]. Some researchers found that the inclusion of high dietary levels of flaxseed (10%) have been associated with a decrease in performance parameters such as reducing feed intake [10,11]. Therefore, research was needed on the effect of duck feed supplementation containing sources of omega-3 fatty acids (Alpha-Linolenic Acid) by using frutescens seed meal (*Perilla Frutescens L.*) as a supplementation of Alpha-Linolenic Acid sources so that there is an increase in carcass quality and nutritional content of duck meat.

2. Materials and methods

2.1. Research materials

A total of 90 ducks were divided into 3 treatment groups, with 5 replicates. The feeding treatments were:
P0: basal diet without supplementation frutescens seeds meal (0%)
P1: treatment diet with 2.5% frutescens seeds meal, and
P2: treatment diet with 5% frutescens seeds meal
The experiment were conducted for 60 days.

2.2. Research methods

The diets were consisted of yellow corn, pollard, rice polish, soy bean meal and vitamin mix. The diets contained crude protein 18-19% and energy 2900-3000 Kcal/kg. Chemical composition of experimental diets were presented in Table 1. Treatment's feedstuffs composition and nutrients content is presented in Table 2.

Table 1. Chemical composition (on dry matter basis) of experimental diets.

Ingredients	Crude protein (%)	ME (Kcal/kg)	Ether extract (%)	Crude fiber (%)	Calcium (%)	Phosphorus av (%)
Yellow corn	8.02	3321	0.97	1.77	0.02	0.069
Pollard	16.85	2446	3.47	9.15	0.09	0.234
Rich polish	11.2	2887	9.49	9	0.04	0.381
Soy Bean Meal	50.94	2577	2.11	3.73	0.29	0.18
Frutescens seed meal	25.01	3457.55	43.1	29.83	2.49	2.031
Premix	0	0	0	0	32.5	22

Table 2. Feedstuffs composition and nutrient content of treatment diets

	Treatment (% of diets)		
	P0	P1	P2
Basal diet (%)	100	97.5	95
Treatment diet (%)	0	2.5	5
Feedstuffs composition			
ME, kcal/kg	2908.09	2994.53	3080.97
Crude protein, %	18.52	19.15	19.77
Crude fiber, %	4.29	5.04	5.79
Extract ether, %	3.13	4.21	5.29
Ca, %	0.90	0.96	1.03
P av, %	0.72	0.77	0.82

P0: basal diet without supplementation frutescens seeds meal (0%); P1: treatment diet with 2.5% frutescens seeds meal, and P2: treatment diet 5% frutescens seeds meal.

2.3. Growth performance
The live weight and feed intake of the ducks were recorded weekly during the experimental period, except for the last period, which lasted seven days. Data on the average daily gain (ADG), and feed intake (FI) were calculated.

2.4. Carcass and non-carcass collection
Data collection of cut weight and carcass weight was carried out when the ducks were eight weeks old. Before slaughtering, two of the ducks were slaughtered in an experimental slaughterhouse with fasting form twelve hours. The process of slaughtering ducks is done by cutting the carotid artery, jugular vein, trachea, and esophagus. The carcass is the body part of ducks after slaughtering based on Islamic Shari'ah, with removal of feathers, and removal of offal, without head, neck and legs. The breast section is taken for meat chemical analysis. In addition, Non carcass weight is obtained by means of life weight reduced by the total weight of the head, neck, blood, hair, skin, legs and also the internal organs (viscera) [12].

2.5. Chemical analyses
The chemical composition of meat (crude protein and ether extract) [13] analysis was conducted in duplicate on freeze-dry samples of the breast and expressed on a fresh basis.

2.6. Data analyses
This research used one way randomized design. The data were analyzed using analysis of variance (ANOVA) and differences between treatment means were further analyzed using Duncan's New Multiple Range Test (DMRT) with significance level of $p<0.05$.

3. Results and discussion

Addition of *Perilla frutescens* meal, as feed supplement up to 5% into the diets have significantly increased ($p<0.05$) average daily gain and non carcass weight (Table. 3). There were no differences among the groups in body weight, feed intake of the diets, slaughter weight and carcass weight ducks with different levels of the diets. Perilla seed and meal are suited as animal feed. Perilla meal in animal feed is still not pervasive to use in animal feed because the limitation of it such as fiber composition is quite high (problem with monogastric animal). It is interesting to use for increase value in meat and egg due to the property is high with protein and omega-3 [5]. The use of plant sources rich in ALA does not negatively influence broiler performance and the sensory qualities of the meat. In addition, the diets enriched with omega-3 fats, alpha-linolenic acid did not change body weight and carcass percentage of broilers. In terms of cuts yield, there was no significant different on the percentage of breast, back and wings by feeding diets supplemented with purslane meal [14].

Table 3. Productive performance and carcass characteristics of ducks fed diets

	Treatments		
	P0	P1	P2
Body weight, g ^{ns}	1183.35	1267.88	1237.40
Feed intake ^{ns}	102.21	103.74	103.51
ADG, g	21.44^{a}	25.49^{b}	26.56^{b}
Slaughter weight, g ^{ns}	1057.88	1115.10	979.38
Carcass weight, g ^{ns}	678.92	755.86	722.23
Non carcass weight, g ^{ns}	475.81^{a}	538.89^{b}	588.34^{b}

P0: basal diet without supplementation frutescens seeds meal (0%); P1: treatment diet with 2.5% frutescens seeds meal, and P2: treatment diet with 5% frutescens seeds meal.
^{a,b} different superscript in the same row indicates significantly ($P<0.05$); ^{ns} = not significant.

There were significant differences in carcass parameters, average daily gain and non carcass weight of the ducks in this study. This result was caused by increased metabolism energy and extract ether accumulation in the pericardic zone as a consequence of the high content of ether extract from the ration with increased perilla seed and digestion supplementation higher than ether extract [6]. Animals fed a diet enriched with animal fat showed higher lumbar circumference, liver weight and carcass and retail cut weights than animals fed linseed or sunflower seed oil supplemented diets [15][16].

Table 4. Chemical composition of meat (on a dry matter basis)

	Treatments		
	P0	P1	P2
Dry matter, % ^{ns}	26.47	25.39	25.17
Crude protein, % ^{ns}	19.22	19.74	19.98
Fat, %	2.51^{a}	2.60^{a}	3.39^{b}

P0: basal diet without supplementation frutescens seeds meal (0%); P1: treatment diet with 2.5% frutescens seeds meal, and P2: treatment diet with 5% frutescens seeds meal.
^{a,b} different superscript in the same row indicates significantly (P<0.05); ^{ns} = not significant.

The dry matter and protein contents were not affected by either the Frutescens level. The impact of the Frutescens enrichment in the diets was significantly affected fat content in the duck meat. The highest fat content of meat has been found in the duck raised with 5% of Frutescens seeds supplementation in the diet (Table 4). The fat content should be emphasized that due to the high percentage of red fibers in duck muscles, their metabolism is higher when compared with chicken or turkey's muscles, and lipids constitute an important component of duck meat [17,18]. In addition, researchers reported that the decrease in lipid content in chickens was very associated with the composition of dietary fatty acid [19]. The fatty acid profiles of breast meat and abdominal fat reflected the fatty acid composition of the diet [20].

4. Conclusions

It was concluded that diets enriched with supplementation Frutescens seeds (*Perilla frutescens* L.) meal up to 5% in the diet effectively increased ADG, non-carcass weight, and fat content of meat, on the other hand, other traits observed in this study were not affected.

Acknowledgements
This research was funded by the research project of Sebelas Maret University (PNBP UNS No. 543/UN27.21/PP/2018).

References

[1] Bosco a D 2011 Performance and meat quality of pure Ancona and Cornish x Ancona chickens organically reared *Arch. Fur* **75** 7–12
[2] Galal A 2011 Performance and Carcass Characteristics of Dumyati , Muscovy , Peking and Sudani Duck Breeds **48** 191–202
[3] Longvah T and Deosthale Y G 1991 Chemical and nutritional studies on hanshi (perilla frutescens), a traditional oilseed from northeast india *J. Am. Oil Chem. Soc.* **68** 781–4
[4] Asif M and Kumar A 2010 Nutritional and Functional Characterisations of Perilla Frutescens Seed Oil and Evaluation of Its Effect on Gastrointestinal Motility *J. Pharm. Sci.* **8** 1–12
[5] Montha N, Yothinsirikul W, Tippaya K, Chompupan K, Lambertz C and Jaturasitha S 2017 Perilla Mint (Perilla frutescens): An Alternative Animal Feed to Enhance Omega- 3 Fatty Acids in Meat and Eggs as Functional Food *J. Agric. Fac. Agric. Chiang Mai Univ.* **33** **(3)** 463–473 (2560)

[6] Peiretti P G, Gai F, Meineri G, Zoccarato I and Gasco L 2010 Apparent digestibility of mixed feed with increasing levels of perilla (Perilla frutescens L.) seeds in rabbit diets *Ital. J. Anim. Sci.* **9** e81

[7] Skowyra M, Falguera V, Azman N, Segovia F and Almajano M 2014 The Effect of Perilla frutescens Extract on the Oxidative Stability of Model Food Emulsions *Antioxidants* **3** 38–54

[8] Gwari G, Lohani H, Haider S ., Bhandari U, Chauhan N and Rawat D S 2015 Chemical and nutritional studies on perilla (perilla frutescens l.) accessions collected... *Int. J. Phytopharm.* **5(5)** 379–82

[9] Asif M 2011 Health effects of omega-3,6,9 fatty acids: Perilla frutescens is a good example of plant oils *Orient. Pharm. Exp. Med.* **11** 51–9

[10] Geflu A, Leinsaat Z Von, Sarı M and Aks M 2002 Effects of addition of flaxseed to diets of laying hens on some production characteristics , levels of yolk and serum cholesterol , and fatty acid composition of yolk **66** 75–9

[11] Bean L D and Leeson S 2003 Long-Term Effects of Feeding Flaxseed on Performance and Egg Fatty Acid Composition of Brown and White Hens *Poult. Sci.* **82** 388–94

[12] Indonesia S N and Nasional B S 2009 Mutu karkas dan daging ayam

[13] Editor A 2006 Of fi cial Methods of Anal y sis *AOAC. 2005. Official Methods Anal. Assoc. Anal. Chem. 18th ed., AOAC, Gaithersburg, Maryland.*

[14] LR Kartikasari B S H and A M N 2018 Carcass and cut yields of broiler chickens fed diet containing purslane meal rich in omega-3 fats Carcass and cut yields of broiler chickens fed diet containing purslane meal rich in omega-3 fats *IOP Conference Series: Earth and Environmental Science (2018) 102(1)* pp 1–6

[15] Pla M, Zomeño C and Hernández P 2008 Effect of the dietary N-3 and N-6 fatty acids on rabbit carcass and meat quality *9th World Rabbit Congress* pp 1425–30

[16] Nuernberg K, Slamecka J, Mojto J, Gasparik J and Nuernberg G 2011 Muscle fat composition of pheasants (Phasianus colchicus), wild ducks (Anas platyrhynchos) and black coots (Fulica atra) *Eur. J. Wildl. Res.* **57** 795–803

[17] Lorenzo J M, Purriños L, Temperán S, Bermúdez R, Tallón S and Franco D 2011 Physicochemical and nutritional composition of dry-cured duck breast *Poult. Sci.* **90** 931–40

[18] Kowalczyk A and Adamski M 2012 Carcass composition and meat characteristics of Pekin ducks in relation to age at slaughter and level of maize distiller ' s dried grains with solubles *J. Anim. Feed Sci.* **21** 157–67

[19] Ferrini G, Baucells M D E-G E and B A C 2008 Dietary Polyunsaturated Fat Reduces Skin Fat as Well *Poult. Sci.* **87** 528–35

[20] Skřivan M, Marounek M, Englmaierová M, Čermák L, Vlčková J and Skřivanová E 2018 Effect of dietary fat type on intestinal digestibility of fatty acids, fatty acid profiles of breast meat and abdominal fat, and mRNA expression of lipid-related genes in broiler chickens *PLoS One* **13** 1–11

IOP Conf. Series: Materials Science and Engineering **633** (2019) 012014 doi:10.1088/1757-899X/633/1/012014

Effect of diets containing fish oil and different levels of purslane meal on production performance and carcass quality of broiler chickens

L R Kartikasari, B S Hertanto, S M Putra, Y Iswara and A M P Nuhriawangsa

Department of Animal Science, Faculty of Agriculture, Universitas Sebelas Maret, Jl. Ir. Sutami No. 36A Surakarta 57126, Indonesia

Email: lilikretna@staff.uns.ac.id

Abstract. The aim of the study was to investigate the effect of dietary inclusion of purslane meal (*Portulaca oleraceae*) and sardines fish oil as a source of omega-3 fats on production performance and carcass quality of broiler chickens. A total of 180 one-day old unsexed Cobb broiler chickens were used in this study. The broilers were randomly allocated into 30 pens with each pen consisted of 6 birds. The pens were randomly assigned to five dietary treatments with 6 replicates (36 birds per treatment). The diets were prepared from a basal diet (P0). The experimental diets were formulated by supplementing the basal diet containing 1.5% sardines fish oil and levels of purslane meal 0 (P1), 6 (P2), 12 (P3), and 18.0% (P4). Water and diets were provided ad libitum for a period of 35 days. Data production parameters were recorded weekly. The data were analysed using analysis of variance (ANOVA). Differences between treatment means were further analysed using Duncan's New Multiple Range Test. Results showed that there were no significant effects on production parameters and carcass weight by adding of dietary purslane meal up to a level of 12%. However, supplementation of 18% purslane meal (P4) resulted in a decrease in body weight from 1718 g (P0) to 1278 g (P4) and in carcass weight from 1139 g (P0) to 811.5 g (P4). It was concluded that dietary levels of purslane meal up to a level of 12% didn't have negative effect on production performance and carcass quality of broiler chickens.

1. Introduction

The effect of supplementation of broiler diets, such as fish meal or fish oil, on broiler performance and the accumulation of meat omega-3 long chain polyunsaturated fatty acids (n-3 LCPUFA) has been investigated by previous researchers [1, 2]. In order to increase the incorporation of n-3 LCPUFA into chicken meat, it is commonly considered to include fish meal or fish oils, alone or in combination with n-3 PUFA (alpha-linolenic, ALA) sources [2]. Unfortunately, there are some negative impacts on the sensory properties of chicken products by using marine sources to manipulate meat fatty acid profiles or meat quality. This is due to the presence of an off-odour or off-flavour [3, 4], and this causes the reduction of consumer acceptability. In order to enhance n-3 LCPUFA deposition in chicken products, the inclusion of plants or seed high in the n-3 PUFA, (ALA, 18:3n-3) which is the n-3 LCPUFA precursor can be an alternative to the supplementation of fish meal or fish oil to chicken diets. Some results from previous studies [5, 6] showed an increase in chicken meat n-3 LCPUFA including eicosapentaenoic acid (EPA), docosapentaenoic acid (DPA, 22:5n-3), and docosahexaenoic acid (DHA) by enriching diets with ALA-rich plant sources. In addition, a study conducted by Kartikasari et al. [7] found that all n-3 fatty acid content and PUFA of meat increased as a result of dietary inclusion of flaxseed oil rich in ALA into broiler diets. Importantly, the use of flaxseed oil does not negatively affect broiler performance and the sensory qualities of the meat. Therefore, the dietary supplementation of plant seeds or oils rich in ALA for broiler chickens can be one of dietary strategy to enhance n-3 fat content in chicken meat and may improve the quality of carcass.

Portulaca oleracea (purslane) is a plant that contains many biologically active compounds, which are β-catotene, folic acid, alpha-tocopherol, ascorbic acid and calcium [8] and high level of n-3 PUFA, ALA [9, 10]. The effect of purslane that is rich in ALA on the accumulation of n-3 fat in eggs, egg quality and production performance of laying hens have been reported by some investigators [10,11,12]. For example, a study conducted by Kartikasari et al. [13] found the use of purslane up to a level of 6% did not influence performance parameters and egg quality such as egg weight, yolk weight, albumen index, yolk index and Haugh Unit. When the level of purslane was increased to 8%, the increase in egg weight, yolk weight, albumen weight, and the intensity of yolk colour was observed [12]. There is little information in relation to the use of purslane meal in performance parameters or carcass quality. In terms of carcass quality, a study conducted by Zhao et al. [14] showed that body weight gain improved and feed conversion reduced by adding purslane extract (0.4%) in the diets for both on day 28 and 42. In addition, there were no effects on slaughter weight, carcass, and carcass cut yields of quails when the diets were supplemented with 10% purslane seed [15]. Therefore, the objectives of the current study were to investigate the effect of diets enriched with purslane meal, as a source of ALA on performance parameters carcass and carcass quality of broiler chickens.

2. Materials and Methods

2.1. Research Materials

In this study, a total of 180 one-day old unsexed Cobb broiler chickens were used. Using 30 pens, broilers were randomly placed in the pens with each pen containing 6 birds. The pens were randomly assigned to five dietary treatments with 6 replicates so each treatment consisted of 36 birds. The basal diet was the corn-soybean based diets containing crude protein 23.43% and energy 3318.94 kkal/kg for starter diet, and crude protein 21.3% and energy 3317.73 kcal/kg finisher diet.

2.2. Research Methods

This experimental design used was a one-way classification design. The variable factor was varying levels of dietary purslane meal, as a source of ALA in the diet. There were a total of five diets comprised of a basal diet (D0) and four experimental diets. The diets were formulated by supplementing a basal diet containing 1.5% fish oil added with 0 (D1), 6 (D2), 12 (D3), and 18% (D4) *Portulaca oleracea* meal (w/w). The level of fat in the experimental diets was held constant at approximately 4% for both starter and finisher diets. Ingredient composition and nutrient content of dietary treatments were presented in Table 1. Upon arrival, the broilers were immediately weighed and housed six birds per cage. There were six replicates for each of the five dietary treatments (n = 6 chickens for each replication). Broiler chickens were fed for 42 days with water and feed provided ad libitum for the duration of period. During the first few days, the chickens were observed frequently to ensure that feed and water were available in adequate amount, and that they were comfortable with the environmental conditions. The data of performance parameters were recorded and measured for a 42-days period. Feed intake (FI) was recorded daily and body weight were determined on weekly basis. Body weight gain and feed conversion ratio (FCR) were calculated and determined throughout the experimental duration (kg feed/kg egg). At day 35 and 42, a total of 30 chickens (n = 6 chickens for each treatment) were slaughtered to evaluate carcass quality and cut yields of chickens.

2.3. Data Analysis
The observed data in this research were the performance data and carcass quality. The obtained data were analysed using analysis of variance (ANOVA). Differences between treatment means were further analysed using Tukey test with significance level of P<0.05.

3. Results and discussion
Adding *Portulaca oleracea* (purslane) meal into the diets influenced production performance including feed intake, weight gain, feed conversion and final body weight (Table 1). Feeding diets enriched with purslane meal at a level of 12 and 18% for 35 days significantly reduced (P<0.01) weight gain. This result was consistent with body weight of chickens, which was diet containing purslane meal 12% resulted in lower body weight compare to those fed diet enriched with 6%. However, while a study found that the use of diets containing 5.0–7.5% flaxseed resulted in significantly lower weight gain [15], in this current study weight gain and body weight of chickens were not different between dietary inclusion of 6% purslane meal and control group. Interestingly, we observed that feed conversion ratio of chickens fed diets supplemented with 12% purslane meal were relatively similar with basal diet (D0).

Table 1. Production performance of broiler chickens fed diets supplemented with Portulaca oleracea meal at day 35.

Parameters	D0	D1	D2	D3	D4	P Value
FC	534.11^{a}	522.35^{ab}	527.97^{a}	496.68^{b}	442.23^{c}	0.000
ADG	46.04^{a}	47.31^{a}	45.592^{a}	41.39^{b}	32.88^{c}	0.000
FCR	1.82^{bc}	1.74^{c}	1.83^{bc}	1.94^{b}	2.13^{a}	0.000
Body weight	1701.0^{a}	1699.0^{a}	1638.8^{a}	1492.4^{b}	1213.4^{c}	0.000

D0: Basal diet, D1: basal diet containing fish oil 1.5%, D2: basal diet containing fish oil 1.5% + 6% purslane meal; D3: basal diet containing fish oil 1.5% + 12% purslane meal; D4: basal diet containing fish oil 1.5% + 18% purslane meal
** significant (P<0.01)

Table 2. Carcass quality of broiler chickens fed diets supplemented with *Portulaca oleracea* meal at day 35.

Parameters	R0	R1	R2	R3	R4	P value	Significance
			(g)				
Slaughtered weight	1718.00^{a}	1720.50^{a}	1783.83^{a}	1553.50^{a}	1278.5^{b}	0.000	***
Carcass weight	1139.00^{a}	1143.17^{a}	1164.30^{a}	1008.17^{a}	811.5^{b}	0.000	***
Carcass percentage	66.29^{ab}	66.43^{a}	65^{ab}	65.23^{ab}	63.47^{b}	0.032	**
Breast weight	441.67^{a}	397.33^{a}	418.50^{a}	359.33^{ab}	288.00^{b}	0.030	**
Thigh weight	161.50^{ab}	183.50^{ab}	188.67^{a}	163.33^{ab}	136.00^{b}	0.000	***
Drumstick weight	158.33^{ab}	172.00^{a}	166.83^{a}	141.17^{b}	117.83^{c}	0.000	***
Back weight	243.17^{a}	266.50^{a}	256.33^{a}	230.33^{a}	172.50^{b}	0.000	***
Wing Weight	121.83^{ab}	126.83^{a}	128.50^{ab}	107.50^{b}	90.50^{c}	0.000	***
Abdominal fat weight	35.84^{a}	35.27^{a}	36.17^{a}	27.29^{ab}	20.27^{b}	0.001	***

D0: Basal diet, D1: basal diet containing fish oil 1.5%, D2: basal diet containing fish oil 1.5% + 6% purslane meal; D3: basal diet containing fish oil 1.5% + 12% purslane meal; D4: basal diet containing fish oil 1.5% + 18% purslane meal
** significant (P<0.01); ***(P<0.001)

Dietary inclusion of Portulaca oleracea (purslane) meal up to a level of 12% into the diets of chickens did not influence the slaughter weight, carcass weight and carcass percentage (Table 3). Similarly, carcass cut yields such as breast weight, thigh weight and back weight tended to decrease when the diets were supplemented with purslane meal beyond 12%. It appears that drumstick weight of chickens fed 12% purslane meal had a lower weight compare to those fed diets containing 6%.

The results of the study were supported by Konca et al. [15] which reported that the dietary inclusion of 10% purslane seed did not influence carcass and carcass cut yields of quails. The percentage of carcass in this study was approximately 65%. Importantly, the inclusion of dietary purslane meal caused a decrease in the weight of abdominal fat. It appears that chickens fed diets containing 12% purslane meal had a lower weight of abdominal fat (20.27 g) compared to those fed control diet (38.84 g). The reduction of lipid content in broiler chickens was strongly associated with the composition of dietary fatty acid [16]. The decreased in abdominal fat weight by feeding diets containing purslane meal is in agreement with the previous studies conducted by Kartikasari et al. [17], who reported that dietary inclusion of 6% purslane meal reduced abdominal fat. In addition, a study conducted by Ferrini et al. [18] showed that abdominal fat significantly decreases in chickens fed diets supplemented with linseed oil rich in n-3 PUFA at a level of 10% ($P < 0.05$).

4. Conclusions

It was concluded that diets enriched with *Portulaca oleracea (*purslane meal*)* up to a level of 12% decreased weight gain and body weight but didn't have negative effects on slaughter weight, carcass quality and cut yields of broiler chickens The dietary inclusion of *Portulaca oleracea* meal up to 6% can be applied without affecting production performance and carcass quality of the chickens.

Acknowledgment

This study was supported by the research grant "Penelitian Dasar Unggulan Perguruan Tinggi (PUPT-DIKTI)" Project Number: 474/UN27.21/PP/2018.

References

[1] Lopez-Ferrer S, Baucells M D, Barroeta A C and Grashorn MA 2001 *Poult. Sci.* **80** 741-52
[2] Hulan H W, Proudfoot F G, Ackman R G and Ratnayake W M 1988 *Can. J. Anim. Sci.* **68** 533-547
[3] Bou R, Guardiola F, Barroeta A C and Codony R 2005 *Poult. Sci.* **84** 1129-1140
[4] Chekani-Azar A, Shahriar H A and Maheri-Sis M 2008 *Asian J. Anim. Vet. Adv.* **3** 62-69
[5] Febel H, Mezes M, Palf T, Herman A, Gundel J, Lugasi A, Balogh K, Kocsis I and Blazovics A 2008 *J. Anim. Physiol. Anim. Nutr.* (Berl) **92** 369-376
[6] Zelenka J, Schneiderova D, Mrkvicova and Dolezal P 2008 *Vet. Med.* **53** 77-85
[7] Kartikasari L R, Hughes R J, Geier M S, Makrides M and Gibson R A 2012 *Prostaglandins Leukot. Essent. Fatty Acids* **87** 103-109
[8] Irawan D, Hariyadi P and Wijaya H 2003 *Indonesian Food and Nutrition Progress* **10** (1)
[9] Uddin M K, Juraimi A S, Hossain M S, Nahar M A U, Ali M E and Rahman M M 2014 *Scientific World J.* 1-6
[10] Aydin R and Dogan I 2010 *J. Sci. Food Agric.* **90** 1759-63
[11] Evaris E, Sarmiento-Franco L A, Segura-Correal J C and Capetillo C M 2015 *Trop. Subtrop. Agroec.* **18** 33- 38
[12] Kartikasari L R, Hertanto H S, Pranoto D, Salim W N and Nuhriawangsa A M P 2017 *Int. Conf. on Food Science and Engineering 2016,* 18-19th October 2016, Surakarta, Indonesia *IOP Conf. Series: Materials Science and Engineering* 193 (2017) 012027 doi:10.1088/1757-899X/193/1/012027
[13] Kartikasari L R, uhriawangsa A M P, Hertanto B S and Swastike W 2015. *J Anim Production* 2015, 41(1) 7-12.
[14] Zhao X H, He X, Yang X F and Zhong X H 2013 *Poult. Sci.* **92** 1343-1347
[15] Konca Y, Beyzi S B, Karabacak M and Yaylak E 2015 *Tavukçuluk Araştırma Dergisi* **12** 1-6
[16] Mridula D, Kaur K, Nagra S S, Barnwal P, Gurumayum S and Singh K K 2015 *J. Applied Anim. Res.* **43** 345-351

[17] Kartikasari L R, Hertanto B S and Nuhriawangsa A M P 2017 *Int. Conf. on Food Science and Engineering 2016*, 18-19th October 2016, Surakarta, Indonesia *IOP Conference Series: Earth and Environmental Science* 102 (2017) 012088 doi:10.1088/1755-1315/102/1/012088
[18] Ferrini G, Baucells M D, Esteve-García E and Barroeta A C 2008 *Poult. Sci.* **87** 528-535

IOP Conf. Series: Materials Science and Engineering **633** (2019) 012015 doi:10.1088/1757-899X/633/1/012015

Physical quality of broiler chickens fed diets containing fish oil and different levels of purslane meal

L R Kartikasari, B S Hertanto, S Y Sutanto and A M P Nuhriawangsa

Department of Animal Science, Faculty of Agriculture, Universitas Sebelas Maret, Jl. Ir. Sutami No. 36A Surakarta 57126, Indonesia

Email: lilikretna@staff.uns.ac.id

Abstract. The aim of the study was to investigate the effect of dietary supplementation of purslane meal (Portulaca oleraceae) and sardines fish oil as a source of omega-3 fats on physical quality of broiler chickens. A total of 180 one-day old unsexed Cobb broiler chickens were used in this study. Chickens were randomly placed into 30 pens with each pen consisted of 6 birds. The pens were randomly assigned to five dietary treatments with 6 replicates (36 birds per treatment). The diets were prepared from a basal diet (P0). The experimental diets were formulated by adding the basal diet containing 1.5% sardines fish oil and levels of purslane meal 0 (P1), 6 (P2), 12 (P3), and 18% (P4). Water and diets were provided ad libitum for a period of 35 days. Six chickens for each treatment were processed as carcass and breast meat (Pectoralis Major) samples were collected for physical quality analysis. The data were analysed using analysis of variance (ANOVA). Differences between treatment means were further analysed using Duncan's New Multiple Range Test. Results showed that the results showed that adding Portulaca oleracea (purslane) meal into the diets at a level of 12 and 18% did not influenced pH value. Feeding diets enriched with purslane meal up to a level of 18% for 35 days did not alter meat WHC with the average of approximately 51%. Diets containing purslane meal decreased meat cooking loss. Meat tenderness of chickens fed diets supplemented with 12% purslane meal (1.12) was higher than those fed 18% purslane meal (1.93). It was concluded that dietary levels of purslane meal up to a level of 12% did not affect the physical quality of broiler chickens.

1. Introduction

The use of marine sources in broiler diets in order to increase broiler performance, meat quality and the accumulation of meat omega-3 long chain polyunsaturated fatty acids (n-3 LCPUFA) has been investigated by previous researchers [1, 2]. However, there are some negative impacts on the sensory properties of chicken products by using fish meal or fish oil to manipulate meat fatty acid profiles or meat quality, as reported by the presence of an off-odour or off-flavour [3,4]. Therefore, an alternative to the supplementation of fish meal or fish oil to chicken diets is needed, for example the use of ALA-rich plant sources [5, 6]. Moreover, a study conducted by Kartikasari et al. [7] found that all chicken meat n-3 LCPUFA including eicosapentaenoic acid (EPA), docosapentaenoic acid (DPA, 22:5n-3), and docosahexaenoic acid (DHA) increased as a result of dietary inclusion of flaxseed oil rich in ALA without negatively affecting broiler performance and the qualities of the meat. Therefore, this dietary strategy can be used to increase n-3 fat content in chicken meat and may improve the quality of meat..

One of plant that contains many biologically active compounds is Portulaca oleracea (purslane). These compounds include alpha-tocopherol, folic acid, β-catotene, calcium, and ascorbic acid [8]. Purslane also contains high level of n-3 PUFA, ALA [9, 10]. Dietary inclusion of purslane rich in ALA on production performance of laying hens, quality of egg, and n-3 fat content has been investigated by some researchers [10, 11, 12]. For example, a study conducted by Kartikasari et al. [13] found the supplementation of purslane up to a level of 6% did not alter performance parameters and egg quality such as yolk weight, egg weight, yolk index, albumen index, and Haugh Unit. Moreover, there was an increase in albumen weight, yolk weight, egg weight, and yolk colour intensity when the level of purslane was increased to 8% [12]. In broiler, a study conducted by Zhao et al. [14] showed that body weight gain improved and feed conversion reduced by adding purslane extract (0.4%) in the diets for both on day 28 and 42. A study in quails reported that there were no effects on slaughter weight, carcass, and carcass cut yields of quails when the diets were supplemented with 10% purslane seed [15]. In addition, Kartikasari et al. [16] reported that diets supplemented with *Portulaca oleracea* up to a level of 6% did not influence carcass quality and carcass cut yields of broilers. There is limited data regarding to the use of purslane meal in meat quality. Therefore, the aims of the current study were to evaluate physical quality of broiler chickens fed diets containing fish oil and purslane meal meat, rich in ALA.

2. Materials and Methods

2.1. Research Materials

A total of 180 one-day old unsexed Cobb broiler chickens were used in this study. The birds were randomly allocated in 30 pens with each pen containing 6 birds. There were five dietary treatments with 6 replicates (n= 36 birds for each treatment). The basal diet was the corn-soybean based diets. Starter diet contained crude protein 23.43% and energy3318.94 kcal/kg and finisher diet contained crude protein 21.3% and energy 3317.73 kkal/kg.

2.2. Research Methods

A one-way classification design was used in this study. The variable factor was purslane meal levels, as a source of ALA in the diet. Five dietary treatments were formulated with a basal diet (D0) supplemented with 1.5% fish oil and added with 0 (D1), 6 (D2), 12 (D3), and 18% (D4) purslane meal (w/w). The fat content in the experimental diets was approximately 4% for both starter and finisher diets. The composition of the diets and nutrient content were presented in Table 1. Upon arrival, the broilers were immediately weighed and housed six birds per pen. There were six replicates for each of the five dietary treatments. The diets were given to the broiler chickens for 42 days. Water and feed were provided ad libitum for the experimental period. To ensure that feed and water were available in adequate amount, the chickens were observed frequently during the first few days. This activity was also to ensure that they were comfortable with the environmental conditions. The data of performance parameters were recorded and measured for a 42-days duration of period. At day 35 and 42, a total of 30 chickens, which were six chickens for each treatment, were slaughtered to evaluate meat quality.

2.3. Data Analysis

All collected data of meat quality in this research were analysed using analysis of variance (ANOVA). Differences between treatment means were further analysed using Tukey test with significance level of $P<0.05$.

3. Results and discussion

The meat qualities of broiler chickens fed diets supplemented with Portulaca oleracea meal for 35 days were presented in Table 1. The results showed that adding *Portulaca oleracea* (purslane) meal into the diets influenced pH value with the diets containing 6% purslane meal increased pH value of the meat from 5.82 for control diet to 6.30 for diets supplemented with 6% purslane meal. Similarly, there was no significant effect on water content observed by feeding purslane meal up to a level of 18%.

Importantly, meat cooking loss of broilers fed diets containing purslane meal was a lower than those fed the control diet however there was no different in cooking loss among dietary treatments. When cooking loss did not change by the increasing levels of dietary inclusion of purslane meal, it appears that fat content of meat increased with the highest level achieved for diets containing 18% purslane meal (5.09%). This could be due to an increase in fat content of the diets by supplementation of dietary purslane meal.

The lipid content of broiler chickens was strongly associated with the composition of dietary fatty acid [17]. This current result was support by Watanabe et al. [18] which reported that there was no significant correlation between intramuscular fat content and meat cooking loss. In addition, a study conducted by Cannata et al. [19] found a negative correlation between cooking loss and fat content of meat. In contrast, a significant positive correlation between lipid content and cooking loss was observed. In this study, Feeding diets enriched with purslane meal up to a level of 18% did not alter meat WHC with the average of approximately 51%. This suggests that there was no relationship between total fat content of meat and WHC as reported by Watanabe et al. [18]. The findings of the study showed that dietary inclusion of Portulaca oleracea (purslane) meal up to a level of 12% into the diets of chickens increased meat tenderness (1.12) compared to those fed 18% purslane meal diets (1.93).

Table 1. Meat quality of broiler chickens fed diets supplemented with Portulaca oleracea meal at day 35.

Parameters	D0	D1	D2	D3	D4	P Value
pH	5.82^{b}	6.27^{a}	6.30^{a}	6.11^{ab}	6.11^{ab}	0.001
WHC (%)	47.39^{a}	43.14^{a}	53.03^{a}	50.93^{a}	50.86^{a}	0.260
Cooking loss (%)	16.77^{a}	10.02^{b}	10.30^{b}	9.07^{b}	10.90^{b}	0.000
Tenderness	1.82^{ab}	1.72^{ab}	1.43^{ab}	1.12^{b}	1.93^{a}	0.036
Water content (%)	75.34^{a}	74.17^{a}	76.15^{a}	75.07^{a}	74.27^{a}	0.464
Fat content (%)	4.15^{bc}	4.23^{bc}	3.62^{c}	4.48^{ab}	5.09^{a}	0.001

D0: Basal diet, D1: basal diet containing fish oil 1.5%, D2: basal diet containing fish oil 1.5% + 6% purslane meal; D3: basal diet containing fish oil 1.5% + 12% purslane meal; D4: basal diet containing fish oil 1.5% + 18% purslane meal
** significant (P<0.01)

4. Conclusions

It was concluded that diets enriched with *Portulaca oleracea* (purslane meal) up to a level of 12% didn't change the meat quality including pH, WHC, tenderness, water content and fat content. Diets supplemented with purslane meal reduced cooking loss of the meat. The dietary inclusion of *Portulaca oleracea* meal up to a level of 12% can be applied without affecting meat quality of the chickens.

Acknowledgment

This study was supported by the research grant "Penelitian Dasar Unggulan Perguruan Tinggi (PDUPT-DIKTI)" Project Number: 474/UN27.21/PP/2018.

IOP Conf. Series: Materials Science and Engineering **633** (2019) 012015 doi:10.1088/1757-899X/633/1/012015

References

[1] Lopez-Ferrer S, Baucells M D, Barroeta A C and Grashorn M A 2001 *Poult. Sci.* **80** 741-52

[2] Hulan H W, Proudfoot F G, Ackman R G and Ratnayake W M 1988 *Can. J. Anim. Sci.* **68** 533-547

[3] Bou R, Guardiola F, Barroeta A C and CodonyR 2005 *Poult. Sci.* **84** 1129-1140

[4] Chekani-Azar A, Shahriar H A and Maheri-Sis N 2008 *Asian J. Anim. Vet. Adv.* **3** 62-69

[5] Febel H, Mezes M, Palf T, Herman A, Gundel J, Lugasi A, Balogh K, Kocsis I and Blazovics A 2008 *J. Anim. Physiol. Anim. Nutr.* (Berl) **92** 369-376

[6] Zelenka J, Schneiderova D, Mrkvicova E and Dolezal P 2008 *Vet. Med.* **53** 77-85

[7] Kartikasari L R, Hughes R J, Geier M S, Makrides M and Gibson R A 2012 *Prostaglandins Leukot. Essent. Fatty Acids* **87** 103-109

[8] Irawan D, Hariyadi P, Wijaya H 2003 *Indonesian Food and Nutrition Progress* **10** (1)

[9] Uddin M K, Juraimi A S, Hossain M S, Nahar M A U, Ali M E and Rahman M M 2014 *Scientific World J.* 1-6

[10] Aydin R and Dogan I 2010 *J. Sci. Food Agric.* **90** 1759-63

[11] Evaris E, Sarmiento-Franco L A, Segura-Correal J C and Capetillo C M 2015 *Trop. Subtrop. Agroec.* **18** 33- 38

[12] Kartikasari L R, Hertanto H S, Pranoto D, Salim W N and Nuhriawangsa A M P 2017 *Int. Conf. on Food Science and Engineering 2016*, 18-19th October 2016, Surakarta, Indonesia *IOP Conf. Series: Materials Science and Engineering* 193 (2017) 012027 doi:10.1088/1757-899X/193/1/012027

[13] Kartikasari L R, Nuhriawangsa A M P, Hertanto B S and Swastike W 2015. *J Anim Production* 2015, 41(1) 7-12

[14] Zhao X H, He X, Yang X F and Zhong X H 2013 *Poult. Sci.* **92** 1343-1347

[15] Konca Y, Beyzi S B, Karabacak M and Yaylak E 2015 *Tavukçuluk Araştırma Dergisi* **12** 1-6

[16] Kartikasari L R, Hertanto B S and Nuhriawangsa A M P 2017 *Int. Conf. on Food Science and Engineering 2016,* 18-19th October 2016, Surakarta, Indonesia *IOP Conference Series: Earth and Environmental Science* 102 (2017) 012088 doi:10.1088/1755-1315/102/1/012088

[17] Mridula D, Kaur K, Nagra S S, Barnwal P, Gurumayum S and Singh K K 2015 *J. Applied Anim. Res.* **43** 345-351

[18] Watanabe G, Motoyama M, Nakajima I and Sasaki K 2018 *Asian-Australas. J. Anim. Sci.* **31** 914-91812

[19] Cannata S, Engle T E, Moeller S J, Zerby H N, Radunz A E, Greens M D, Bass P D and Belk K E 2010 *Meat Sci.* **85** 428-434

IOP Conf. Series: Materials Science and Engineering **633** (2019) 012016 doi:10.1088/1757-899X/633/1/012016

Process optimization for producing pumpkin (*Cucurbita moschata* D) and arrowroot (*Marantha arundinaceae* L) starch-based instant porridge

A Slamet[1,2], D Praseptiangga[3], R Hartanto[3] and Samanhudi[4]

[1]Doctoral Program of Agricultural Science, Graduate School of Sebelas Maret University (UNS), Jl. Ir. Sutami 36 A, Kentingan, Surakarta 57126, Central Java, Indonesia
[2]Department of Food Science and Technology, Faculty of Agroindustry, University of Mercu Buana Yogyakarta, Jl. Wates Km 10 Yogyakarta , Indonesia
[3]Department of Food Science and Technology, Faculty of Agriculture, Sebelas Maret University (UNS), Jl. Ir. Sutami 36 A, Kentingan,Surakarta 57126, Central Java, Indonesia
[4]Department of Agrotechnology, Faculty of Agriculture, Sebelas Maret University (UNS), Jl. Ir. Sutami 36 A, Kentingan, Surakarta 57126, Central Java, Indonesia

Email: agusumby@yahoo.com

Abstract. Pumpkin is a potential source of vitamin A and β-carotene as an antioxidant, thus it is promising to be developed as functional food ingredients. However, pumpkin-based manufactured food is still limited. Hence an experiment is initiated to produce pumpkin and arrowroot starch-based instant porridge which is easy to prepare and people ranging from children to adults fond of it. Pumpkin was whisked and mixed with arrowroot starch in the ratio of 1:1, 2:1, 3:1, 4:1, and 5:1 until it became an instant porridge. The porridge was then put into a drum dryer (1 bar) and it was rotated at the speed of one or one and a half round per minute. Color and viscosity of the produced instant porridge were determined. The results indicated that physical characteristics of instant porridge were affected by ratio of pumpkin with arrowroot starch and the rotation speed of the drum dryer. Instant porridge made from pumpkin-arrowroot starch in the ratio of 5:1 and produced at the speed of one and a half round per minute had a color value of L: 74.43, a: 9.33, b: 32.50, and viscocity 47.4 cP.

1. Introduction

Pumpkin (*Cucurbita moschata* D.) is a type of annual plant from the *Cucurbitaceae* family. Availability of pumpkin in Indonesia is relatively high. The three most common varieties are *Cucurbita pepo, Cucurbita maxima* and *Cucurbita moschata*. Pumpkin is also rich in carotene, vitamins, minerals and fiber [1]. The chemical composition as well as the antioxidants content makes pumpkin an important food product for human consumption and industrial utilisation [2]. Just as in most fruits, pumpkins are very sensitive to microbial spoilage, even under refrigerated conditions. Thus, it is best if they are preserved in order to increase shelf life [3]. This is especially true as the perishable nature of pumpkin tends to limit its utilisation, hence the need to be processed by drying [4]. Pumpkin contains high carotenoids reaching 160 mg/100 g. Pumpkin comes from the *Cucurbitaceae* family and *Cucurbita* genus. Common cultivars (*Cucurbita moschata* D.) for human consumption are called pumpkins [5].

Pumpkin names are vary between different countries, among others, pumpkins are called Kabocha (Japan), buttercup (New Zealand). Pumpkin has a yellow-orange color, sweet taste and sweeter than other pumpkin varieties [6]. Pumpkin is stable during storage about 3 to 4 months after harvesting. However, after the skin is peeled off, the pumpkin is in a state that is susceptible to softening, microbes, discoloration and decay. Pumpkin is rich in fiber (dietary fiber), contains bioactive compounds, β-carotene, vitamin A and tocopherol [7]. Besides these vitamins, pumpkin also contains other vitamins, namely B6, K, C, thiamine and riboflavin. The mineral contents in the pumpkin include K, P, Mg, Fe and Se [8] Pumpkin is rich in fiber (dietary fiber), contains bioactive compounds, β-carotene, vitamin A and tocopherol [9]. In addition to these vitamins, pumpkin also contains other vitamins, namely B6, K, C, thiamine, and riboflavin.

Arrowroot (*Maranta arundinacea* L.) is a starch carbohydrate source that has the advantage of having a low glycemic index, making it suitable for people with diabetes mellitus. Arrowroot tubers have health benefits because of the lower glycemic index (14), compared to other tubers, such as gembili (90), kimpul (95), canna (105), and sweet potato (179). The glycemic index states a measure of an increase in a person's blood sugar level after consuming the food in question [10].

The optimal drying process with drum dryer which produces pumpkin powder with good aroma uses a vapor pressure of 313.54 kPa with a 1.27 rpm drum dryer rotation. Drum dryer at a vapor pressure of 313.54 kPa and with drum rotation of 1.27 rpm 7.5% in the process of processing the noodles with the highest fiber and β-carotene content [11].

The purpose of this study is to determine the optimization of the process of making instant porridge of pumpkin and arrowroot starch to produce instant porridge of pumpkin and arrowroot starch which has optimal physical properties.

2. Materials and methods

2.1. Materials

The main ingredients used in this study were pumpkin (*Cucurbita moschata D.)* and arrowroot starch. Pumpkin was obtained from the Ungaran area of Central Java. The specifications of the pumpkins that were used were already ripe optimally at the age of 3 months of harvest, had orange skin and flesh. The pumpkins used were weighing between 8-10 kg/fruit. Arrowroot starch was obtained from a group of arrowroot processed craftsmen from Mekar Jaya in Pajangan Area, Bantul, Yogyakarta. The tool used in this study was drum dryer for instant porridge of pumpkin and arrowroot starch production.

2.2. Methods

The skin of the pumpkin was peeled, then the fruit was cut into pieces with a size of 2 x 2 x 2 cm^3. Pumpkin pieces were included in the grinding machine with arrowroot starch added in the ratio of each of: 1:1, 2:1, 3:1, 4:1, and 5:1. Each mixture of pumpkin and arrowroot starch was added with 200 ml of distilled water. The resulting mixture of pumpkin and arrowroot starch was processed using a drum dryer with a variation of the drum dryer rotation speed of 1 and 1.5 rpm at a pressure of 1 bar. The resulting drum dryer process was in the form of a thin slurry of instant slurry, then it was ground and sieved with an 80 mesh sieve.

2.3. Color Testing

The color test in this study used a Chromameter CR 300 Minolta (Japan). Testing was conducted using the Hunter L *, a * and b * color systems. Chromameter before use was calibrated first with the white standard on the device.

2.4. Viscosity Testing

The viscosity test was carried out in this study by using a Brookfield Viscometer tool. The working principle of liquid viscosity testing is to measure the degree of viscosity of liquid samples. The measurement of the thickness of the liquid is useful to know that the frictional force experienced by an object moving in a fluid is related to the viscosity of the liquid. The thicker a liquid means that it will inhibit the flow of the liquid to move.

2.5. Statistical Analysis

The data were analyzed statistically by one-way variance analysis (ANOVA) at a significance level of 0.05 and the difference in mean values was determined by Duncan's Multiple Range Test (DMRT) ($p<0.05$) to determine the difference between concentration treatments. For statistical analysis using the SPSS 16, one way anova program.

3. Results and Discussion

3.1. Color

The color of instant porridge of pumpkin and arrowroot starch with variations of pumpkin- arrowroot starch and rotation speed of drum dryer are presented in Table 1.

Table 1. Color of Instant Porridge of Pumpkin and Arrowroot Starch

Pumpkin : arrowroot starch	Rotation speed of drum drier (rpm)	L	a	b
1:1	1	78.66 ± 0.23^{d}	4.36 ± 0.57^{a}	22.46 ± 0.11^{d}
2:1	1	78,70 ± 0.17^{d}	6.36 ± 0.57^{c}	29.10 ± 0.26^{e}
3:1	1	73.70 ± 0.00^{a}	6.13± 0.57^{c}	28.43 ± 0.05^{d}
4:1	1	79.36 ± 0.04^{e}	7.83 ± 0.57^{e}	28.20 ± 0.51^{d}
5:1	1	78.00 ± 0.00^{c}	9.00 ± 0.00^{g}	32.70 ± 0.00^{g}
1:1	1.5	80.03 ± 0.46^{f}	5.33 ± 0.57^{b}	25.63 ± 0.28^{b}
2:1	1.5	77.50 ± 0.00^{b}	6.10 ± 0.17^{c}	29.13 ± 0.32^{e}
3:1	1.5	78.23 ± 0.32^{c}	6.73± 0.20^{d}	26.70 ± 0.20^{c}
4:1	1.5	80.30 ± 0.00^{f}	8.20 ± 0.00^{f}	29.80 ± 0.00^{f}
5:1	1.5	77.43 ± 0.15^{b}	9.33 ± 0.00^{h}	32.50 ±0.20^{g}

Values are mean ± standard deviation. Means within the same column with different letters are significantly different ($p<0.05$).

Based on the color Table 1, it shows that the more the pumpkin ratio added, the more the brightness value decreased. This is because the more pumpkin added the instant porridge color was produced, the color turned into yellow-orange, so the white degree decreased. The lowest level of instant porridge brightness in the ratio of pumpkin and arrowroot starch was 5:1 at 1 or 1.5 rpm drum dryer rotation speeds. As for the redness value (a) shows that there were more pumpkin produced instant slurry which was getting red. This was because the pumpkin contained carotene which gave the red pigment. The yellowish value (b) also had the same tendency as the reddish value, that was, the more pumpkin would produce instant porridge which was getting yellowish. The effect of drum dryer rotation speed on the value of redness and yellowish indicated that the drum drier speed of 1.5 rotations per minute increased the value of redness and yellowish in the instant porridge produced. This was because the faster the rotation of the drum drier did not cause much decrease in the value of the reddish and yellowish color. Carotenoids would decrease due to longer heat contact. Drum dryer rotation at 1.5 rpm was longer than 1 rpm. [11] reported that the higher drying temperature and long drying time caused a decrease in beta carotene levels to reach 56%. [12] stated that the temperature and time of the process of making instant porridge using a drum dryer affected the product produced.

3.2. Viscosity of Slurry Instant Porridge of Pumpkin and Arrowroot Starch

Table 2 shows the viscosity of instant porridge of pumpkin and arrowroot starch before being processed using a drum dryer. The more pumpkin, the viscosity increased. At the pumpkin and arrowroot starch ratio of 5:1, the highest viscosity was 970.8 cP. This was because the water content of pumpkin was greater than arrowroot starch. This was consistent with what was stated by [13] that the total solids in pumpkin was ranged between 7-10%, which meant the water content was between 90-93%. Based on the best instant porridge sensory testing produced by the drum dryer process at a ratio of 5:1 at 1.5 rpm rotation speed, it had a viscosity value of 47.4 cP.

Table 2. Viscosity of Slurry Instant Porridge of Pumpkin and Arrowroot Starch

Pumpkin : arrowroot starch	viscosity (cP)
1:1	140.50 ± 0.70^{a}
2:1	212.00 ± 2.12^{b}
3:1	338.80 ± 1.41^{c}
4:1	549.90 ± 0.00^{d}
5:1	970.80 ± 1.41^{e}

Values are mean ± standard deviation. Means with different letters are significantly different ($p<0.05$).

4. Conclusion

The treatment ratio of pumpkin and arrowroot starch and rotation speed of drum dryer affected the physical properties of instant porridge of pumpkin and arrowroot starch produced. Variation of pumpkin and arrowroot starch by 5:1 in the drum dryer process in a pressure of 1 bar and a speed of 1.5 rotations per minute produced instant porridge with colors: L: 74.43, a: 9.33, and b: 32.50, and viscosity 47.4 cP.

References

[1] Krokida, M. K., Karathanos, V. T., Maroulis, Z. B. and Marinos-Kouris, D. 2003. *Journal of Food Engineering*. **59**(4) 391– 403.
[2] Guiné, R. P. F. and Barroca, M. J. 2012. *Food and Bioproducts Processing* **90** (1) 58–63.
[3] Doymaz, I. 2007. *Journal of Food Engineering*. **79** 243–248.
[4] Onwude, D. I., Hashim, N., Janius, R. B., Nawi, N. and Abdan, K., *2016 International Food Research Journal*, **23** (3) 1173-1181
[5] Sudarto Y., 1993. *Budidaya Waluh*. (Yogyakarta Kanisius)
[6] Loy, J.B., 2004. Critical reviews in plant sciences, **23** (4) 337-363.
[7] Wang, J., Wang, J. S., and Yu, Y. 2007. *Archivos Latinoamericanos de Nutrición* **51** 395–399
[8] Nawirska, A., Figiel, A., Kucharska, A.Z., Sokół-Łętowska, A. and Biesiada, A., 2009. Journal. Food Engineering, 94 (1) 14-20
[9] Wang, Y.M., Zhang, B., Xue, Y., Li, Z.-J.,Wang, J.F., Xue, C.-H., and Yanagita, T. 2010. The mechanism of dietary cholesterol effects on lipids metabolism in rats. Lipids in Health and Disease, **9** (1) 4.
[10] Marsono, Y. 2002. *Journal Agritech*, **22** (1) 13-16.
[11] Đào 2015 *Journal of Science* , **3** (3) 149 – 160.
[12] Roongruangsri, W. and Bronlund, J.E., 2016. *International Food Research Journal,* **23** (3) 962-972
[13] Arévalo-Pinedo, A. and Murr, F.E., 2006. *Journal of Food Engineering,* **76** (4) 562-567.

Acknowledgement
This research is fully supported by using the scholarship funds of the Government of the Republic of Indonesia through the Indonesian Institute of Education Scholarship Fund Management Institution (LPDP BUDI-DN) Indonesia. Number: PRJ-4952/LPDP.3/2016.

IOP Conf. Series: Materials Science and Engineering **633** (2019) 012017 doi:10.1088/1757-899X/633/1/012017

Efficacy of betaine as carcass modifier in female broiler chickens to produce leaner carcass for human consumption

A Ratriyanto, S Prastowo and N Widyas

Department of Animal Science, Faculty of Agriculture, Universitas Sebelas Maret, Surakarta, Indonesia

E-mail: ratriyanto@staff.uns.ac.id

Abstract. Betaine is a methyl group donor which involved in protein and energy metabolism, offering an opportunity to the poultry producers to satisfy consumer needs by producing leaner carcass. This study was conducted to investigate the efficacy of betaine as carcass modifier in female broiler chickens fed diet containing sufficient methyl group donor. Two hundred female day-old chicks (DOC) were allotted to four assay diets with 5 replicates of 10 birds. The starter and finisher diets contained 0.50 and 0.38% methionine as methyl group donor, respectively. The basal diets were fed without supplementation or supplemented with 0.10, 0.20 and 0.30% betaine. The assay diets were fed from the age 8–42 days. Two birds per replicate were slaughtered at the age of 42 days to measure the carcass characteristics. The fat content was derived from breast meat samples. Supplementation of betaine did not influence the slaughter weight, carcass and breast yield, indicating similar growth and carcass production. Furthermore, administration of betaine produced 29.8–42.5% lower abdominal fat pad deposition and 9.2–35.9% lower meat fat content than those without betaine administration ($P<0.05$). It can be concluded that betaine can modify carcass characteristic in female broiler chickens by lowering fat deposition.

1. Introduction

Broiler chickens have a high growth rate, good feed efficiency and can be slaughtered at a relatively young age between 35–42 days. To obtain optimal growth performance and good carcass quality, the broilers have to be provided with sufficient nutrient content in their diet. The nutritional requirements in intensive livestock farming are well known and can be met through a precise formulation [1]. However, there is a room to improve the growth rate or carcass characteristics by supplementation with a specific feed additive which influence the nutrient digestion or metabolism. For instance, betaine (trimethyl glycine) has been considered to improve growth rate and carcass characteristics in poultry [2] and pigs [3].

Due to its role as methyl group donor, betaine is involved in protein and energy metabolisms and has a potential to produce leannes carcass with low fat content [4]. In this regard, betaine is categorized as “carcass modifier” which refers to a substance that can modify and improve carcass composition and quality through specific metabolic pathway [4]. Therefore, it is expected that betaine may offer an interesting opportunity to the poultry producers to satisfy consumers need in term of leaner carcass for human consumption [5]. However, provious studies revealed that responses to the mode of actions of betaine as carcass modifier are still inconsistent [6,7]. Several studies reported improvement in carcass yield or breast yield in poultry [6,8] and reduction in carcass fat [9]. In contrast, other investigations indicated less pronounced responses of carcass traits or fat deposition following addition of betaine in the diet [7,10].

It is well known that the role of betaine as methyl group donor is associated with the present of methyl group originated form other sources in the diet. Thus, the level of other methyl group donors, such as methionine and choline, would affect its efficacy [4,11]. Furthermore, growth rate and carcass production in poultry are categorized as sex-related traits and exist a sexual dimorphism between male and female. Investigation with a certain gender would result a more accurate response rather than using unsexed birds. Therefore, this study was performed to investigate the efficacy of betaine as carcass modifier in female broiler chickens fed diet containing sufficient methyl group donor.

2. Materials and methods

2.1. Animals and diets

This study used 200 female day-old chicks (DOC) of broiler chickens of New Lohmann (MB 202) strain produced by PT. Japfa Comfeed Indonesia with average body weight of 44.4±0.8 g. During adaptation for 7 days after arrival, the birds were kept with a standard brooding management according to the breeder. At 8 days, the birds with an average body weight of 149.5±2.5 g were allotted to 4 dietary treatments and 5 replicates with 10 birds per replicate. The 4 dietary treatments included a basal diet and a basal diet supplemented with 0.1, 0.2, and 0.3% anhydrous betaine (Betafin 96%, Danisco-Finnland). The diets were formulated based on maize and soybean meal to meet the nutrient requirements of broiler chickens according to Nutrition Research Council [12]. The starter and finisher diets contained sufficient level of methionine as methyl group donor amounting 0.50 and 0.38% methionine, respectively (**Table 1**). Betaine was supplemented to the diet at the expense of maize [13].

Table 1. Nutrient composition of the basal diets.

Nutrients	Starter Diet	Finisher Diet
Metabolizable energy (kcal/kg)	3200.51	3226.38
Crude protein (%)	22.00	20.06
Calcium (%)	1.10	1.14
Phosphorus (%)	0.47	0.57
Lysine (%)	1.19	1.10
Methionine (%)	0.50	0.38

2.2. Data collection

The birds were raised on litter cages with the size 1.0 × 1.0 m for each replicate. The assay diets were given *ad libitum* from the age of 8 to 21 days during starter period, while the assay diets for finisher period were fed thereafter until 42 days. Accordingly, the birds have free access to water during the experiment. The ND B1, Gumboro and ND La Sota vaccines were applied according to standard from the producers.

At the end of experiment, 2 boiler chickens from each replicate (10 birds per treatment) were slaughtered after an overnight (12 hours) fast to measure the carcass characteristics [14]. Prior to slaughtering, the birds were individually weighed. Eviscerated carcass was obtained after the birds being sacrificed by removing the blood, feather, head, neck, shank and internal organs. Carcass and breast yield were determined as percentage of live body weight at slaughtering. The fat present in abdomen and attached at the gizzard was removed and weighed to measure abdominal fat deposition [14]. Samples of breast meat were collected and analized to measure the fat content according to the procedure of AOAC [15].

2.3. Statistical analysis

The data were checked for homogeneity and then analyzed with analysis of variance the effect of treatments. If the treatments showed significant effect, it was continued to Duncan's test to analyze the difference between treatments [18]. The following mathematical model was applied: $Y_{ij} = \mu + \alpha_i + \epsilon_{ij}$

where μ was the grand mean, α_i was the deviation from the grand mean due to the treatment levels and ϵ_{ij} was the error term. The R program was applied for statistical analyses [16].

3. Results and discussion

3.1. Slaughter weight, carcass and breast yield

Supplementation of betaine in the diet containing sufficient methionine did not affect slaughter weight, carcass and breast yield indicating similar growth and carcass production (**Table 2**). This result indicated that betaine is less efficient in improving protein deposition during carcass production when supplemented in the diet containing high level of methyl group donor. Efficacy of betaine as a carcass modifier is still inconsistent, which may be influenced by several factors such as methyl group donor level in the diet, betaine level and source, as well as age and species of the animal [4].

Betaine as a methyl group donor is expected to increase the availability of methionine for protein synthesis to produce optimal growth [17]. During the transmethylation reaction, betaine donates its labile methyl group to form methionine or other substances involved in protein and energy metabolism [4], as has been shown by Apicella et al. [18] that betaine enhanced protein synthesis. It is hypothesized; however, the betaine effect in this study may be masked by the high methionine content in the diet. Previous observation revealed that betaine showed higher efficacy under sub-optimal conditions [19], although Wang et al [8] did not find any effect of betaine supplementation in low-methionine diet on carcass yield in meat duck. In addition, Matthews et al. [20] reported that response to betaine was higher in broilers received coccidiosis challenge.

This study is in line with observation of Waldroup and Fritts [7], in which, supplementation of 0.1% betaine from ages 0 to 49 days to the diet containing sufficient methionine did not affect the body weight gain and carcass characteristics of male broilers. Moreover, other authors showed that body weight of broilers at 42 days was lower when the diet supplemented with 0.2% betaine [21], which may be due to lower availability of methyl acceptor than methyl group donor, resulting in lower efficacy of betaine [11]. Different to this study, previous observation showed that supplementation of 0.14% betaine to the diet containing 20, 21.5 and 23% protein enhanced carcass and breast yield without affecting body weight of unsexed broilers [2]. In addition betaine improved carcass composition, especially breast yield in broilers fed diet deficient in methionine [6].

Table 2. Slaughter weight, carcass and breast yield of female broiler chickens fed diet containing sufficient methyl group donor and supplemented with betaine.

Parameters	Betaine Supplementation (%)			
	0.0	0.1	0.2	0.3
Slaughter weight (g)	2002.42	2048.20	2012.19	2008.80
Carcass yield (%)	67.83	69.91	68.46	68.25
Breast yield (%)	36.31	37.49	36.09	36.14

3.2. Abdominal fat deposition and meat fat content

Supplementation of betaine in the diet containing sufficient methionine produced produced 29.8–42.5% lower abdominal fat pad deposition and 9.2–35.9% lower meat fat content than those without betaine administration ($P<0.05$). Supplementation of 0.1 and 0.2% betaine generated in a similar efficacy while 0.3% betaine showed the highest efficacy in lowering fat deposition (**Table 3**). This finding supports the hypothesize in which betaine has a potential role as carcass modifier [4,11]. Zhan et al. [22] reported that supplementation of betaine in the diet deficient in methionine decreased abdominal fat deposition in broiler chickens. A lower in abdominal fat deposition was also reported in broiler chickens and turkey due to betaine supplementation [9,23].

The role of betaine as carcass modifier in lowering fat carcass content can be attributed to several mechanisms associated with lipotropic activity. For instance, methyl group donated by betaine improves the synthesis of lecithin, carnitine, creatine and hormone-sensitive lipase activity which function to prevent fat deposition [22,24]. Furthermore, since betaine is an oxidation product of choline, provision of betaine in the diet would improve choline availability. Choline is a precursor of very-low-density lipoprotein which can prevent deposition of fat in the body [25]. The more recent observation revelaed that betaine has antioxidant function which prevent lipid peroxidation, associated with a better meat quality of broilers [26].

Table 3. Abdominal fat and meat fat content of female broiler chickens fed diet containing sufficient methyl group donor and supplemented with betaine

Parameters	Betaine Supplementation (%)			
	0.0	0.1	0.2	0.3
Abdominal fat percentage (%)	1.81^{a}	1.27^{b}	1.24^{b}	1.04^{c}
Meat fat content (%)	2.06^{a}	1.87^{b}	1.74^{b}	1.32^{c}

^{a,b,c} Means within a row followed by different superscripts differ significantly (P<0.05)

4. Conclusion

Supplementation of betaine in the diet containing sufficient methionine showed a high efficacy in decreasing fat deposition although less efficient in improving carcass yield, which associated with a high methyl group donor availability in the diet. This finding meets the consumers need in providing the leannes carcass. Betaine effect would be more pronounced when supplemented in the diet containing low level of methyl group donor.

References

[1] Li Y X, Wang Y Q, Pang Y Z, Li J X, Xie X H, Guo T J and Li W Q 2011 *Int. J. Poult. Sci.* **10** 110–2
[2] Ratriyanto A, Indreswari R and Sunarto 2014 *Int. J. Poult. Sci.* **13** 575–81
[3] Huang Q C, Xu Z R, Han X Y and Li W F 2008 *Anim. Feed Sci. Technol.* **140** 365–75
[4] Ratriyanto A, Mosenthin R, Bauer E and Eklund M 2009 *Asian-Australasian J. Anim. Sci.* **22** 1461–76
[5] Hassan R A, Attia Y A and El-Ganzory E H 2005 *Int. J. Poult. Sci.* **4** 840–50
[6] McDevitt R M, Mack S and Wallis I R 2000 *Br. Poult. Sci.* **41** 473–80
[7] Waldroup P W and Fritts C A 2005 *Int. J. Poult. Sci.* **4** 442–8
[8] Wang Y Z, Xu Z R and Feng J 2004 *Anim. Feed Sci. Technol.* **116** 151–9
[9] Amerah A M, Gimenez-Rico R D and Ravindran V 2013 *Worlds. Poult. Sci. J.* **69** 1–3
[10] Esteve-Garcia E and Mack S 2000 *Anim. Feed Sci. Technol.* **87** 85–93
[11] Ratriyanto A and Mosenthin R 2018 *J. Anim. Physiol. Anim. Nutr. (Berl).* In Press
[12] Nutrition Research Council 1994 *Nutrient Requirements of Poultry* (Washington DC: National Academic Press)
[13] Ratriyanto A, Mosenthin R, Jezierny D, Sauer N and Eklund M 2009 *J. Anim. Feed Sci.* **18** 453–64
[14] Sun H, Yang W R, Yang Z B, Wang Y, Jiang S Z and Zhang G G 2008 *Am. J. Anim. Vet. Sci.* **3** 78–84
[15] AOAC 2001 *Official Methods of Analysis of the Association of Official Analytical Chemists* (Washington DC)
[16] R Core Team 2015 *R: A Language and Environment for Statistical Computing* (Vienna: R Foundation for Statistical Computing)
[17] Eklund M, Bauer E, Wamatu J and Mosenthin R 2005 *Nutr. Res. Rev.* **18** 31–48
[18] Apicella J M, Lee E C, Bailey B L, Saenz C, Anderson J M, Craig S A S, Kraemer W J, Volek J S and Maresh C M 2013 *Eur. J. Appl. Physiol.* **113** 793–802
[19] Spreeuwenberg M, Hees H M J, Smits C H M, Paterson J E and Barker J A 2007 *Manip. pig Prod. XI* **14** 186
[20] Matthews J O and Southern L L 2000 *Poult. Sci.* **79** 60–5

IOP Conf. Series: Materials Science and Engineering **633** (2019) 012017 doi:10.1088/1757-899X/633/1/012017

[21] Konca Y, Kirkpinar F, Mert S and Yaylak E 2008 *J. Anim. Vet. Adv.* **7** 930–7
[22] Zhan X A, Li J X, Xu Z R and Zhao R Q 2006 *Br. Poult. Sci.* **47** 576–80
[23] Al-Shukri A Y, Kaab H T and Abdulwahab H M 2012 *Kufa J. Vet. Med. Sci.* **3** 12–20
[24] Saunderson C L and Mackinlay J 1990 *Br. J. Nutr.* **63** 339–49
[25] Yao Z and Vance D E 1989 *J. Biol. Chem.* **264** 11373–80
[26] Alirezaei M, Gheisari H R, Ranjbar V R and Hajibemani A 2012 *Br. Poult. Sci.* **53** 699–707

IOP Conf. Series: Materials Science and Engineering **633** (2019) 012018 doi:10.1088/1757-899X/633/1/012018

Predicting yolk and albumen weight using egg weight of quails (*Coturnix coturnix japonica*) fed diet supplemented with betaine

A Ratriyanto, Y H Amirullah, S Prastowo and N Widyas

Department of Animal Science, Faculty of Agriculture, Universitas Sebelas Maret, Surakarta, Indonesia

E-mail: ratriyanto@staff.uns.ac.id

Abstract. Betaine as a methyl donor is proven to increase the synthesis of egg components which presence in poultry diet is suspected to modify the composition of the eggs. Eggs are multi-purpose animal product. It is utilized in various ways; either as a whole or separated between its yolk and albumen. For the later purpose, the knowledge about egg composition is essential. This study aimed to build models to predict the yolk and albumen weight using egg weight of quails (*Coturnix coturnix japonica*) fed diet supplemented with betaine. In total, 239 data obtained from 3 assay diets were subjected to analysis of variance, correlation, regression and cross validation. The assay diets consisted of a basal diet containing 16.5% protein fed without or with supplementation of 0.06 and 0.12% betaine. The supplementation of 0.12% betaine increased egg, yolk and albumen weight ($p<0.01$). The correlation of egg weight~yolk weight and egg weight~albumen weight ranged between 0.65–0.69 and 0.69–0.79, respectively. The regression models showed high predictability between 92.7–94.5% for yolk weight prediction and 96.0–97.2% for egg weight prediction based on egg weight. Thus, egg weight is a good predictor to predict the yolk and albumen weight of quail eggs.

1. Introduction

Coturnix coturnix japonica quails start to lay eggs at the age of 35–42 days [1] and produce 200–300 eggs in a year with average egg weight of 10 grams [2]. The interior quality of the egg is affected by nutrient content in the diet [3]. The yellow yolk is composed of lipoproteins while albumen composed of proteins. Protein requirements for laying quails according to recommendation of Nutrition Research Council [4] is 20% while Indonesian National Standard recommends 17% protein. Due to high temperature in Indonesia, it is required to adjust protein content in the diet because the too high protein stimulates heat stress, leads to reduces protein synthesis [5]. An alternative to avoid heat stress is feeding low-protein diet and supplementation with feed additive that play a role in protein metabolism, for example betaine.

Betaine donates its methyl group to synthesize various important metabolically substances such as carnitine, creatine and phosphatidylcholine, thus betaine plays a role in energy and protein metabolism [5]. Betaine increases the efficiency of energy and protein utilization, leads to improve egg quality [6]. In addition, betaine supplementation in the diet has been shown to increase yolk and albumen weight in quails [7,8].

Eggs are multi-purpose animal product. It is utilized in various ways; either as a whole or separated between its yolk and albumen. For the later purpose, the knowledge about egg composition is essential. Yolk and albumen weight are generally determined by breaking the eggs [9]. This method is complicated, takes a lot of time and causes economic losses due to broken eggs [10]. Predictions of yolk and albumen weight with the regression equation using egg weight as predictor allow to measure these variables without breaking the eggs [11]. Several studies have been conducted to determine the correlation between egg weight and yolk weight as well as albumen weight. Kul and Seker [9] showed a positive correlation between egg weight and yolk or albumen weight in quails. However, there is a lack of information on the predictability with cross validation method, particularly in quails received betaine supplementation. Therefore, this study aimed to build models to predict the yolk and albumen weight using egg weight of quails (*Coturnix coturnix japonica*) supplemented with betaine.

2. Materials and methods

2.1. Experimental design and diet formulation

In total, 225 laying quails (*Coturnix coturnix japonica*) were distributed to 3 dietary treatments with 5 replicates of 15 quails each. The basal diet was formulated to meet the nutrient requirements of laying quails according to Nutrition Research Council [4] except the protein which was formulated below the standard (16.5%), as presented in **Table 1.** The other 2 assay diets were obtained by supplementing 0.06 and 0.12% betaine. Betaine supplementation was performed according to the procedure of Ratriyanto et al. [8]. The dietary treatments were given for two periods (2×28 days). The data were collected at last 3 days of the first and second periods, namely days 26, 27 and 28. The eggs were individually weighed and then cracked. The yolks were separated from the albumen using yolk separator. The albumen weight was calculated by reducing the egg weight with the yolk and shell weight [8].

Table 1. Nutrient composition of the basal diet.

Nutrients	Content
Metabolizable energy (kcal/kg)	2,800
Crude protein (%)	16.50
Crude fat (%)	5.09
Crude fiber (%)	4.58
Crude ash (%)	5.00
Calcium (%)	3.40
Phosphorus (%)	0.61
Lysine (%)	1.13
Methionine (%)	0.41

2.2. Management and analysis of the data

The data used in the analyses included egg, yolk and albumen weight. The data were separated between control (without supplementation) and those supplemented with 0.06 and 0.12% betaine. The data were subjected to analysis of variance, correlation, regression and cross validation. All data analyses were performed using custom scripts of R program [12].

2.2.1. Analyses of variance

Analysis of variance was performed to determine the effect of treatments on egg, yolk and albumen weight. If the treatments showed significant effect, it was continued with Duncans test [13].

2.2.2. Correlation

Correlation was calculated to determine the relation between egg weight and yolk or albumen weight, and to determine the regression model. The correlation of egg weight~yolk weight and egg weight~albumen weight were calculated according to Ott et al. [14] as follows:

$$R_{EW,YA} = \frac{\text{Cov (EW, YA)}}{\sigma_{BT}\,\sigma_{YA}}$$

$$\text{cov (EW, YA)} = \frac{\sum_1^n (\text{EW. YA})}{n}$$

$$\sigma = \sqrt{\sigma^2}$$

$$\sigma^2 = \frac{\sum_1^n (x-\mu)^2}{n}$$

Where: $R_{EW,YA}$ = correlation of egg weight~yolk weight or egg weight~albumen weight; Cov (EW,YA) = covariant between egg weight~yolk weight or egg weight~albumen weight; σ_{BT} = standard deviation of egg weight; σ_{YA} = standard deviation of yolk or albumen weight; n = number of data; x = data; μ = average.

2.2.3. Regression
Regression model was built to predict yolk and albumen weight using egg weight data according to Gujarati [15]:

$$y_i = \alpha + \beta x_i$$

Where: Y_i = response (yolk weight, albumen weight); α = intercept; β = slope of independent variable; x_i = egg weight.

2.2.4. Cross validation
The model built was compared to the predictor's ability by using the accuracy of the prediction model. Predictability was obtained by applying the cross validation method, which determine the performance of a prediction model [11] by using most of the data to create a prediction model and save the rest for validation process [16]. Cross validation applied a regression formula obtained in another data to measure the accuracy of the formula to predict another set of data. The cross validation was tested by taking a certain amount of data (test set) and the remaining data are training set (**Figure 1**).

Figure 1 shows the data 2–10 were the training set that will be used to create a regression equation, then to predict the test set of the data 1. The actual value of the test set data and predicted value based on the regression was then compared to determine the predictability of the regression equation.

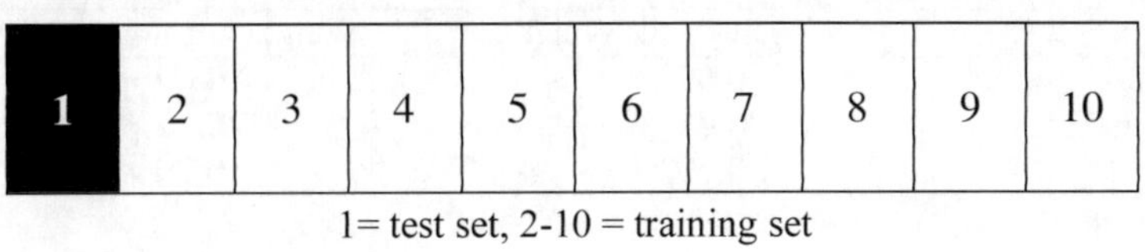

Figure 1. Illustration of cross-validation method.

3. Results and discussion

3.1. Actual weight of the egg, yolk and albumen
Average weight of the egg of all treatments ranged from 8.77 to 9.35 grams (**Table 2**), which were categorized normal [1]. Yolk and albumen weight in this study are also in the normal range between

2.50–2.74 and 5.37–5.74, respectively. Supplementation of betaine enhanced the egg weight associated with enhanced in yolk and albumen weight. Betaine has been shown to improves nutrient digestion and involved in protein and energy metabolism, led to improvement in egg weight [8]. According to Horhoruw et al. [17] yolk and albumen weight increased as egg weight increased. Yolk composed mainly of lipids while albumen composed of protein, thus involvement of betaine in the metabolism increase the synthesis of yolk and albumen precursors. In this study, each variable (egg, yolk and albumen weight) was affected by betaine supplementation, thus further analyses were performed separately based on the treatment.

Table 2. Description of data.

Variables	Treatments	Total Data	Min.	Max.	Mean±SD
Egg weight	Control	80	7.63	10.38	8.77 ± 0.55^b
	Betaine 0.06%	79	7.78	9.97	8.99 ± 0.52^b
	Betaine 0.12%	80	8.50	10.32	9.35 ± 0.48^a
p value					< 0.01
Yolk weight	Control	80	1.76	3.47	2.50 ± 0.32^b
	Betaine 0.06%	79	2.13	3.23	2.55 ± 0.26^b
	Betaine 0.12%	80	2.21	3.27	2.74 ± 0.25^a
p value					< 0.01
Albumen weight	Control	80	4.38	6.41	5.37 ± 0.38^b
	Betaine 0.06%	79	4.68	6.53	5.53 ± 0.43^{ab}
	Betaine 0.12%	80	5.00	6.50	5.73 ± 0.34^a
p value					< 0.01

[a,b] Different superscript in the same column and variable indicated significant difference ($p < 0.01$).

3.2. Correlation of egg weight~yolk weight and egg weight~albumen weight

Egg weight is a good predictor for some characteristics of egg quality such as yolk and albumen weight [9]. Kul and Seker [9] showed that quails egg weight had a positive correlation with yolk and albumen weight. Predicting the interior quality of the egg using egg weight data is considered easier and without breaking the egg.

There were positive correlations of egg weight~yolk weight and egg weight~albumen weight (**Table 3**). Correlations of egg weight~yolk weight are classified as moderate correlation (0.3–0.7) while correlations of egg weight~albumen weight are classified as high correlation with the correlation value above 0.7 [18]. Control treatment and supplementation of 0.12% betaine showed highest correlation of egg weight~yolk weight while and supplementation of 0.12% betaine showed highest correlation of egg weight~albumen weight. This result was in accordance with previous studies which observed positive correlation of weight~yolk weight and egg weight~albumen weight in quails and chickens [9,19].

Table 3. Correlation of egg weight~yolk weight and egg weight~albumen weight.

Treatments	Egg Weight~Yolk Weight	Egg Weight~Albumen Weight
Control	0.69	0.69
Betaine 0.06%	0.65	0.76
Betaine 0.12%	0.69	0.79

3.3. Linear regression model and cross validation

The model built in this study was simple linear regression. The regression used dependent variable (y), that is yolk weight or albumen weight and independent variable (x), that is egg weight. Since there were 3 treatments in each variable, 6 regression models were built in this study (**Table 4**).

The R^2 measures the capability of the model in explaining variations in the dependent variable [15]. R^2 ranged between 0 and 1, with the higher value indicating the better the independent variable in providing all the information needed to predict the dependent variable. All the 6 models were then subjected to K–fold cross validation resulting the predicted values of yolk and albumen weight. The prediction accuracy were obtained by comparing the actual values with the prediction values [11] as presented in **Table 5**. Regression models had predictability ranged between 92.7–94.5% for yolk weight and 96–97.2% for albumen weight. The highest predictability for both yolk and albumen weight obtained by the supplementation with 0.12% betaine treatment. The predictability of all treatments was above 90%, thus its classified as very high accuracy.

Table 4. Linear regression model and R^2.

Variables	Treatments	Regression Models	R^2
Yolk weight	Control	YW= –0.04+0.41.EW	0.47
	Betaine 0.06%	YW= –0.35+0.32.EW	0.42
	Betaine 0.12%	YW= –0.68+0.37.EW	0.48
Albumen weight	Control	AW= 1.23+0.47.EW	0.47
	Betaine 0.06%	AW= –0.06+0.62.EW	0.58
	Betaine 0.12%	AW= 0.45+0.56.EW	0.63

AW = albumen weight; EW = egg weight; YW = yolk weight; R^2 = coefficient of determination.

Table 5. Accuracy of prediction model.

Variables	Treatments	Actual Value	Prediction Value	Accuracy (%)
Yolk weight	Control	2.50±0.32	2.50±0.22	92.70
	Betaine 0.06%	2.55±0.26	2.55±0.17	94.10
	Betaine 0.12%	2.74±0.25	2.74±0.18	94.50
Albumen weight	Control	5.37±0.38	5.37±0.26	96.00
	Betaine 0.06%	5.53±0.43	5.53±0.33	96.00
	Betaine 0.12%	5.73±0.34	5.73±0.27	97.20

4. Conclusion

Linear regression model with egg weight as a predictor was considered as good model for predicting yolk and albumen weight in quails fed diet supplemented with betaine. All models for predicting yolk and albumen weight generated very high predictability. Thus, egg weight is a good predictor to predict the yolk and albumen weight of quail eggs.

5. References

[1] Ratriyanto A 2018 *Caraka Tani J. Sustain. Agric.* **2570** 1–7
[2] Kaye J, Luka S J, Akpa G N and Adeyinka I A 2017 *Int. J. Innov. Res. Adv. Stud.* **4** 93–7
[3] Shanaway M M 1994 *Quail Production Systems: A Review* (Rome: FAO)
[4] Nutrition Research Council 1994 *Nutrient Requirements of Poultry* (Washington DC: National Academic Press)
[5] Ratriyanto A, Mosenthin R, Bauer E and Eklund M 2009 *Asian-Australasian J. Anim. Sci.* **22** 1461–76
[6] Ratriyanto A, Indreswari R, Dewanti R and Wahyuningsih S 2018 *IOP Conference Series: Earth and Environmental Science* vol 142 p 012002
[7] Ratriyanto A, Indreswari R, Nuhriawangsa A M P and Haryanti A E 2015 *Proc. 6th International Seminar on Tropical Animal Production* ed C T Noviandi (Yogyakarta: Faculty of Animal Science, Universitas Gadjah Mada) pp 118–22
[8] Ratriyanto A, Indreswari R and Nuhriawangsa A M P 2017 *Rev. Bras. Cienc. Avic.* **19** 445–54
[9] Kul S and Seker I 2004 *Int. J. Poult. Sci.* **3** 400–5
[10] Kurshid, A., M. Farooq, F.R. Durrani K S and N C 2003 *Int. J. Puoltry Sci.* **2** 164–167
[11] Zhang Y and Yang Y 2015 *J. Econom.* **187** 95–112
[12] R Core Team 2015 *R: A Language and Environment for Statistical Computing* (Vienna: R Foundation for Statistical Computing)
[13] Steel R G D and Torrie J H 1995 *Prinsip dan Prosedur Statistik* (Jakarta: Gramedia Pustaka

Utama)
[14] Ott R L, Longnecker M, Taylor M and Payumo M A 2010 *An Introduction to Statistical Methods and Data Analysis Sixth Edition* (USA: Brooks/Cole, Cengage Learning)
[15] Gujarati D 2012 *Econometrics by Example* (New York: Palgrave MacMillan)
[16] Kohavi R 1995 *Int. Jt. Conf. Artif. Intell.* **14** 1137–43
[17] Horhoruw W M, Wihandoyo and Yuwanta T 2009 *Bul. Peternak.* **33** 8–16
[18] Gerstman B B 2008 *Basic biostatistics: Statistics for the public health practice* (USA: Jones and Bartlett Publishers)
[19] Udoh U H, Okon B and Udoh A P 2012 *Int. J. Poult. Sci.* **11** 696–9

IOP Conf. Series: Materials Science and Engineering **633** (2019) 012019 doi:10.1088/1757-899X/633/1/012019

Egg production patterns of quails given diets with different nutrient density

N Widyas, T Nugroho, B F Hidayat, A Masykur, S Prastowo and A Ratriyanto

Department of Animal Science, Sebelas Maret University, Surakarta INDONESIA

E-mail: ratriyanto@staff.uns.ac.id

Abstract. Energy and protein ratio in the diet plays an important role in laying quails production. High nutrient density could expose the quails into heat stress due to inefficient metabolic process and nutrient absorption. On the other hand, lower density possesses the risk of nutrient insufficiency. Both cases are potential in affecting the egg production. This study aimed to investigate the egg production patterns of quails fed different nutrient density with constant ratio of metabolizable energy (ME) and crude protein (CP). Egg production data was obtained from two hundred laying quails (*Coturnix coturnix japonica*) which were randomly assigned into two distinct experimental diets. Dietary treatments were given for eight weeks and replicated five times each with 20 quails. Two diets were formulated to contain High ME-CP (2,800 kcal/kg ME and 18.7% CP) and Low ME-CP (2,600 kcal/kg ME and 17.3% CP). The data were analysed using t-test and egg production patterns were plotted. Quails fed with different nutrient density yielded similar egg production values and patterns. We fit non-linear mathematical model into the datasets; similar trends and fitness were also observed ($R^2 = 0.96$). It is concluded that quail fed with different nutrient density exhibited minor difference in egg production pattern.

1. Introduction

Feed formulation for laying quails in Indonesia is still based on the nutrient requirements published by other countries with different climates, thus it is not necessarily ideal for tropical conditions in Indonesia. The nutrient requirement for quails during production period according to Nutrition Research Council [1] is 20% crude protein (CP) and 2,800 kcal/kg metabolisable energy (ME). This condition provides an opportunity to adjust the nutrient content to the hot climate regions to produce optimal performance [2]. It is well known that environment is a fundamental factor that limits poultry production in the tropics due to excessive metabolic heat production leads to heat stress [3].

Excessive content of protein and energy in the diet would not have a positive impact on improving the productivity, but rather reduce the performance of poultry in hot climates. This condition also causes inefficiency due to high undigested nutrients, thus, the diet has to be formulated precisely since egg production is affected by nutrient balance in the diet [4]. In feed formulation, the emphasis is placed on protein and energy content which holds 70% of feed costs [5]. A decrease in CP or ME in the diet can reduce feed costs [3,6] and minimize metabolic heat produced by protein and energy [7]. Therefore we need to provide a precise feed formulation for tropical regions with lower CP and ME content in accordance with local climate in Indonesia.

Poultry egg production follows specific pattern. Deviation from the pattern can indicate anomalies, thus the information on egg production pattern is very useful for control and evaluation mechanism [8–11]. The objective of this paper is to investigate the egg production patterns of quails fed different nutrient density with constant ratio of metabolizable energy (ME) and crude protein (CP).

2. Materials and Methods

2.1. Experimental design and diet formulation
Two diets with different nutrient densities were formulated: High ME-CP (2,800 kcal/kg ME and 18.7% CP) and Low ME-CP (2,600 kcal/kg ME and 17.3% CP) as in Table 1.

Table 1. Nutrient content of basal diets with constant energy to protein ratio

Nutrien dalam pakan	Metabolizable Energy : Crude Protein	
	2.800 : 18.7	2.600 : 17.3
Metabolisable energy (kcal/kg)	2800	2601
Crude protein (%)	18.70	17.30
Calcium (%)	3.38	3.36
Phosphorus (%)	0.49	0.49
Lysine (%)	1.06	0.96
Methionine (%)	0.40	0.41

The experiment was started with female quails at the age of 21 days. The quails were randomly allotted into two diets. Each diet was replicated five times with twenty quails per replication. The quails were then immediately adapted to the experimental diets from age 41 to 60 and the full experimental diets were fed from age 61. Quails started to lay eggs around 42 days of age and the daily egg production was recorded. The experiment lasted until 49 days after the experimental diets were given.

2.2. Data analyses
Egg production data were obtained as hen day average which is the percentage of daily eggs produced by each flock relative to the number of birds in the corresponding flock. All data analyses were performed using custom scripts of R program [12].

2.2.1. Comparing treatments
To observe the difference of eggs production between high and low-density diets, independent sample t-test was performed under 5% significance level. The test was done for both the adaptation (41-60 days) and treatment period (61-109 days). Later, we visualize the actual egg production data into a trend line plot.

2.2.2. Logistic regression
The production plot of real data contained fluctuation due to uncontrolled effect. To obtain the idealized sigmoid curve which is identical with eggs production pattern can be achieved by utilizing non-linear mathematical models. We employed logistic regression to build the hypothesized production pattern with the following formula:

$$Y_t = \frac{\alpha}{1+\beta \exp[kt]}$$

Where :
Yt = production at time-t
α = peak of production
β = carrying capacity
k = production rate
t = time of production
The fitness of the models were then assessed through the coefficient of determination parameter.

3. Results and Discussion

3.1. Comparing experimental diets

The egg production of two diets with different nutrient densities were compared. The comparation results can be seen in Table 2.

Table 2. Mean, standard deviation and t-test results

Diets	Observation days	Egg production (%)			p-value
		Min	Max	Mean±Sd	
Adaptation period					
High	20	0.00	68.00	23.32±19.74	0.54
Low	20	0.00	72.00	21.64±18.99	
Treatment period					
High	49	28.00	92.00	64.44±11.33	0.92
Low	49	39.13	92.00	64.35±10.69	
Whole period					
High	69	0.00	92.00	52.52±23.50	0.75
Low	69	0.00	92.00	51.97±23.70	

Results showed that whether it is during the adaptation or treatment period, the egg production were relatively consistent. This finding means that there was no significant difference ($p<0.05$) of eggs production between different diets

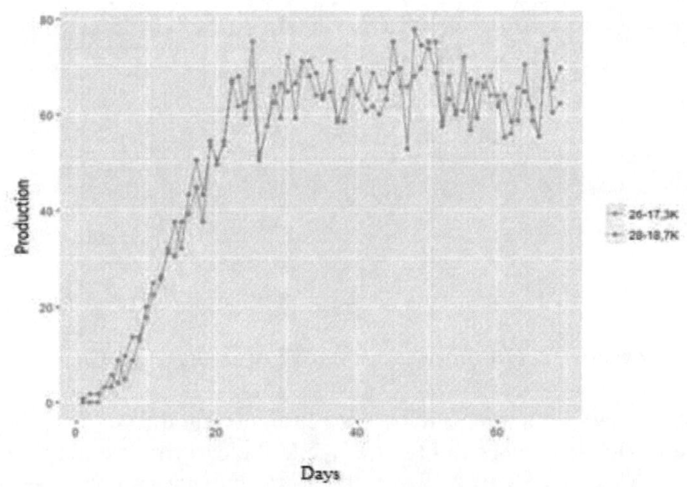

Figure 1. Egg production pattern

The results from t-test in Table 2 were supported by the eggs production pattern in Figure 1. Although the eggs productions were fluctuating among days, but the pattern for both diets were similar. The low-density diet which contained less nutrients (metabolizable energy and crude protein) stimulated the quails to eat more; and thus yielding in higher feed intake [13]. With this mechanism, the nutrients utilization is optimized while the metabolic heat production is minimized [14]. The positive side effect of high feed intake was that the intake of the other nutrients such as minerals, amino acids and vitamins were also higher and supporting the productivity.

3.2. Logistic Regression

We employ logistic regression to obtain the hypothesized pattern of eggs production for low diet (Figure 2) and high diet (Figure 3)

IOP Conf. Series: Materials Science and Engineering **633** (2019) 012019 doi:10.1088/1757-899X/633/1/012019

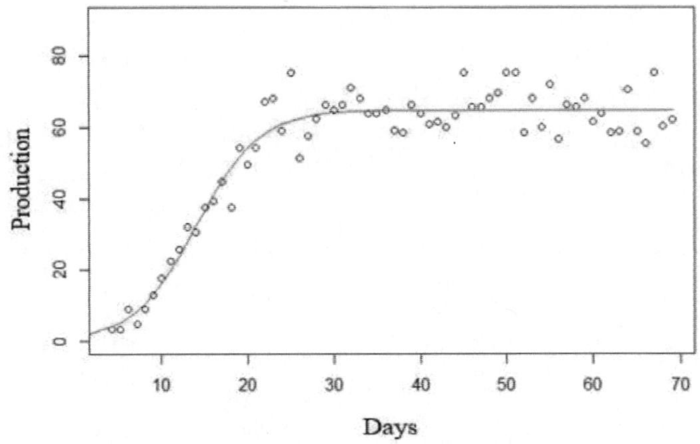

Figure 2. Non-linear curve of high density group

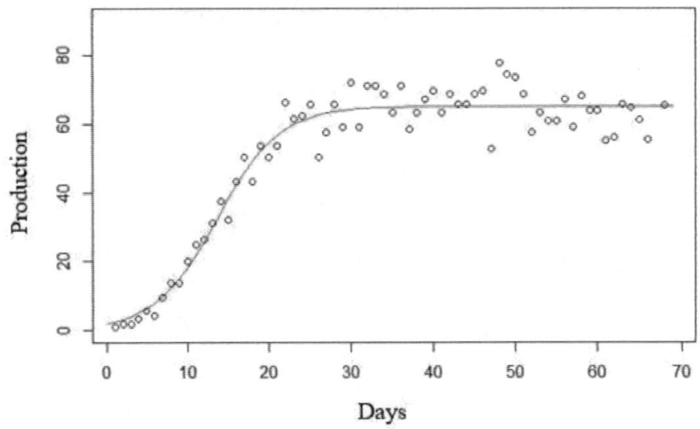

Figure 3. Non-linear curve of low density group

The patterns and the parameters estimate from logistic regression model were very similar for both diets. The peak production was slightly higher in low diet while the carrying capacity of the high diet was a bit higher. Logistic regression models fit well with the data showed with the coefficient determination (R^2) of 0.96 for both diets.

Table 3. Parameter estimates of the logistic regression model

Parameters	High	Low
Peak production	64.92	65.14
Production rate	0.27	0.26
Carrying capacity	46.10	35.21
R^2	0.96	0.96

4. Conclusion

Diets with high and low protein density showed similar egg production and egg production pattern.

References

[1] Nutrition Research Council 1994 *Nutrient Requirements of Poultry* (Washington DC: National Academic Press)

[2] Ghazvinian K, Irani M, Jamshidi R and Mirzaei-aghsaghali A 2011 *Ann. Biol. Res.* **2** 122–8
[3] Ratriyanto A, Indreswari R and Nuhriawangsa A M P 2017 *Rev. Bras. Cienc. Avic.* **19** 445–54
[4] Li Y X, Wang Y Q, Pang Y Z, Li J X, Xie X H, Guo T J and Li W Q 2011 *Int. J. Poult. Sci.* **10** 110–2
[5] Moosavi M, Chaji M, Rahimnahal S and Kazemi A R 2012 **3** 2485–8
[6] Kamran Z, Sarwar M, Nisa M, Nadeem M A, Mahmood S, Babar M E and Ahmed S 2008 *Poult. Sci.* **87** 468–74
[7] Furlan R L, Faria Filho D E de, Rosa P S and Macari M 2004 *Brazilian J. Poult. Sci.* **6** 71–9
[8] Savegnago R P, Cruz V a. R, Ramos S B, Caetano S L, Schmidt G S, Ledur M C, El Faro L and Munari D P 2012 *Poult. Sci.* **91** 2977–87
[9] Bain M M, Nys Y and Dunn I C 2016 *Br. Poult. Sci.* **57** 330–8
[10] Grossman M, Gossman T N and Koops W J 2000 *Poult. Sci.* **79** 1715–24
[11] Gavora J S, Liljedahl L E, McMillan I and Ahlen K 1982 *Br. Poult. Sci.* **23** 339–48
[12] R Core Team 2015 R: A language and environment for statistical computing
[13] Ferket P R and Gernat A G 2006 *Int. J. Poult. Sci.* **5** 905–11
[14] Li Y X, Qang Y Q, Pang Y Z, Li J X, Xie X H, Guo T J and Li W Q 2011 *Int. J. P* **10** 110–2

IOP Conf. Series: Materials Science and Engineering **633** (2019) 012020 doi:10.1088/1757-899X/633/1/012020

The consistency of quail's egg production supplemented with vitamin C in the diets

A Ratriyanto, T Nugroho, A Masykur, B F Hidayat, S Prastowo, Sunarto and N Widyas

Department of Animal Science, Sebelas Maret University, Surakarta, Indonesia

E-mail: ratriyanto@staff.uns.ac.id

Abstract. In tropical area, animals are prone to stress due to high environmental temperature hence affecting the production. Vitamin C has been offered as an agent to help animal to cope with heat stress. This paper aimed to estimate the consistency of quail's egg production supplemented with vitamin C through repeatability estimates. A total of 136 quails (*Coturnix coturnix japonica*) were randomly allotted into 2 dietary treatments with 4 replicates each containing 17 quails. The dietary treatments were basal diet (T0) and basal diet + 250 mg/kg vitamin C (T1). Dietary treatments were given for 8 weeks in laying phase. In total we obtained 56 days of egg production data. Data were analysed using t-test to infer the differences between the treatments and analysis of variance (ANOVA) to obtain the repeatability value. The results showed that the supplementation of vitamin C increased egg production up to 14.3% ($P<0.01$). Repeatability value of T0 was higher (0.33) than T1 (-0.13). Quails given T0 has a high egg production consistency but produce lower egg production than T1. It is concluded that supplementation of vitamin C increased egg production but were not yet able to produce the optimal consistency.

1. Introduction

Egg production as an important trait in poultry industry is affected by both genetic and environmental factors [1,2]. Genetic effect contained information regarding the potential of the animals in producing eggs while the environment served as the supporting factor. Hence, no matter how superior an individual genotype is, it will never reach its optimum condition without the support of the environment. Environmental factor can be theoretically divided into two parts: the permanent environment and the temporary environment which roles in an individual's phenotype differed [1,3]. The environment which an individual experienced during early life such as non-direct maternal effects will leave a permanent influence on its phenotype. Meanwhile, other types of environment such as weather and diets will only affect an individual's phenotype temporarily. The change in the later effect will be followed by the change in phenotype as a response.

High ambient temperature in the tropics becomes a significant challenge in poultry production and impairs feed intake, physiology and performance of poultry, including laying quails [4]. Poultry are susceptible to heat stress, particularly when kept above the thermoneutral zone of 18–22°C [5,6]. Various techniques can be practiced to alleviate heat stress in poultry, including many nutritional strategies [4,7]. One of the nutritional strategies to alleviate stress in laying quails is the use of Vitamin C in the diet [8].

IOP Conf. Series: Materials Science and Engineering **633** (2019) 012020 doi:10.1088/1757-899X/633/1/012020

Vitamin C is one of the important antioxidants and plays a role in immune formation [9]. Vitamin C can be synthesized in the body of poultry, however, high ambient temperature can cause a decrease in nutrient absorption and a decrease in the concentration of nutrients in the blood, including vitamin C [8]. In stressful conditions (for example due to density, high temperature, humidity), the synthesis of vitamin C in the body decreases or the vitamin C requirement increases, leads to insufficiency of vitamin C. Therefore, it is necessary to provide vitamin C in the diet or drinking water [9] as an anti-stress to maintain or improve the performance of poultry [5].

By adding the vitamin C, we wished to manipulate the temporary environmental effect. It is interesting to see how the quail will response this supplementation from the production point of view. The aim of this article is to present the difference in egg production consistency of quails fed with and without vitamin C supplementation in tropical climate and to observe how much egg production trait is affected by the temporary environmental effect which is vitamin C supplementation.

2. Materials and Methods

2.1. The feeding experiments

A population of 136 quails were used in the experiment. The quails were divided into two distinct treatments: basal diet as control (T0; Table 1) and basal diet + 250mg vitamin C supplementation (T1). Each diet treatment was replicated four times with 17 individual quails per flock (as the replicate).

Table1. Nutrient content of the basal diet

Nutrient	Content
Metabolizable energy (kcal/kg)	2,800.00
Crude protein (%)	18.00
Lysine (%)	1.04
Methionine (%)	0.45
Calcium (%)	3.40
Phosphorus (%)	0.50

Quails were obtained from a village breeding center at the age of 23 weeks and were immediately given the basal diet. The quails were slowly adapted into the experimental diet; this adaptation period lasted for 4 weeks. The experimental diets were given from age 28 and the egg production data was recorded for 8 weeks.

2.2. Data analyses

Egg production data was recorded daily as the percentage of eggs produced by each flock relative to the number of birds in the flock at each day. However, the data used in the analysis was weekly averaged egg production. Data analyses were performed using custom scripts of R programming language [10].

Given that we intended to compare the difference of eggs production between two diets, two-samples independent t-test was performed under 5% significance level. The hypothesis was then formulated as $H_0: \mu_1 = \mu_2$ where μ_1 was the mean of control group and μ_2 was the mean of group with vitamin C supplementation. We assumed that the datasets were ~iid, N(μ, σ^2); no collinearity as well as autocorrelations were found among analysed variables. A common two sample t-test formula was utilized to test the hypothesis as in the following model:

$$y_{ij} = \mu + \tau_i + e_{ij}$$

We denoted that y was the observed variables, which was the egg production, μ was the general mean, τ_i was the effect of the i-th treatment and e_{ij} was the random residual term of diet i and replicate j where $e \sim N(0, \sigma_e^2)$.

The consistency of eight consecutive weeks egg production between treatments were observed by employing intraclass correlation procedure for both diets. Datasets were partitioned into weekly egg production. Intra class correlation procedure was conducted for each diet following a linear model with week as the systematic factor.

$$y_{ij} = \mu + w_i + e_{ij}$$

Where y was the egg production of a diet, μ was the general mean, w_i was the effect of i-th week and e was random residual. Intraclass correlation was estimated as:

$$\hat{\rho} = \frac{\sigma_w^2}{\sigma_w^2 + \sigma_e^2}$$

With $\hat{\rho}$ was the estimated intraclass correlation; σ_w^2 is the between week variance which the difference between Mean squares (MS) between week and MS residual. Whereas σ_e^2 was residual variance which nominal was (k-1)MS residual; k = 8 which was the number of weeks.

3. Results and Discussion

3.1. Explorative data analysis

The comparison of egg production (averaged for eight week) between diets is presented in Table 2. The vitamin C supplementation has significant effect (P<0.05) on quails' egg production. This signify our hypotheses that vitamin C helped the quails to cope with the environmental pressure and thus the egg production is higher.

Table 2. Descriptive statistic and t-test for egg production data

Treatments	Minimum	Maximum	Mean±sd	P-value
Control	41.18	100	73.82±12.48	P<0.01
Vit C	52.94	100	84.38±9.79	

We later analyse the data each week and plot the egg production trends for both diets as in Figure 1. It can be seen that quails supplemented with vitamin C produce remarkable more eggs compared to the control and this trend is consistent throughout the experimental period.

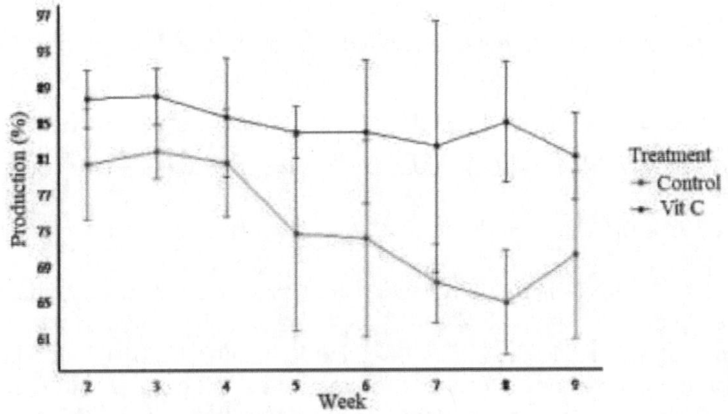

Figure 1. Production pattern

3.2. Intraclass correlation

We run single score intraclass correlation to check the consistencies of egg production for each treatment. The analysis was done under the null hypothesis that there were no correlations among egg productions over different weeks. The results were presented in Table 3.

Table 3. Intraclass correlation parameters

Parameters	Treatment	
	Control	Vit C
Number of repeated measure	8	8
Intraclass correlation	0.33	-0.13
P-value for hypotheses	0.02	0.81

The records were available for eight consecutive weeks during the experiment each with four replicates. We subjected the intraclass correlation to observe the consistency off egg production across this time period. From the p-value we can deduce that the on the control treatment, there is correlation among the eight week of egg production with a value of 0.33. This value represented the repeatability estimate of egg production of the control quails. Biologically speaking, 33% of the variation in egg production of the control population were affected by genetic and permanent environmental effect [2,11] while the rest were affected by temporary environment during the data collections. On the other hand, there is almost no noticeable correlation among observations in egg production of quails with vitamin C supplementation ($P>0.05$) with intraclass correlation value of -0.13. The low value of intraclass correlation of the quails supplemented with vitamin C suggested that the egg production in this population over eight weeks were more affected by the specific condition of each observation time point instead of the permanent effects. Heat exposure increased serum corticosterone levels and Hsp70 expression. Serum corticosterone level was significantly decreased by vitamin C or E supplementation in the groups for quail [12]. In the previous study, vitamin C or E supplementation significantly improve feed intake, improve growth, and egg production by buffering heat stress effect [7,8,13]. This finding justified that vitamin C supplementation played the role of temporary environment for the quails during their production.

IOP Conf. Series: Materials Science and Engineering **633** (2019) 012020 doi:10.1088/1757-899X/633/1/012020

4. Conclusion
It is concluded that supplementation of vitamin C increased egg production but was not able to produce the optimal consistency.

References
[1] Falconer D S and Mackay T F C 1996 *Introduction to Quantitative Genetics* (Edinburgh: Prentice Hall)
[2] Hardjosubroto W 1994 *Aplikasi Pemuliaan Ternak di Lapangan* (Jakarta: Gramedia Widiasarana Indonesia)
[3] Lynch M and Walsh B 1998 *Genetics and Analysis of Quantitative Traits* (Sunderland, MA: Sinauer Associates, Inc)
[4] Chand N, Muhammad S, Khan R U, Alhidary I A and Rehman Z U 2016 Ameliorative effect of synthetic γ-aminobutyric acid (GABA) on performance traits, antioxidant status and immune response in broiler exposed to cyclic heat stress *Environ. Sci. Pollut. Res.* **23** 23930–5
[5] Khan R U, Naz S, Nikousefat Z, Selvaggi M, Laudadio V and Tufarelli V 2012 Effect of ascorbic acid in heat-stressed poultry *Worlds. Poult. Sci. J.* **68** 477–89
[6] Khan R U, Naz S and Dharma K 2014 Chromium: Pharmacological applications in heat-stressed poultry *Int. J. Pharmacol.* **10** 213–7
[7] Farooqi H A G, Khan M S, Khan M A, Rabbani M, Pervez K and Khan J A 2005 Evaluation of betaine and vitamin C in alleviation of heat stress in broilers *Int. J. Agric. Biol.* **7** 744–6
[8] Sahin K, Sahin N, Onderci M, Gursu M F and Issi M 2003 Vitamin C and E can alleviate negative effects of heat stress in Japanese quails *Food, Agric. Environ.* **1** 244–9
[9] Sahin K, Onderci M, Sahin N, Gursu M F and Kucuk O 2003 Dietary Vitamin C and Folic Acid Supplementation Ameliorates the Detrimental Effects of Heat Stress in Japanese Quail K. *J. Nutr.* **133** 1882–6
[10] R Core Team 2015 R: A language and environment for statistical computing
[11] Bourdon R M 2000 *Understanding Animal Breeding* (Prentice Hall)
[12] Sahin N, Tuzcu M, Orhan C, Onderci M, Eroksuz Y and Sahin K 2009 The effects of vitamin C and E supplementation on heat shock protein 70 response of ovary and brain in heat-stressed quail *Br. Poult. Sci.* **50** 259–65
[13] Abdulrashid M, Agwunobi L and Hassan M 2004 *Ascorbic acid and heat stress in domestic chicken nutrition: A review* vol 8 (Department of Animal Science, University of Benin)

IOP Conf. Series: Materials Science and Engineering **633** (2019) 012021 doi:10.1088/1757-899X/633/1/012021

Milk production of imported Holstein cows over different environment

S Prastowo[1], T Nugroho[1], N Mahfudhoh[1], F Y Putra[1], Subiakti[2], A Ratriyanto[1], A Susilowati[1], Sutarno[1], N Widyas[1]

[1]Department of Animal Science, Sebelas Maret University, Surakarta Indonesia
[2]Baturaden Dairy Cattle Breeding Centre, Banyumas, Central Java, Indonesia

Email: prastowo@staff.uns.ac.id

Abstract. Indonesia is importing genetically superior Friesian Holstein (FH) cows as breeding stocks for domestic dairy cattle population. The milk production of these imported cows, however, are very much affected by temporary environmental effect due to the habitat changes and transportation. This study aimed to observe the consistency of milk production of imported cows viewed from the repeatability estimates. In total, milk yield records on lactation one (L1), two (L2) and three (L3) of 70 imported FH cattle were obtained from Baturaden Dairy Cattle Breeding Centre. The repeatability values were estimated with intraclass correlation method utilizing information from three lactation periods. Milk yield decreased over lactation periods which were 6077.55±1239; 4821.79±888 and 4374.6±1012 kg for L1, L2, and L3 respectively. The repeatability of milk yield trait in this population was 0.41. It indicates 59% of the variation in milk yield is controlled by the unknown error term, including temporary environmental effects. Imported cows undergone extreme adaptation processes to be able to live optimally in a new environment which would modify their physiology. We suspect environmental changes and resource differences between tropical and subtropical climates triggers different trait expression as the explanation of the phenomenon.

1. Introduction

Milk is food rich in nutrition which is commonly harvest from dairy cattle. In Indonesia, there were no native dairy cattle breeds, therefore importing dairy genetic is the strategy to overcome the national milk demand. Friesian Holstein (FH) is the breed which selected and imported from both Australia and New Zealand to establish dairy cattle breeding stock [1]. This breed carries superior genetic potential for milk production by able to produce 5217 kg milk per year. Also, FH is able to adapt with tropical and sub-tropical countries under temperatures range 5-25°C [2].

As generally known, milk production is the result of interaction between genetic and environment. Genetic itself is traits information written in the DNA sequence, inherited from both parents to the offspring [3]. However, its expression to phenotype depends on its interaction with environment. Due to the environmental different between Indonesia and Australia or New Zealand, imported Holstein cows would facing climate adaptation process, feed sources, and dairy management. This situation leads to the different of milk resulted. Different environment change the set of genes expression controlling milk production to be different from its origin environment [4] in different time point such as lactation period.

For that, this study aims to evaluate the consistency of milk yield of imported Friesian Holstein which reared in Indonesia in different lactations period. The result could be useful for rearing management, specially to provide better environment to support milk production potential from superior genetic of imported dairy cattle.

2. Materials and methods

The materials in this study were 70 cows which imported from either Australia or New Zealand on 2015 when they are having their first pregnancy. All the cows were housed in Dairy Breeding Center Baturaden, Purwokerto at Limpakuwus farm under intensive production system to minimize variation due to management aspects. Milk yield data were obtained from production sheet at year 2015 to 2017, then standardized to 305 days from three consecutive lactations.

In order to separate the variance components due to genetic and permanent environmental effect versus temporary environmental effect, intraclass correlation was done with individuals as the main factor. The repeatability estimate was the outcome from the intraclass correlation. It represents the proportion of variation in milk yield due to genetic and permanent environmental effect. Whereas the residual represents the variation due to unknown aspect which we hypothesized as due to temporary environmental effect. The linear model for intraclass correlation is $Y_{km} = \mu + \alpha_k + e_{km}$, where Y is the lactation of the individual, μ is general mean, α_k is the effect of the individual, and e_{km} is random residual. To obtain the parameter estimates, ANOVA table was built according to the intraclass correlation model (Table 1). The components of this table were then extracted to estimate the repeatability value.

Table 1. ANOVA table

Source	df	SS	MS	EMS
Between id	(N-1)	SS_w	MS_w	$\sigma_e^2 + k_1\sigma_w^2$
Within id	N(M-1)	SS_e	MS_e	σ_e^2

where,
df = Degrees of freedom
SS = Sum of squares
MA = mean squares
EMS = expected mean of squares
N = number of observations per individuals
M = number of individuals
k_1 = M if the number of observations is equal
m_k = number of repeated measurements of the k individual
σ_w^2 = variance among individuals
σ_e^2 = residual variance

The variance component estimated from Table 1 were used to obtain the repeatability estimate according to the following formula:

$$r = \frac{\sigma_w^2}{\sigma_w^2 + \sigma_e^2}$$

where, r is repeatability estimate, σ_w^2 is variance among individuals, and σ_e^2 is residual variance.

IOP Conf. Series: Materials Science and Engineering **633** (2019) 012021 doi:10.1088/1757-899X/633/1/012021

3. Results and discussions

In this study, the average of milk yield shows decline trend over lactations period. Highest average of milk yield was found in lactation 1, followed with lactation 2 and 3 (6077.55±1239; 4821.79±888 and 4374.6±1012 Kg). This result was in contrary with previous study, where milk yield tends to increase from first to the following lactations [6,7]. The different with our result is the milk yield obtained in these studies were obtained from the original environment as the dairy cattle origin.

As mentioned, female dairy cattle in this study were imported either from Australia or New Zealand, shipped to Indonesia in pregnant condition (4-6 months pregnancy). The change of environment later changes its effect on different milk yield over lactation period / number. In the 1st lactation, we found highest milk yield average which could explained that cow still carry environment effect from its original environment. Moreover, in the 2nd and 3rd lactation, cow start to adapt with Indonesian environment which might cannot support the genetic potential of milk production.

In the view of udder development physiology, our result also shows contrary concept. Normally, cattle udder will develop it size in term of its lobule, lumen, alveoli, and secretory cell capacity over lactation number [8,9]. In this study, it seems there is no development of udder size according to the milk yield resulted. There could be different expression of related hormone controlling udder development, such as bST (bovine somatotropin), in different environment. As explained, bST is a hormone which control the mammary gland development to support milk production [10].

The repeatability of milk yield trait in this study was found at 0.41. This mean there were 59% of the variation in milk yield is controlled by the unknown error term, including environmental effects. Considering the milk yield trend over lactations period, the repeatability value could explain that milk yield mostly affected by the environment. Our previous study also support this result [11], when we found a significant effect ($p<0.05$) of both lactation period and calving month to milk yield. In this regard, calving month represent different season that directly impact to the quality of feed supplied. Genetic and environment are the factors which interact to form the animal phenotype. Mostly environment shared bigger part in quantitative trait such as milk yield. Unsuitable environment would impact to the decrease of animal genetic potential expressed in phenotype. Therefore, an effort to provide better environment to the imported dairy cattle would improve the milk yield resulted.

4. Conclusions

It is concluded, there was a decrease of milk yield over lactation 1 to 3 of imported Holstein cows in Indonesia. First lactation shows highest milk yield followed with lactation 2 and 3 respectively. According to that, we consider environment factors is the most plausible explanation according to the repeatability estimation value in this study. In the effort to improve milk yield, serving better environment in term of feed, housing, and farm management could be a way to try.

5. References

[1] Pangestu M, Subagyo Y, Yuwonol P and Rustomo B 2000 *Asian-Australasian J. Anim. Sci.* **13** 505–8

[2] Coffey E L, Horan B, Evans R D and Berry D P 2016 *J. Dairy Sci.* **99** 5681–9

[3] Bourdon R M 2014 *Understanding Animal Breeding* (United State of Amerika: Pearson)

[4] Cromie A R, Kelleher D L, Gordon F J and Rath M 1998 *Interbull Bulletin 17* **17** 100–5

[5] Hardjosubroto W 1994 *Aplikasi Pemuliabiakan Ternak di Lapangan* (Jakarta: PT. Gramedia Widiasarana Indonesia)

[6] Domecq J J, Skidmore A L, Lloyd J W and Kaneene J B 1997 *J. Dairy Sci.* **80** 101–12

[7] Pösö J and Mäntysaari E A 1996 *J. Dairy Sci.* **79** 1284–91

[8] Senger P 2005 *Pathway to Pregnancy and Parturition* (Washington: Pullman)

IOP Conf. Series: Materials Science and Engineering **633** (2019) 012021 doi:10.1088/1757-899X/633/1/012021

[9] Shennan D B and Peaker M 2000 *Physiol. Rev.* **80** 925–51
[10] Mellado M, Antonio-Chirino E, Meza-Herrera C, Veliz F G, Arevalo J R, Mellado J and de Santiago A 2011 *J. Dairy Sci.* **94** 4524–30
[11] Widyas N, Putra F Y Y, Nugroho T, Pramono A, Susilowati A, Sutarno and Prastowo S 2018 *IOP Conference Series: Earth and Environmental Science* vol 142 p 012005

IOP Conf. Series: Materials Science and Engineering **633** (2019) 012022 doi:10.1088/1757-899X/633/1/012022

Yield and anthocyanin content of M1 generation of black rice induced by gamma rays

E Purwanto, Z Nahdhiana, M B Yunindanova and Nandariyah

Department of Agrotechnology, Faculty of Agriculture, Sebelas Maret University, Jl. Ir. Sutami 36 A, Kentingan, Surakarta, 57126, Indonesia

Email: edipurwanto@staff.uns.ac.id

Abstract. Black rice is rice that has a dark purple to blackish pigment and has high antioxidant properties. The gamma rays can induce mutation in plant. The purpose of this research was to understand the yield and anthocyanin of black rice with gamma ray irradiation using M1 generation from Bantul and Boyolali varieties. This experiment employed 0, 100, 200 and 300 Gy gamma ray irradiation. The effects of gamma rays on Boyolali variety differed from Bantul variety. Bantul variety was more sensitive to gamma rays irradiation than Boyolali variety. The application of gamma rays did not significantly change the flowering age Boyolali variety. However, on Bantul variety, gamma rays reduced flowering age. Application 100 Gy irradiation yielded early maturing plant on Bantul variety. This result revealed that the gamma ray had potential for decreasing the harvest age in black rice. The application of irradiation could enhance the size of grain of black rice indicated by higher weight of 1000 seed in Boyolali variety. Irradiation 100 Gy also boosted yield per plant in Boyolali variety. In Boyolali variety, gamma ray boosted the yield per plant by producing bigger seed rather than increased the number of productive tillers. The increase of the irradiation doses did not always correlate with the changes of plant phenotype. To conclude, application 100 Gy was the most effective dose to produce higher anthocyanin content compared to another doses.

1. Introduction

Black rice is rice that has a dark purple to blackish pigment on the pericarp, aleuron, endosperm and other parts such as leaf petals. Black rice is a good nutritional source for health due to anthocyanin content. The anthocyanin content in black rice is higher than in the red rice reaching 200-400 milligrams per 100 grams [1]. Black rice is rarely cultivated by farmers because farmers are less interested to cultivate it, mainly due to the long harvest age.

Researches to improve the rice varieties performances using gamma-ray irradiation have been widely conducted. Atomita-4 rice variety with 200 Gy irradiation resulted in a shorter plant mutant [2]. Mutation induction is directed to enlarge plant genetic diversity by altering one or more important characters and retaining most of the original characters. Gamma-ray irradiation aims to obtain new plant variants through high mutation frequencies. Therefore, this experiment aimed to evaluate the yield and anthocyanin content of M1 generation of black rice by gamma-rays induction.

2. Materials and Methods

2.1. Rice Seeds Irradiation
This study used Boyolali and Bantul black rice varieties, from Bantul, Yogjakarta dan Boyolali, Central Java, Indonesia. This experiment employed irradiation doses of 0, 100, 200 and 300 Gy of gamma-rays. Each variety consisted of 2400 mutant seeds; $A_{1\text{-}50}$, $B_{1\text{-}50}$, $C_{1\text{-}50}$, $D_{1\text{-}50}$, $E_{1\text{-}50}$, $F_{1\text{-}50}$, $G_{1\text{-}50}$, $H_{1\text{-}50}$, $I_{1\text{-}50}$, $J_{1\text{-}50}$, $K_{1\text{-}50}$, and $L_{1\text{-}50}$. The irradiation induction was conducted at the Isotope and Irradiation Application Center (PAIR) of the National Atomic Energy Agency (BATAN) Jakarta, Indonesia.

2.2. Rice Cultivation
The rice seeds were cultivated on a rice field at Gotanon Village, Jaten Sub-district, Karanganyar Regency, Central Java Province, Indonesia from December 2017 to April 2018. Plants were cultivated with a spacing of 25cm x 25cm.

2.3. Anthocyanin Analysis
Anthocyanin content was analyzed by differential pH method in the Laboratory of Food Technology and Agricultural Products, Faculty of Agriculture, Sebelas Maret University. It was detected using UV-VIS spectrophotometer by measuring the light absorption of solutions with pH 1.0 and 4.5 at 510 nm and 700 nm, respectively. The preparation of buffer solution at pH 1.0 was by diluting 1.86 g KCl in 980 ml distilled water and added by HCl until reached pH 1.0. While the buffer solution at pH 4.5, was prepared by diluting 54.43 g $CH_3CO_2Na.3H_2O$ in 950 ml distilled water and added by HCl until reached pH of 4.5. The results were calculated referring to Giusti and Wrolstad [3] as follows:

$$Asp = (A510 - A700)pH1.0 - (A510 - A700)pH\ 4.5$$

While the content of total anthocyanin (TA) was calculated as follows:

$$TA = \frac{A}{\varepsilon \times L} \times MW \times DF \times \frac{V}{Wt}$$

Where A is the absorption, ε is the molar absorptivity coefficient (26900 $M^{-1}cm^{-1}$), L is the cuvette optical path length (1 cm), MW is the molecular weight (449.2 g mol^{-1}), DF is the dilution factor, V is the volume and Wt is the weight of the sample (g). The total anthocyanin content was expressed as mg anthocyanin 100 g^{-1} fresh weight.

2.4. Observation
Variables observed were the number of productive tillers, flowering age, harvest age, the weight of 1000 seeds, yield per plant and anthocyanin content. This study compared plants at each dose of irradiation with control.

3. Result and Discussion

3.1. Yield and Agronomic Traits
Currently, radiation technology is widely utilized to produce changes in plant performances. Gamma irradiation is one of electromagnetic radiation that is capable of hydrolyzing chemical bonds [4]. The result of this study indicated that the effects of gamma rays on Boyolali variety differed from Bantul variety (Table 1). The results showed that number of productive tillers of Boyolali variety was less than Bantul variety. Application 100-200 Gy irradiation did not change the number of productive tillers of Boyolali variety. However, the usage of 300 Gy increased number of productive tillers of Boyolali variety 25% compared to control (0 Gy). In the contrary, on Bantul variety, gamma rays 100-300 Gy decreased the number of productive tillers. This finding exhibited Bantul variety was more sensitive to gamma rays irradiation than Boyolali variety. The influence of gamma rays on a plant depends on the dose and many factors such as species, cultivars, physiology, plant age, the state and size of the plant genome [5]. Exposure of plants to ionizing irradiation potentially produces negative effects, but effects may vary with irradiation dose and species. Ionizing irradiation could induce DNA breaks that may lead to chromosomal and genomic abnormalities [6].

Gamma rays irradiation also could modify the flowering age. The application of gamma rays did not significantly change the flowering age Boyolali variety. However, on Bantul variety, gamma rays reduced flowering age (Table 1). It meant gamma rays gave positive impact on black rice growth since the problem in black rice cultivation is the long flowering age. The shortest flowering age was Bantul variety under 100 Gy irradiation. Shorter flowering age meant shorter harvest age. Application 100 Gy irradiation also yielded early maturing plant. This result revealed that the gamma ray had potential for decreasing the harvest age in black rice but the dosage did not always correlate with the level of reduction in harvest age. This finding, however, was slightly different from previous research that stated that 300 Gy showed more influence in decreasing flowering age [7]. Gamma irradiation mostly affects the plant growth by variation in production through cytology, biochemistry, physiology, and morphogenetic of the cells [8]. These changes might be beneficial and possess higher economical values [9].

Table 1. Average values of agronomic trait of rice Boyolali variety dan Bantul variety

Boyolali Variety					
Irradiation Dosage (Gy)	Number of Productive Tillers	Flowering Age (day)	Harvest Age (day)	Weight of 1000 seed (g)	Yield per Plant (g)
0	12	70	115	31.0	32.14
100	12	69	114	33.8	36.47
200	12	69	115	33.1	28.28
300	15	70	115	33.4	26.04
Bantul Variety					
Irradiation Dosage (Gy)	Number of Productive Tillers	Flowering Age (day)	Harvest Age (day)	Weight of 1000 seed (g)	Yield per Plant (g)
0	17	68	113	24.9	27.78
100	15	65	110	31.6	30.45
200	15	66	111	24.4	25.88
300	15	67	112	29.9	33.18

Irradiation of gamma rays produced higher weight of 1000 seed in Boyolali variety (Table 1). The highest weight of 1000 seed was detected under 100 Gy reaching 33.8 g or increased 9% compared to 0 Gy. This result exhibited that the application of irradiation could enhance the size of grain of black rice. Genetic diversity can occur through mutations [10]. This result was in contrary with previous experiment [11] that stated in M1 generation of NERICA-1 rice, the weight of 1000 seeds gradually decreased with application of gamma ray. Genetic changes in somatic cells can occur due to gamma ray irradiation. In addition, it can be derived and cause changes in plant phenotypes [12]. Irradiation 100 Gy also boosted yield per plant. This dosage could improve 13.5% yield per plant compared to control. On the other hand, application 200-300 Gy significantly reduced the yield per plant. Moreover, 300 Gy decreased 18.9% yield per plant compared to control. This result indicated that in Boyolali variety, gamma ray boosted the yield per plant by producing bigger seed rather than increased the number of productive tillers.

Although the weight of 1000 seed of Bantul variety was lower than Boyolali variety, there was a significant increase in seed weight of Bantul variety compared to controls. Application 100 Gy also generated the highest weight of 1000 seed of Bantul variety reaching 31.6 g or 26.9% higher than control. However, in this variety, the yield per plant did not correlate with weight of 1000 seed. Application 300 Gy composed the highest yield per plant reaching 33.18 g. Gamma ray is one of the significant physical agents to enhance the characters and productivity of several plants [13]. In Bantul variety, the addition dosage of irradiation did not correlate with the increment of yield per plant. This pattern could be seen in weight of 1000 seed and yield per plant in which application 100 Gy increased weight of 1000 seed and yield per plant followed by a decline under 200 Gy and continued increase under 300 Gy.

3.2. Anthocyanin

Bantul variety contained greater anthocyanin content than Boyolali control (Figure 1). Gamma-ray irradiation yielded 15% higher anthocyanin content at Bantul variety under 100 and 300 Gy. This finding is in accordance with previous research [14] that irradiation increased the levels of anthocyanin content in buckwheat (*Fagopyrum esculentum* Moench) sprouts. The anthocyanin pigment in rice acts as an antioxidant compound. Conversely, 200 Gy lowered anthocyanin content in Bantul variety.

The application of irradiation significantly increased the anthocyanin content in Boyolali variety. The highest anthocyanin content in Boyolali variety was found under 100 Gy irradiation. Irradiation 100 Gy produced 1.6 times higher anthocyanin content than control. Similarly, 200 and 300 Gy increased 0.8 and 0.14 times higher anthocyanin content, respectively. It can be noticed that 100 Gy was the most effective dose to produce higher anthocyanin content compared to another doses. However, the level of doses did not correlate with the anthocyanin content increases. The greater exposure of gamma irradiation were usually inhibitory by decreasing mitotic activity. On the contrary, low doses could be utilized as safer and more stimulatory tools to improve variations [15, 16]. In addition, low doses of irradiation stimulate growing by altering the hormonal effect in the plant cells [17].

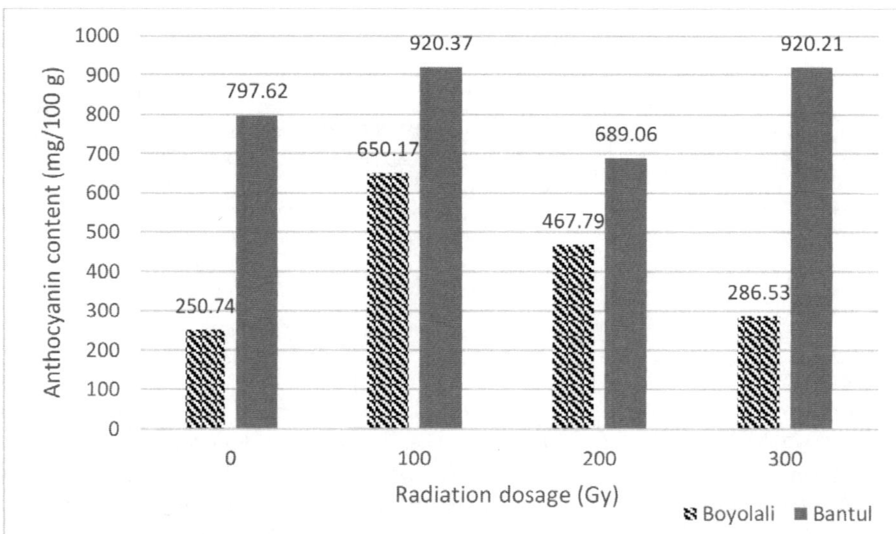

Figure 1. Anthocyanin content of Boyolali and Bantul varieties on different irradiation doses

4. Conclusion

The effects of gamma rays on Boyolali variety differed from Bantul variety. Bantul variety was more sensitive to gamma rays irradiation than Boyolali variety. The application of gamma rays did not significantly change the flowering age Boyolali variety. However, on Bantul variety, gamma rays reduced flowering age. Application 100 Gy irradiation yielded early maturing plant on Bantul variety. This result revealed that the gamma ray had potential for decreasing the harvest age in black rice. The application of irradiation could enhance the size of grain of black rice indicated by higher weight of 1000 seed in Boyolali variety. Irradiation 100 Gy also boosted yield per plant in Boyolali variety. In Boyolali variety, gamma ray boosted the yield per plant by producing bigger seed rather than increased the number of productive tillers. The increase of the irradiation doses did not always correlate with the changes of plant phenotype. To conclude, application 100 Gy was the most effective dose to produce higher anthocyanin content compared to another doses.

Acknowledgements
We thank to the Ministry of Research, Technology and Higher Education for funding this research through PUPT Program.

References

[1] Kristamtini T, Basunanda P and Murti R H 2014 Keragaman genetik dan korelasi parameter warna beras dan kandungan antosianin total sebelas kultivar padi beras hitam lokal *Ilmu Pertanian* **17** 57-70

[2] Sobrizal, Sutisna S, Carkum and Ismachin M 2004 Mutan padi pendek hasil iradiasi sinar gamma 0.2 Kgy pada varietas Atomita 4 *Risalah Pertemuan Ilmiah Penelitian dan Pengembangan Aplikasi Isotop dan Tadiasi* (Jakarta: Badan Tenaga Nuklir-Puslitbang Teknologi Isotop dan Radiasi)

[3] Giusti M M and R E Wrolstad 2001 Characteristic and measurement of anthocyanins by UV-visible spectroscopy *Current Protocols in Food Analytical Chemistry* (Hoboken-New Jersey: John Wiley & Sons, Inc.)

[4] Piri I, Babayan M, Tavassoli A and Javaheri M 2011 The use of gamma irradiation in agriculture *African J. of Microb. Res.* **5** 5806-11

[5] De Micco V, Arena C, Pignalosa D and Durante M 2011 Effects of sparsely and densely ionizing radiation on plants *Radiat. Environ. Biophys.* **50** 1-19

[6] Blagojević D 2015 Effects of gamma radiation on growth and development in Arabidopsis thaliana *Master Thesis* (Norway: Department of Plant Sciences-The Norwegian University of Life Sciences) 1-59

[7] Hartanti R S, Putri T A N, Zulfa F, Sutarno and Suratno 2017 Identification of morphological character and esterase isozyme pattern in second-generation black rice plant irradiated to gamma rays *Int. Conf. on Food Sci.and Engine. 2016* **193** 012038

[8] Rahimi M M and Bahrani A 2011 Effect of gamma irradiation on qualitative and quantitative characteristics of canola (Brassica napus L.) *Middle-East J. of Sci. Res.* **8** 519–25

[9] Majeed A, Khan A U R, Ahmad H and Muhammad Z 2010 Gamma irradiation effect on some growth parameters of Lepidium sativum L *ARPN J. Agric. Biol. Sci.* **5** 39-42

[10] Sasikala R and Kalaiyarasi R 2010 Sensitivity of rice varieties to gamma irradiation *Electronic J. of Plant Breeding* **1** 885-9

[11] Islam F, Azad M A K and Nath U K 2014 Effect of gamma ray on Nerica-1 rice and selection of desirable mutants *Bangladesh J. Nuclear Agric.* **30** 13-20

[12] Haris A, Abdullah, Bakhtiar, Subaedah, Aminah and Jusoff K 2013 Gamma ray radiation mutant rice on local aged dwarf *Middle-East J. of Sci. Res.* **15** 1160-64

[13] Sharma D K and Rana D S 2007 Response of castor (Ricinus communis) genotypes to low doses of gamma irradiation *Indian J. Agric. Sci.* **77** 467-9

[14] Tsurunaga Y, Takahashi T, Katsube T, Kudo A, Kuramitsu O, Ishiwata M and Matsumoto S 2013 Effects of UV-B irradiation on the levels of anthocyanin, rutin and radical scavenging activity of buckwheat sprouts *Food Chem.* **141** 552-6

[15] Kumari R and Singh Y 1996 Effect of gamma rays and EMS on seed germination and plant survival of Pisum sativum L. and Lens culinaris *Med. Neo. Botanica* **4** 25–29

[16] Khalil S J, Rehman S, Afridi K and Jan M J 1986 Damage induced by gamma irradiation in morphological and chemical characteristics of barley *Sarhad J. of Agric.* **2** 45–54

[17] Moghaddam S S, Jaafar H, Ibrahim R, Rahmat A, Aziz M A and Philip E 2011 Effects of acute gamma irradiation on physiological traits and flavonoid accumulation of Centella asiatica *Molecules* **16** 4994–5007

IOP Conf. Series: Materials Science and Engineering **633** (2019) 012023 doi:10.1088/1757-899X/633/1/012023

Egg quality in F_1 cross between brown and black lines of Japanese quail

R Dewanti, P Fadilla, M Cahyadi
Department of Animal Science, Faculty of Agriculture, Universitas Sebelas Maret
Jl. Ir. Sutami No. 36A Surakarta 57126, Indonesia

E-mail: mcahyadi@ uns.ac.id

Abstract. The objective of this study was to determine both external and internal egg quality in F_1 cross between Brown and Black lines of Japanese quail. A total of 186 eggs given basal diets have been collected at the age of ten weeks and those eggs were then tested to evaluate their external and internal qualities. They were produced by four group of F_1 cross between two lines of Brown male and Black female of Japanese quail and vice versa. Cross between Brown male and Black female lines produced Brown layer lines (B1 and B2 lines, respectively) while cross between Black male and Brown female lines was called L1 and L2 lines, respectively. The data was analyzed by using one-way analysis of variance (ANOVA) with an alpha level of 5%. Furthermore, the pairwise differences among populations were tested using Duncan's multiple range test (DMRT). The results showed that egg weight, shell weight and egg interior traits including yolk index, albumin index, Haugh unit (HU), and yolk color score were significantly difference among F_1 populations ($P<0.05$). Egg and shell weights were highest in B1 line, on the other hand, albumin index, yolk index, HU, yolk color score were found highest in B2 line. In addition, no differences among lines have been observed for shell thickness, egg index, and air cavity of egg. These results suggested that Brown quail line was better both exterior and interior qualities of egg.

1. Introduction

Quail is poultry species raised for egg and meat productions due to easy to be raised, low feed consumption (about 20 g/head/day), and growing quickly. It is able to produce egg at 42 days of age. Quail egg contains 13.1% of proteins, 11.1% of fats and 168 kilocalories. In addition, quail predominantly lays eggs at 3.00 to 6.00 p.m. and sometime at night. In ideal environment, laying quail can produce until 250 eggs/year [1]. In Indonesia, the most popular quail raised by small holder farmers is Japanese quail (*Coturnix coturnix japonica*), especially Brown and Black lines of Japanese quail due to sexing reason.

Generally, cross-breeding among different quail lines is conducted to prevent inheritance of negative traits due to inbreeding. It also facilitates sexing at one day old quail (DOQ) by using criss-cross inheritance among them [2]. Moreover, performance of individual quail will be better than average performance of its parents [3].

One of the parameters which is possible to determine the quality of F_1 offspring is egg quality traits including both external and internal quality of egg [4]. Exterior quality consists of egg shape (egg length and width), egg weight, egg shell thickness and shell weight, whereas interior quality consists of albumin length and height, yolk diameter and height, and yolk color. This study was aimed to determine both external and internal egg quality in F_1 cross between Brown and Black lines of Japanese quail.

2. Materials and Methods

2.1. Quail population and management

A total of 186 eggs were produced by four Japanese quail lines. Eggs were produced by four group of F_1 cross between two lines of Brown male and Black female of Japanese quail and vice versa. They were raised in colony cage sized 50×30×20 cm (width×length×height) with 15 to 16 quails per cage. Commercial feed for broiler starter was given to growing starter of quail (0-30 days old). Furthermore, commercial laying quail was started to be given to quail at the age of 31 days old. The nutrient content of commercial feed is showed in Table 1.

Table 1. Nutrient content of commercial feed for laying quail

Nutrient	Content (%)
Crude protein	20.0-22.0
Crude fat	4.0-7.0
Crude fiber	Max 6.0
Calcium	3.0-3.5
Phosphor	0.4-0.6

Source: PT. Japfa Comfeed Indonesia (2018)

2.2. Measurement of egg quality traits

Both exterior and interior quality of egg has been conducted in this study. Exterior quality included egg weight, shell weight, shell thickness, egg index, and air cavity, on the other hand, interior quality included index of albumen, yolk index, HU and yolk color score. Egg and shell weight were measured using digital scales. Shell thickness was obtained by measuring the blunt, equator and taper area after the shell was cleaned and aired. The measurement of the yolk color score was determined using yolk colour fan with 1-15 scale. Data was calculated using the formula below:

1. $\text{Egg index (\%)} = \frac{\text{egg width (mm)}}{\text{egg lenght (mm)}} \times 100\%$
2. $\text{Egg shell weight (\%)} = \frac{\text{Egg shell weight (g)}}{\text{Egg weight (g)}} \times 100\%$
3. $\text{Yolk Index} = \frac{\text{yolk height (mm)}}{0{,}5\ (\text{yolk max.(mm)} + \text{yolk min.(mm)})} \times 100\%$
4. $\text{Albumen Index} = \frac{\text{albumen height (mm)}}{0{,}5\ (\text{albumen max.(mm)} + \text{albumen min.(mm)})} \times 100\%$
5. $\text{HU} = 100 \log [(\text{albumen height (mm)} + 7.57) - (1.7 \times \text{egg weight (g)} \times 0.37)]$

2.3. Experimental design and data analysis

This study used completely randomized design (CRD). The data obtained was analysed using one-way analysis of variance (one-way ANOVA) and pairwise differences among lines was tested using Duncan's multiple range test (DMRT) with 5% alpha. Mathematic model used in this study was as follows:

$$Y_{ij} = \mu + L_i + \varepsilon_{ij}$$

where, Y_{ij} is the phenotype of the 1th quail, μ is population mean, L_i is line effect, and ε_{ij} is random error.

3. Results and Discussion

3.1. External quality of quail egg

This study was conducted on 28°C average environment temperature and 61% average humidity. Average of feed consumption was 8 g/quail/day in the morning, 7 g/quail/day in the afternoon, and 10 g/quail/day in the evening, respectively. Data of external quality of quail eggs were egg weight, shell weight, shell thickness, egg index and air cavity which are presented in Table 2.

Plumage color differences of Japanese quail affected egg and shell weight, but no effect on shell thickness, egg index and air cavity were observed. The average value of egg weight was significantly higher in brown lines (10.66-11.8 g) than black lines (9.30-10.09 g). The shell weight of brown lines was 1.32-1.44 g while the black line was 1.33-1.37 g.

Table 2. External quality of egg in four different Japanese quail lines

Variable	B1	B2	L1	L2	*P-Value*
Egg weight (g)	11.08 ± 0.88^a	10.66 ± 0.63^b	10.09 ± 0.68^c	9.30 ± 0.82^b	P<0.050
Shell thickness (mm)	0.29 ± 0.19	0.36 ± 0.24	0.30 ± 0.24	0.26 ± 0.17	P=0.097
Shell weight (g)	1.44 ± 0.14^a	1.32 ± 0.19^b	1.33 ± 0.11^b	1.37 ± 0.12^b	P<0.050
Egg index (%)	79.00 ± 2.99	78.86 ± 2.65	78.10 ± 2.26	78.94 ± 2.33	P=0.404
Air cavity (mm)	0.62 ± 0.51	0.78 ± 0.49	0.58 ± 0.50	0.64 ± 0.54	P=0.238

B1 (Brown 1) and L1 (Black 1) are from VBC 1, B2 (Brown 2) and L2 (Black 2) are from VBC 2
^{a,b,c} Superscript on the same line shows a significant difference (P<0.05)

Egg weight is influenced by genetic inheritance [5]. Egg weight is influenced by quail line [6] and age [7]. This study showed that egg weight from different line had congruency with the results. Previous study reported that Brown quail (base population) has 10.88 g of egg weight while black quail has 10.74 g of egg weight in average [8]. Egg weight and most qualified characteristic are dominantly possessed by Brown quail [9].

Shell weight on Brown quail was better than Black quail. Shell weight contributes on 10% of whole egg weight [10]. Shell weight ranged from 0.56-0.9 g. Shell weight is influenced by shell and membrane thickness [7]. Calcium (Ca) and Phosphor (P) contents in feed affected shell thickness [5,11]. The shell weight difference in this study was due to different quail line having different genetic trait, instead of having same feed nutrition.

Shell thickness, egg index and air cavity difference were not observed on this study. Height of air cavity was ranged from 0.58 to 0.78 mm. Shell thickness was ranged from 0.26 to 0.36 mm and egg index was ranged from 78.10 to 79.00%. Previous study measured shell thickness average in 0.234 mm containing 2.3% of calcium [12]. Shell thickness average generated by seven weeks old quail fed on low protein fodder fortified by commercial enzyme was 0.302 mm [13].

3.2 Internal quality of quail egg
The results showed that differences plumage color of Japanese quail lines significantly effected egg interior quality represented by albumen index, yolk index, HU and yolk colour as displayed in Table 3. Brown quail albumen index is higher than Black quail (13.20 vs 13.10%) [14]. Additionally, higher albumen index value in Brown quail (13.10 – 15.81%) than black quail (13.6-14.73%) was observed in this present study. Different albumen index may be caused by the influence of various factors including egg weight, diameter of albumen, line and age. Thicker and higher egg albumen represents better egg albumen quality [14].

Table 3. Internal quality of quail egg from four different Japanese quail lines

Variable	B1	B2	L1	L2	*P-Value*
Albumin Index (%)	13.10 ± 2.83^c	15.81 ± 2.74^a	13.65 ± 2.59^{bc}	14.73 ± 2.55^{ab}	P<0.05
Yolk index (%)	40.13 ± 3.88^b	42.75 ± 5.05^a	39.85 ± 5.21^b	39.24 ± 2.93^b	P<0.05
Haugh Unit	76.92 ± 8.42^c	84.48 ± 5.92^a	80.45 ± 6.78^b	84.87 ± 5.01^a	P<0.05
Yolk Color	4.89 ± 0.89^{ab}	5.15 ± 0.85^a	4.63 ± 0.66^b	5.14 ± 0.86^a	P<0.05

B1 (Brown 1) and L1 (Black 1) are from VBC 1, B2 (Brown 2) and L2 (Black 2) are from VBC 2
^{a,b,c} Superscript on the same line shows a significant difference (P<0.05)

Yolk index represents freshness of the yolk which is measured by yolk height and diameter [15]. This study showed significant difference yolk index among quail lines. Brown quail yolk index was better than black quail's yolk index (Table 3). Yolk index of quail egg fed by 18% protein in the diet was 40.6% and increasing until 20.1% crude protein in the diet also increased yolk index 42.2% [13]. Yolk index is varied from 30-50% [16] and it even can reach 52% [17].

Haugh Unit (HU) represents quail egg's quality. Previous study reported that HU value of quail egg was 84.49 [18]. According to HU value, egg is classified into four groups, i.e. HU > 79 belongs to AA class, HU 79 > U > 55 belongs to A class, HU 55 > U > 31 belongs to B class, and HU < 31 belongs to C class [19]. The HU value calculated in this study was ranged from 76.92–84.48 for Brown quail population, on the other hand the HU value of Black line was ranged from 80.45– 84.87. The HU may be affected by egg weight and albumen height.

Yolk color of Brown and Black lines were ranged from 4.89 to 5.15 and 4.63 to 5.14, respectively. Pigment on egg yolk is carotene and riboflavin which are classified as lipochrome (xanthophyll) colouring egg yolk into reddish orange. Orange on yolk presents carotenoid rich-containing zeaxantin, criptoxantin and lutein or xantofil [7]. Factors affecting yolk color were breed, genetic, environment condition, disease, heat stress, xanthophyll oxidation and egg production rate. Previous studies reported that yolk color was 4 in accordance to yolk color fan [21, 22]. Orange yolk color represented presence of carotenoids containing a lot of zeaxanthin, cryptoxanthine, and lutein (xanthophyll) [7].

4. Conclusions

Different plumage color lines affected on the internal and external quality of the Japanese quail eggs. Brown quail has better quality than Black quail lines in both external and internal quality of egg of F_1 cross between brown and black lines of Japanese quail, thus brown quail is more appropriate to be layer quail.

Acknowledgments

This research was supported and funded by the Research Scheme of the "Penelitian Unggulan–Universitas Sebelas Maret (PU-UNS)" with the contract number: 543/UN27.21/PP/2018.

References

[1] Woodard A, Abplanalp H, Wilson W O and Vohra P 1973. *Japanese quail Husbandry in the Laboratory (Coturnix-coturnix japonica)* (California: University of California Davis)
[2] Rifki M, Dewanti R, Widyas N and Cahyadi M 2018 *J. Anim. Breed. Genomics* **2** DOI: 10.12972/jabng.20180023
[3] Warwick E J, Astuti J M dan Hardjosubroto W 1990 *Pemuliaan Ternak* (Yogyakarta: Gadjah Mada University Press)
[4] Kul S and Seker I 2004 *Int. J. Poult. Sci.* **3** 400-405
[5] Ensminger M E 1992 *Poultry Science* (Danville, IL: Interstate Pub)
[6] Santos T C, Murakami and Fanhani J C 2011 *Brazilian J. Poul. Sci.* **13** 9-14
[7] Yuwanta, T 2010 *Telur dan Kualitas Telur* (Yogyakarta: Gadjah Mada University Press)
[8] Sujana E, Tuti W and Asep A 2014 *Characteristics of Quail Eggs Exterior Quality Basic Population on Brown and Black Feather Color Strains at the Center for Nursery Quality of Derivative Quail Eggs The Result of Quail Crosses* (Bandung: Padjajaran University)
[9] Chimezie V O, Fayeye T R, Ayorinde K L and Adebunmi A 2017 *J. Agrosearch* **17** 44– 53
[10] Amrullah, I K 2003 *Nutrisi Ayam Petelur* (Bogor: Lembaga Satu Gunung Budi)
[11] Wahju, J 1997 *Ilmu Nutrisi Unggas* (Yogyakarta: Gadjah Mada University Press)
[12] Hazim J A, Razuki W M, Al-Hayani W K and Al-Hassani A S 2011 *J. Poultry Sci.* **9** 584-590
[13] Suprijatna E, Kismiati S, Furi N R 2008 *J. Indonesian Trop. Anim. Agri.* **33** 66-71
[14] Jessy B and Panigrahi B 2016 *Veter. World J.* **9** 832-836
[15] Sudaryani T 2006 *Kualitas Telur* (Jakarta: Penebar Swadaya)
[16] Romanoff A L and Romanoff A 1963 *The Avian Egg* (New York: John Wiley and Sons)
[17] Imai C, Mowlah A and Saito J 1984 *J. Poultry Sci.* **65** 474-480
[18] Song K T, Choi S H and Oh H R 2000 *Asian–Aus. J. Anim. Sci.* **13** 986–990
[19] Yuwanta T 2007 *Telur dan Produksi Telur* (Yogyakarta: Gadjah Mada University Press)
[20] North M O and Bell D D 1992 *Commercial Chicken Production Manual* (New York: An AVI Book Published)

[21] Sipayung, P P 2012 Performa Produksi dan Kualitas Telur Puyuh (*Coturnix-Coturnix Japonica*) pada Kepadatan Kandang yang Berbeda (Bogor: Bogor Agricaltural University)
[22] Nagarajan S, Narahari D, Jayaprasad I A and Thyagarajan D 1991 *British Poul. Sci.* **32** 243-248

The effects of plumage color lines and sex on slaughter weight and carcass parts of Japanese quail

R Dewanti, I A Harminanti, N Widyas, M Cahyadi

Department of Animal Science, Faculty of Agriculture, Universitas Sebelas Maret
Jl. Ir. Sutami No. 36A Surakarta 57126, Indonesia

E-mail: mcahyadi@uns.ac.id

Abstract. This study was conducted to evaluate the effects of plumage color lines and sex on slaughter weight and carcass parts of Japanese quail. A total of 80 birds consisting 40 Brown (20 males and 20 females) and 40 Black lines (20 males and 20 females) have been raised under the same condition and they were grouped into four groups, Brown 1 line (B1), Brown 2 line (B2), Black 1 line (L1), and Black 2 line (L2), respectively. After 14 weeks of age, they were slaughtered. Slaughter weight, carcass weight, thighs, drumsticks, wings, breasts, and back were calculated. Furthermore, the data were analyzed using the general linear model (GLM) and the differences among lines have been tested using Duncan's multiple range test (DMRT) with 5% of the alpha level. The results found that slaughter weight, thighs, and back were significantly different among quail lines ($P<0.05$). Slaughter weight was lowest in B1 line and highest in L1 line (135.67 vs 149.85 g). The highest thigh percentage was found in B1 line population while L1 line was the lowest thigh percentage (14.29 vs 13.46%). The B2 line has the highest back part percentage. In addition, sex was significantly affecting carcass parts of Japanese quail ($P<0.05$). Female quail has more breast, drumstick, and wing parts compared to male quail. On the other hand, male quail has higher thigh percentage than female quail. This study concluded that plumage color affecting slaughter weight, thigh and back of Japanese quail carcass and female quail produced more breast, drumstick, and wings parts of quail carcass.

1. Introduction

In Indonesia, *Coturnix coturnix japonica*, which is well known as Japanese quail, is quail species mostly raised by farmers. It is the most popular quail to be farmed for producing eggs and meats [1]. It has two lines based on its plumage colors, i.e. black and brown plumage lines, respectively. Besides producing the egg, Japanese quail is also raised to produce meat. Quail carcass can be derived from male quail or female quail that already had low egg production due to economically inefficient to be kept.

Carcass quality may be affected by various factors such as slaughter age and genetics [2]. Indicators determining carcass quality is meat and skin colors, and also thickness of meat fibers. In addition, factors such as sex, strain, body size, age of slaughter, and inheritance factor may affect carcass quality of quail [3, 4]. Previous study reported that different plumage color line significantly affected body weight, heart weight, and fatness of quail. Black plumage quail line has higher body weight than Brown plumage quail line [3]. Therefore, this study was carried out to evaluate the effects of plumage color lines and sex on slaughter weight and carcass parts of Japanese quail.

2. Materials and Methods

2.1. Quail population

IOP Conf. Series: Materials Science and Engineering **633** (2019) 012024 doi:10.1088/1757-899X/633/1/012024

The population used in this study was 80 parents stock quail consists of 20 Brown male and 20 Brown female quails and also 20 Black male and 20 Black female quails. They were fed with commercial broiler starter diet started from 0 to 30 days of age. After that, they were fed by using commercial laying quail feed until 14 weeks of age.

2.2. Slaughtering and carcass processing

Quails were fasted for 4 hours before they were slaughtered. Furthermore, they were dipped into warm water at 50 to 55°C [5]. Then, feather was removed and evisceration was also conducted to remove internal organs. The head, neck, shank were cut [6]. The wings were separated by the shoulder joint and cut circularly. The thigh was separated from the acetabulum with pelvic muscles and without pelvic bones. The intact back separated from the pelvic bone, scapula part vertebrae from the posterior neck to tail [5].

The data collected in this study were slaughter weight, carcass weight, carcass percentage, breast, thighs, drumstick, back, and wings. Slaughter weight and carcass weight are obtained by weighing using a digital scale with the capacity of 500 g and sensitivity of 0.01. Carcass percentage and proportion of carcass parts are measured by using these following formulas:

$$\text{Carcass Percentage} = \frac{\text{Carcass Weight (g)}}{\text{Slaughter Weight (g)}} \times 100\%$$

$$\text{Carcass Parts Percentage} = \frac{\text{Carcass Parts Weight (g)}}{\text{Carcass Weight (g)}} \times 100\%$$

2.3. Experimental design and analysis of data

This present study was using completely randomized design (CRD) and general linear model (GLM) had been applied to evaluate significant effect of plumage color line and sex of quail to carcass traits. If there was a significant effect then it would be continued with Duncan's multiple range test (DMRT) with 5% of alpha.

3. Results and Discussion

3.1. The effect of plumage color line to slaughter weight and carcass proportions

Statistical analysis showed that slaughter weight, thigh, and back were significantly different among plumage color lines (Table 1). Slaughter weight of Black line was much higher than Brown line, 149.85 vs 135.67 g (P<0.05). Previous study reported that difference of slaughter weight might be caused by different genetic potency among quail lines [7]. Selection and age of slaughter were also affected slaughter weight of quail [8]. Moreover, higher back proportion of Black plumage line was found in this study (Table 1). On the other hand, thigh proportion was significantly higher in Brown plumage line than Black one, 14.29 vs 13.46%. This result was much lower than previous study that reported thigh percentage were ranged from 16.03 to 22.21% [3, 9]. In addition, no significance effect of plumage color line to carcass weight, breast, drumsticks, and wings were observed in this current study. This result was different with previous study reported that drumstick weight, carcass weight, the percentage of carcass, breast, and thigh weight were significantly different among strains [10, 11].

Table 1. Carcass Weights Based on Black and Brown Plumage

Variable	B1	B2	L1	L2	*P-Value*
Slaughter weight (g)	135.67^{b}	140.12^{ab}	149.85^{a}	145.99^{ab}	<0.05
Carcass (g)	86.33	88.65	93.30	87.97	0.18
Breast (%)	39.93	38.70	40.08	39.05	0.35
Thigh (%)	14.29^{a}	13.65^{ab}	13.46^{b}	14.14^{a}	<0.05
Drumstick (%)	10.70	10.44	10.23	9.83	0.09
Backs (%)	26.16^{c}	28.51^{a}	27.87^{b}	28.43^{a}	0.05
Wing (%)	8.88	8.53	8.24	8.47	0.32

L1 is Black 1 line; B1 is Brown 1 line; L2 is Black 2 line, B2 is Brown 2 line.

^{a,b} Different superscript on the same row shows a significant difference among lines (P<0.05).

3.2. The effect of sex on slaughter weight and proportions of carcass part

The effects of sex on slaughter weight and carcass parts are presented in Table 2. The statistical analysis showed that breast, thigh, drumstick, and wings were significantly different between sex (P<0.05). Breast, drumsticks, and wings of female quail were higher than male quail, on the other hand, male quail had higher thigh percentage than female.

Table 2. The effect of sex on slaughter weight and proportions of carcass part of Japanese quail

Variable	Male	Female	*P-Value*
Slaughter weight (g)	135.01	150.88	0.83
Carcass (g)	92.83	85.55	0.54
Breast (%)	38.64^{b}	40.07^{a}	<0.05
Thigh (%)	14.30^{a}	13.54^{b}	<0.05
Drumstick (%)	10.09^{b}	10.75^{a}	<0.05
Backs (%)	28.88	26.77	0.21
Wings (%)	7.98^{b}	8.81^{a}	<0.05

^{a,b} Different superscript on the same row shows a significant difference between sex (P<0.05).

The percentage of female breasts was higher than males. It may be caused by the slaughter weight in females is higher than males so that the percentage of breasts can also be higher. The percentage of breasts in this study was higher than previous study [3]. The percentage of female drumstick and wings were found higher than male that similar to previous report [11, 12, 13]. However, thigh percentage in male sex was 1% higher than female (Table 2). Previously, it is known that thigh length of male quail is higher than female which is affecting the proportion of thigh [14].

Additionally, slaughter and carcass weights and back part were not different between sex in this study. This results were different with previous studies that reported male quail had higher backs [13], on the other hand slaughter weight were higher in female quail population due to heavier female internal organ weights than males [8, 11, 12, 15] impacting to higher carcass weight in female [16].

4. Conclusions

This study concluded that Black plumage line had higher slaughter weight and backs. Brown plumage color had higher thigh percentage. In addition, female quail had higher breast, drumstick, and wing percentages. Male quail had higher thigh percentage than female quail.

Acknowledgments

This research was supported by the Research Scheme of the "Penelitian Unggulan - Universitas Sebelas Maret (PU-UNS)" with the contract number: 246 623/UN27.21/PP/2017.

References

[1] Subekti E dan Hastuti D 2013 *Jurnal Mediagro* **9** 1-3
[2] Genchev A G, Ribarski S S, Afanasjev G D and Blohin G I 2005 *J. Cent. Eur. Agric.* **6** 501-505
[3] Raji A O, Girgiri A Y, Alade N K and Jauro S A 2015 *Trakia J. Sci.* **2** 159-165
[4] Resnawati H 2004 *Proc. Seminar Nasional Teknologi Peternakan dan Veteriner* (Bogor: Balai Penelitian Ternak Ciawi)
[5] Soeparno 2005 *Ilmu dan Teknologi Daging* (Yogyakarta: Gadjah Mada University Press)
[6] Hadiwiyoto S 1992 *Kimia dan Teknologi Daging Unggas* (Yogyakarta: Universitas Gadjah Mada)
[7] Arora K L and Samples O 2011 *Int. J. Poult. Sci.* **8** 640-643
[8] Narinc D, Karaman E and Aksoy T 2014 *J. Poult. Sci.* **93** 762-760
[9] Genchev A, Mihaylova G, Ribarski S, Pavlov A and Kabakchiev M 2008 *Trakia J. Sci.* **6** 72-82
[10] Shokoohmand M, Kashan N E J and Maybody M A E 2007 *Int. J. Agric. Biol.* **9** 945-947
[11] Vali N, Edriss M A and Rahmani H R 2005 *Int. J. Poult. Sci.* **4** 296-300
[12] Alamuoye O F and Ojo J O 2015 *Scholars J. Agric. Vet. Sci.* **5** 342-344

IOP Conf. Series: Materials Science and Engineering **633** (2019) 012024 doi:10.1088/1757-899X/633/1/012024

[13] Banerjee S 2010 *World App. Sci. J.* **2** 174-176
[14] Fachri B I, Asmara I Y dan Garnida D 2016 *Students e-J.* **5** 1-10
[15] Lotfi E, Zerehdaran S and Azari M A 2011 *J. Poult. Sci.* **90** 2202-2208
[16] Minvielle F 1998 *Proc. 6th Asian Pacific Poult. Cong.* (Nagoya-Japan/Japan Branch of the World Poultry Science Association) p. 122-127

Production of antioxidant C-phycocyanin using extraction process of *Spirulina platensis* in large scale industry

Wachda[1,2], H Hadiyanto[2], G D Harjanto[1], M L Huzain[1], R W Aji[3]

[1] Neoalgae Indonesia Makmur, Jl. Slamet Riyadi 20, Sukoharjo, Indonesia, 57561

[2] Center of Biomass and and Renewable Energy, Faculty of Engineering, Diponegoro University, Jl. Prof Soedarto, SH, Tembalang Semarang, Indonesia, 50275

[3] Algaepark Indonesia Mandiri, Jalan Raya Puspiptek - Gunung Sindur, Bogor, Indonesia

E-mail: hady.hadiyanto@gmail.com

Abstract. *Spirulina platensis* is one of the popular species that containing C-Phycocyanin. C-Phycocyanin has been used in many applications such as antioxidant, anti-inflammatory, and natural blue colorant. The increase of phycocyanin production from *S. platensis* is influenced by extraction method and solvent ratio. The phycocyanin pigment can be massive extracted using mechanical cell rupture like freezing thawing method and distillated water as solvent. This research aimed to produce high concentration of antioxidant phycocyanin from *S. platensis* using simple and efficient method for large scale industry and develop a method for primary extraction from dry biomass of *S. platensis*. There were 3 treatments and each treatment was conducted in 3 replications. Standardization of process treatments such as biomass : solvent ratio (1:6, 1:50 and 1:100) and processing time were carried out at thawing time (1 hour), freezing time (2 hours) and stirring time in water bath (0,5 hour). The phycocyanin concentration and yield were determined by using spectrophotometer. The result showed that at biomass : solvent ratio of 1:50, resulted C-PC concentration of 0.179 mg/ml and its yield of 17.944 mg/g.

1. Introduction

Spirulina platensis is a rich and inexpensive source of natural blue pigment [1]. The phycobiliprotein in cyanobacteria *S. platensis* produces allophycocyanin (APC) and C-phycocyanin (C-PC), both blue colored. C-PC as the major is a photosynthetic pigment of the phycobiliprotein family and is the major light-harvesting protein pigment present in the cells of *S. platensis*, whereas APC is a minor component [2].

C-PC has many applications in pharmaceutical (fluorescent marker [3], the neuroprotective [3], hepatoprotective [3], anti-inflammatory [4] and antioxidant [5]), cosmetic and also in food industry such as (as natural blue colorant). In general, the extraction method is the key for maximum recovery of phycobiliproteins in the natural state from algae [6]. The extraction of phycobiliproteins involves the cell rupture and release of these proteins from within the cell, and thus, the use of variations in the osmotic pressure, abrasive conditions, chemical treatment, freezing-thawing and sonication, among others, are necessary. Its extraction is difficult as the cell wall of the cyanobacteria is quite resistant [7], which is made of four layers i.e. fibril, peptidoglycan, proteins, and analogous to gram negative bacteria [8].

The extraction process should be effective in terms of high extraction yield and must be environmental friendly. However, all of these cell disruption methods lack specificity as cell debris and other unwanted impurities are also released. Nevertheless, phycocyanin is sensitive to light, oxygen, moisture, and temperature. Therefore, it is able to readily degrade under certain physicochemical conditions [9]. The most important are the cell disruption method, type of solvent, solvent-to-biomass ratio and extraction time [10, 11, 20].

Phicobiliprotein is a polar compound that dissolves in polar solvents. The use of water solvents in extracting phycocyanin in *S. platensis* is considered safe and can attract the active substances. The use of water as a solvent can more easily dissolve the biomass of *Spirulina* cells. Unlike methanol which includes organic solvents and is toxic, water is not an organic solvent and is not toxic. Because water does not have toxic power, no preliminary test for solvent toxicity is necessary. In addition, the economic value of water is cheaper when compared to other types of solvents [2].

This study was primarily aimed to determine the most effective solvent concentration to obtain optimize the process, minimizing costs and maximizing yield. Several factors can influence C-PC extraction. However, it is important to study the factors affecting extraction as a group in order to optimize the process, minimizing costs and without the use of chemical agents, in order to obtain this natural blue dye from *S. platensis* with a high phycocyanin yield and concentration.

2. Materials and Methods

2.1. Materials

Dry biomass of *S. platensis* was procured from PT. Neoalgae Indonesia Makmur, Sukoharjo, Indonesia which stored in aluminum phthalate pouches at room temperature (20 ± 5°C) and distillated water as solvent.

2.2. Extraction procedures

The dry biomass underwent the drying step that was carried out at 40°C for 1 hour and the freezing step at -18°C for 2 hours. Mechanical rupture was carried out by stirring in a water bath for 0.5 hour then settled for 24 hours. The extraction of C-PC was carried out according to Silveira [11], which used water as solvent under the following conditions: a biomass-to-solvent ratio of 1:6, 1:50, 1:100. Each analysis was conducted with 3 replications. The crude extract was separated from debris using a set of sieves.

2.3. Analytical procedures

2.3.1 C-PC Concentration. The C-PC concentration in mg/mL was calculated and anlyzed by the optical densities of spectrophotometry at 652 nm and 620 nm, using Eq. (1) [12]:

$$\text{C-PC} = \frac{(OD620 - 0.474 \times (OD652))}{5.34} \tag{1}$$

2.3.2 Extraction Yield. The extraction yield was calculated using Eq. (2) [12]. Yield is the extraction yield of phycocyanin in mg of C-PC/ dry biomass (g), V is the solvent volume (mL) and DB is the dry biomass (g).

$$\text{Yield} = \frac{(C - PC \times V)}{DB} \tag{2}$$

C-Phycocyanin concentration and yield were presented in table form then also analyzed for significance level using SPSS software based on oneway Analysis of Variance (ANOVA) and Duncan's Multiple Range (DMRT) tests at a significance level of 5%.

3. Result and Discussion

The effects of the different treatments on the biomass of *S. platensis* were investigated individually (Table 1):

Table 1. Concentration and Yield of C-Phycocyanin

Treatments		OD 620 (nm)	OD 652 (nm)	C-PC (mg/ml)	C-PC average (mg/ml)	V/DB (1000 ml/10 gr)	Yield (mg/g)	Yield average (mg/g)
A (1:6)	1	1.849	1.948	0.173		100	17.334	
	2	1.843	1.948	0.172	0.172	100	17.222	17.243
	3	1.838	1.943	0.171		100	17.173	
B (1:50)	1	1.888	1.958	0.179		100	17.976	
	2	1.888	1.958	0.179	0.179	100	17.976	17.944
	3	1.883	1.958	0.178		100	17.882	
C (1:100)	1	1.878	1.943	0.179		100	17.922	
	2	1.878	1.943	0.179	0.178	100	17.922	17.884
	3	1.872	1.943	0.178		100	17.809	

The table of C-Phycocyanin concentration and yield showed that ratio 1:50 was the highest among the other treatments. It resulted the concentration of of 0.179 mg/ml and yield of 17.944 mg/g. On the otherhand ratio 1:6 achieved the lowest concentration of 0,172 mg/ml and yield of 17.243 mg/g. Phycocyanin is a water-soluble protein that can be released simply by the destruction [13]. The analysis of variance (ANOVA) test results showed that the use of water solvent with different concentrations had no significant effect ($P < 0.05$) on the C-Phycocyanin concentration and yield.

According to the literature, freeze-thawing method is widely used for the extraction of phycocyanin from fresh biomass that improves the efficiency of extraction [1, 10, 14, 15]. The main purpose of the freeze-thaw process is for stressing cells to accelerate the release of pigments from cells after the collision process. The cell breakage that is caused by freezing can improve the phycocyanin extraction yield [16]. The impact of the freezing process is the occurrence of cell swelling and cell damage caused by the formation of sharp crystals that form during clotting. The next step with the thawing process will cause cellular contraction due to the piercing of the cell due to the formation of sharp crystals, so that a cell leak will occur which results in cellular pigment discharge.

The solvent used in the extraction process must be able to pull the active component from the mixture in the sample [17, 21]. Extraction of phycocyanin with water does not differ much from that of sodium phosphate buffer (pH 7.4), it is found twofold higher extraction yield of phycocyanin in frozen biomass using distilled water as compared to phosphate buffer [11]. Through the sample immersion process, there is a breakdown of cell walls and membranes due to differences in pressure inside and outside the cell. The bioactive compounds in the cytoplasm will dissolve in the solvent and extract the compound will be perfect. Solvent volume affects extraction yield. This is in accordance previous report who extracted dyes on suji leaves with ethanol solvent, stating that extraction with a stirring speed of 300 rpm with a ratio of 5 grams of suji in 200 ml of solvent resulted in the greatest extract yield [18]. Water is not an organic solvent and is not toxic. Because water does not have toxic power, no preliminary test for solvent toxicity is necessary. In addition, the economic value of water is cheaper when compared to other types of solvents. Increasing the biomass / solvent ratio can increase effectiveness solvent into the cell and give more extract. However, an advantage solvents have been reported to absorb cavitation energy from extraction system, much of the other contaminant proteins from cytoplasm and other organelles resulting in lower extraction results, In the case of a smaller amount of solvent, biomass cannot find sufficient blood solvent, resulting in insufficient duplicate phycocyanin from the cell [19, 11].

4. Conclusion

The biomass to solvent ratio showed effect on the C-PC concentration and extraction yield. Moreover, this methodology does not employ chemical products, using only water to make the extraction, and is a simple procedure. Using dried biomass from S. platensis, in a 1:50 biomass to solvent ratio, a C-PC concentration of 0.179 mg/mL was obtained with yield of 17.944 mg/g.

Acknowledgement
This work was financially supported by research project of “PT. Neoalgae Indonesia Makmur, Center of Biomass and and Renewable Energy, Universitas Diponegoro, and PT. Algaepark Indonesia Mandiri”.

References

[1] Minkova K M, Tchernov A A, Tchorbadjieva M I, Fournadjieva S T, Antova R E, Busheva M C 2003 Purification of C-phycocyanin from *Spirulina (Arthrospira) fusiformis J. Biotechnol* **102** pp 55–59.
[2] Boussiba S, Richmond A E 1979 Isolation and characterization of phycocyanins from the blue-green alga *Spirulina platensis*, *J. Arch. Microbiol* **120** pp 155–159.
[3] Romay C, Gonzalez R, Ledon N, Remirez D, Rimbau V 2003 C‐Phycocyanin: a biliprotein with antioxidant, anti-inflammatory and neuroprotective effects *J. Curr. Protein Pept. Sci.* **4** pp 207–216.
[4] Reddy C M, Bhat V B, Kiranmai G, Reddy M N, Reddanna P, Madyastha K M 2000 Selective inhibition of cyclooxygenase‐2 by C‐Phycocyanin, a biliprotein from *Spirulina platensis J. Biochem. Biophys. Res. Commun* **277** pp 599–603.
[5] Bhat V B, Madyastha K M 2000 C-Phycocyanin: a potent peroxyl radical scavenger in vivo and in vitro *J. Biochem. Biophys. Res. Commun* **275** pp 20–25.
[6] Niu J F, Wang G C and Tseng C K 2006. Method for large scale isolation and purification of R-phycoerytrin red alga Polysiphonia urceolata *J. Grev. Protein Expr. Purif.* **49** pp 23–31.
[7] Wyman M 1992 An in vivo method for the estimation of phycoerythrin concentrations in marine cyanobacteria. *J. Limnol. Ocean* **37** pp 1300–1306.
[8] Van E C 1977 On the morphology and ultrastructure of the cell wall of *Spirulina platensis J. Antonie Van Leeuwenhoek* **43** pp 89–99.
[9] Kannaujiya V K, Sinha R P 2016 Thermokinetic stability of phycocyanin and phycoerythrin in food-grade preservatives *J. Appl. Phycol.* **28** pp 1063–1070.
[10] Abalde J 1998 Purification and characterization of phycocyanin from the marine cyanobacterium *Synechococcus sp. J. Plant Sci* **136** pp 109–120.
[11] Silveira S T, Burkert J F M, Costa J A V, Burkert C A V, Kalil S J 2007 Optimization of phycocyanin extraction from *Spirulina platensis* using factorial design *J. Bioresour. Technol* **98** pp 1629–1634.
[12] Bennett A, Bogorad L 1973 Complimentary chromatic adaptation in a filamentous blue-green alga. *J. Cell Biol.* **58** pp 419–435.
[13] Antelo F S, Anschau A, Costa J A V, Kalil S J 2010 Extraction and Purification of C-phycocyanin from *Spirulina platensis* in Conventional and Integrated Aqueous Two-phase Systems *J. Braz. Chem. Soc.* **21** pp 921–926.
[14] Doke J M 2005 An improved and efficient method for the extraction of phycocyanin from *Spirulina sp. Int. J. Food Eng* **1** pp 1–11.
[15] Soni B, Kalavadia B, Trivedi U, Madamwar D 2006 Extraction, purification and characterization of phycocyanin from Oscillatoria quadripunctulata–isolated from the rocky shores of Bet-Dwarka, Gujarat, India *J. Process Biochem* **41** pp 2017–2023.
[16] Nakagawa K, Ritcharoen W, Sri-Uam P, Pavasant P, Adachi S 2016 Antioxidant properties of convective-air-dried Spirulina maxima: evaluation of phycocyanin retention by a simple mathematical model of air-drying *J. Food Bioprod* **100** pp 292–302.
[17] Ansel H C 1989 Pengantar Bentuk Sediaan Farmasi. Diterjemahkan oleh Farida Ibrahim. UI-Press. Jakarta.
[18] Yuniwati M, Iskarima F, Padulemba A 2012 Optimasi kondisi proses pembuatan kompos dari sampah organik dengan cara fermentasi menggunakan EM4 *Jurnal Teknologi* **5** pp 172- 181.
[19] Samavati V dan Manoochehrizade A 2013 Dodonaea viscosa Var. angustifolia Leaf: New Source of Polysaccharide and Its Anti-Oxidant Activity *J. Carbohydrate Polymers* **98** pp 199-200.

[20] Hrishikesh A T, Rochak M, Jampani C 2018 Simple and efficient method for extraction of C-Phycocyanin from dry biomass of *Arthospira platensis* *J. Algal Research* **31** pp 239-251.

[21] Isil I, Saniye A, Zeliha D, Mehmet K, Meltem C D, Figen K E 2018 Optimization of phycocyanin extraction from *Spirulina platensis* using different techniques *J. Food Composition and Analysis* **70** pp 78-88.

IOP Conf. Series: Materials Science and Engineering **633** (2019) 012026 doi:10.1088/1757-899X/633/1/012026

Zeolite and manure treatment on the increase of N soil, N absorption and soybean production in alfisols

V Triatmoko[1], P Alvernia[1], I L Haniati[2], S Minardi[1], Suntoro W[1], and D P Ariyanto[1]

[1]Department of Soil Science, Faculty of Agriculture, Sebelas Maret University, Surakarta, Indonesia
[2]Postgraduate of Soil Science, Sebelas Maret University, Surakarta, Indonesia

Email: sminardi@staff.uns.ac.id

Abstract. Soybean production in Indonesia could be improved by utilizing Alfisols as potential land. Zeolites and manure are used in this study as slow release materials to reduce nitrogen losses from soil. The aim of this research is reviewing the effect of zeolites and manure in improving N total contents, N absorption, and growth and production of soybean in Alfisols. Factorial randomized complete block designed of two factors (zeolite doses; 0 ton / ha, 2.5 ton / ha, and 5 ton / ha, and type of manure; P0: 0 ton / ha, P1: cow manure, P2: quail manure) were used and field experiments were conducted in this study. Obtained results showing zeolite with dose of 5 ton/ha gave the highest influence indicated by the increase of N total 35.13% and the increase of N absorption up to 53.37%. Cow manure gave the highest efficiency with total N increase of 40.54% and an increase in N absorption of 48.93%. Zeolite 5 ton/ha with cow manure gave the highest yield with an increase of total pod content 129.41%, and the increase of seed weight 35.39 %. Finally, zeolite treatment with cow manure at dose 5 ton/ha is the best treatment in this research.

1. Introduction

A very wide gap can still be found in the production and consumption of soybean in Indonesia. Increase in domestic soybean consumption potentially increase soybean imports, because soybean production has not been able to meet the needs of the community as soybean production in the country tends to decrease [1]. The low production of soybean is caused by the decreasing of fertile land caused by continuous use of inorganic fertilizer [2]. Efforts to increase soybean production in Indonesia to suppress imports are done through intensification, extensification, diversification, and rehabilitation. Extensification efforts are done by utilizing marginal lands, one of them is Alfisols [3].

Alfisols is a soil that experienced intensive weathering and further development, resulting in leaching of nutrients, especially N, P, K. This type of soil generally have low chemical fertility. If the nitrogen element is present in a low state, then the growth and production of the plant will be disrupted [4]. Nitrogen is the most important element for growth and filling of soybean seeds. However, the availability of nitrogen in the soil is generally very low. Whereas the quantity and quality of high soybean seed yield require a high N supply as well.

Manure is a potential source of organic material with its presence available in the environment and it is affordable. Manure is able to increase soil CEC and can form complex compounds with metal ions such as Al, Fe, and Mn so as not to poison its environment [5].

IOP Conf. Series: Materials Science and Engineering **633** (2019) 012026 doi:10.1088/1757-899X/633/1/012026

Zeolite as a soil enhancer is a mineral of a hydrogenated aluminosilicate compound with a hollow structure and contains interchangeable alkali cations. The cations in the zeolite are bonded by a weak bond and can be easily replaced without altering the aluminosilicate structure [6]. Zeolite is a natural material that has a high CEC (120-180 meq/100g) and is hollow with a cavity size corresponding to the size of the ammonium ion. This causes the zeolite to absorb the ammonium ion before it turns into nitrate [7].

2. Materials and Methods

This study was carried out on the dry land, Jumantono, located in central Java Indonesia. Laboratory analysis was done in Soil Chemistry and Soil Fertility Laboratory of FP UNS to analysis begins before planting (initial soil), at the time of maximum vegetative and generative.

The research was conducted by using a complete Randomized Complete Block Design factorial with two factors, namely zeolite and manure. The dose of zeolite consists of 3 levels (Z0: 0 t ha^{-1}, Z1: 2.5 t ha^{-1}, and Z2: 5 t ha^{-1}). Type of manure consisting of 3 types, namely (P0: Without Manure, P1: Quail Manure (5 t ha^{-1}), and P2: Cow Manure (5 t ha^{-1}). Each treatment combination was repeated 3 times so that 27 experimental units were obtained. The implementation of the study included plant preparation, land preparation, initial soil sampling, zeolite incubation and manure, planting, maintenance, harvesting (maximum and generative vegetative), and laboratory analysis.

The soil pH was determined in 1:2.5 w/v soil: water mixture. Particle-size distribution was determined by sedimentation method [8], organic carbon concentration by Walkley and Black [9] rapid chromic acid titration, calcium carbonate ($CaCO_3$) by Bundy and Bremner method [10], cation exchange capacity (CEC) by the neutral normal ammonium acetate method [11]. N (Kjeldahl method) was determined according to Jackson [12] and for analysis crop yield include plant height, fresh weight, dry weight, total pod content, and seed weight. The observed data were analyzed using multiple analysis or ANOVA, Duncan Multiple Range Test (DMRT) and correlation test.

3. Result and Discussion

The soil fertility level in Jumantono area is included in the low category of pH 5.4 (acid), organic c-1.15% (low), 1.93% KTK 18.93 me/100g (medium), KB 14.32% (very low), total N 0.26% (medium), P available 4.45 ppm (very low), K available 0.22 me/100g (Low), Ca 1.72 me/100g, Mg 0.56 me/100g (low), and Na 0.20 me/100g (low). Thus it is necessary to provide manure and zeolite so that plants can grow well and produce a high harvest and profitable.

Table 1. The result of initial soil analysis

Observation Variable	Result	Unit	Rating
pH H_20	5.4		Acid
C-organic	1.15	%	Low
Organic Material	1.93	%	Low**
CEC	18.93	me/100g	Medium
Base Saturation	14.32	%	Low***
Total N	0.26	%	Medium**
P available	4.45	Ppm	Very low**
K available	0.22	me/100g	Very Low
Ca	1.72	me/100g	Very low
Mg	0.56	me/100g	Low
Na	0.20	me/100g	Low

Description *): Rating according to FAO (2006) [13]
**) : Rating according to Tekalign (1991) [14]
***): Rating according to Landon (1991) [15]

IOP Conf. Series: Materials Science and Engineering **633** (2019) 012026 doi:10.1088/1757-899X/633/1/012026

Table 2. Results of analysis of quail manure, cow manure, and zeolite

Variable	Quail	Cow	Zeolite	Unit
N total	1.32	1.40	-	%
P_2O_5	4.54	4.41	-	%
K_2O	1.51	1.40	-	%
C-organic	17.56	19.94	-	%
C/N ratio	13.31	14.30	-	
KTK	-	-	128.20	me/100g

C/N ratio of quail manure is 13.31 and C/N of cow manure is 14,30. C/N ratio is high then the nutrient content is slightly available for plants, otherwise, C/N ratio is low then the availability of high nutrients and available for plants [16]. The value of CEC in the zeolite which is 128.20% me/100g CEC (Table 2). Values has met the criteria of technical requirement of the regulation of the ministry of agriculture number 70 with KTK zeolite which is at least 120 me/100g.

Applying zeolite dose of 5 t ha^{-1} increased pH value by 5.26% compared without applying zeolite. Soil pH increased due to the zeolite undergoes hydrolysis process of silicates which produces OH^- ion leading to an increase in soil pH [17]. Application of manure type has no significant effect on soil pH since the organic material has large buffer capacity that if enough soil contains this component, soil pH will be relatively stable [18].

Table 3. Effect of dosing of zeolite on soil pH, N total, and c-organic

Dose of Zeolite (t ha^{-1})	pH	N total (%)	C-organic (%)
0	5.7a	0.37a	1.5a
2.5	5.78a	0.45b	1.73b
5	6.0b	0.50b	1.86b

Description: The numbers followed by the same letters show no significant difference in the DMRT test of 5%

Dose of zeolite and manure type gave a real effect to the increase of total N of soil, while the interaction between them gave no significant effect. Zeolite with a dose of 5 t ha^{-1} giving the highest total N total value, with increase of 35.13% compared with without zeolite. The increase in total soil N can occur because the zeolite content is capable of absorbing NH_4^+ [19].

Application of cow manure increased N total by 40.54% (Table 4). Manure can increase the total N-content in the soil. Organic material from manure plays a role in inhibiting nitrification, so that N elements are not easily lost so that it can be used by plants. Organic matter is important for retaining water for supplying and retaining nutrients [20]. Zeolite gave a real effect to c-organic soil (Table 3). Zeolite with a dose of 5 t ha^{-1} increased the c-organic by 24%. The use of amelioran in acid soils can increase the soil C organic [21]. Treatment of cow manure 5 t ha^{-1} yields the highest C-organic soil value which is 1.87%. Cow manure 5 t ha^{-1} treatment resulted the increase of C-organic soil by 26.35% (Table 4).

C-organic is positively correlated with total soil N, and CEC, this shows soil organic matter has a positive effect, when the content of organic matter is high then the value of N total is also high. C-organic has high CEC which is derived from functional groups of organic acids such as carboxyl and hydroxyl. When the clusters are hydrolyzed it will produce a lot of negative charges. The increase in negative charge due to the process will lead to an increase in CEC.

Table 4. Effect of giving different types of manure to N-total, and C-organic

Types of Manure	N total (%)	C-organic (%)
Control	0.37a	1.48a
Quail	0.41a	1.73b
Cow	0.52b	1.87b

Description: The numbers followed by the same letters show no significant difference in the DMRT test of 5%

Dose of zeolite and type of manure had significant effect to soil CEC (Table 5). An increase of value of soil CEC due to the applied zeolite based on laboratory analysis has been known that zeolite minerals have very high. Zeolites have high CEC values useful as cation adsorbers and binder and exchangers [22].

Table 5. Effect of Interaction of Zeolite Dosage and Manure on Soil CEC.

Dose of Zeolite (t ha^{-1})	Type of Fertilizer			Mean
	Control	Quail	Cow	
0	23.56 a	46.74 b	47.44 b	39.25
2.5	48.49 b	49.78 b	51.24 b	49.83
5	52.48 b	46.49 b	49.16 b	49.38
Mean	41.51	47.67	49.28	

Description: The numbers followed by the same letters indicate different yet unreal on the DMRT test 5%

Current study indicated soil CEC correlated positively with soil c-organic and N total of soil. Type of manure provides real effect to fresh weight of plant, manure gives high N nutrients that soybean crops can absorb N. High N uptake of nutrients can make maximum plants productivity, as plants growth and development will be optimal. Applying a single zeolite gave a significant effect on the nutrient uptake (Table 6). The highest N uptake in zeolite treatment with a dose of 5 t ha^{-1} is 9.54%. Zeolite's hollow structure channeled in all directions that it can store ammonium ions (NH_4^+) and other gases as well as limiting volatilization and washing N from fertilizer [22].

Type of manure gives real effect on N uptake of plants (Table 6), with the highest nutrient uptake in cow manure was 9.8%. The increase in cow manure was 48.93% compared with control treatment. Amount of N absorbed by plants is closely related to the resulting dry weight of the plant. The increase of N nutrient uptake can stimulate better plant growth (plant dry weight), and vice versa. Correlation test between CEC and N uptake of a plant indicates positive interaction. This represents that increase in soil CEC will be followed by increased N uptake of plants. In addition, there was a positive correlation between total N soil with N uptake by soybean crops.

Table 6. Effect of dosing of zeolite and types of manure on N uptake

Dose of Zeolite (t ha^{-1})	N uptake (%)	Types of Manure	N uptake (%)
0	6.22a	Control	6.58a
2.5	8.36ab	Quail	7.72b
5	9.54b	Cow	9.8b

Description: The numbers followed by the same letters indicate different yet unreal on the DMRT test 5%

Interaction between dose of zeolite and type of manure has a significant effect on a number of pods (Table 7). The highest increase of pod content was obtained at zeolite 5 t ha^{-1} with a type of cow manure which increased by 129% compared to treatment without manure and zeolite. Cow manure in addition to improve soil conditions is also able to supply nutrients required by plants [23]. Production component is determined by the number of pods and weight of the pod content. The higher is a value of the component, the higher is its productivity.

Table 7. Effect of Interaction of Zeolite Dose and Type of Manure on Number of Soy Pods Contents

Dose of Zeolite	Type of Manure			Mean
	Control	Quail	Cow	
0	102 a	124 c	193g	139
2.5	113 b	145e	193g	150
5	132d	187f	234 h	184
Mean	119	175	180	

Description: The numbers followed by the same letters indicates no significant difference in the DMRT test of 5%

Amount of pod content with N uptake has a positive correlation. At the time of pod filling process, nitrogen is highly required as it is the main constituent of protein in a seed. Sufficient nitrogen availability will provide high-quality seed formation [24].

Table 8 shows that zeolite and manure gave significant effect on seed weight. An increase of seed weight was obtained at 5 t ha^{-1} zeolite and 5 t ha^{-1} cow manure of 56.01% compared with zeolite and quail manure treatment. The highest seed weight is generated by application of cow manure. Cow manure has high N, P, K nutrients absorbed by plants in sufficient quantities.

Table 8. Effect of Interaction of Zeolite Dose and Type of Manure at Weight of whole seeds

Dose of Zeolite	Type of Manure			Mean
	Control	Quail	Cow	
0	316 a	345 c	416 d	359
2,5	332 b	449e	451 e	410
5	334d	458f	493 g	429
Mean	327	429	442	

Description: The numbers followed by the same letters indicates no significant difference in the DMRT test of 5%

There was a strong positive correlation between seed weight and N tissue. Amount of nitrogen absorbed by plants through the soil is initially buried in the stems and leaves, after forming pods, it is then collected in the pod skin, the older is the pod, then most of nitrogen (80-85%) is absorbed into the seeds.

4. Conclusion

An interaction between applying zeolite and manure increases the value of CEC, total pod content and weight of whole batch. Giving dose of zeolite by 5 t ha^{-1} with cow manure gave the highest yield, total pod content, and seed weight. Zeolite with dose of 5 t ha^{-1} gives highest influence indicated by increased N total of 35.13% and increase of N uptake by 53.37%. Applying manures gives effect to total soil N and N uptake with cow manures give the highest influence. Based on the results of this study, the application of zeolite and cow manures with dose 5 t ha^{-1} increases N total soil and soybean yield.

Acknowledgements

The authors wish to express their gratitude for PNPB research grant 2018 of Universitas Sebelas Maret, Indonesia and everyone involved in helping this content and for their valuable discussions.

References

[1] Sari P M, Aimon H, and Syofyan E 2014 Analysis Factors Affecting Production, Consumption and Import of Soybean in Indonesia *J. Eco* 3(5)
[2] Sinuraya M A, Barus A and Hasanah Y 2015 Respons Growth and Prodution of Soybean (Glycine max (L.) Meriil) to Concentration and Ways of Application of Liquid Organic Fertilizer *J. Agroeco* (562) pp 1721 – 1725 E-ISSN No. 2337- 6597
[3] Yuniati R 2004 Screening of Line Soybean Glycine max (L.) Merrill Tolerant to NaCl for Planting in Saline Land Makara Sains 8(1) pp 21-24

[4] Armiadi. 2009 Nitrogen Uptake by Biologically on Leguminosae Plants. Wartazoa 19 (1)
[5] Siregar and Hartatik 2010 Application of Organic Fertilizer in Increasing Efficiency of Anorganic Fertilizer on Paddy Field *Proc. Nat. Seminar*: BALITTAN
[6] Andrius J, Subacius B and Rasa Slinksiene. 2015. Adsorption of Potassium Ions on Natural Zeolite: Kinetic and Equilibrium Studies. Chemija 26(2): 69-78
[7] Suwardi 2009 Use of Zeolit to Improve Soil Characteristic and Increasing Agriculture Production. Departement of Agriculture: Jakarta. ISSN : 1411-6723
[8] Day P R 1965 Particle fractionation and particle size analysis. In: Black, C.A. et al. (Eds.), Methods of Soil Analysis, Part 1. Agronomy Monograph No. 9 American Society of Agronomy, Madison, WI, pp. 545–567
[9] Walkley A and Black I A 1934 An examination of the Degitjareff method for determining soil organic matter, and a proposed modification of the chromic acid titration method. *J. Soil Sci.* 37, 29–38
[10] Bundy L G and Bremner J M 1972 A single titrimetric method for determination of inorganic carbon in soils. *Soil Sci. Soc. Am. Proc.* 36 pp 273–275
[11] Richards L A 1954 Diagnosis and Improvement of Saline and Alkali soils. USDA Agricultural Handbook No. 60. US Government Printing Office, Washington, DC
[12] Jackson, M.L. 1967 Soil Chemical Analysis. Prentice-Hall of India Pvt. Ltd., New Delhi
[13] FAO (Food and Agriculture Organization) 2006 Plant Nutrition for Food Security: A guide for Integrated Nutrient Management. FAO, Fertilizer and Plant Nutrition Bullet in 16, Rome, Italy
[14] Tekalign T 1991 Soil, Plant, Water, Fertilizer, Animal Manure and Compost Analysis. Working Document No. 13. International Livestock Research Center for Africa (ILCA), Addis Ababa, Ethiopia.
[15] Landon J 1991 Booker Tropical Soil Manual: Handbook for Soil Survey and Agricultural Land Evaluation in the Tropics and Subtropics. John Wiley and Sons Inc., New York
[16] Surtinah 2013 Test on Nutrient in Sweet Corn (Zea mays saccharata) Compost. *J. Agri* 11 (1)
[17] Juliana E, Sariffudin, and Jamilah 2015 Use of Zeolit and Rice Charcoal on Contaminated Pb Paddy Field and Paddy Plants. *J. Agrotech* 3 (2) pp 703-709 ISSN: 2337- 6597
[18] Utami and Handayani 2003 Soil Chemical Characteristic of Entisol on Organic Agriculture. *J. Agri. Sci.* 10 (2) pp 63-69
[19] Sudirja R, Joy B, Rosniawaty S, Setiawan A, and Yunianto R 2016 Effect of Formula Urea-Zeolit-Activate Charcoal to pH, N-total, CEC and Pb residue in Contaminated Soil. *Soilrens* 14 (1)
[20] Grifal F F and Vance E D 2000. Influence of Soil Organic Matter on Forest Productivity. *New Zealand J. For. Sci* 30 (1) pp 169-205
[21] Jabri A M 2009 Increase Production of Crops with Zeolite. Sinar Tani Magazine, 7 January 2009. Indonesia Soil Research Centre, Indonesia Agriculture Research Centre
[22] Eghball B 2002 Soil properties as influenced by phosphorus- and nitrogen-based manure and compost applications. *J. Agron.* 94 pp 128-135
[23] Muzammil, Rusmawan D, and Asmarhansyah 2012 Effect of Nitrogen Dosage to Growth and Soybean Production in Ex- Lead Mining Area Central Bangka, Bangka Belitung. Agriculture Technology Centre in Bangka Belitung
[24] Nurlisan, Rasyad A, Yoseva S 2013 Effect of Anorganic Fertilizer to Growth and Soybean Production (Glycine max (L.) Merril). *J. Agrotech of Riau University*

IOP Conf. Series: Materials Science and Engineering **633** (2019) 012027 doi:10.1088/1757-899X/633/1/012027

Effect of arbuscular mycorrhiza and organic matter type toward growth and yield of sorghum on Oxisol Tuntang

V R Cahyani[1,3], J Muntoyib[2] and S Hartati[1]

[1] Department of Soil Science, Faculty of Agriculture, Sebelas Maret University, Indonesia 57126
[2] Department of Agrotechnology, Faculty of Agriculture, Sebelas Maret University, Indonesia 57126

Email: vitaratri@staff.uns.ac.id

Abstract. Utilization Oxisol as plant growth media for sorghum is still limited. The main problems of the cultivation in Oxisol are low pH, high soluble form of Al, Mn, Fe, low organic matter (OM) and lack of nutrients especially P. The purpose of the present research was to examine the effect of arbuscular mycorrhiza (AM) and OM type toward the growth and yield of sorghum on Oxisol Tuntang. The pot experiment was conducted in greenhouse using factorial completely randomized design with two factors of AM inoculation and OM type with three replications. The first factor was inoculation of AM: without and with AM inoculation. The second factor was OM type: without OM, elephant grass compost, cow dung compost, worm castings and goat manure. The inoculation of AM in interaction with OM significantly increased AM colonization, spore density, leaf number, dry weight, water uptake, P concentration, P uptake and grain mass. Interaction of AM with worm castings and interaction of AM with goat manure indicated the same higher effect on sorghum growth and yield, compared with the interaction of AM with elephant grass compost and interaction of AM with cow dung compost. The control treatment showed the lowest plant performance.

1. Introduction

Sorghum is one of important cereal crops in the world which placed fifth after rice, wheat, maize and barley [1]. Sorghum is also the second important crop in semi-arid tropics because of its tolerant nature for dry soil condition and low water requirement compared to other cereal crops such as maize and wheat [2]. The nutrient composition of sorghum was similar to rice and wheat, such as: 72.09% of carbohydrate (79.34% rice, wheat 76.31%), and 10.62% of protein (6.61% rice, wheat 10.33%) which make it very potential as substitute food source [3].

Most lands in the tropics have low fertility level and have limitations in their intensive use, especially Oxisols dry land (covering 22.5% of the total land area in the tropics) [4]. Oxisols are soil type with adequate physical properties for plant growth but has low chemical fertility such as high acidity and an imbalance nutrient content. Oxisols are easily eroded which brings P and K nutrient are easily bounded by Al, Mn and Fe and make these nutrients to be unavailable for plants that causing P and K deficiency. Al, Mn and Fe will dissolve easily in acidic conditions and they are toxic to plants [5].

Sorghum cultivation on Oxisol is limited by many limitation factors resulting in low sorghum productivity. The strategy to overcome the problems is by inoculating arbuscular mycorrhiza (AM) and applying organic matter (OM) to the soil. AM has mutualistic symbiosis with plant roots by its role in nutrients uptake especially P and also water uptake [8-10], while plants provide organic C for AM for its growth and development [11].

OM is able to support soil chemical fertility, stabilizing soil pH due to its buffer properties [6]. OM is also proven to improve the quality of soil biology by providing soil organic C [7]. The present study aimed to examine the effect of AM and OM type toward the growth and yield of sorghum on Oxisol Tuntang.

2. Materials and Methods

2.1 Soil preparation
Oxisol Soil from Tuntang, Semarang Regency, by geographical location of $7^{0}21'08''$ S - $110^{0}28'43''$ E was used as a planting medium. Soil samples were taken randomly from non-cultivated land then air-dried and sieved using a 2 mm diameter soil filter and put into a 15×25 polybag.

2.2 OM preparation
OM used was elephant grass compost, cow dung compost, worm castings and goat manure. Each OM was analyzed for its chemical characteristics before application.

2.3 AM inoculum source
The source of mycorrhizal inoculum used was AM culture with zeolite media which was dominated by the genus *Glomus* sp. with spore density of 182 – 238 100 g^{-1} media. AM density analysis was carried out by wet filtration and centrifugation method which continued by AM genus analysis using Melzer staining.

2.4 Pot experiment
This study used a completely randomized design with a factorial pattern. The first factor was AM (M) with two levels: without AM (M0) and with AM (M1) inoculation. The second factor was OM types (P) with 5 treatments: without OM input (P0), 248.16 g elephant grass compost (P1), 195.17 g cow dung compost (P2), 223.09 worm castings (P3) and 235.43 g goat manure (P4). From two factors above, 10 combinations of treatments were obtained then repeated three times to obtain 30 experimental units. Calculation of OM dosage was based on the requirement to achieve to the level of 2% of the total organic C content, by considering the total C content in soil and the total C in the each type of OM used in the present study.

2.5 OM incubation
OM was mixed with soil until it is homogeneous in a polybag. This medium was then moistened with tap water to about 70% of field capacity and incubated for 10 days before planting.

2.6 Planting and AM inoculation
Planting was done after 10 days of OM incubation by making a hole as deep as ± 3 cm filled with 2 seeds. The hole was covered with thin soil layer and then watered using a sprayer. After 7 days, thinning was done by leaving one plant per polybag. AM inoculation was done by making a hole as deep as ± 7 cm around and under the root of the plant and then added AM inoculum 25 g / pot culture (± 50 spores / pot culture). Soil moisture was managed around 70% field capacity during plant growth. Plants were harvested on 126 days after planting (DAP).

2.7 Variable observation
The observed variables were soil pH, AM infectivity (the percentage of root infection and spore density), AM effectiveness (plant height, shoot fresh and dry weight, root fresh and dry weight, flowering time, P concentration, P uptake) and the production of Sorghum grains (total grain weight and 100 grains weight).

3. Results and Discussion

3.1 Characteristics of Oxisol Tuntang and organic matters type

Oxisol Tuntang used as planting media in the present study based on the criteria of the assestment of soil analysis according to Balai Penelitian Tanah Indonesia (Indonesian of Soil Research Center, 2009) had slightly acidic pH, low total N, available P, available K and very low total organic C (Table 1).

The four kinds of OM used in this study had slightly alkaline pH (Table 1). The C/N ratio of all OM ranged from 15.11 – 16.03. According to Peraturan Menteri Pertanian Republik Indonesia (Regulation of the Minister of Agriculture of the Republic of Indonesia) No.70/Permentan/SR.140/10/2011, the minimal requirement for quality standart of C/N ratio for OM is around 15-25. Table 1 showed that elephant grass compost has the lowest total N and total P.

Table 1. Soil characteristics and organic matters analysis

Parameter	Soil	Different type of organic matters			
		EGC	CDC	WC	GM
pH H_2O	5.8	7.5	7.8	7.9	7.7
Total N (%)	0.12	1.26	1.51	1.33	1.31
Total P (%)	-	1.48	1.61	1.81	1.53
Available P (ppm)	1.82	-	-	-	-
Total K (%)	-	1.47	1.55	1.78	1.79
Available K (me%)	0.70	-	-	-	-
Organic C (%)	0.65	19.04	24.21	21.18	20.07
Ratio C/N	5.42	15.11	16.03	15.92	15.32

EGC (elephant grass compost), CDC (cow dung compost), WC (worm castings), GM (goat manure)

3.2 Soil pH

Based on ANOVA results, it was found that the OM treatment had a very significant effect on soil pH, while AM and its interaction with OM had no significant effects. Figure 1 shows that the treatments of all OM had statisticaly the similar effect in increasing pH to the range of 6.43 - 6.51.

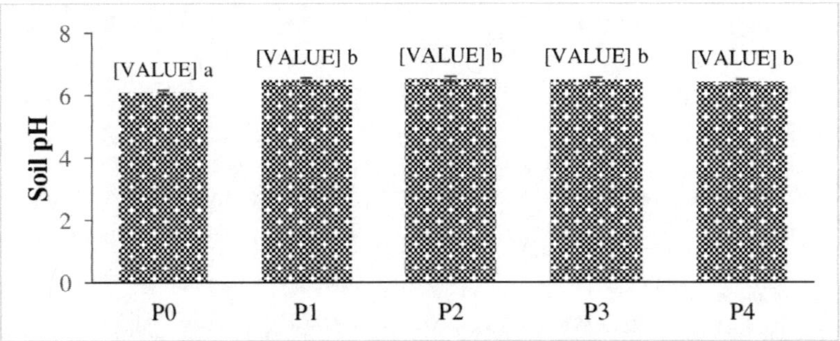

Figure 1. The effect of organic matters on soil pH. P0 (without OM), P1 (elephant grass compost), P2 (cow dung compost), P3 (worm castings), P4 (goat manure)

3.3 Mycorrhizal infectivity (percentage of root infection and spore density)

ANOVA results showed that the interaction of AM and OM had significant effect on root infection and spore density of AM. AM colonization in plant root and AM spores were found in all treatments with inoculation of AM and were not found in all treatment without AM (Figure 2).

The highest infection was obtained in the treatment of M1P3 (23.33%), followed by the treatment of M1P4 (20%), M1P1 (18.33%), M1P2 (15%) and the lowest M1P0 (11.67%). The highest spore density was indicated in the treatment of M1P3 (60.00 spores/100g soil), M1P1 (57.33 spores/100g soil) and M1P4 (56.33 spores/100g soil) followed by the treatments of M1P2 (50.67 spores/100g soil) and M1P0 (49.33 spores/100g soil). According to Gryndler et al. [12] and Leigh et al. [13], AM mycelium activity is influenced by the decomposition of organic matter through compounds released during the decomposition process and also by the secondary metabolites of microorganisms involved in the decomposition of organic matter.

Correlation analysis showed that spore density was positively correlated (r=0.991**) with root infection by AM. The higher percentage of AM infection was followed by higher spores density. This is in accordance with the report by Douds and Schenck [14] and Hindumathi and Reddy [15] which stated that sporulation increased along with increasing AM infection or colonization in the root system.

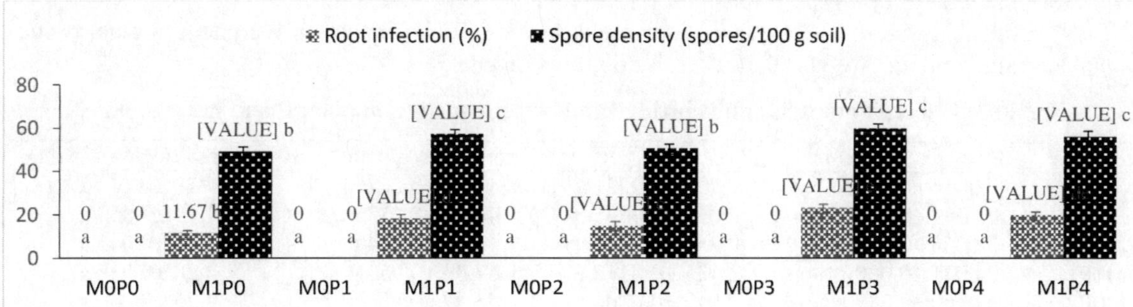

Figure 2. The effect of arbuscular mycorrhiza and organic matter on root infection and spore density. M0 (without AM), M1 (with AM inoculation), P0 (without OM), P1 (elephant grass compost), P2 (cow dung compost), P3 (worm castings), P4 (goat manure)

3.4 Plant growth

ANOVA results showed that AM and OM interactions significantly affected the number of leaves, shoot fresh and dry weight, root fresh and dry weight but did not significantly affect the plant height and flowering time (Table 2).

Table 2. The effect of arbuscular mycorrhiza and organic matter on plant growth of sorghum

T	PH (cm)	LN	SFW (g)	SDW (g)	RFW (g)	RDW (g)	FT (dap)
M0P0	215.9 ±5.18	14.7 ±0.58 a	263.9 ±7.98 a	94.1 ±3.28 a	94.2 ±3.87 a	35.8 ±1.83 a	84.3 ±3.06
M1P0	233.0 ±21.34	15.7 ±0.58 ab	292.4 ±6.79 bc	103.5 ±1.46 b	113.0 ±2.60 b	42.9 ±0.88 b	86.7 ±10.26
M0P1	260.4 ±16.85	16.7 ±0.58 b	302.4 ±10.17 cd	109.3 ±2.34 c	117.7 ±1.28 b	44.6 ±0.54 bc	84.0 ±1.73
M1P1	255.4 ±2.90	18.7 ±0.58 cd	316.7 ±5.17 e	112.9 ±2.69 c	123.8 ±1.44 c	47.0 ±1.01 c	84.7 ±1.53
M0P2	235.1 ±32.46	17.0 ±1.00 bc	283.8 ±6.78 b	101.1 ±1.45 b	131.1 ±3.97 d	49.7 ±0.98 d	80.0 ±5.00
M1P2	223.0 ±18.21	19.3 ±0.58 d	310.3 ±2.16 de	110.6 ±1.75 c	140.3 ±1.64 e	53.2 ±0.60 e	77.3 ±2.08
M0P3	246.8 ±29.90	17.3 ±0.58 bc	334.3 ±9.07 f	119.1 ±2.17 d	181.2 ±2.46 h	68.7 ±0.75 h	78.7 ±3.51
M1P3	240.9 ±28.18	22.7 ±0.58 e	345.5 ±5.51 f	123.1 ±2.98 d	198.8 ±2.80 i	75.4 ±1.04 i	78.0 ±1.73
M0P4	257.1 ±24.46	17.3 ±2.52 bc	336.6 ±2.51 f	120.9 ±1.40 d	164.7 ±2.65 f	62.5 ±1.66 f	81.7 ±3.51
M1P4	246.3 ±5.03	21.3 ±0.58 e	343.5 ±6.82 f	122.4 ±2.01 d	171.9 ±7.61 g	65.2 ±2.94 g	78.7 ±1.15

Numbers followed by the same letter show no significant difference in 5% DMRT, T (treatments), PH (plant height), LN (leave numbers), SFW (shoot fresh weight), SDW (shoot dry weight), RFW (root fresh weight), RDW (root dry weight), FT (flowering time) , M0 (without AM), M1 (with AM inoculation), P0 (without OM), P1 (elephant grass compost), P2 (cow dung compost), P3 (worm castings), P4 (goat manure)

As presented in Table 2, the lowest number of leaves obtained in the control treatment (14.7), while the highest number of leaves was obtained in the treatment of M1P3 (22.7) and M1P4 (21.3) that were 44.9 - 54.4% higher compared to control. The lowest shoot dry weight obtained in the control treatment (94.1 g), while the highest shoot dry weight obtained in the treatment of M1P3 (123.1 g), M1P4 (122.4 g), M0P4 (120.9 g) and M0P3 (119.1 g) that were 26.6 - 30.9% higher compared to the control. The lowest root dry weight obtained in control treatment (35.8 g), while the highest root dry weight was obtained in M1P3 (75.4 g) that was 110.9% higher compared to the control. The AM inoculation showed the higher shoot and root fresh weight than non-inoculation treatment, these results indicated the improvement of water uptake. The enhancement of water uptake was mediated by the role of AM fungal hyphae [16]. The highest shoot and root fresh weight was indicated by MIP3 which explained that worm castings gave higher effect compared with the other OM types. This result was supported by Hameeda et al [17].

The treatment of AM inoculation showed higher sorghum growth responses than without AM inoculation (Table 2). Correlation analysis showed that AM infectivity was positively correlated with the number of leaves (r=0.671**), shoot fresh weight (r=0.390*), shoot dry weight (r=0.371*), root fresh weight (r=0.264) and root dry weight (r=0.263). These results were similar with other research results that showed the roles of mycorrhiza for increasing plant growth [18-19].

3.5 P concentration, P uptake and sorghum yield

ANOVA results showed that AM and OM treatment had significant effect on P content in plant tissue, P uptake, grain yield and weight of 100 sorghum grains (Table 3).

Table 3. Effect of arbuscular mycorrhiza and organic matter on plant P and grain yield.

Treatment	P concentration (%)	P uptake (g pot^{-1})	Grain yield (g/plant)	100 grains weight (g/plant)
M0P0	0.0054 ±0.000062 a	0.701 ±0.0221 a	41.22 ±0.267 a	4.18 ±0.134 a
M1P0	0.0059 ±0.000045 b	0.857 ±0.0095 b	41.55 ±0.213 a	4.19 ±0.059 a
M0P1	0.0055 ±0.000089 a	0.853 ±0.0210 b	47.45 ±0.484 c	4.27 ±0.045 a
M1P1	0.0071 ±0.000075 e	1.135 ±0.0112 e	67.08 ±0.063 g	4.28 ±0.012 a
M0P2	0.0060 ±0.000045 b	0.910 ±0.0106 c	53.89 ±0.487 e	4.25 ±0.056 a
M1P2	0.0065 ±0.000136 d	1.073 ±0.0218 d	64.57 ±0.579 f	4.32 ±0.100 ab
M0P3	0.0059 ±0.000165 b	1.113 ±0.0216 de	44.56 ±2.246 b	4.24 ±0.088 a
M1P3	0.0076 ±0.000091 f	1.502 ±0.0130 f	70.64 ±0.599 h	4.43 ±0.067 bc
M0P4	0.0063 ±0.000107 c	1.146 ±0.0263 e	51.63 ±0.260 d	4.25 ±0.044 a
M1P4	0.0081 ±0.000327 g	1.514 ±0.0723 f	71.79 ±1.376 h	4.50 ±0.076 c

Numbers followed by the same letter show no significant difference in 5% DMRT, M0 (without AM), M1 (with AM inoculation), P0 (without OM), P1 (elephant grass compost), P2 (cow dung compost), P3 (worm castings), P4 (goat manure)

As shown in Table 3, the lowest P concentration was obtained in the control treatment (0.0054%), while the highest one was obtained in M1P4 (0.0081%) that was 50% higher compared to control. The lowest P uptake was obtained in the control treatment (0.701 g / plant), whereas the highest P uptake was obtained in M1P4 (1.514 g/plant) and M1P3 (1.502 g/plant) that were 114.3 - 115.9% higher compared to control.

The lowest grain yield was obtained in the control treatment (41.22 g), while the highest grain yield was obtained in the treatment of M1P4 (71.79 g) and M1P3 (70.64 g) which were 71.4 - 74.2% higher compared to control. The highest 100 grains weight obtained in M1P4 (4.50 g) which was 7.7% higher compared to control.

Table 3 shows that the use of AM gave higher P concentration, P uptake and grain yield than without AM. AM can accelerate the movement of P to the root by increasing P affinity, increasing phosphatase activity and increasing phytohormone production which has an impact on changes in root functions to increase nutrient uptake [20]. Correlation analysis showed that P concentration and P uptake were positively correlated with grain yield and 100 grains weight. Phosphorus is needed for plant growth in the whole life cycle including vegetative to generative phases. Phosphorus nutrient plays important role in storing and transferring energy (ADP and ATP) obtained from photosynthesis and carbohydrate metabolism for the process of growth, grain production and accelerates maturity [21].

4. Conclusion

Among the four types of OM that were applied in this study, the highest effect of OM on AM infectivity in plant root was worm castings, followed by goat manure, elephant grass compost and cow dung compost. Three types of organic matter namely worm castings, elephant grass compost and goat manure showed similar effect on AM spore density, while cow dung compost showed a lower effect.

The treatment of AM inoculation showed better sorghum growth than without AM inoculation, which indicated by the variables of the number of leaves, shoot dry weight and root dry weight. The treatments of worm castings and goat manure showed higher plant growth compared with cow dung and elephant grass compost. The AM and OM interaction showed higher results on shoot and root fresh weight which also meant higher plant water uptake. The highest P uptake in sorghum was shown by goat manure and worm castings treatment followed by elephant grass compost and cow dung compost. The treatments of goat manure and worm castings showed higher grain yield compared with elephant grass compost and then following by cow dung compost. The main findings of the present study indicated that the interactions of AM with worm castings and AM with goat manure resulted in higher effectiveness on sorghum growth, P uptake and grain yield compared with the interactions of AM with elephant grass compost and AM with cow dung compost.

References

[1] Leslie J F 2014 *Mycotoxins in the sorghum grain chain* (Mycotoxin reduction in grain chains) ed Leslie J F and Logrieco A (Oxford: John Wiley & Sons, Inc) chapter 20 pp 282-295
[2] Krupa K N, Dalawai N, Shashidhar H E, Harinikumar K M, Manojkumar H B, Bharani S and Turaidar V 2017 *Int. J. Pure App. Biosci.* **5** 221-237
[3] USDA (2018) *USDA Food Composition Databases* https://ndb.nal.usda.gov/ndb/
[4] Lal R 1997 *Soil of the tropics and their management for plantation forestry* (Management of soil, nutrient and water in tropical plantation forest) ed Nambiar E K S and Brown A G (Canberra: Australian Centre for Agricultural Research) chapter 4 pp 97-123
[5] Fölster H and Khanna P K 1997 *Dynamics of Nutrient Supply in Plantation Soils* (Management of soil, nutrient and water in tropical plantation forest) ed Nambiar E K S and Brown A G (Canberra: Australian Centre for Agricultural Research) chapter 10 pp 339-371
[6] Fulekar M H 2010 *Composting* (Environmental biotechnology) ed Fulekar M H (Boca Raton, Florida: Taylor and Francis Group) chapter 10 pp 285-297
[7] Guo L, Wu G, Li Y, Li C, Liu W, Meng J, Liu H, Yu X and Jiang G 2016 *Soil & Tillage Research* **156** 140-147
[8] Ruiz-Lozano J M and Azcon R 1995 *Physiologia Plantarium* **95** 472-478
[9] Guo W, Zhao R, Zhao W, Fu R and Guo J 2013 *Appl. Soil Ecol.* **72** 85-92
[10] Smith S E, Jakobsen I, Gronlund M and Smith F A 2011 *Plant Physiol.* **156** 1050-1057
[11] Cahyani VR 2009 Proc. of Int. Seminar of Upland for Food Security. Faculty of Agriculture Jenderal Soedirman University, Purwokerto, Indonesia. pp254-260.
[12] Gryndler M, Hrselova H, Cajthaml T, Havrankoca M, Rezacova V, Gryndlerova H and Larsen J 2009 *Mycorrhiza* **19** 255-266
[13] Leigh J, Fitter A H and Hodge A 2011 *FEMS Microbiol. Ecol.* **76** 428-438
[14] Douds D D and Schenck 1990 *New Phytol.* **116** 621-627
[15] Hindumathi A and Reddy B N 2011 *J. Mycol. Plant Pathol.* **41** 537-542
[16] Jiang X Y and Huang Y 2003 *Ecol. Environ.* **12** 353-356
[17] Hameeda B, Harini G, Rupela O P and Reddy G 2007 *Afr. J. Biotechnol.* **6** 009-012
[18] Raju P S, Clark R B, Ellis J R and Maranville J W 1990 *Plant and Soil* **121** 165-170
[19] Auge R M 2004 *Can. J. Soil Sci.* **84** 373-381
[20] Orcutt M D and Nilsen E T 2000 *Nutrient defiency stress and plant growth and development* (The physiology of plants under stress: soil and biotic factors) (New York: John Wiley & Sons, Inc) chapter 3 pp 46-107
[21] Campbell C A, Lafond G P and Zentner R P 1991 *Fertilization and Legumes Influence Spring Wheat Yield Trends* (Better Crops with Plant Food Vol. 77 No. 4) ed Armstrong D L (Georgia, USA: Potash & Phosphate Institute (PPI)) pp 6-15

IOP Conf. Series: Materials Science and Engineering **633** (2019) 012028 doi:10.1088/1757-899X/633/1/012028

Postharvest application of an edible coating based on chitosan and gum Arabic for controlling respiration rate and vitamin C content of chilli (*Capsicum frustecens* L.)

N Muthmainnah, Suratman, and Solichatun

Department of Biology, Faculty of Mathematics and Natural Sciences, Universitas Sebelas Maret, Ir. Sutami Street No.36A, Surakarta, Central Java, Indonesia 57126

Email : suratman@staff.uns.ac.id

Abstract. Chili (*Capsicum frutescens* L.) is the most important vegetable cum spice grown in tropical and sub tropical regions of the world. The post-harvest handling of chilli must be done well so that the quality is maintained. One way of handling post-harvest is to use edible coating such as chitosan and gum Arabic. The research was aimed to find out the effects of chitosan and gum Arabic coating to control of respiration rate and vitamin C content of post-harvest chilli. The research was conducted using a Completely Randomized Design (CRD) factorial consisting of two factors. The first factor consisted of various concentration of chitosan (0%, 1%, 1.5%, 2%) and the second factor consisted of various concentration of gum Arabic (0%,5%,10%,15%) with three replications for each treatment. The coated fruits were then stored at room temperature (28°C) for 15 days. The observed data included respiration rate and vitamin C content. Data were analyzed using Analysis of Varians (ANOVA) and followed by Duncan's Multiple Range Test (DMRT) at a 5% confidence level. The results showed that the application of chitosan and gum Arabic coating significantly affected the respiration rate and maintained vitamin C content.

1.Introduction

Chili (*Capsicum frutescens* L.) is the most important vegetable cum spice grown in tropical and sub tropical regions of the world. Chilli fruits are widely used as a flavoring and a spice throughout the world due to their color, taste, aroma and characteristic pungency [1,2]. The main factor associated with chilli postharvest shelf-life, particularly in tropical regions where the temperature is high, is increased respiration which results in faster fruit ripening and deterioration of fruit quality. The chilli fruits are usually used for fresh consumption and raw material for the food processing industry. Threfore, chilli fruits must be available in fresh condition at any time because it cannot be stored for a relatively long time. This causes demand and the need for chilli is always high at all times. The quality of chilli will drop after 2-3 days if stored at room temperature. Chilli can last for 10-20 days if stored in a cold temperature room (5-7 ° C) and humidity 90-95% [3]. Storage of chilli fruits in cold temperatures will certainly require equipment that is not cheap, especially if it has to be marketed in areas far from the harvesting location. Therefore, post harvest handling is needed to extending postharvest life and keeping production costs low.

One of such promising postharvest treatment used for extending the market life of lively respiring horticultural produce is edible coating technology. Edible coatings indicate the application of commercial food grade waxes or films to protect the loss of natural glossiness during the postharvest period. The best part of the edible coating is that these are edible, healthy and biodegradable in nature, unlike other chemical postharvest treatments which leave a residue [4,5].

The use of edible coating is another safe method of prolonging storability of perishable crops [6]. Edible coatings are mostly derived from biological sources. Application of edible coating on fresh fruits is able to reduce quality changes and slow down quantity losses, for example, moisture loss by controlling and modifying the internal atmosphere of the individual fruits [7]. The edible coating material forms thin layer on the surface of food and gives a selective barrier against moisture, oxygen, and carbon dioxide [6]. If a film or coating with the appropriate permeability is chosen, a controlled respiratory exchange can be established and thus the preservation of fresh produce can be achieved [8]. Many materials such as polysaccharides, proteins, essential oils, may economically serve as edible coatings [9,10,11]. Chitosan and gum Arabic are classified as polysaccharides and can be used as an edible coating materials.

Chitosan has strong antimicrobial and antifungal activities that could effectively control fruit decay [12,13]. It could easily form coating on fruit and vegetable, and the respiration rate of fruit and vegetable was reduced by adjusting the permeability of carbon dioxide and oxygen [14]. Considering the superior properties of chitosan, it has been successfully used in many postharvest fruits and vegetables, such as grape, strawberries, jujube, fresh-cut lotus root and guava[11,15,16,17,18].

Gum Arabic is a soluble fiber obtained from acacia trees grown in sub-Saharan Africa (*Acacia senegal* (L.) Willd. or *Acacia seyal* Delile). Gum Arabic has been widely used as an emulsifier, thickening agent and film forming [19]. Coating with gum Arabic has been reported to be effective in increasing the shelf life of cucumber and tomatoes [20,21].

Chitosan and gum Arabic coating can maintain the permeability of CO_2 and O_2 of banana which will form the desired modified atmosphere that can increase the shelf life [22]. The use of edible coatings combination of chitosan and gum Arabic has not been applied to chilli fruits yet. Therefore in the present study was mainly focused to examine the effect of treatment variations in chitosan and gum Arabic concentration to improving shelf life of chilli fruits, especially for controlling respiration rate and vitamin C content.

2. Materials and Methods

2.1 Plant material preparation

Chilli fruits (*Capsicum frutescens* L.) at commercial maturity were harvested from an agricultural nursery in Klaten, Central Java, Indonesia and transported to the Universitas Sebelas Maret within 2 hours. Chilli fruits that are of uniform in size, shape, and color and are also free from any blemish or defects were selected and then washed with distilled water.

2.2 Experimental design

The research was conducted using a Completely Randomized Design (CRD) factorial consisting of two factors. The first factor consisted of various concentration of chitosan (0%, 1%, 1.5%, 2%) and the second factor consisted of various concentration of gum Arabic (0%,5%,10%,15%) with three replications for each treatment.

2.3 Preparation of edible coating

Edible coating solution from chitosan (1%, 1.5%, 2%) were prepared by dissolving 1 g, 1.5 g and 2 g of chitosan powder in 100 ml of 1% acetic acid. Edible coating from gum Arabic (5 %, 10 %, 15%) were prepared by dissolving 5 g, 10 g and 15 g of gum Arabic powder in 100 ml aquades. The solutions were stirred with low heat (40 ◦C) for 60 minute on a magnetic stirrer, then filtered to remove any undissolved impurities using a vacuum flasks [22].

IOP Conf. Series: Materials Science and Engineering **633** (2019) 012028 doi:10.1088/1757-899X/633/1/012028

2.4 Application of chitosan and gum Arabic coating

The coating solution was applied uniformly on the whole surface of fruits. Chili fruits were dipped in the chitosan solution (0%; 1%; 1.5% and 2%) for 5 minutes and then allowed to dry for 30 min at room temperature (28^0 C). After the chilli fuits were dry, the fuits then dipped in the gum Arabic solution (0%, 5%, 10% and 15%) for 5 minutes. After that, the fruits were dried again to ensure that the coating material was applied thoroughly on surface of the fruits. The fruits were then stored at room temperature (28°C) for 15 days for further analysis.

2.5 Characterization of applied edible coating

2.5.1 Analysis respiration rate

The respiratory rate was determined using a Lutron GCH 2028 carbon dioxide analyzer (Lutron Electronic Enterprise Co., Ltd., Taipei, Taiwan). Three coated chilli fruits (approximately 30 g) were placed in glass jar and incubated at room temperature (28°C) for 1 h. The glass jar had air-tight screw caps and rubber septum to allow headspace sampling. Percentages of carbon dioxide were recorded when the readings were stabilized. The results were calculated and expressed as ml CO_2/kg day [23].

2.5.2 Analysis of vitamin C content

The vitamin C content of the samples was estimated based on absorbance in conjunction with spectrophotometer at 264 nm [24]

2.6 Statistical analysis

The data was expressed as the means value and standard deviations. The IBM SPSS software 17.0 was used to perform analysis of variance (ANOVA). Significance differences between the means were set at a 5% confidence level. Duncan's Multiple Range Test (DMRT) was used to compare the mean values of multiple treatments.

3. Result and Disccussion

3.1 Respiration Rate

Fruits and vegetables are living commodities and their rate of respiration is of key importance to maintenance of quality. It has been commonly observed that the greater the respiration rate of a fruit, the shorter the postharvest life [25]. In the present study, the respiration rate of chilli fruits decreased until the end of the storage day. Chitosan and gum Arabic treated chilli fruits exhibited a significantly lower respiration rate than control during the storage period (Table 1).

Edible coating has the potential of reducing respiration rate of fruits. This is associated with a decrease in the metabolic rate. At 15 day, respiration rate decreased until 40-62%. Application of chitosan as edible coating showed that the lowest decreased respiration rate occured on the use of chitosan 1.5%, with respiration rate of 51.78% while application of gum Arabic showed that the lowest decreased respiration rate occured on the use of gum Arabic 10%, with respiration rate of 41.38%.

In the combination of chitosan and gum Arabic as coating material, the lowest decrease in respiration rate occurred in the treatment of chitosan 1.5% and gum Arabic 10% with decreased in respiration rate of 40% that capable of producing thick and effective layers to reduce the respiration rate compared than other treatments. In the treatment of gum Arabic 10%, the results were not significantly different with those obtained in the use of chitosan 1.5% and gum Arabic 10%.

The coating materials which play more important role in inhibiting the respiration rate of chilli fruits is application of chitosan 1.5% and gum Arabic 10%. The presence of these coating materials were able to maintain the surface of the chili fruits so that the outside air cannot enter through the chili pores. Similar report has showed that the application of gum Arabic 10% in coating of tomato fruits has been found to delay the ripening procces by slowing down the rate of respiration. The other result from Scanning Electron Microscope (SEM) observations on bananas showed that coating material such as chitosan and gum Arabic can cover the cuticle and block pores on the surface of the fruits. Therefore, the amount of oxygen that can be accessed by cells to be reduced and the amount of cabondioxide increases so that the respiration process can be inhibited [21]. Reduction of the respiration rate as a result of chitosan coating also has been reported in longan fruits [26] and plum fruits coated [27].

Table 1. The effect of different edible coating combinations of chitosan and gum Arabic on respiration rate of chilli fruits.

Treatment		Respiration rate (ml CO2/kg·day)		Decreased respiration rate (%)
Concentration of chitosan (%)	Concentration of gum gum Arabic (%)	Day 0	Day 15	
0	0	93.43^{a}	36.23^{a} ± 3.13	60.75^{efg} ± 4.09
	5	96.35^{a}	43.24^{bcd} ± 2.37	55.12^{cdef} ± 2.46
	10	95.32^{a}	55.87^{e} ± 4.62	41.38^{a} ± 4.85
	15	102.09^{ab}	40.51^{abc} ± 3.73	60.31^{efg} ± 3.65
1	0	101.98^{ab}	38.35^{ab} ± 2.48	62.40^{g} ± 2.45
	5	102.22^{ab}	43.29^{bcd} ± 0.65	57.66^{defg} ± 0.64
	10	94.26^{a}	43.53^{bcd} ± 1.43	53.82^{cde} ± 1.52
	15	95.09^{a}	40.88^{abc} ± 2.15	57.01^{cdefg} ± 2.27
1.5	0	95.29^{a}	45.95^{cd} ± 4.25	51.78^{bcd} ± 4.46
	5	101.85^{ab}	46.13^{cd} ± 2.80	54.70^{cdef} ± 2.83
	10	94.11^{a}	55.92^{e} ± 4.97	40.57^{a} ± 5.46
	15	99.96^{ab}	42.30^{bcd} ± 0.85	57.68^{defg} ± 0.84
2	0	94.38^{a}	41.74^{bc} ± 0.44	55.77^{cdefg} ± 0.47
	5	110.13^{b}	40.77^{abc} ± 1.29	61.61^{ef} ± 8.22
	10	95.59^{a}	47.50^{d} ± 4.05	50.31^{bc} ± 4.21
	15	102.04^{ab}	54.93^{e} ± 2.08	46.17^{ab} ± 2.04

Note = Different lowercase letter indicates significant difference at $p < 0.05$ in each storage day; different uppercase letter indicates significant difference at $p < 0.05$ in each treatment

3.2 Effect of chitosan and gum Arabic coating on vitamin C content

Vitamin C, also known as ascorbic acid, is easily destroyed by oxidation especially at high temperature and easily lost during processing for preservation. Based on the results of ANOVA, various variations of chitosan and gum Arabic treatment effected on vitamin C content significantly (table 2). A continuous decline in ascorbic acid of chilli fruit throughout the storage period was observed during storage time. After 15 days storage, vitamin C content was decreased until 72-98%.

Decrease in vitamin C content can be caused due to the occurrence of oxidative processes and the ongoing process of respiration. Decrease in vitamin C content because oxidative processes can occur due to stimulation by light, oxygen, heat, peroxide and enzymes [28]. The higher the oxidative process takes place, the more vitamin C is damaged. In this study, degradation of vitamin C was caused by the presence of contact with surrounding air. The mechanism is the presence of oxygen molecules causing ascorbic acid monoanion to produce ascorbic anion radicals and H_2O followed by dehydro formation of ascorbic acid and hydrogen peroxide. Dehydro of ascorbic acid (L-dehydroaskorbat acid) is a form of oxidation of L-ascorbic acid which still has activity as vitamin C. L-dehydroaskorbat acid which is very unstable can change to 2.3-L-diketogulonat (DKG) which has no activity vitamin C again. If DKG has been formed, it will reduce and even eliminate the vitamin C content available [29]. The process of respiration that is still ongoing during storage in chilli fruits also causes a decrease in vitamin C content.

The process of respiration will produce water then water evaporates from the fruit tissue. Evaporation of water causes the cell structure that originally intact to wither so that the enzyme ascorbate oxidase will be released by direct contact with ascorbic acid. As a result, vitamin C content is damaged [30]. Similar reports have been published for other horticultural crops (such as banana, green peppers, red peppers, etc) indicate that loss of vitamin C is also accelerated by longer storage time [31]. Our results are also indicated that concentration of vitamin C in chilli fruits declined gradually during storage at room temperature (28 °C) and with increasing duration of storage time after 15 days storage.The gradually declining in ascorbic acid in coated fruit with gum Arabic suggests that the coating slowed down but did not prevent the synthesis of ascorbic acid during ripening [5,21]. Similar gradually slowing down of ascorbic acid content during ripening has been reported in the the application of chitosan as coating of strawberry fruits after 9 days of storage, longer than uncoated fruits [32].

Table 2. The effect of different edible coating combinations of chitosan and gum Arabic on vitamic c content of chilli fruits.

Treatment		Vitamin C content (mg/100ml)		Decreased Vitamin C content(%)
Concentration of chitosan (%)	Concentration of gum Arabic (%)	0 Day	15 Day	
0	0	15.30^{b}	1.54^{c} ± 0.000	89.88^{h} ± 0.10
	5	14.82^{ab}	0.88^{b} ± 0.000	94.06^{i} ± 0.02
	10	14.96^{ab}	0.57^{ab} ± 0.011	96.17^{ij} ± 0.11
	15	15.10^{ab}	1.82^{c} ± 1.020	88.49^{h} ± 6.78
1	0	15.30^{b}	0.18^{a} ± 0.029	98.88^{j} ± 0.19
	5	15.00^{ab}	2.78^{de} ± 0.000	81.47^{fg} ± 0.03
	10	14.98^{ab}	3.65^{f} ± 0.135	75.63^{e} ± 0.88
	15	14.79^{a}	6.76^{j} ± 0.000	54.29^{a} ± 0.03
1.5	0	15.04^{ab}	4.19^{g} ± 0.111	72.12^{d} ± 0.10
	5	14.99^{ab}	6.09^{i} ± 0.000	59.37^{b} ± 0.05
	10	15.02^{ab}	2.47^{d} ± 0.009	83.55^{g} ± 0.05
	15	15.10^{ab}	5.30^{h} ± 0.040	64.31^{c} ± 1.15
2	0	14.79^{a}	2.92^{de} ± 0.011	80.27^{f} ± 0.02
	5	14.85^{ab}	0.98^{b} ± 0.061	93.38^{i} ± 0.43
	10	15.00^{ab}	0.87^{b} ± 0.058	94.17^{i} ± 0.36
	15	14.79^{a}	2.99^{e} ± 0.029	79.80^{f} ± 0.18

Note = Different lowercase letter indicates significant difference at $p < 0.05$ in each storage day; different uppercase letter indicates significant difference at $p < 0.05$ in each treatment

4. Conclusion

The application of chitosan and gum Arabic as edible coating combination could be an alternative in the preservation of fruits and vegetables. The combination of chitosan and gum Arabic has been able to maintain respiration rate and gradually slowing down vitamin C content of chilli fruits for 15 day storage.

Acknowledgment

This study was partly supported by a grant from Universitas Sebelas Maret (UNS) through Research Group of Plant Biomaterial, Faculty of Mathemathics and Natural Sciences (No. Grant: 543 /UN27.21/PP/2018)

References

[1] Barbero GF, Liazid A, Azaroual L, Palma M and Barroso CG 2014 *Int. J. Food Prop.* **19** 485–493
[2] Giuffrida D, Dugo P, Torre G, Bignardi C, Cavazza A, Corradini C and Dugo G 2014 *Food Res. Int.* **65** 163–170
[3] Syukur M, Yunianti R and Dermawan R 2016 *Cultivation of Chilli Fruits, Harvesting Every Day* (Jakarta : Penebar Swadaya) [in Indonesian]
[4] Kumar R and Kapur S 2016 *Int. Journ. of Sci. Tech. Eng.* **2 (11)** 694-697.
[5] K Prasad K, Guarav AK, Preethi P and Neha P 2018 *Acta Scientific Agriculture* **2 (5)** 55-64.
[6] Dhall RK 2013 *Critical Reviews in Food Science and Nutrition* 53 **(5)** 435–450.
[7] Turhan KN 2009 *International Controlled and Modified Atmosphere Research Conference* **876** 299-305
[8] Ayranci E and Tunc S 2003 *Food Chemistry* 80 **(3)** 423–431
[9] Vanzela ESL, Nascimento PD, Fontes EAF, Mauro MA and Kimura M 2013 *Food Sci. Technol.* **50** 420-425.
[10] Lima MA, Cerqueira MA, Souza BWS, Santos ECM and Teixeira JA 2010 *J. Food Eng.* **97** 101-109.
[11] Santos NSD, Aguiar AJA, Oliveira CED, Sales CVD and Silva SDM 2012 *Food Microbiol.* **32** 345-353.
[12] Aider M 2010 *Food Sci. Technol-LEB* **43** 837-842
[13] Shiekh RA, Malik MA, Al-Thabaiti SA and Shiekh MA 2013 *Food Sci. Technol. Res.* **19(2)** 139-155.
[14] Elsabee MZ and Abdou ES 2013 *Eng C Mater. Biol. Appl.* **33** 1819-41
[15] Perdones A, Sánchez-González L, Chiralt A and Vargas M 2012 *Postharvest Biol. Tech.* **70** 32-41.
[16] Yu YW, Zhang SY, Ren YZ, Li H and Zhang XN 2012 *J. Food Eng.* **113** 408-414
[17] Xing Y, Li X, Xu Q, Jiang YH and Yun J 2010 *Innov. Food Sci. Emerg.* **11** 684-689
[18] Hong K, Xie J, Zhang L, Sun D and Gong D 2012 *Scientia Horticulturae* **144** 172-178
[19] Motlagh S, Ravines P, Karamallah KA and Ma Q 2006 *Food Hydrocolloids* **20(6)** 848-854
[20] Al-Juhaimi F, Ghafoor K and Babiker EE 2012 *Pakistan Journal of Botany* **44(4)** 1439–44.
[21] Ali A, Maqbool M, Anderson PG and Zahid N 2013 *Postharvest Biol.Tech.* **76** 119-124
[22] Maqbool M, Ali A, Alderson PG, Zahid N and Siddiqui Y 2011 *J. Agric. Food Chem.* **59** 5474-82
[23] Azarakhsh N, Osman A, Ghazali HM, Tan CP and Mohd Adzahan N 2012 *Int. Food Res. J.* **19(1)** 279-285
[24] AOAC (Association of Official Analytical Chemist) 1990 *Official Method of Analyst* (15th Ed) (Washington DC : AOAC)
[25] Aked J 2002 Maintaining the Postharvest Quality of Fresh Fruits and Vegetables *Fruit and Vegetable Processing: Improving Quality* ed W Jongen (Cambridge : Woodhead Publishing) pp 119-149
[26] Lin MG, Lasekan O, Saari N and Kahirunniza-Bejo S 2017 *Journal of Food Quality* **2017** 1-11
[27] Bal E 2013 *J. Agr. Sci. Tech.* **15** 1219-30
[28] Chitravathi K, Chauchan OP and Raju PS 2015 *Food Packaging and Shelf Life* **45** 1-9
[29] Sine HMC 2013 *Partner* **20(2)** 165-171 [in Indonesian]
[30] Rahayu ES and Pribadi P 2012 *Biosaintifika* **4(2)** 89-97 [in Indonesian]
[31] Lee SK and Kader AA 2000 *Postharvest Biology and Technology* **20** 207-220
[32] Petriccione M, Mastrobuoni F, Pasquariello MS, Zampella L, Nobis E, Capriolo G and Scortichini M 2015 *Foods* **2015**, 4, 501-523

IOP Conf. Series: Materials Science and Engineering **633** (2019) 012029 doi:10.1088/1757-899X/633/1/012029

Color and texture analyses of meatballs made from beef, pork, rat, dog meats, and their mixtures

F R Tathma,T Wibowo, I M Taufik and M Cahyadi
Department of Animal Science, Faculty of Agriculture, Universitas Sebelas Maret, Surakarta 57126, Indonesia

E-mail: mcahyadi@uns.ac.id

Abstract. The aim of this study was to evaluate color and texture of meatballs made from beef, pork, rat, dog meats, and their mixtures. A total of 32 meatballs have been made and they were grouped into eight treatments containing four meatballs, respectively. Color intensity consisting of lightness(L*), redness (a*), and yellowness (b*) has been tested using chromameter CR-400. Additionally, texture of meatballs consisting hardness, gumminess, springiness, chewiness was analyzed using AMETEX test and calibration instrument. The data obtained from this study was furthermore analyzed using one-way analysis of variance (ANOVA) and the pairwise differences among treatments were tested by Duncan's multiple range test (DMRT). The results showed that L*, a*, b* were significantly different among treatments ($P<0.01$). The L* value of dog meat was different with beef, pork, and rat meat. Also, the a* value of beef was similar to pork and significant different with dog and rat meats, while b* value of dog meat was different with other meats. The L*, a*, b* scores in mixture samples consistently represented the average value of color scores of samples consisting various species. Moreover, texture analysis indicated highly significant effect of treatments on hardness, gumminess, springiness, and chewiness ($P<0.05$). Highest hardness, gumminess, and chewinessscores were found in meatballs made from beefand rat, respectively. In addition, springiness of meatballs was relatively similar among treatments. This study concluded that meatballs made from different species and mixtures had different color and texture. Further study should be conducted to test whether color and texture analysis could be utilized for checking meat-based food adulteration and halalness.

1. Introduction

Nowadays, there are many meat processed products derived from chopped meat, one of them is meatball. Meatball is one of popular meat processed products in the world [1]. According to SNI 1-3818-1995, meatball is round shaped food derived from meat mixture (not less than 50%) and starch or other cereals with/out allowed additional component.Meatball is widely spread and favored in Indonesia. Several factors affectingconsumer preferencesof meatball are appearance, juiciness, taste and texture [2]. Texture includes hardness, springiness, cohesiveness, gumminess and chewiness[3]. Hardnessis the most important factor determining meatball quality [4]. Texture identification is performed by several ways [5]. The most common method is texture profile analysis (TPA) [3]. Numbers of texture assessment have previously been established [6].

Color of food product also determines consumer choice because it is one of freshness indicators which is able to be assessed bychromameter test.Differences of color intensity among species have been previously reported due to different concentrations of myoglobin [7]. Beef contains highest myoglobin concentration followed by lamb, and pork [7]. Additionally, texture and meat colorevaluations have been reported to describe transformation of meat color during storage [8]. Generally, species identification in meat and meat processed products is conducted by DNA analysis of mitochondrial genome using polymerase chain reaction (PCR) [9,10,11]. However, different color intensity among meats from different animal species may be useful to check species origin of meat based products, especially meat processed products such as meatball, sausage, and etc. Utilizations of texture and color analyses to distinguish and to trace meat origin in the processed products have not been reported elsewhere.Thus, the aim of this study was to evaluate color and texture of meatballs made from beef, pork, rat, dog meats, and their mixtures.

2. Materials and Methods

2.1.Meatball production

Meatballs were made by using meats from different animal species. They were beef (S), pork (B), and dog meats (A) boughtfrom traditional market, whereas rat (T) meat was obtained from wild rat in the rice field. Right thigh muscles were selected to be used as main raw material to produce meatballs. A total of 32 meatballs were made that divided into eight kinds of meatball based on the origin of meat used, i.e. beef, dog, pork, rat, beef-dog, beef-pork, beef-rat, and beef-dog-pork-rat meatballs, respectively. The ingredients and method to produce meatballs were according previous study [10]. The proportion of meat in the meatball was set according to Table 1.

Table 1. Ingredients to produce meatball from different animal species

Ingredients	S(g)	A(g)	B(g)	T(g)	SA(g)	SB(g)	ST(g)	SABT(g)
Beef (S)	100	0	0	0	50	50	50	25
Dog meats (A)	0	100	0	0	50	0	0	25
Pork (B)	0	0	100	0	0	50	0	25
Rat (T)	0	0	0	100	0	0	50	25
Garlic	2	2	2	2	2	2	2	2
Tapioc flour	15	15	15	15	15	15	15	15
Salt	2	2	2	2	2	2	2	2
Pepper	1	1	1	1	1	1	1	1
Total	120	120	120	120	120	120	120	120

S is beef meatball; A is dog meatball; B is pork meatball; T is rat meatball; SA is meatball containing beef and dog meat; SB is meatball containing beef and pork; ST is meatball containing beef and rat meat; and SABT is meatball containing meat from four species.

2.2.Texture and color analyses

Hardness, cohesiveness, gumminess, resilience, springiness and chewiness of meatballs were analyzed using AMETEX Test and Calibration Instrument. Meatball with 2 cm of diameter were placed on the instrument and compressed with 10.000 mm/s of pressure. Also, meatball samples were analyzed using Chromameter CR-400 to determine L*, a*, and b* scores.

2.3. Experimental design and data analysis

This study applied completely randomized experimental design (CRD) with eight treatments and four replications. The data observed in this study was then analyzed using one-way analysis of variance (ANOVA) to evaluate significant effect of treatments to the parameters observed with 5% alpha and Duncan's multiple range test was also performed to evaluate pairwise differences among meatballs.

3. Results and Discussion

3.1.Texture and color of meatball made from different meat sources

Both texture and color analysis of meatballs indicated inconsistent results. The ingredients used to produce meatballs were similar for each treatments. The only different thing in the ingredients was meat sources that used asdifferentiator among treatments, therefore ingredients and procedure to produce meatballs in this study should notbe taken into account in discussionsince they were exactly the same. Texture differences among meatballs from different species are presented in Table 2.Hardness is the maximum strength on the first bite [12]. Meatballs made from pork and dog meats were significantly different with meatballs made from beef and rat meat. In addition, meatballs containing more than one species were not consistently different. The highest hardness score went to beef and rat meatballs. Beef and rat meatballs showed no gumminess difference but they were highest score compared to other meatballs with different meat sources. Gumminess is product of hardness and cohesiveness [12]. Springiness represents recovery period between first and second bite [12]. Moreover, springiness of meatballs were relatively similar among meatballsamples. Meatball originated from rat meat had highest chewiness score than other samples ($P<0.01$). Chewiness is product of gumminess and springiness [12]. Additionally, no significant difference among treatments was found for cohesiveness and resilience.

Colorsof meatball were objectively analyzed usingChromameter CR-400 for parameters oflightness (L^*) or white within range 0 to 100,score of a^* and b^*, respectively. Score of a^* and b^* represent redness and yellowness that havepositiveand negative values. Score of a^* is ranged from0 to 60 for red and 0 to -60 for green. Furthermore, score of b^* is ranged from0 to 60 for yellow and 0 to -60 for blue [13]. Dog meatball had lowest L^* score compared to meatball made from other species while meatballs made from meat mixtures were inconsistent (Table 2). No difference L^* score between beef and pork and rat meat was detected in this study. This result suggested that it could not be used to differentiate species origin in meatball. Redness(a^*) score of beef meatball was different with dog and rat meatballs, however it was not different with pork meatball. In addition, yellowness score (b^*) of beef was significantly higher than dog meatball, on the other hand it was not different with pork and rat meatballs. These results suggested that meatballs could not differentiate based on L^*, a^*, b^* scores due to inconsistent results.Flour and spices addition were able to be reason of this result[14]. Moreover, using DNA analysis of mitochondrial genome is much more accurate and effective to identify species origin in meat and meat products to date compared to physical analyses due to lack of consistent results [10, 11].

Table 2. Texture and color scores of meatballs made from different sources of meat

Parameter	S	A	B	T	SA	SB	ST	SABT	*P-Value*
Lightness (L^*)	44.25^{bcd}	38.60^{a}	44.91^{cd}	46.16^{d}	42.19^{abc}	45.53^{cd}	47.42^{d}	41.17^{ab}	0.000
Redness (a^*)	8.11^{a}	11.35^{c}	7.98^{a}	9.29^{b}	9.26^{b}	8.56^{ab}	10.94^{c}	9.53^{b}	0.000
Yellowness (b^*)	11.51^{c}	10.02^{a}	11.21^{bc}	11.09^{bc}	11.24^{bc}	12.55^{d}	10.16^{a}	10.51^{ab}	0.000
Hardness (N)	18.06^{e}	10.41^{abc}	7.67^{a}	17.99^{e}	13.17^{cd}	11.76^{ab}	9.62^{ab}	15.97^{de}	0.000
Cohesiveness	0.38	0.38	0.38	0.44	0.41	0.38	0.40	0.41	0.206
Gumminess (N)	6.68^{cd}	3.99^{ab}	2.97^{a}	7.87^{d}	5.39^{bc}	4.44^{ab}	3.83^{a}	6.06^{c}	0.000
Resilience	0.51	0.49	0.47	0.62	0.55	0.49	0.53	0.51	0.504
Springiness	0.91^{abc}	0.89^{ab}	0.86^{a}	0.94^{c}	0.91^{abc}	0.91^{bc}	0.89^{ab}	0.93^{bc}	0.027
Chewiness (N)	6.05^{d}	3.56^{ab}	2.61^{a}	7.39^{e}	4.88^{cd}	4.05^{bc}	3.43^{ab}	6.08^{d}	0.000

S is beef meatball; A is dog meatball; B is pork meatball; T is rat meatball; SA is meatball containing beef and dog meat; SB is meatball containing beef and pork; ST is meatball containing beef and rat meat; and SABT is meatball containing four species of meat. Different superscript in similar row shows significant differences.

4. Conclusions

Texture and color analyses on meatballs were conducted in this study, however, they could not be used to distinguish original material of meatballs due to inconsistent results. Further study should be taken into account to get more consistent result in order to be an alternative to check food adulteration and halalness of animal based foods.

Acknowledgments
This research was partially funded by research scheme of *Penelitian Berbasis Kompetensi (PBK)* with contract number: 089/SP2H/LT/DRPM/2018, Directorate of Research and Community Service, and supported by *Program Kreativitas Mahasiswa Penelitian (PKM) 5 Bidang* with contract number: 1020/B3.1/KM/2018, Ministry of Research, Technology and Higher Education, Republic of Indonesia.

References
[1] Eaqub Ali M, Hashim U, Mustafa S and Che Man Y B 2012 *Food Anal. Method.* **5** 613-623
[2] Barton-Gade P A, Cross H R, Jones J M and Winger R J 1988 *Factor Affecting Sensory Properties of Meat. In Meat Science, Milk Science and Technology* (Amsterdam-The Netherlands: Elsevier Science Publishers)
[3] Szczesniak A 1963 *J. Food Sci.* **29** 385-389
[4] Chambers E N, and Bowers J R 1993 *Food Tech.* **47** 116-120
[5] Bourne M C 2002 Principles of objective texture measurement. In M. C. Bourne (Ed.), *Food texture and viscosity: Concept and measurement* 107–188.
[6] Tabilo G, Flores M, Fiszman S M and Toldra F 1999 *Meat Sci* .**51** 255-260
[7] Seideman S C, Cross H R, Smith G C and Durland P R 1984 *J. Food Qual.* **6** 211-237
[8] Olivera D F, Bambicha M, Laporte G, Cárdenas F C and Nora Mestorino 2012 *J. Food Sci. Technol.* **50** 821-825
[9] Ni'mah A, Kartikasari Y, Pratama A D, Kartikasari L R, Hertanto B S and Cahyadi M 2016 *J. Indonesian Trop. Anim. Agric.***41** 7-12
[10] Novianty E, Kartikasari L R, Lee J H and Cahyadi M 2017 *IOP Conf. Ser.: Mater. Sci. Eng.***193** 012002
[11] Cahyadi M, Puruhita, Barido FH andHertanto BS 2018 *IOP Conf. Ser.:Earth Environ. Sci.* **102** 012038
[12] Indiarto R 2012 *Jurnal Teknologi Hasil Pertanian* **5** 106-116
[13] Mc Williams 1997 *FOODS: Experimental Perspectives* **3** 41-47
[14] Garcia M L, Caceres E and Selgas M D 2006 *IJFST* **4 1**1207-1215

IOP Conf. Series: Materials Science and Engineering **633** (2019) 012030 doi:10.1088/1757-899X/633/1/012030

Functionality of xanthan and almond gum in colloidal shellac nanoparticles containing cinnamon

D R A Muhammad[1, 3], VGupta[1], A S Doost[2], K Dewettinck[1]
[1] Laboratory of Food Technology and Engineering, Department of Food Technology, Safety and Health, Ghent University, Coupure links 653 Gent 9000 Belgium.
[2]Particle and Interfacial Technology Group, Department of Green Chemistry and Technology, Ghent University, Coupure links 653 Gent 9000 Belgium
[3] Department of Food Science and Technology, UniversitasSebelasMaret, Jl. IrSutami 36A Surakarta 57126 Indonesia.

E-mail: DimasRahadianAji.Muhammad@ugent.be, rahadiandimas@uns.ac.id

Abstract. Instability of shellac nanoparticles at acidic pH is the main challenge of its use as an oral delivery system. This study aims to investigate the functionality of xanthan and almond gum in shellac nanoparticles containing cinnamon prepared by anti-solvent precipitation. The gums were added into the anti-solvent phase while the cinnamon extract was incorporated into the solvent phase. The results show that the minimum concentration of xanthan and almond gum to prevent shellac aggregation at simulated gastric pH (1.2) was 0.3% and 0.6%, respectively. Higher concentration of gums resulted in a bigger particle size and more negative ζ-potential, regardless the type of the gums. Cinnamon loading in the shellac-xanthan gum complex resulted in nanoscale sized particles, while the complex system consisted of shellac, almond gum 0.6% and cinnamon extract created a network entrapping the individual particle. In conclusion, xanthan gum incorporation was better than almond gum to stabilise shellac nanoparticle containing cinnamon in simulated gastric pH. This study confirms that shellac-xanthan gum complex can be potentially used for an oral delivery system of bioactive compounds.

1. Introduction

Nanoscience and nanotechnology are the new frontiers of the century and are considered as the "next big thing" in many industries, including in food manufactures [1]. Nanoparticulate delivery systems (e.g. nanoparticles, liposomes, nanocapsules and nanospheres) are some of the examples of nanotechnology applications in food industry [2]. Nanotechnology has been successfully applied to mask undesirable odours and flavours. In addition, nanotechnology can be used to improve bioavailability as well as to control the release of bioactive compounds [3, 4, 5]. Shellac, a nontoxic and physiologically harmless material secreted by lac insect (*Kerrialacca*), can be potentially used as a nanocapsule wall-material. It has tremendous film forming and protecting properties [6, 7]. However, the use of shellac as an oral delivery system of bioactive compounds in a nanoparticulate form has a limitation as it forms aggregate at gastric pH [8]. Nanoparticulate delivery system of bioactive compounds targeting the colon has to be stable in the harsh acidic conditions of the stomach [9, 10]. Therefore, it is important to create a system which can prevent the aggregation of shellac at acidic pH. Various natural polymers has been reported to have a great potency as a stabiliser in many systems; therefore, it might be an alternative to stabilise shellac colloidal particle in the gastric pH condition.

In our preliminary study, guar gum, K-carrageenan, Na-alginate, xanthan gum and almond gum were used (data not shown). Only the last two polymers showed a positive effect in preventing shellac aggregation at acidic pH. Thus, in this study, xanthan gum and almond gum were further investigated to prepare stable shellac nanoparticles. Xanthan gum is a natural biopolymer produced by fermentation by bacteria *Xanthomonas campestris* [11]. Xanthan gum has been extensively used as a stabilising agent in many products. It has pseudo plastic rheological properties, temperature stability and compatibility with many food ingredients [12]. Almond gum, a biopolymer derived from the tree *Amygdalus communis* L., is another type of gum which has a high potency as a stabilising agent in food. Almond gum is classified as arabinogalactan polysaccharide with arabinose, galactose, xylose, and uronic acid as its major constituents [13].

Cinnamon extract was used as the model of the core material of the nanoparticles in this study. This is based on the fact that synthesis of nanoparticles containing spice extract are still scarce until to date. Cinnamon extract contains a wide range of bioactive compounds acknowledged to have benefits for human health, such as an antioxidant, anti-inflammatory, antitumor, anticancer, antidiabetic, and anti-hypertriglyceridemia agent due to its phenolic and volatile constituents [14]. The long list of beneficial effect of cinnamon for health as well as its high antioxidant activity indicates that cinnamon extract can be potentially used in pharmaceutical and food industry [15]. For its application, nano-sized particles may have significant impact to improve its compatibility to food or beverage matrix. In this study, anti-solvent precipitation method was selected to fabricate the nanoparticles due to its low cost and easy operation. Thus, this study aims to investigate the functionality of xanthan and almond gum in stabilising shellac nanoparticles containing cinnamon extract.

2. Experimental

A traditional method described by Muhammad et al [16] was followed to prepare the cinnamon extract. In brief, 5 grams of cinnamon powder was extracted in 50 ml ethanol for 48 hours at 20°C. Afterwards, the cinnamon residue was separated using a vacuum filter and then the cinnamon extract was collected (Laboport, KNF Neuberger, Inc, USA)

Raw almond gum (containing both water soluble and insoluble substances) was obtained from Sepahan Nano-Food Co. (Isfahan, Iran). For the separation of the water-soluble from the insoluble part, the almond gum was firstly pulverised using a ball mill (Retsch, PM400, UK) for 30 min at 14 x g. Then, the powdered almond gum was mixed with distilled water in the concentration of 4% and left on stirring for 4 hours at room temperature (20°C). The mixture was kept at 4°C overnight for complete hydration. To separate the water-soluble part, the mixture was centrifuged at 20000 rpm for 30 min at 25°C, and then the clear solution on top was gently collected. The dry matter concentration of this solution was determined from the mass obtained by placing 10 ml at 105°C overnight [17].

The shellac colloidal nanoparticles were fabricated using an anti-solvent precipitation method [8]. The first solution (solvent phase) was made by solubilising 2% (w/w) g of shellac fine powder (SSB 55 Astra FP, SSB Stroever GmbH & Co. KG (Bremen, Germany)) in ethanol. The second solution (anti-solvent phase) was made by solubilising xanthan gum (Satiaxane CX 931, Cargill France SAS) in distilled water in different concentration. The first solution was injected by using a syringe to the anti-solvent phase with the ratio of 1:3 (w/w) in order to obtain homogeneous mixture. Afterwards, a vacuum rotary evaporator was used to remove ethanol from the mixture; thereby, the colloidal shellac nanoparticles were formed. A freeze-dryer (Vaco 5-D Zirbus technology GmbH) was used to obtain dry particles. To prepare shellac nanoparticles containing cinnamon, the cinnamon extract was subjected to replace the ethanol in the first solution in multilevel proportion (12.5% to 37.5% (w/w)).

Photon Correlation Spectroscope (PCS100M, Malvern Instrument Ltd, UK) and Zetasizer IIC (Malvern Instrument Ltd, UK), with the temperature setting of 25°C, were used to measure the particle size and the surface charge of the shellac nanoparticles, respectively. Before the measurement, a few drops of the colloidal shellac nanoparticles was appropriately diluted in distilled water and KCl 0.01 M for the particle size determination and the surface charge analysis, respectively.

To observe the shape of the nanoparticles, a JEOL JSM 7100F SEM equipped with a PP3010T Cryo-SEM Preparation System (Oxford Instruments) was used. The microstructural observations were conducted in the dried nanoparticles form and rehydrated nanoparticles form. After placed on the cryospecimen holder the sample was then cryo-fixed in slush nitrogen (–210°C) and transferred to the cryo-unit in the vitrified state. Subsequently, the sample was fractured well, sublimated for 20 min at -140°C and sputter coated with platinum (4 min, 0.5 mbar). The microstructural properties of the nanoparticles were observed in the microscope with the temperature setting of 140°C.

The measurement of apparent viscosity was conducted by an AR2000-ex rheometer (TA Instruments) equipped with concentric DIN cylinder (cylinder: 42.00 mm; rotor outer radius: 14.00 mm; stator inner radius: 15.00 mm; geometry gap: 5920 mm). Approximately, 20 g of samples was placed at the cylinder and then the analysis was performed at shear rate level 50 s^{-1}. A cross-hatched 40 mm stainless-steel plate was used for the determination of yield stress of shellac-gum colloidal system loaded with cinnamon extract. The measurement was conducted by increasing the shear stress from 0.01 to 50 Pa. All the tests were carried out at 20°C.

SPSS Statistics 23 was used to perform analysis of variance (one way ANOVA), Duncan's Multiple Range Test (DMRT) and T-test. The differences were considered significant at $p<0.05$. In addition, Pearson's test was also carried out to determine the correlation coefficient between gums concentration and the physicochemical characteristic of the nanoparticles.

3. Results and discussions

The use of xanthan and almond gum was intended to prevent shellac aggregation at acidic pH. Interestingly, an instant aggregation of some parts of shellac was found in the step of injection of the solvent phase into the anti-solvent phase when the gums was absent. The instant aggregation can be prevented by the incorporation of either 0.1% of xanthan gum or 0.5% of almond gum. At the pH 1.2, the lowest concentration of xanthan and almond gum to inhibit shellac nanoparticles aggregation were 0.3% and 0.6%, respectively (Figure 1, Table 1). At the lower concentration of the gums, the repulsion was not adequate to prevent aggregation. Prevention of shellac aggregation at simulated gastric pH is by the electrosteric stabilisation [8]. The electrosteric stabilisation is gained by the adsorption of the gums on the surface of the colloidal particles. Hydrogen bonding between shellac and xanthan gum were reported by Patel et al. [18] on the contrary, until to date none of literature has discussed a possible interaction between shellac and almond gum [8]. There are some factors affecting the ability of the gums to stabilise shellac nanoparticles in addition to its concentration, such as composition, molecular weight and chain length [19]. This may be one of the reason why both gums have different stabilisation effect.

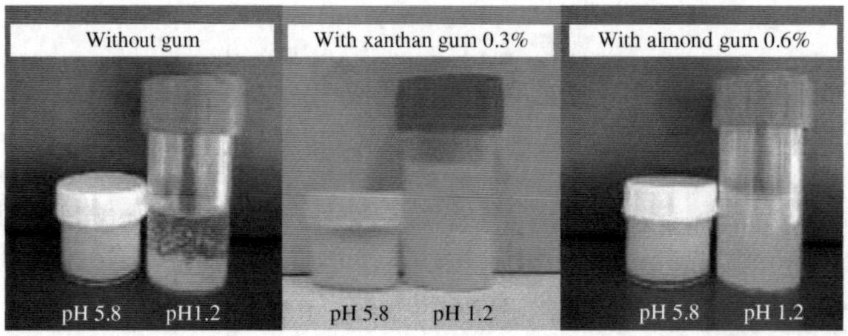

Figure 1. Influence of xanthan and almond gums incorporation on the visual appearance of the colloidal shellac nanoparticles

The gum concentration has a significant influence on the particle size of the nanoparticles when observed at pH 5.8 (the natural pH of the shellac colloidal system), regardless the type of the gum. It indicates that more gum molecules were adsorbed onto the shellac particles surface [8]. The higher concentration of gums can also increase the viscosity. A high viscosity can play a significant role to inhibit diffusion between solvent and anti-solvent, consequently, resulted in the formation of agglomeration [19]. Figure 2A confirms that a higher concentration of the gums led to a higher apparent viscosity. This manifestation is mainly because of the pseudoplastic effect of the gums which have been widely reported in previous studies [11]. Interestingly, it was observed that at the same concentration (e.g. 0.3% and 0.5%), xanthan gum is more viscous than almond gum. Therefore, it can be understood that the particle size of colloidal particles made by the combination of shellac-xanthan gum were relatively bigger than that of shellac-almond gum at the same concentration of the gums.

Table 1. Particle size of shellac colloidal dispersion stabilised by xanthan gum and almond gum

Gum concentration	Xanthan gum		Almond gum	
	Stability at pH 1.2	Particle size (nm) at pH 5.8	Stability at pH 1.2	Particle size (nm) at pH 5.8
0.0%	aggregation•	66.5 ± 1.9^a		
0.1%	aggregation	99.4 ± 1.6^b		
0.2%	aggregation	127.0 ± 4.5^c		
0.3%	stable	147.1 ± 3.9^d	aggregation•	109.5 ± 0.9^a
0.4%	stable	153.2 ± 1.2^e	aggregation•	137.1 ± 3.2^b
0.5%	stable	207.5 ± 4.2^e	aggregation	195.7 ± 2.5^c
0.6%			stable	381.4 ± 3.2^d
0.7%			stable	428.4 ± 12.5^e
0.8%			stable	625.8 ± 17.0^f
Pearson's correlation coefficient with gum concentration		0.975**		0.967**

Mean values with the same letter do not differ significantly ($p>0.05$) in the same column. Notation (•) indicates that the nanoparticle was formed, but aggregation of some parts of the particle was observed in the injection step. Notation (**) indicates significant at $p<0.01$.

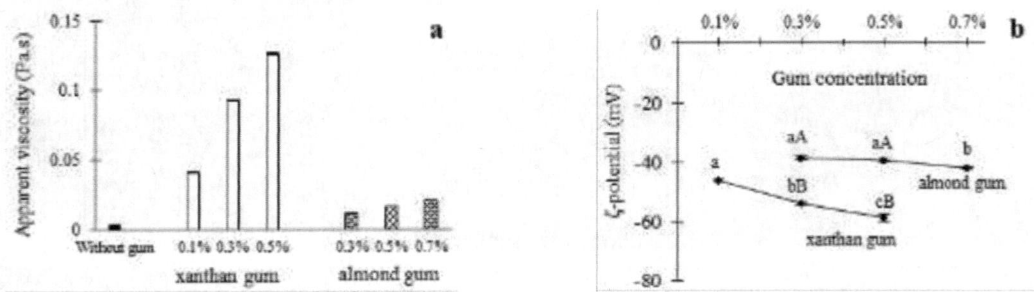

Figure 2. (a) Apparent viscosity of xanthan and almond gum in different concentration at the shear rate of 50 s^{-1}; (b) The effect of gum incorporation on the zeta potential of shellac colloidal dispersion. Mean values with the different lowercase and uppercase letter differ significantly ($p<0.05$) in the same type of gum and the same gum concentration, respectively.

Gums incorporation resulted in the more negative surface charge, regardless the type of the gums. Both gums have been reported as a negatively charged polymer in previous studies [8, 19]. However, when comparing both gums at the same concentration, for instances 0.3% and 0.5%, it was noticeably shown that the surface charge of xanthan gum had higher negative values than that of almond gum; and the statistically significant difference was observed between the samples (Figure 2B). The fact that the rise of almond gum concentration gave a limited effect on the surface charge of the nanoparticle may be because the almond gum has a lower number of carboxylate groups. As mentioned by Cerrutti et al [20], a mutual repulsion of electrical charges is the basis of electrostatic stabilisation. In the electrostatic stabilisation, the carboxylic group can act as a proton donor or acceptor in hydrogen bonds, while the carboxylate group can act as a proton acceptor. Since xanthan gum contributes with appreciably more negative charges than almond gum, the incorporation of xanthan gum induces more repulsive effect. It might be the reason why a less concentration of xanthan gum (0.1%) was required to prevent instant aggregation of shellac during injection step than the almond gum (0.5%).

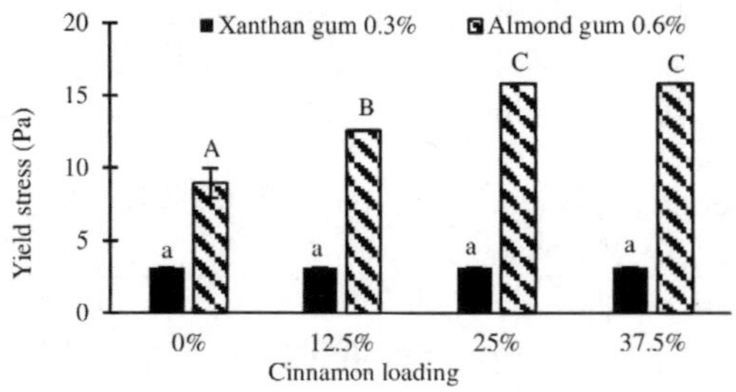

Figure 3. The effect of cinnamon extract loading on the yield stress of the shellac-gum complex. Statistical analysis was carried out separately for each type of gum. Mean values within each gum type with the same notation do not differ significantly ($p>0.05$).

The minimum concentration of xanthan and almond gum to stabilise shellac nanoparticles was found at the concentration of 0.3% and 0.6%, respectively. Therefore, the study was continued by incorporating cinnamon extract in the shellac nanocapsule templates. In Figure 3, it was shown that the blank nanoparticles made of shellac-almond gum had a higher yield stress than shellac-xanthan gum. The incorporation of cinnamon extract resulted in a more viscous colloidal system in the shellac-almond gum system, but not in the shellac-xanthan gum system. It is interesting since the apparent viscosity of almond gum was lower than that of xanthan gum before the nanoparticles fabrication as previously shown in Figure 2A. To get better understanding on the network formation of the colloids, the microstructural properties of the colloidal nanoparticles were then investigated.

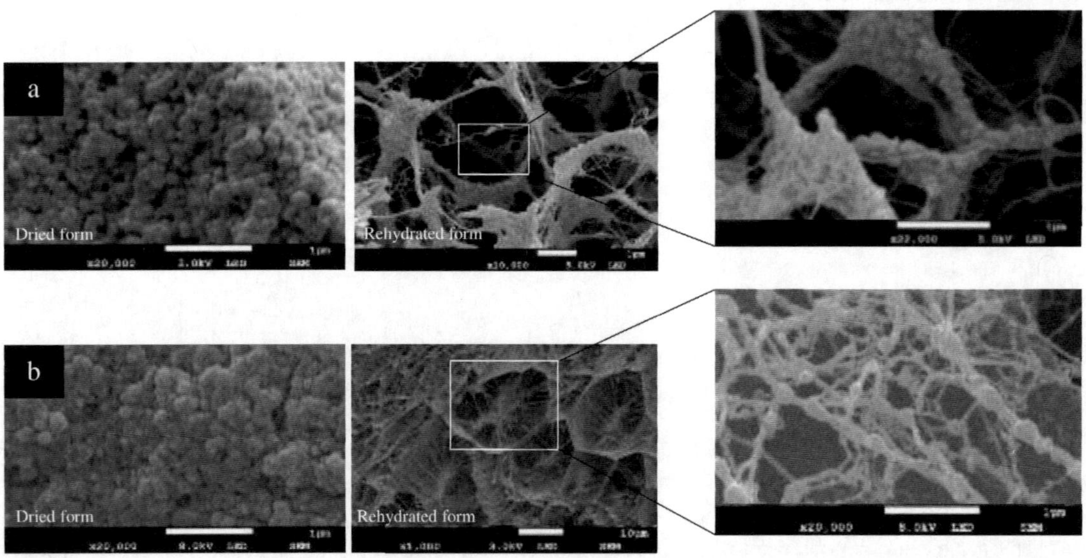

Figure 4. Typical morphology of shellac nanoparticle incorporated with xanthan gum 0.3% (a) and almond gum 0.6% (b) loaded with cinnamon extract 12.5%.

Figure 4 shows that the individual nanoparticle has spherical shape with a smooth outer surface agreeing a report by Joye & McClements [19]. The difference on the microstructure of shellac colloidal nanoparticles loaded with cinnamon stabilised using xanthan gum and almond gum was observed. As shown, shellac-xanthan gum nanoparticles containing cinnamon were in the nanoscale size with spherical shape. PCS measurement confirmed that the particle size of the shellac-xanthan gum nanoparticles containing cinnamon was in the range of 100-200 nm. Shellac-almond gum system loaded with cinnamon also have spherical shape when observed in dry form. It was also perceived that the individual particle has nanoscale size. However, by the PCS measurement, it was found that the particle size of shellac-almond gum incorporated with varied level cinnamon extract were surprisingly in the range of 3 to 9 μm. When observed in the rehydrated form, it was discovered that the complex system consisted of shellac, almond gum 0.6% and cinnamon extract created a network entrapping the individual particle. The nanoparticles were flocculated forming a bigger particle and even formed a gel-like matrix when observed visually.

4. Conclusions

In conclusion, xanthan and almond gum have an ability to stabilise shellac colloidal nanoparticles through electro-steric mechanism. However, almond gum needs a higher concentration than xanthan gum to prevent shellac nanoparticles aggregation at acidic pH. The shellac-xanthan gum complex was more suitable for cinnamon extract carrier than the shellac-almond gum complex since the gel-like structure of the shellac-almond gum complex were formed after the incorporation of the cinnamon. Further study is required to investigate the encapsulation efficiency and particle loading of the nanoparticles made from shellac-xanthan gum-cinnamon complex.

References

[1] Sozer N and Kokini J L 2009 *Trends Biotechnol.* **27** 82-89
[2] Chau C F, Wu S Hand Yen G C 2007 *Trends Food Sci. Technol.***18** 269-280
[3] Esfanjani A F and Jafari S M 2016 *Colloids Surf. B Biointerfaces* **146** 532-543
[4] Handford C E, Dean M, Henchion M, Spence M, Elliott C T and Campbell K 2014 *Trends Food Sci. Technol.***40**(2) 226-241
[5] Katouzian I and Jafari S M 2016 *Trends Food Sci. Technol.***53** 34-48
[6] Byun Y, Ward A and Whiteside S 2012 *Food Hydrocol.***27** 364-370
[7] Farag Y and Leopold C S 2011 *Eur. J. Pharm. Sci.***42** 400-405
[8] Patel A, Heussen P, Hazekamp J and Velikov K P 2011 *Soft Matter* **7** 8549-8555
[9] Castangia I, Nácher A, Caddeo C, Merino V, Díez-Sales O, Catalán-Latorre A, Fernàndez-Busquets X, Fadda A M andManconi M 2015 *Acta Biomater.***13** 216-227
[10] Park J H, Saravanakumar G, Kim K and Kwon I C 2010 *Adv. Drug Deliv. Rev.***62** 28-41
[11] Habibi H and Khosravi-Darani K 2017 *Biocatal. Agric.Biotechnol.***10**130-140
[12] Garcıa-Ochoa F, Santos V E, Casas J A and Gomez E 2000 *Biotechnol. Adv.***18** 549-579
[13] Rezaei A, Nasirpour A and Tavanai H 2016 *Food Hydrocol.* **60** 461-469
[14] Muhammad DRA and Dewettinck K 2017 *Int. J Food Prop* **20**:sup2 2237-2263
[15] Muhammad D R A, Saputro A D, Rottiers H, Van de Walle D and Dewettinck K 2018 *Eur. Food Res. Technol.***244** 1185–1202
[16] Muhammad DRA, Praseptiangga D, Van de Walle D and Dewettinck K 2017 *Food Chem* .**231** 356-364
[17] Doost A S, Muhammad D RA, Stevens CV, Dewettinck K and Van der Meeren P 2018 *Food Hydrocol.***83** 190-201
[18] Patel A R, Drost E, ten Hoorn J S and Velikov K P 2013 *Soft Matter* **9** 6747-6751
[19] Joye I Jand Mc Clements D J 2013 *Trends Food Sci. Technol.***34** 109-123
[20] Cerrutti B M, de Souza C S, Castellan A, Ruggiero R and Frollini E 2012 *Ind. Crops Prod.***36** 108-115

Acknowledgements

This work was supported by the Ministry of Research, Technology, and Higher Education, the Republic of Indonesia (BPPLN Scholarship No. 15.1/E4.4/2015). Hercules Foundation is acknowledged for its financial support in the acquisition of the cryo-SEM JEOL JSM-7100F (grant number AUGE-09-029).

IOP Conf. Series: Materials Science and Engineering **633** (2019) 012031 doi:10.1088/1757-899X/633/1/012031

Panelist acceptance level and characterization of physical and chemical properties on dark chocolate bar with addition of kaffir lime (*Citrus hystrix* DC.) leaf essential oil

D P Yoriska, D Praseptiangga and L U Khasanah

Department of Food Science and Technology, Universitas Sebelas Maret (UNS) Jl. Ir. Sutami 36A Surakarta 57126, Central Java, Indonesia

E-mail: dpraseptiangga@yahoo.com

Abstract. In this study, panelist acceptance level, physical, and chemical properties of dark chocolate bar enriched with kaffir lime (*Citrus hystrix* DC.) leaf essential oil (0.25%, 0.5%, and 0.75%) were studied. The addition of kaffir lime leaf essential oil in dark chocolate bar is aimed to enrich the content of functional compounds in dark chocolate and is expected to create an Indonesian chocolate signature. The results showed that panelist acceptance level decreased with increasing concentration of kaffir lime leaf essential oil added in parameters of aroma, taste, and overall, thus, the highest panelists acceptance was dark chocolate bar with the addition of 0.25% kaffir lime leaf essential oil. The addition of various formulations of kaffir lime leaf essential oil showed a decrease in hardness significantly and the color differences between dark chocolate with and without addition of kaffir lime leaf essential oil were very small. The addition of 0.25% kaffir lime leaf essential oil increased the moisture content and antioxidant activity in dark chocolate bar significantly. Major compounds detected in GC-MS analysis of dark chocolate bar with the addition of 0.25% kaffir lime leaf essential oil were theobromine, citronellal, caffeine, and vanillin.

1. Introduction

Indonesia is the third biggest cocoa producer in the world, after Ivory Coast and Ghana. Cocoa production in Indonesia has increased from 593,331 tons to 656,817 tons in 2015 and 2016 respectively and estimated to reach 688,345 tons in 2017 [1]. Cocoa contains large concentrations of flavonoids, even greater than tea and wine. Dark chocolate contains considerably higher amounts and also has greater biological effects of flavonoids than milk chocolate because the milk in milk chocolate may slow down the intestinal absorption of flavonoid. Aside from antioxidant, cocoa also contains minerals such as potassium, phosphorus, copper, iron, zinc, and magnesium [2].

Citrus hystrix DC., commonly known as kaffir lime or makrut lime, is a common tropical herb in the family Rutaceae found everywhere in Southeast Asia [3]. The production of kaffir lime leaf essential oil in Indonesia in 2011 reached 2-3 million tons and has a stable trend [4]. The main component in kaffir lime leaf essential oil is citronellal. Several studies have revealed that citronellal has antioxidant activity and also antibacterial activity that can inhibit Salmonella and other Enterobacteria [5][6]. Apart from that, the essential oil extracted from kaffir lime has been demonstrated to reduce blood pressure and relieve depression in human studies [7].

The research trend of chocolate products innovation is growing quite rapidly in the community. Several studies have been undertaken to investigate the interaction between cocoa and some herbs and spices such as aniseed, lemon, ginger [8], and cinnamon [9][10] on sensory [11][12], physical [8], and chemical properties [13]. However, the interaction between cocoa and kaffir lime is still poorly mapped and not well understood. This research aims to study the panelist acceptance level, physical properties, and chemical properties of dark chocolate bar with the addition of kaffir lime leaf essential oil. This study also aimed to enrich the content of functional compounds in dark chocolate and is expected to create an Indonesian signature chocolate.

2. Experimental

2.1 Chocolate preparation

Couverture Dark Chocolate 65% "Easimelt Dark - Aura" from Tulip and kaffir lime (*Citrus hystrix* DC.) leaf essential oil (CV. Orizho Indonesia, Bantul) were used in this study. Couverture Dark Chocolate was first melted using chocolate melter at 50°C for about 45 minutes until chocolate completely melted. The mixing process of the melted dark chocolate and kaffir lime leaf essential oil (0.25%, 0.5%, 0.75% (v/w); ρ = 0,8507 g/ml at 27°C) was done with a ball mill for 10 minutes with 90 rpm at 50°C. Afterwards, the chocolate mixture was molded into dark chocolate block and cooled in a cooling chamber at 11°C for about 30 minutes. The tempering process was conducted by melted down the dark chocolate block with a temperature less than 45°C. Approximately 3/4 part of the melted dark chocolate was then poured on the marble table, flattened, and inverted by using a scrapper and a chocolate knife until the chocolate dough temperature was stabilized at 25°C. Once stabilized, the tempered chocolate was then poured back into the bowl containing 1/4 part of untempered chocolate, then stirred for a few minutes until the temperature stable at ±30°C, and then followed by moulding into chocolate plastic moulds and cooled in a refrigerator at 8°C. After 15 minutes, the moulded chocolates were de-moulding and were wrapped in the aluminum foil and stored at refrigerator temperature until used for further analysis.

2.2 Sensory evaluation

Kaffir lime leaf essential oil-enriched dark chocolates were then evaluated using Hedonic methods [14]. The experimental chocolates were subjected to sensory evaluation by a group of panelists, comprising 30 random panelist members varying in age between 20-22 years. Five attributes (color, aroma, taste, appearance, overall acceptability) were evaluated. Three samples of chocolate with different codes were presented in a single serving plate with crackers and mineral water for rinsing between samples. Panelists were asked for not comparing the samples with other samples. The sensory properties were presented on a five point scoring scale 1) Dislike, 2) Dislike moderately, 3) Neutral, 4) Like moderately, 5) Like.

2.3 Physical and chemical analyses

Color measurements of dark chocolates were conducted by using Chromameter CR-400, Konica Minolta Optic inc. [15]. °Hue was calculated based on the formula: °Hue = $\tan^{-1} b^*/a^*$ [16]. Total Color Difference (TCD*) was calculated based on the values of L*, a*, and b* obtained by the formula: $TCD^* = [(L_{treatment} - L_{control})^2 + (a_{treatment} - a_{control})^2 + (b_{treatment} - b_{control})^2]^{1/2}$ [15]. The hardness (texture measurement) of dark chocolate samples (3 x 2 cm, depth 1 cm) were measured using Zwick Universal Testing Machine SA/0.5 [15] at pre-load 0,02 N, pre-load speed 50 mm/min, and test speed 10 mm/min. Moisture content of chocolate was determined using Moisture Meter CA-200 Mitsubishi Chemical Analytech by Karl Fischer titration method [17]. The measurement for antioxidant activity were investigated using DPPH (1,1-diphenyl-2-picryhydrazyl) method [18]. GC-MS analysis was conducted based on [19] with slight modification to identify the active compounds and citronellal content in the sample. The control and the best formula samples were extracted by dissolving samples in methanol solvent, and melted in water bath at 110°C for about 2 minutes, mixed by vortex, and then centrifuged for 2 minutes until separated. One μL of the sample was injected into the Gas Chromatography Mass Spectrometry (GC-MS) Agilent Technologies Model 6890N (GC) 5973 inert (MS). Component separation was performed with helium as carrier gas, sample injected in split mode. The initial oven temperature was programmed at 60°C, kept constant for 3 minutes, with an increase in temperature at a rate of 20°C/min until it reached 100°C, 7°C/min until 240°C, kept constant for 5 minutes. The total program time was 30 minutes. Components of the samples were identified by comparing each peak in the chromatogram with the WILEY 7 database.

IOP Conf. Series: Materials Science and Engineering **633** (2019) 012031 doi:10.1088/1757-899X/633/1/012031

2.4 Statistical analysis
Data analysis was performed by SPSS Statistics 20 using One-Way Analysis of Variance (ANOVA) followed by DMRT (Duncan's Multiple Range Test) with a significance level of $\alpha = 5\%$. From the sensory, color, and texture test results, the best formula will be obtained. Afterward, Independent-Samples T Test was used to evaluate moisture content and antioxidant activity results for the best formula with a significance level of $\alpha = 5\%$.

3. Results and Discussion

3.1 Panelist acceptance level (sensory properties)
Table 1 shows that the variation of kaffir lime leaf essential oil addition in dark chocolate gave no significant effect on panelist acceptance in color and appearance parameters, however, all samples received score above 4, suggesting a good acceptability for the color and appearance parameters. The appearance parameter, particularly the glossiness of the samples, is greatly influenced by the tempering process. Tempering process is one of the important steps in producing chocolate as it is affecting the quality of chocolate which were the glossiness of the chocolate, good snap, and chocolate shelf life [21]. There were significant differences at panelist acceptance between F1 sample and the other samples (F2 and F3) in the parameter of aroma and taste. F1 received the highest score from the panelist in both aroma and taste parameters among the samples. F1 sample scored closed to "like moderately" in aroma parameter, while the other samples were both scored "neutral". This is allegedly due to the increased of kaffir lime leaf essential oil concentration that strengthen the aroma of kaffir lime leaf, thus defeating the aroma of the dark chocolate itself, and affecting the panelist acceptance of the samples. Citronellal, the major compound of kaffir lime leaf essential oil, consisted of terpenoid mixture that can give a special scent on kaffir lime leaf essential oil [20]. In the taste parameter, F1 also scored near "like moderately", while F2 scored near "neutral" and F3 "neutral". Panelists tend to dislike the sample with higher concentration of kaffir lime leaf essential oil addition. This might be due to the higher concentration of kaffir lime leaf essential oil addition gave the more bitter taste in the sample. Kaffir lime leaf essential oil has a strong bitter taste as it is consisted of esters, phenols, carbonyls, terpenes, and others that have bitter taste [22]. The result in overall parameter showed significant differences between F1 and the other samples (F2 and F3). F1 received the highest panelist acceptance and was scored "like moderately", while F2 and F3 were "neutral". From the five parameters tested, the F1 sample has the highest acceptance level. Therefore, it can be concluded that the best formula based on the panelists acceptance level is dark chocolate with 0.25% kaffir lime leaf essential oil addition.

3.2 Color measurement
This study was also supported by the result of color measurement. There were significant differences between dark chocolate control and the other three formulas in °Hue value (Table 1). Thus, the addition of kaffir lime leaf essential oil to dark chocolate gave statistically significant effect on °Hue of dark chocolate. However, °Hue value in each sample showed the same color range, which is in red color range [23]. The color difference level of Total Color Difference (TCD*) value obtained between control and F1, F2, F3 were very small, medium, and small, respectively [24]. The result of color measurement showed that F1 (0.25% kaffir lime leaf essential oil addition) has the slightest difference with the control. Therefore, F1 was also selected as the best formula based on the color measurement.

3.3 Texture measurement (hardness)
Table 1 shows that increasing concentration of kaffir lime leaf essential oil caused significant reduction in F max (hardness) in dark chocolate. The decrease in F max may caused by the addition of essential oil which in the form of liquid. Small changes in composition can give effects on texture in chocolate products [17]. Besides, the addition of the kaffir lime leaf essential oil can increase the particle size of dark chocolate. Increase in particle size resulted in linear decreases in textural properties (hardness) in dark chocolate [25]. According to [17], a good chocolate product is a chocolate that has a good snap. The best formula from the texture measurement was determined by selecting the formula with the biggest F max value among the formulations, thus F1 was selected.

Referring to the selected formula on the color and texture measurements, as well as the selected formula from the panelist acceptance level evaluation, F1 (dark chocolate with 0.25% kaffir lime leaf addition) was selected as the best formula. The best formula and control were then tested for the following chemical properties (moisture content, antioxidant activity, and GC-MS) evaluation.

Table 1. Panelist acceptance level, color, and texture measurement of the dark chocolate bars

	Control	F1 (0.25%)	F2 (0.5%)	F3 (0.75%)
Panelist acceptability[a]				
Color	-	$4.50^{a} \pm 0.68$	$4.63^{a} \pm 0.77$	$4.57^{a} \pm 0.73$
Aroma	-	$3.93^{b} \pm 0.94$	$3.03^{a} \pm 1.30$	$3.03^{a} \pm 1.40$
Taste	-	$3.90^{b} \pm 1.13$	$2.80^{a} \pm 1.06$	$3.03^{a} \pm 1.38$
Appearance	-	$4.43^{a} \pm 0.86$	$4.57^{a} \pm 0.77$	$4.53^{a} \pm 0.97$
Overall	-	$4.00^{b} \pm 0.95$	$3.07^{a} \pm 1.11$	$3.10^{a} \pm 1.40$
Color Measurement				
L*	$29.09^{a} \pm 0.45$	$29.17^{a} \pm 0.39$	$28.42^{a} \pm 0.27$	$28.78^{a} \pm 0.61$
a*	$9.29^{a} \pm 0.17$	$9.26^{a} \pm 0.18$	$8.98^{a} \pm 0.07$	$8.94^{a} \pm 0.27$
b*	$6.92^{c} \pm 0.18$	$6.66^{bc} \pm 0.06$	$6.14^{a} \pm 0.35$	$6.32^{ab} \pm 0.27$
°Hue	$36.68^{c} \pm 0.23$	$35.74^{b} \pm 0.29$	$34.34^{a} \pm 0.56$	$35.25^{b} \pm 0.34$
TCD	-	0.27	1.08	0.76
Texture Measurement (Hardness)				
F max (N)	$85.99^{c} \pm 2.34$	$73.52^{b} \pm 0.71$	$72.28^{b} \pm 4.08$	$64.54^{a} \pm 0.99$

[a] Hedonic scale: 1) Dislike, 2) Dislike moderately, 3) Neutral, 4) Like moderately, 5) Like
[b] Different letters in the same row indicate significant differences ($\alpha < 0.05$)

3.4 Moisture content

There was a significant difference between dark chocolate control and F1. The result of the moisture content test in this study was contradicting with the similar research that has been done. In the research by [26] there was a significant decrease in moisture content in 0.25% and 0.5% of cinnamon essential oil and a significant increase at 0.75% compared to the control. In the research by [27] noted that with the addition of various essential oils (lime leaf, bay leaf, and lemongrass leaf) into edible film of fish gelatin decrease the moisture content of the edible film. This contradicting result with the similar previous researches is suspected due to very small water vapor migration during the cooling process of the dark chocolate block after the mixing process. The occurrence of heat shock (from extreme temperature changes) from the ball mill process to cooling process in the cooling chamber causes condensation. The dew is allegedly reentered into the chocolate due to being retained by the plastic used as a cover during the cooling process of dark chocolate blocks. According to [28], temperature, solid fat content, crystal structure, and composition of the chocolate are some of the factors affecting moisture migration in chocolate. The result showed that the addition of kaffir lime leaf essential oil can increase the moisture content of dark chocolate significantly, however, it was still good and acceptable, because the results were still below 2%. According to [17], 2% moisture can give a poor shelf life as well as an inferior texture on the chocolate.

Table 2. Moisture content and antioxidant activity of the dark chocolate bars

	Control	F1 (0.25 %)
Moisture content (%)	$1.19^{a} \pm 0.00$	$1.74^{b} \pm 0.06$
Antioxidant activity (%)	$57.73^{a} \pm 1.46$	$80.41^{b} \pm 1.46$

[a]Different letters in the same row indicate significant differences ($\alpha < 0.05$)

3.5 Antioxidant activity

Antioxidants have caught the attention of biochemists and health experts for its ability to protect the body against damage by reactive oxygen, thus avoiding some degenerative diseases. Table 2 shows a significant increase in antioxidant activity by the addition of 0.25% of kaffir lime leaf essential oil to dark chocolate. Citronellal, one of the main active compounds in kaffir lime leaf essential oil, has been known of its high antioxidant activity. From the past studies, Kaffir lime leaf essential oil is known to has antioxidant activity of 17.65 μmol TE/g [5], while the total antioxidant capacity of dark chocolate is 227 μmol TE/g [29]. Aside from citronellal content, kaffir lime leaf essential oil also contained other compounds that play a role in increasing antioxidant activity in F1 sample, such as citronellyl acetate, geranyl acetate, and β-citronellol.

3.6 Active compounds and citronellal contents

In this study, theobromine has been detected as the major compound in both control and F1 (figure not shown) with relative content 59.27% (in control) and 48.31% (in F1). In dark chocolate bar control, vanillin, caffeine and squalene were also detected with relative content 12.78%, 13.77%, and 14.18%, respectively. While F1 was consisted of theobromine (48.31%), citronellal (18.35%), caffeine (12.81%), vanillin (11.58%), palmitic acid (3.36%), and citronellyl acetate (0.82%). Theobromine is the main compound in cacao and gave distinct bitter taste in cocoa products along with a small contribution from caffeine. It is used as a vasodilator, diuretic, and cardiac stimulant [30]. Caffeine had effects on cardiovascular parameters and mood including feelings of alertness. Caffeine was reported to increase self-alertness and improve attention and psychomotor performance [31]. Citronellal was found in GC-MS test results in dark chocolate with the addition of 0.25% kaffir lime leaf essential oil. Citronellal is a monoterpene that has fungicidal ability, antifungal [32], antibacterial, and bactericidal activity [33], as well as depressant and antinociceptive activity and potentially useful in treating inflammatory and damage caused by oxidant agents [34]. The presence of squalene in dark chocolate control is suspected to be due to the use of vegetable emulsifier in the form of soybean oil in dark chocolate couverture used. Interestingly, squalene was not found in the dark chocolate with the addition of 0.25% kaffir lime leaf essential oil. This is allegedly due to the advanced process that uses heat. [35] reported that squalene is thermolabile due to its unsaturated linear chain. Squalene is commonly found in many plant oils such as olive oil, soybean oil, and sunflower oil. According to [36], squalene acts as an anticancer, has antihypertensive effects, and cardioprotective activity. Vanillin commonly used as a flavoring agent in chocolate. Citronellyl acetate was derived from the kaffir lime leaf essential oil, plays a role in creating a distinctive aroma on the kaffir lime leaf essential oil. It has antinociceptive, fungicidal, bactericidal, and potentially antihepatoma activity [37]. As the relative content of palmitic acid in this study showed a relatively small percentage, which is 3.53%, palmitic acid detected in this study is believed to be fatty acids derived from cocoa fat. [38] reported that the content of palmitic acid which exceeds 20% is strong evidence of the addition of coconut oil. However, palmitic acid was only detected in F1 allegedly due to the impurities from the ball mill process performed on the dark chocolate bar with the addition of kaffir lime leaf essential oil, since cooking oil was used in cleaning process of the ball mill. The small remnants of cooking oil is possibly become the impurities that causing the detection of palmitic acid in F1.

4. Conclusion

Dark chocolate bar with 0.25% kaffir lime leaf essential oil addition has the highest panelist acceptance. There were significant difference between dark chocolate control and the other formulas (0.25%, 0.5%, and 0.75%) at °Hue value. The increasing concentration of kaffir lime leaf essential oil caused significant reduction in F max (hardness) in dark chocolate. Moisture content and antioxidant activity in dark chocolate bar were significantly increased by the addition of the best formula (0.25% kaffir lime leaf essential oil) in comparison to the control. Theobromine, citronellal, caffeine, and vanillin were detected as major compounds in GC-MS analysis of dark chocolate bar with the addition of 0.25% kaffir lime leaf essential oil.

References

[1] Directorate General of Estate Crops 2016 *The Crop Estate Statistics of Indonesia for Cocoa 2015-2017* (Jakarta: Directorate General of Estate Crops) p 3

[2] Latif R 2013 *J. Med.* **71(2)** 64

[3] Srisukh V, Tribuddharat C, Nukoolkarn V, Bunyapraphatsara N, Chokephaibulkit K, Phoomniyom S, Chuanphung S and Srifuengfung S 2012 *Sci. Asia* **38** 212

[4] Rusli M S 2012 *Proc. IFEAT Int. Conf. in Singapore 'Essential Asia'* **4** 163

[5] Ayusuk S, Siripongvutikorn S, Thummaratwasik P and Usawakesmanee W 2009 *Kasetsart J. (Nat. Sci.)* **43** 305–12
[6] Nanasombat S and Lohasupthawee P 2005 *KMITL Sci. Tech. J.* **5(3)** 527–38
[7] Yee L P 2014 *Study on the Effect of Hydrodistillation on Yield and Limonene Composition of Various Citrus Essential Oil and its Application as Natural Aromatic Gel (Thesis)* (Pahang: Universiti Malaysia Pahang)
[8] Albak F and Tekin A R 2014 J. *Food Sci. Tech.* **1(7)** 51–9
[9] Ilmi A, Praseptiangga D and Muhammad D R A 2017 *IOP Conf. Ser.: Mater. Sci. Eng.* **193**
[10] Dwijatmoko M I, Muhammad D R A and Praseptiangga D 2016 *Nusantara Biosci.* **8(2)** 301–5
[11] Praseptiangga D, Nabila Y and Muhammad D R A 2018 *Caraka Tani: J. Sustainable Agri.* **33(1)** 78–88
[12] Rasuluntari I N, Muhammad D R A and Praseptiangga D 2016 *Nusantara Biosci.* **8(2)** 297–300
[13] Muhammad D R A, Praseptiangga D, Walle D V and Dewettinck K 2017 *Food Chem.* **231** 356–64
[14] Setyaningsih D, Apriyantono A and Sari M P 2010 *Analisis Sensori untuk Industri Pangan dan Agro* (Bogor: IPB Press)
[15] Misnawi 2011 *Pelita Perkebunan* **27(3)** 218–9
[16] Andarwulan N, Kusnandar F and Herawati D 2011 *Analisis Pangan* (Jakarta: Dian Rakyat)
[17] Beckett S T 2008 *The Science of Chocolate 2nd Edition* (UK: RSC Publishing)
[18] Molyneux P 2004 *Songklanakarin J. Sci. Tech.* **26(2)** 211–9
[19] Kawiji, Khasanah L U, Utami R and Aji Y M 2015 *J. Aplikasi Tek. Pangan* **4(2)** 48–55
[20] Ketaren S 1993 *Pengantar Minyak Atsiri Jilid II* (Jakarta: Balai Pustaka)
[21] Afoakwa E O, Paterson A, Fowler M and Viera J 2008 *J. Food Eng.* **89** 128–36
[22] Meyerhof W 2013 *Neurosciences - From Molecule to Behavior: A University Textbook* (Berlin: Springer Berlin Heidelberg)
[23] Hutchings J B 1999 *Food Color and Appearance 2nd Edition* (Gaithersburg: Aspen Publishers)
[24] PCE Instruments 2010 *Color Meter PCE-TCD 100 Instruction Manual* (Hampshire: PCE Inst.)
[25] Afoakwa E O, Paterson A, Fowler M and Viera J 2008 *J. Food Eng.* **87** 181–90
[26] Dwijatmoko M I 2014 *Tingkat Penerimaan Panelis Serta Karakterisasi Sifat Fisik dan Kimia Dark Chocolate Bar dengan Penambahan Minyak Atsiri Kayu Manis (Cinnamomum burmannii) (Thesis)* (Surakarta: Universitas Sebelas Maret)
[27] Tongnuanchan P, Benjakul S and Prodpran T 2013 *Int. J. Food Sci. Tech.* **48** 2143–9
[28] Hondo H 2013 *Modeling of Moisture Migration through Chocolate (Thesis)* (Gothenburg: Chalmers University of Technology)
[29] Gu L, House S E, Wu X, Ou B and Prior R L *J. Agri. Food Chem.* **54** 4057–61
[30] Haryadi and Supriyanto 2012 *Teknologi Cokelat* (Yogyakarta: Gadjah Mada University Press)
[31] Baggott M J, Childs E, Hart A B, Bruin E, Palmer A A, Wilkinson J E and Wit H 2013 *Psychopharmacology (Berl.)* **228(1)** 109–18
[32] Zore G B, Thakre A D, Jadhav S and Karuppayil S M 2011 *Phytomedicine* **18** 1181–90
[33] Romero J C L, Rios H G, Borges A and Simoes M 2015 *Hindawi* **2015** 6–7
[34] Melo M S *et al* 2011 *Biol Res* **44** 363–8
[35] Popa O, Babeanu N E, Popa I, Nita S and Parvu C E D 2015 *BioMed Research Int.* **2015** 1–11
[36] Martinez C S, Ribotta P D and Leon A E 2016 *Cogent Food Agri.* **2** 2
[37] Rios E R V *et al* 2013 *Chemico-Biological Interactions* **203** 573–9
[38] Suzuki R M, Montanher P F, Visentainer J V and Souza N E 2011 *Ciencia e Tecnologia de Alimentos* **31(2)** 541–6

Acknowledgements
The authors would like to express their gratitude for PNBP research grant of Universitas Sebelas Maret (UNS), Indonesia.

IOP Conf. Series: Materials Science and Engineering **633** (2019) 012032 doi:10.1088/1757-899X/633/1/012032

Chitosan-garlic essential oil incorporation on beef meatball edible coatings as antioxidant-based functional food

S Ariviani[1], M Kusumawati[1] and W Atmaka[1]

[1]Department of Food Science and Technology, Faculty of Agriculture, Universitas Sebelas Maret, Jl. Ir. Sutami 36 A, Kentingan, Surakarta 57126, Indonesia

Email: setya_ariviani@yahoo.com; setyaningrum_ariviani@staff.uns.ac.id

Abstract. In addition to providing protection, edible coatings can also improve the functional properties of the coated food through the incorporation of bioactive compounds, such as antioxidants. The study aimed to investigate the potential of chitosan and/or garlic essential oil (EO) incorporation into the edible coatings of beef meatballs in enhancing the antioxidant properties. The potency of chitosan, garlic EO and its combination in increasing the antioxidant properties of meatballs were examined by comparing them toward meatballs without bioactive compounds incorporation (control) for 18 days of storage at 4±1°C. This study used a completely randomized design and the data were analyzed using an analysis of variance at $p > 0.05$. Results showed that incorporation of chitosan and garlic EO into edible coatings increasing both either RSA or RP of coated meatballs. RSA of meatballs increased 3.7 times (chitosan), 4.6 times (garlic EO) and 7 times (chitosan-garlic EO), while the RP increased 2.6 times (chitosan), 2.9 times (garlic) and 3.0 times (chitosan-garlic EO). During storage, all of the meatballs tested showed a reduction in both RSA and RP. Incorporation of chitosan-garlic EO into the edible coating produces a synergistic effect in RSA but it showed antagonist effect in RP.

1. Introduction

Beef meatballs are one of the popular and favorite beef products in all classes of Indonesian society especially the youngsters. Dewi and Mahmudiono [1] reported that beef meatballs consumption of Indonesian people reaches 1-2 times a week on average. Akcan et al. [2] proved that beef meatballs coated with edible coatings prepared using whey protein isolate incorporated with *Laurus nobilis* and/or *Salvia officinalis* extracts were increases the antioxidant activity of beef meatballs during frozen storage.

Active packaging (AP) is a packaging which was incorporated with bioactive compounds so that in addition to imparting product protection its also improves the product functional values [3]. One of the bioactive compounds that can be incorporated into AP is antioxidants [4]. Antioxidants are compounds which are capable to prevent oxidative stress through their abilities to neutralize reactive oxygen species (ROS), therefore provide various health benefits such as anti-diabetic, anti-cancer, and anti-atherosclerotic activities [5,6]. Antioxidants define as "any substance that delays, prevents, or removes oxidative damage to a target molecule"[7].

Chitosan and essential oil of garlic are natural ingredients which have been proven to exhibit antioxidant capabilities [8,9,10,11,12,13]. The combination of two or more antioxidant may show synergistic, antagonistic or additive effects [14]. Incorporation of essential oil (EO) in chitosan edible coatings increased the antioxidant capacity, such as EO of *Zataria multiflora* Boiss [15], EO of Thymus [16], EO of cinnamon [17], and EO of caraway [18].

Previous researchers have proven that chitosan and/or EO added to meat products inhibits the oxidative deterioration [19,20,21,22,23]. The addition of antioxidant directly into food products decreases its antioxidant capacity due to interactions with the food matrix. Incorporation of antioxidants into edible coatings is one simple strategy to minimize the decrease in antioxidant effectiveness. Ponce et al. [24] reported that the direct addition of chitosan did not show antioxidant effects but in the form of chitosan edible coatings showed inhibition of polyphenol oxidase and peroxidase enzymes.

Kanatt et al. [25] showed that application of chitosan coatings on chicken meatballs capable to inhibit the lipid oxidation and microbial growth. Yuan et al. [26] reported that chitosan coatings have the potential to extend the shelf life of meat and fish products. Several studies have examined the incorporation of EO in edible coatings to inhibit oxidation and microbial growth in meat and meat products, such as Ginger EO [27], *Ziziphora clinopodioides* EO [28], *Origanum virens* EO [23], *Ziziphora persic*a EO [22], and *Satureja thymbra* EO [29]. To the best of our knowledge, study on the incorporation of chitosan and/or garlic EO in meatballs edible coatings to improve the antioxidants properties have not been reported..

This study examined the potential of chitosan and/or garlic EO incorporated in beef meatballs edible coatings to improve the antioxidant properties as an alternative of antioxidant-based functional food development. The potential of the coated beef meatballs as functional food was determined by measuring its antioxidant activity for 18 days of storage at a temperature of 4 ± 1°C. The potential of chitosan and garlic EO as antioxidant ingredients were evaluated by measuring the radical scavenging activity and reducing power.

2. Methods

2.1 Beef Meatballs Preparation
Beef meatballs were prepared refers to Komariah and Hendrawati [30]. The silverside beef minced using a meat chopper (Philips HR2939) with the addition of ice water (20%). Salt (3%) was added to pulverized meat, followed by addition of tapioca flour (20%), it was then mixed until homogenous. The resulting dough was molded into balls (15g of each ball). The raw meatballs was boiled until cooked (for about 30 minutes) and then drained.

2.2 Preparation of Edible Coatings
Edible coatings prepared by the method previously described by Amalia and Putri [31] with slight modifications. Briefly, dispersion of cornstarch in water at a concentration of 5% was added with 1% (v/v) glycerol and then heated and stirred using hotplate magnetic stirrer (Heidolph MR 3001 R) for 30 minutes at 75°C and 50 rpm. This edible coatings solution is further cooled until the temperature reaches 37°C. In this study, edible coatings were prepared using 4 formulas, namely control (without any addition), chitosan addition, garlic EO addition, and the addition of chitosan-garlic EO mixture. Chitosan was purchased from CV. Bio Chitosan Indonesia (Jakarta, Indonesia), whereas garlic EO was obtained from CV. Happy Garden (Jakarta, Indonesia). Chitosan solution of 1% (in 1% acetic acid solution) was added as much as 1% (v/v) after glycerol added (before heating treatment). Incorporation of garlic EO of 0.5% (v/v) was carried out after the temperature of the edible coating solution reached 37°C.

2.3 Application of Edible Coatings on Beef Meatballs
The method used for edible coating applications on beef meatballs as previously described [32]. Meatballs are dipped into edible coatings solutions for 60 seconds, lifted and air dried. The coated beef meatballs then stored at a controlled temperature of 4±1°C. Antioxidant activity evaluation

(radical scavenging activity and reducing power) of beef meatballs were conducted at 0, 5, 10, 14 and 18 days of storage.

2.4 Radical Scavenging Activity (RSA) Analysis

RSA of beef meatballs was determined using DPPH (2,2-Diphenyl-1-picrylhydrazyl) Antioxidant assay [33]. The RSA was expressed as Gallic acid equivalent activity (GAEA) (mg GA/ kg sample)

2.5 Reducing Power (RP)Analysis

RP test was conducted using the FRAP (ferric-reducing antioxidant power) method [34]. The RP of beef meatballs was expressed as Ascorbic acid equivalent activity (AAEA) (mg AA/kg sample)

2.6 Statistical Analysis

Data were presented as mean and standard deviations. The data of radical scavenging activity and the reducing power were analyzed using the program of IBM SPSS Statistics 22 (SPSS Inc., Chicago, USA) by analysis of variance (ANOVA). Independent sample t-test was used to evaluate the significant differences ($p<0.05$) on the potency of chitosan and garlic EO as antioxidant agent. Duncan's multiple range test (DMRT) was used to evaluate the significant differences ($p<0.05$) on the antioxidant activity among samples (chitosan, garlic EO, chitosan-garlic EO, and control).

3. **Results and Discussions**

3.1 The potential of chitosan and garlic EO as antioxidant agent

Antioxidant potential of chitosan and garlic EO which were used in this study was evaluated by measuring its radical scavenging activity (primary antioxidant) and reducing power (secondary antioxidant), the results were presented in Table 1. Primary antioxidant retards oxidation by radical scavenging, whereas secondary antioxidant inhibits oxidation by a mechanism that does not involve direct scavenging of free radicals[35].

Tabel 1. Radical scavenging activity (Gallic acid equivalent activity) and reducing power (Ascorbic acid equivalent activity) of chitosan and garlic essential oil.

Bioactive compounds	RSA (mg GA/kg)	RP(mg AA/kg)
Chitosan	2867.00 ± 5.19^a	51452.00 ± 51.50^a
Garlic EO	3820.16 ± 5.19^b	56477.00 ± 51.50^b

Different letter within a similar column means significant differences ($p<0.05$)

Garlic EO showed higher antioxidant activity rather than chitosan, in term of both radical scavenging activity (RSA) and reducing power (RP) (Table 1). Previous studies reported similar results. The RSA of garlic EO and chitosan at a concentration of 1 mg/ml were 64.1% and 38.03 % respectively [10, 36]. Lawrence dan Lawrence [10] dan Wu et al. [37] indicated that garlic EO exhibited higher reducing capacity rather than chitosan. The reducing power of garlic EO at a concentration of 2.5 ppm equal to that of 500 ppm chitosan.

3.2 Antioxidant activity of coated meatballs with various edible coatings formulas during storage of 4±1°C.

3.2.1 Radical scavenging activity (RSA)

Tabel 2. Radical scavenging activity (Gallic acid equivalent activity) of coated beef meatballs with various edible coatings formulas during storage at 4±1°C.

Edible Coatings Formulas	RSA of coated meatballs during storage (mg GA/kg)				
	0 day	5 days	10 days	14 days	18 days
Control	$103{\pm}0.01^{b}_{A}$	$9{\pm}0.02^{a}_{A}$	$9{\pm}0.02^{a}_{A}$	$0{\pm}0.00^{a}_{A}$	$0{\pm}0.00^{a}_{A}$
Chitosan	$370{\pm}0.01^{c}_{B}$	$136{\pm}0.01^{b}_{B}$	$127{\pm}0.01^{b}_{B}$	$73{\pm}0.01^{a}_{B}$	$70{\pm}0.01^{a}_{B}$
Garlic essential oil	$463{\pm}0.01^{d}_{C}$	$202{\pm}0.01^{d}_{C}$	$193{\pm}0.01^{bc}_{C}$	$181{\pm}0.01^{ab}_{C}$	$175{\pm}0.01^{a}_{C}$
Chitosan-garlic essential oil	$700{\pm}0.01^{c}_{D}$	$670{\pm}0.01^{b}_{D}$	$667{\pm}0.02^{b}_{D}$	$586{\pm}0.01^{a}_{D}$	$586{\pm}0.01^{a}_{D}$

Different superscript in the same rows and different subscript in the same columns mean significant differences ($p< 0.05$)

Data presented in Table 2 indicated that incorporation of chitosan and/ or garlic EO into beef meatballs edible coatings significantly improve RSA of the coated meatballs reach to 3.7 times (chitosan), 4.6 times (garlic EO) and 7 times (chitosan-garlic EO) respectively. Beef meatballs coated with either chitosan or garlic EO edible coatings have significantly higher RSA than that of control in all storage periods. Both chitosan and garlic EO proved to be good radical scavenging agents (Table 1), thus its incorporation into edible coatings improved RSA of the coated meatballs. During storage, chitosan edible coatings produce coated meatballs with a lower RSA rather than the garlic EO ones. This result was consistent with the RSA data in Table 1 which shows that garlic EO exhibits significantly higher RSA than chitosan. Beef meatballs coated using chitosan-garlic EO edible coatings provide the highest RSA during storage periods. The sum of the RSA value of chitosan edible coatings treatment and garlic EO edible coatings treatment was lower than the RSA value of chitosan-garlic EO edible coatings treatment, indicated that there was a synergistic interaction between chitosan and garlic EO on RSA.

Synergistic interaction of chitosan and garlic EO on RSA related to the different radical scavenging mechanisms in both bioactive compounds so that there is no competition in scavenging free radicals. DPPH classified as a method utilizing both Hydrogen Atom Transfer (HAT) and Single Electron Transfer (SET) mechanisms [38]. The radical scavenging capacity of chitosan via the electrons donor, whereas garlic EO through the protons (hydrogen) donor. The antioxidant activity of chitosan as a free radical scavenger generally due to the presence of nitrogen groups in C-2 which can eliminate various types of free radicals [39]. Pasanphan and Chirachanchai [40] stated that the mechanism of chitosan in scavenging free radicals is related to the electrons donor ability. The hydrogen donor ability of chitosan is relatively low due to strong hydrogen bonds both intermolecular and intramolecular. Diallyl disulfide was reported as one of the major compounds of garlic EO [12,41]. The level of this organosulfur compound in commercial garlic EO reached 25.3% [42] dan 25.9% [43]. Colin-Gonzales et al. [44] stated that antioxidant capacity of organosulfur compounds related to the thiol groups which can easily donate its proton to an electrophilic species, thereby neutralizing them. Radical scavenging mechanism of garlic organosulfur compound involves H-atom transfer from the methylene of the allyl group on the divalent sulfur to the radicals [45].

Table 2 showed a decreasing in radical scavenging activity along with increasing storage time. This is related to the stability of the chitosan and garlic EO during storage. Dhawade and Jagtap [46] reported increasing of chitosan molecular weight during storage due to the oligomers changes. Chien et al. [47] stated that high molecular weight chitosan (HMWC) showed a lower DPPH radical scavenging activity than low molecular weight chitosan (LMWC). The decrease in RSA of coated meatballs with incorporation of garlic EO is related to evaporation of the EO volatile compounds during storage [48]. Politeo et al. [49] stated that the decrease in volatile compounds of essential oil resulted in a decrease in its radical scavenging capacity. Kim and Kobayasi [50] explained that the concentration of diallyl disulfide decreases during storage. Diallyl disulfide is one of the major compounds of organosulfur antioxidant in garlic EO [51].

3.2.2 Reducing Power

Tabel 3. Reducing power (Ascorbic acid equivalent activity, AAEA) of coated beef meatballs with various edible coatings formulas during storage at 4±1°C.

Edible Coating	Reducing Power AAEA (mg//kg sampel)				
	0 day	5 days	10 days	14 days	18 days
Control	$3701.72\pm10.30^{b}_{A}$	$3695.77\pm10.30^{b}_{A}$	$3677.93\pm10.30^{a}_{A}$	$3671.98\pm10.30^{a}_{A}$	$3671.98\pm10.30^{a}_{A}$
Chitosan	$9759.70\pm51.50^{c}_{B}$	$9551.60\pm68.13^{b}_{B}$	$9521.60\pm77.25^{b}_{B}$	$9343.50\pm77.25^{a}_{B}$	$9343.50\pm77.25^{a}_{B}$
Garlic EO	$10651.00\pm25.75^{c}_{C}$	$10071.00\pm25.75^{b}_{C}$	$10057.00\pm44.60^{b}_{C}$	$9566.50\pm44.6^{a}_{C}$	$9536.7\pm68.13^{a}_{C}$
Chitosan-garlic EO	$11172.00\pm0.00^{c}_{D}$	$11082.00\pm89.20^{bc}_{D}$	$10993.00\pm44.60^{b}_{D}$	$10413.00\pm44.04^{a}_{D}$	$10354.00\pm92.84^{a}_{D}$

Different superscript in the same rows and different subscript in the same columns mean significant differences ($p < 0.05$)

Table 3 indicated that the control sample exhibits reducing power. According to Udenigwe and Aluko [52], amino acids with sulfur groups (methionine and cysteine) have the higher reducing capacity than other amino acids. Beef has methionine and cysteine contents of around 1.99% and 1.21% respectively [53]. Meatball samples with the incorporation of chitosan or garlic EO showed higher AAEA than the control sample. The AAEA of beef meatballs coated with chitosan, garlic EO, and the chitosan-garlic EO edible coatings were 2.6 times, 2.9 times and 3.0 times higher than that of the control ones. This is related to the reducing power of chitosan and garlic EO (Table 1). The reducing power of chitosan due to its amine group (-NH2) [54]. EO showed reducing capability [10] related to the organosulfur compounds. Meatballs coated with chitosan edible coatings exhibited a lower reducing power than that of garlic EO edible coatings ones. This result was in line with the reducing power data of garlic EO and chitosan presented in Table 1.

The value of RP in meatballs with chitosan-garlic EO treatment showed an antagonistic effect, indicated by its lower RP value compared to the sum of RP values of the individual treatments (chitosan treatment and garlic EO treatment). Antagonistic effects that occur may be attributed to the competition between the two compounds to reduce Fe^{3+} ion or the limited of potassium ferricyanide available as reduced substance in the reaction system.

Similar to the RSA, the RP also decreased with increasing storage time. As explained in the RSA section, this result also related to an increase in chitosan molecular weight as well as reduced of garlic EO volatile compounds during storage. The RP of LMWC was higher than that of HMWC [55,37]. The decrease in the concentration of essential oil volatile components also resulted in a decrease of reducing power [49].

4. **Conclusions**

At all storage periods, meatballs coated with edible coatings incorporated with chitosan, garlic EO, and chitosan-garlic-EO showed significantly higher antioxidant capacity than control (without any incorporation). Incorporation of chitosan-garlic EO garlic into edible coatings provides the highest antioxidant activity of coated meatballs. The antioxidant activity of beef meatballs coated with chitosan-EO garlic edible coatings showed a synergistic effect as a primary antioxidant (radical scavenging activity) and antagonistic effect as a secondary antioxidant (reducing power). It can be concluded that the incorporation of chitosan-garlic EO into beef meatballs edible coatings provide an alternative antioxidant-based functional food development, especially as the radical scavenger.

References

[1] Dewi A C N and Mahmudiono T 2013 *MGI* **9** (1) 42-8
[2] Akcan T, Estevez M and Serdaroglu M 2017 *LWT - Food Science and Technology* **77** 323-31
[3] Bourtoom, T. 2008. *IFRJ* **15** (3) 237-48
[4] Kontominas, M. 2016. Bioactive Good Packaging (USA: DEStech Publications, Inc)
[5] Carocho M and Ferreira I C F R 2013 *Food and Chem. Toxicol.* **51** 15–25
[6] Huang D, Ou B and Prio R *2005 J. Agric. Food Chem.* **53** 1841-56
[7] Halliwell B and Gutteridge J M C 2015 *Free Radicals in Biology and Medicin*e Fifth Edition (UK: Oxford University Press).
[8] Kim K W and Thomas R L 2007 *Food Chem.* **101** (1) 308-13
[9] Yen M T, Yang J H and Mau J L 2008 *Carbohydrate Polymers* **74** (4) 840–44
[10] Lawrence R and Lawrence K 2011 *Asian Pac J Trop Biomed* **1** (1) S51-4
[11] Chekki R Z, Snoussi A, Hamrouni I and Bouzouita N 2014 *Med J Chem* **3** (4) 947-56
[12] Mnayer D, Fabiano-Tixier A-S, Petitcolas E, Hamieh T, Nehme N, Ferrant C, Fernandez X and Chemat F 2014 *Molecules* **19** 20034-53 doi:10.3390/molecules191220034
[13] Abdel-Salam A F, Elaby S M and Ali J B 2014 *Afr. J. Microbiol. Res.* **8** (27) 2574-82.
[14] Choe E and Min D B 2009 *CRFSFS* **8** 345-58.
[15] Moradi M, Tajik H, Rohani S M R, Oromiehie A R, Malekinejad H, Aliakbarlu J and Hadian M 2012 LWT- Food Science and Technology 46 (20) 477–84.
[16] Ruiz-Navajas Y, Viuda-Martos M, Sendra E, Perez-Alvarez J A and Fernandez-Lopez J 2013 Food Control 30 (2) 386–92
[17] Lopez-Mata M A, Ruiz-Cruz S, Silva-Beltran N P, Ornelas-Paz J D, Ocano-Higuera V M, Rodriguez-Felix F and Shirai, K 2015 International Journal of Polymer Science 1 – 8.
[18] Hromis N M, Lazic V L, Markov S L, Vastag Z G, Popovic S Z, Suput D Z, Džinić N R, Velićanski A S and Popovic L M. 2015 Journal of Food Engineering 158 86–93
[19] Georgantelis D, Ambrosiadis I, Katikou P, Blekas G and Georgakis S A 2007 Meat Science 76 172–81
[20] Shaheen M S, Shaaban H A, Hussein A M S, Ahmed M B M, El-Massry K and El-Ghorab A 2015 Pol. J. Food Nutr. Sci 66 (4): 295-02.
[21] Ghaderi-Ghahfarokhi M, Barzegar M, Sahari M A and Azizi M H 2016 Food Bioprocess Technol. 9 (7) 1187–201
[22] Hamedi H, Kargozari M, Shotorbani P M, Mogadam N B and Fahimdanesh M. 2017 Food Hydrocolloids 72 35-46
[23] Catarino M D, Alves-Silva J M, Fernandes R P, Gonçalves M J, Salgueiro L R, Henriques M F and Cardoso S M 2017 Food Control 80 273-80
[24] Ponce A J, Roura S I, del Valle C E and Moreira M R 2008 Postharvest Biology and Technology 49 294–300
[25] Kanatt S R, Rao M S, Chawla S P and Sharma A 2013 LWT - Food Science and Technology 53 321-26
[26] Yuan G, Chen X and Li D. 2016. Food Research International 89 117–128
[27] Noori S, Zeynali F and Almasi H 2018 Food Control 84 312 -20.
[28] Rezaei F and Shahbazi Y 2018 LWT - Food Science and Technology 87 122-33
[29] Choulitoudi E, Bravou K, Bimpilas A, Tsironi T, Tsimogiannis D and Taoukis P 2016 Food and bioproducts processing 100 570–77
[30] Komariah N U and Hendrarti E N 2005 Sifat Fisik Bakso Daging Sapi dengan Jamur Tiram Putih sebagai campuran Bahan Dasar (Skripsi) (Bogor: Institut Pertanian Bogor)
[31] Amalia RR and Putri W D R 2014 Jurnal Pangan dan Agroindustri 2 (3) 43-53
[32] Hadi H N S S 2008 Aplikasi Kitosan dengan Penambahan Ekstrak Bawang Putih sebagai Pengawet dan Ed*ible Coating Bakso Sapi* (Tesis) (Bogor: Institut Pertanian Bogor)

[33] Sharma O P and Bhat T K 2009 Food Chem. 113 1202–05 doi:10.1016/j.foodchem.2008.08.008
[34] Berker KI, Güçlü K, Tor I, Apak R Apak 2007 Talanta 72 (3) 1157-65. doi: 10.1016/j.talanta.2007.01.019
[35] Pokorny J, Yanishlieva N and Gordon M 2001 Antioxidants in Food: Practical Applications. (England: Woodhead Publishing Limited)
[36] Kuppusamy S K and Karrupaiah J 2013 Int. J. of Pharm. & Life Sci. 4 (5) 2685-90
[37] Wu C, Wang L, Fang Z, Hu Y, Chen S, Sugawara T and Ye X 2013. Mar. Drug 14 (95) 1-18.
[38] Prior R L, Wu X and Schaich K. 2005 J. Agric. Food Chem. 53 (10): 4290-4302. doi: 10.1021/jf0502698
[39] Wei L, Li Q, Tan W, Dong F, Luan F and Guo Z 2017 Molecules 22 (501) 1-11.
[40] Pasanphan W and Chirachanchai S 2008 Carbohydrate Polymers 27 169-77.
[41] Satyal P, Craft J D, Dosoky N S and Setzer W N Foods 6 (63) 1-10. doi:10.3390/foods6080063
[42] Zhao N N, Zhang H, Zhang X C, Luan X B, Zhou C, Liu Q Z, Shi W P, Liu Z L 2013 J Econ Entomol. 106 (3) 1349-54.
[43] Lawson L D, Zhen-Yu J W and Bronwyn G H 1991. Planta Med 57 363-70.
[44] Colin-Gonzales A L, Santana R A, Silva-Islas C A, Chanez-Cardenas M E, Santamaria A and Maldonado P D 2012 Oxidative Medicine and Cellular Longevity 1-16. doi:10.1155/2012/907162
[45] Nimse SB and Pal D 2015 RSC Adv. 5 27986-28006. doi: 10.1039/C4RA13315C
[46] Dhawade P P and Jagtap R N 2012 Adv. Appl. Sci. Res. 3 (3) 13-72-82.
[47] Chien P-J, Sheu F, Huang W-T and Su M-S 2007 Food Chem. 102 (4) 1192-98
[48] Turek C and Stintzing F C 2013. CRFSFS 12 40-53.
[49] Politeo O, Jukic M and Milos M 2007 Food Chem. 101 (1) 379-85.
[50] Kim S M and Kobayashi A 1997 Biosci. Biotech. Biochem 61 (9) 1482-85.
[51] Atmaca G. 2004. Antioxidant effects of sulfur-contaaining amino acids. Yonsei Medical Journal 45 (6) 776-88
[52] Udenigwe C C and Aluko R E 2011 International Journal of Molecular Science 12 3148-61.
[53] Hollo G, Csapo J, Szucs E, Tozser J, Repa I and Hollo I 2001 Asian-Australasian Journal of Animal Sciences 14 (11) 1555-59
[54] Sarbon N M, Sandanamsamy S, Kamaruzaman S F, Ahmad F. 2014 J. Food Sci. Technol. 52 (7) 4266-75
[55] Xing R, Liu S, Guo Z, Yu H, Wang P, Li C, Li Z and Li P 2005 Bioorg Med Chem 13 (5) 1573-77.

IOP Conf. Series: Materials Science and Engineering **633** (2019) 012033 doi:10.1088/1757-899X/633/1/012033

The use of solid phase spectrophotometry for control toxic metal ion content in milkfish cultivation

S Saputro, L Mahardiani and Ninda DCD

Chemistry Education, Faculty of Teacher Training and Education, Sebelas Maret University, Jl. Ir. Sutami 36A Surakarta, Central Java-57126, Indonesia

E-mail: sulistyo_s@staff.uns.ac.id

Abstract. Solid-Phase Spectrophotometry (SPS) has been widely used in the study of trace element speciation in water. This spectrophotometer has advantages that have a high sensitivity, a small device (portable), and easy to use in the analysis process. This study aims to: (1) increase the use of SPS values in controlling toxic ions in water used for milkfish cultivation, and (2) determine SPS sensitivity for the determination of Cd(II) in pond water samples analyzed. The method used is laboratory experiment. The result showed that the SPS method can be used in the analysis of the content of Cd(II) in the pond water with such a high enough sensitivity results obtained LoD of 0.51 µg/L, RSD < 5% and recovery of 95-105%.

1. Introduction

Fish has a high nutrient content, especially protein and unsaturated fats. Milkfish is one of the cultivated fish commodities that are popular with people because of its high protein content (20-24%) and other ingredients (amino acids, fatty acids, minerals and vitamins) that are high. Milkfishes cultivation is a lucrative industry sector. The development of milkfish cultivation has a good trend with the yield of milkfish production from milkfish ponds in Central Java reaching 64,305 tons with a production value of Rp. 643,052,700,000.- in 2011[1].

Ponds Milkfish is mostly in coastal areas (estuary water) and most are mixed brackish water ponds from sea water and river water (downstream of the river). Heavy metals are increasingly found in natural water sources and rivers and sea water. Based on previous research on the content of Cd(II), many seawater showed high yields of pollution. Downstream has a wide possibility of pollution from upstream, midstream and downstream (the pollution accumulates downstream (as long as there is no metal sedimentation). This is consistent with Harsono's research about the content of heavy metals in the river in Juwana which arefamous for producing milkfish has a highly toxic heavy metal content [3]. The location of milkfish cultivation that is susceptible to pollution causes periodic monitoring of the content of toxic metal in pond water to ensure the quality of the milkfish produced.

The toxic ion content is usually a very small concentration. Atomic absorption spectroscopy, a device commonly used to measure metal content for the µg/L level, so it needs a more accurate method. In addition, a more practical analysis method is needed so that it can be used to periodically control heavymetal content. In 1976 a new analytical method was reported, namely Solid-Phase Spectrophotometry [6]. After the analytical method, more advanced research and development has emerged with various chemical species. Solid-Phase Spectrophotometry has been Solid-Phase

Spectrophotometry (SPS) is a sensitive measurement method in the determination of trace element metal ion content that has been proven in previous studies to examine Cr (VI), Pb (II) and As III) ion contentand also SPS is a sensitive method used in pharmaceuticals for example in determining phenol concentrations [9] [8] [7] [5].SPS is also a practical method with a relatively small tool size and does not require preconcentration of the sample to be tested so that it can be used for in-situ analysis [10]. Therefore, in this study, the authors tried to use SPS to determine theCd (II) in the water of the ponds milkfish.

2. Methods

2.1 Tools and Materials

The tools used are a Solid-Phase Spectrophotometry (SPS) K-MAC Lab Junior (figure 1), Satoda Science, Japan; analytic balance; timer; Magnetic stirrer; pipette drops; volumetric equipment; measuring flask; micro pipette; volumetric pipette aliquotting device assembled using a syringe, teflon capillary pipe and other devices used to extract 0.06 ml of resin and beaker glass. Material used are the standard solution of $CdNO_3$ 1000 mg/L; 1 M NaOH; ditizon powder (Merck); muromac AG 50W X2 100-200 mesh resin and aquadest.

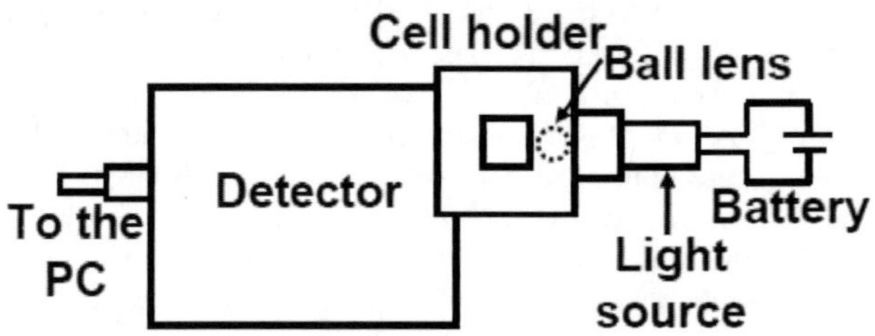

Figure1. Simple Spektrofotometer SPS [10]

2.2 Methods

2.2.1 *Sample Collection and Preparation.* Pond water is measured DO, pH and temperature first. Samples of pond water were taken by dipper (figure 2) at a depth of about 30 cm, then filtered with a cellulose acetate filtering unit, after that, put the sample into a PTFE bottle.

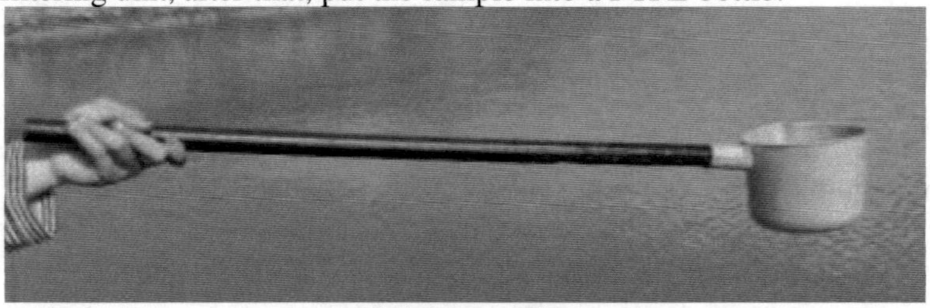

Figure 2. Dipper

2.2.2. *Making Resin solutions.*Muromac AG 50 W X2 100-200 mesh resin was dissolved in a distilled water and left for a while until the resin becomes more bubbly.

2.2.3 *Manufacture of Ditizon Complexing Solution.*Ditizone powder weighing 5 mg was dissolved into 100 ml of chloroform

2.2.4 *Preparation of Cd(II) Standard solution of 100 mg/L.*Cd(II) stock solution 1000 mg/L 2.5 mL was dissolved in a distilled water until the volume is 25 ml.

2.2.5 *Preparation of standard solution of Cd(II) 1 mg/L (1000 µg/L).*Cd(II) solution of 100 mg/L weighing 5 grams was added to aquadest until weighing 50 grams.

IOP Conf. Series: Materials Science and Engineering **633** (2019) 012033 doi:10.1088/1757-899X/633/1/012033

*2.2.6 Preparation of Cd(II) Standard Solution 100 µg/L.*Cd(II) 1000 µg/L 10 mL in volume was dissolved in a distilled water until the volume is 100 ml

2.2.7 Preparation of standard solution of Cd(II) 0 µg/L, 20 µg/L, 40 µg/L, and 80 µg/L.Standard solutions of Cd(II) 0µg/ L, 20µg/ L, 40µg/ L, and 80µg/L, were prepared by diluting Cd 100µg/L solution with distilled water.

*2.2.8 Procedure for Determining Cd(II) Calibration Curves.*Into 20 mL standard Cd (II) solution 0 µg / L, 20 µg / L, 40 µg / L, and 80 µg / L, 0.2 mL 1M NaOH was added. Then, 1 mL of ditizone was added. After that, 0.06 mL resin was added and stirred using a stirrer for 20 minutes and the samples were analyzed using Solid Phase Spectrophotometry with a wavelength of 485 nm and 615 nm. Then the difference in absorbance of two wavelengths is taken, namely ΔA = A 485nm - 615nm where where ΔA is used in making the standard curve Cd (II) (ΔA vs. concentration).

*2.2.9 Procedure for Determining Cd(II) Species in a Sample.*Into 20 mL standard Cd (II) solution 0 µg / L, 20 µg / L, 40 µg / L, and 80 µg / L, 0.2 mL 1M NaOH was added. Then, 1 mL of ditizone was added. After that, 0.06 mL resin was added and stirred using a stirrer for 20 minutes and the samples were analyzed using Solid Phase Spectrophotometry with a wavelength of 485 nm and 615 nm. Then the difference in absorbance of two wavelengths is taken, namely ΔA = A 485nm - 615nm where ΔA obtained has been substituted into the standard curve equation Cd (II) so that the species Cd (II) in the sample is known.

*2.2.10 Determination of Detection Limit.*The limit of detection is calculated using the method 3σ (standard deviation), which is taken 20 ml of aqueous solution 5 times, then 0.2 mL of 1M NaOH is added, 1 mL of ditizone solution and 0.06 ml of resin has been stirred for 20 minutes and the samples were analyzed with Solid Phase Spectrophotometry with a wavelength of 485 nm and 615 nm. Then the difference in absorbance of two wavelengths is taken, namely ΔA = A 485nm - A 615nm, where ΔA is substituted into the standard Cd curve, then the concentration and detection limits can be calculated.

*2.2.11 Determination of recovery.*Recovery was determined by making a standard solution of Cd(II) 0µg/L, 40µg/ L, 80µg/L, and 160µg/L with solvent is one sample solution. Then 0.2 mL of NaOH 1 M, 1 mL dithizone and 0.06 mL resin were added. After that the solution was stirred using a stirrer for 20 minutes and the samples were analyzed using Solid Phase Spectrophotometry with a wavelength of 485 nm and 615 nm. Then the absorbance difference from the two wavelengths is taken, namely ΔA = A 485nm - 615nm where ΔA is substituted into the standard Cd(II) curve.The obtained concentration then compared between the level of readability of the instrument and the results of the calculation.

3. Result and discussion

3.1 Collecting samples

Sample P1 is taken from the pond with the closest distance from the residential area, the P2 is taken out of the pond with a distance of 2 plots of P1, P3 is taken out of the pond with a distance of 2 plots of P2, P4 is taken out of the pond with a distance of 2 plots of P3.

Table 1 Data of Temperature, pH and Do water sample

Sample	Temp °C	pH	DO
P1	37.8	6.64	10.20
P2	38.5	6.10	9.85
P3	39	5.90	10.40
P4	38.8	6.14	10.84

Table 1 contains data of temperature, pH and DO water sample. The data demonstrate that the temperature of the pond water when sampling is quite hot. This is explained by the fact that the sampling time is during the day. pH from 4 pond water shows a value of <7. This shows that pond water is slightly acidic. The pH of pond water should be neutral, so that pH neutrality is needed. The value of dissolved oxygen (DO) of all pond water is very high. This is not suitable for normal conditions at a pressure of 1 atm with an averagetemperature of 38°C where normal DO in this condition are around 6.62 mg/L while the DO content in the sample ranges from 10 mg/L or in other words there is a condition of supersaturation. This is influenced by samples carried out during the day. According to Jeffries and Mills, this is because during the day the release of oxygen in the photosynthetic process which runs intensively and dissolved oxygen levels can exceed saturated oxygen levels and the water experience supersaturation [2].

3.2 *Determination of Cd(II) calibration curve*

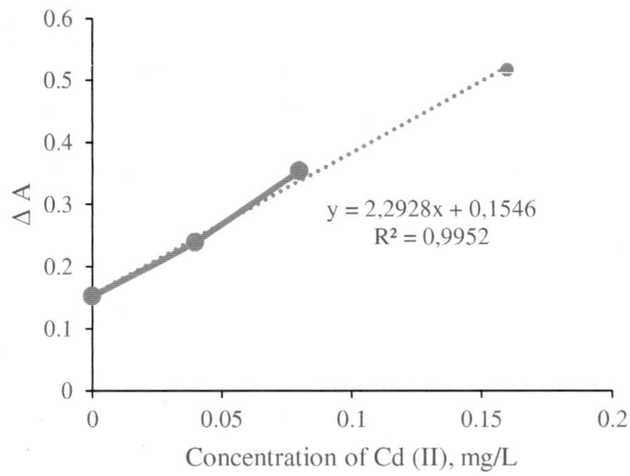

Figure 3. Standard Curve of Cd(II)

From the standard curve (figure 3) is obtained an equation that is y= 2.2928x + 0.1546 where y is ΔA and x are the concentration. These equations will be used to determine the metal concentration of Cd (II) metal ions in samples and test solutions

3.3 Determination of Cd(II) levels in water sample

Based on data in table 2, the concentration of Cd (II) in pond water samples ranged from 0.044 mg/L (sample P4) to 0.10 mg/L (sample P2). The difference in the concentration of Cd (II) in each sample can be seen from the physical condition of the sample where the P2 sample with the largest Cd (II) content has a more dirty water color and there are even some milkfish that die and float on the surface of the water. P2 sample has the smallest DO value. It is appropriate that at relatively low DO water organism will have difficulty surviving.

Table 2. Results of Determination of Cd (II) levels in the sample

Sample	**ΔA**	**Concentraton of Cd(II) (mg/L)**
Sample P1	0.357456	0.089
Sample P2	0.386113	0.10
Sample P3	0.318853	0.072
Sample P4	0.256329	0.044

3.4 Determination of detecting limit

One way to verify the analysis method is by determining the detection limit (limit of detection) and recovery. Detection limit (LoD) is the smallest limit test parameter owned by a tool/instrument. The valuation and determination of detection limits depend on the analysis carried out or what tools/ instruments are used in the analysis. In the experiments that have been carried out, the detection limit value is 0.00051 mg/L or 0.51 μg/L. This shows that at a concentration of Cd (II) of 0.51 μg/L the sample is still legible. From the experiment, it was proved that Solid-Phase Spectrophotometry (SPS) was still able to measure the sample levels up to a concentration of Cd (II) 0.51 μg/L. When compared with the results of a study on verification of the AAS (Atomic Absorbance Spectroscopy) method for determination of cadmium metal that has been carried out by Sasongko, Yulianto, and Sarastri which obtained the detection limit of 0.0045 mg/L, the SPS analysis capability is still superior to AAS [11].

3.5 Determination of Precision

Precision tests were carried out to measure the level of accuracy of the test method. Precision is measured through the distribution of individual results from the average if the procedure is applied repeatedly in samples taken from a homogeneous mixture [4]. Precision testing is performed by calculating RSD from the detection limit data. Based on the calculation results obtained RSD amounted to 3.68%. These results have met the required accuracy of RSD <5%[9].

3.6 Determination of recovery

The accuracy of a measurement can be defined as the suitability between the determination and the actual value. Accuracy is stated by recovery calculation. Recovery is expressed as the ratio between the results obtained and the actual results. The results of this study to comply with the requirements recovery value 95-105%.

4. Conclusion

Based on the results of the research conclusions that can be taken, solid phase spectrophotometry can be used to determine the concentration of Cd(II) in pond water samples with a low detection limit (0.51 μg/L) and have an accuracy and precision that meet the criteria of measurement instruments. Solid phase spectrophotometry is a suitable method to control Cd(II) levels in milkfish ponds regularly because of the ease in the analysis.

References

[1] Andriyanto S 2013 Kondisi terkini budidaya ikan bandeng di Kabupaten Pati, Jawa Tengah *Media Akuakultur 8*(**2**) 139-144

[2] Effendi, Hefni 2003 *Telaah kualitas air: bagi pengelolaan sumber daya dan lingkungan peraian*Yogyakarta Penerbit Kanisius

[3] Harsono D R S 2002 Analisis tingkat pencemaran air sungai di daerah estuari Jawa Tengah*Badan Penelitian dan Pengembangan Propinsi Jawa Tengah*

[4] Harmita H 2012 Petunjuk pelaksanaan validasi metode dan cara perhitungannya*Pharmaceutical Sciences and Research (PSR) 1*(**3**) 117-135

[5] J Vucovic, S Matsuoka, K Yoshimura, R J Grubecic, D Kremer, N Santic 2012 *Analytical Sciences* ***28*(4)** 397-402

[6].S Matsuoka K Yoshimura 2010 Recent trends in solid phase spectrometry: 2003–2009. A Review*Analytica Chimica Acta* **664** 1-18

[7] S Saputro, M Masykuri, L Mahardiani, B Mulyani, I Qorina, K Yoshimura, K Takehara, S Matsuoka 2017 The usage of activated carbon from teak sawdust *(tectona grandis* l.f.*)* and zeolite for the adsorption of Cr(VI) and its analysis using solid-phase spectrophotometry (sps).In IOP Conf. Series: Materials Science and Engineering Vol 176 No 1 p 012019

[8] S Saputro, K Yoshimura, S Matsuoka., K Takehara, J Aizawa and Y Tennichi 2014 *Chemical Geology* **364** 33-42

[9] S Saputro, K Yoshimura, K Takehara, S Matsuoka, and Narsito 2009 *Analytical Sciences* **25** 1445-1450

[10] Sarenqiqige, S Saputro, S KAI, M Satoda, S Matsuoka and K Yoshimura 2013 *Analytical Sciences* ***29*** 677-680
[11] A Sasongko, K Yulianto, D Sarastri 2017 Verifikasi metode penentuan logam kadmium (Cd) dalam air limbah domestik dengan metode spektrofotometri serapan atom *Jurnal Sains dan Teknologi.* **6(2)** 228-237

Acknowledgement
The author would like to thank the Chemistry Laboratory of Faculty of Teacher and Training Education and the Integrated Center Laboratory of Sebelas Maret University for giving permission to conduct this research.

IOP Conf. Series: Materials Science and Engineering **633** (2019) 012034 doi:10.1088/1757-899X/633/1/012034

Elicitation under salinity stress increases flavonoid content and antioxidant activity in cowpea (*Vigna unguiculata*) sprouts

F M Rajendra, L S Kristiani and S Ariviani

Department of Food Science and Technology, Faculty of Agriculture, Universitas Sebelas Maret, Jl. Ir. Sutami 36 A, Kentingan, Surakarta 57126, Indonesia

Email: setya_ariviani@yahoo.com; setyaningrum_ariviani@staff.uns.ac.id

Abstract. Legumes are often used as a source of natural antioxidant. Elicitation is a promising alternative way of improving antioxidant compounds in legumes sprouts, such as flavonoid compound. NaCl stress can be used as one of abiotic elicitation that induced non-enzymatic defense in a plant, thus increases secondary metabolites which enhance the antioxidant capacity. However, its effects on cowpea (Vigna unguiculata) germination need to be more studied. In this study, we germinated cowpea under increasing salinity (0, 50, 100, 150mM NaCl) to investigate its effect on the total flavonoid content and antioxidant activity (radical scavenging activity and reducing power). Total flavonoid content and radical scavenging activity of cowpea sprouts increase along with increasing NaCl concentration. Meanwhile, only 150mM NaCl showed significantly higher reducing power among other concentrations. Total flavonoid content have a high correlation with radical scavenging activity ($r=0,962$; $p<0,01$) but not correlated with reducing power ($r=0,137$; $p>0,05$). This research results proved that elicitation using 150mM NaCl could be used as one of the strategies to enhance bioactive compound and antioxidant activity in legumes, thus increasing its potential to be developed as an antioxidant-based functional food.

1. Introduction

Reactive Oxygen Species (ROS) is one type of free radical compound [1]. In a normal cell, there is a balance condition between ROS production and antioxidant [2]. Excessive ROS production in the human body can cause oxidative stress which leads to oxidative damage towards fat, protein, and DNA [3,4]. Therefore, the intake of rich antioxidant food is needed to keep the balance of ROS and antioxidant.

Legume can be used as one of the natural antioxidant sources. Legume contains several substances that can improve health, such as phenolic and flavonoid. Several studies have proved that cowpea (*Vigna unguiculata*) contains phenolic and flavonoid compounds [5,6].

Germination process is a common way to increase antioxidant capacity in legume [7–9]. There are several studies that report the enhancement of antioxidant capacity in cowpea during the germination process [10,11]. Elicitation during germination process is one of the factors which affect the increase of antioxidant compound in legume sprouts.

Elicitation is an efficient technique to increase phenolic and other antioxidant compounds of plants as the stress response [12,13]. Elicitation can be applied using both of biotic and abiotic elicitor. The

IOP Conf. Series: Materials Science and Engineering **633** (2019) 012034 doi:10.1088/1757-899X/633/1/012034

advantages of using abiotic elicitor are cheaper and relatively easy to be applied [12]. Salt stress is one of abiotic elicitation that can be applied to plants [14,15]. Na^+ ion gives toxicity to the plant, while Cl^- is anion which most frequently found in soil or water which has high salt content. Na^+ and Cl^- ions are easier to get inside a cell compared to the other ions. An excess of Na and Cl ions induce plant's cell damage. Moreover, the high NaCl concentration outside the plant's cell caused stress due to water deficit as a result of osmotic pressure [16]. Swieca [12] reported that elicitation using NaCl in lentil sprouts produced higher antiradical activity and reducing power rather than other abiotic elicitation.

Previous studies reported improvement of the antioxidant capacity of legumes sprouts elicited with NaCl along with increasing of NaCl concentration, such as lentil [12], mungbean [17,18], and common bean [19]. To the best of our knowledge, the study on the antioxidant capacity enhancement using NaCl elicitation with a various concentration in cowpea sprouts hasn't been reported. Thus we hypothesized an increase in flavonoid levels and antioxidant activity (radical scavenging and reducing power) are going to be observed in cowpea sprouts as a response to NaCl stress. Therefore, the present research aims to evaluate the effect of increasing NaCl levels (0, 50, 100, 150mM) on the improvements of total flavonoid content as well as antioxidant activity (radical scavenging activity and reducing power) of cowpea sprouts. The correlation between the total flavonoid content and the antioxidant activity was also analyzed.

2. Methods

2.1 Preparation of Sprouts

Cowpea (*Vigna unguiculata*) seeds were purchased from the local market in Surakarta, Indonesia. Selected seeds (not wrinkled and not hollow seeds) were rinsed thrice with distilled water. Afterward, the seeds were soaked in NaCl solution (0, 50, 100, 150mM) with a ratio of 1:3 w/v for 8 hours at room temperature. The elicited seeds were germinated for 48 hours on bamboo weaved (15 cm x 15 cm). During germination, seeds were sprayed with distilled water every 12 hours.

2.2 Extract Preparation

Dehulled cowpea sprouts (2 grams) were extracted with 20 ml of the methanol: water (80:20, v/v) at 50°C using a water bath shaker (SWB 20, Fisher Scientific Haake, Germany) at 200 rpm for 2 hours. The extract was separated by centrifugation (PLC-05 Sentrifuge, Gemmy, Taiwan) at 10,000 rpm for 15 minutes. The clear supernatant was collected and stored in an amber bottle at 10°C until further analyzed.

2.3 Total Flavonoid Content Analysis

Total flavonoid content was determined following the method described by Pękal and Pyrzynska [19]. Total flavonoid content was expressed as QE (µM Quercetin Equivalent/g sample dry weight).

2.4 Radical Scavenging Activity Test

Radical scavenging activity was analyzed using $ABTS^{\bullet+}$ free radical method [20]. The free radical scavenging activity was expressed as TEAC (Trolox Equivalent Antioxidant Capacity/g sample dry weight).

2.5 Reducing Power Test

Determination of reducing power was conducted using the method described by Berker et al [21]. The reducing power was expressed as AAEA (µM Ascorbic Acid Equivalent Activity/g sample dry weight).

2.6 Statistical Analysis

Data were presented as mean and standard deviations. The data of total flavonoid content, radical scavenging as well as the reducing power of cowpea sprouts were analyzed using the program IBM

SPSS Statistics 22 (SPSS Inc., Chicago, USA) by analysis of variance (ANOVA). Duncan's multiple range test (DMRT) was used to evaluate the significant differences between mean ($p<0.05$). Pearson correlation analysis was used to determine the correlation between total flavonoid content and antioxidant activity.

3. Results and Discussion

Salt stress increases ROS accumulation due to the increase of the respiration rates as a response to the stress [22]. Plants will develop a defense mechanism to reduce ROS excess, through secondary metabolite production such as phenolic and flavonoid compounds [12]. Total flavonoid content and radical scavenging activity, as well as the reducing power of cowpea sprouts which prepared with elicitation using various NaCl concentrations, were presented in Table 1.

Table 1. Flavonoid total content and antioxidant activity of NaCl elicited cowpea sprout with increasing NaCl concentration (0, 50, 100, 150mM).

NaCl concentration (mM)	Total flavonoid content (μM QE/g sprout dry weight)	ABTS•$^+$ radical scavenging activity (mM TEAC/g sprout dry weight)	Reducing power (μM AAEA/g sprout dry weight)
0	250.41 ± 2.42^a	952.16 ± 26.47^a	6790.20 ± 43.02^a
50	398.49 ± 8.72^b	1130.11 ± 43.01^{bc}	6484.93 ± 226.25^a
100	418.01 ± 7.19^c	1192.89 ± 41.54^{cd}	6539.56 ± 188.21^a
150	444.98 ± 8.71^d	1259.51 ± 29.72^d	7237.60 ± 283.53^b

Different superscript within similar column means significant differences ($p<0.05$)

NaCl concentration significantly affects the total flavonoid content of cowpea sprouts. The flavonoid content increased as the NaCl concentration raised (Table 1). Taïbi et al [18] that studied the effect of salt stress on antioxidant defense systems in *Phaseolus vulgaris* reported a similar result. Total flavonoid content in *Phaseolus vulgaris* sprouts increase along with increasing NaCl concentration. Study of Świeca [12] showed that lentil sprouts elicited with higher NaCl concentration showed significantly higher ($p<0.05$) total flavonoid content.

The results in Table 1 also exhibited that the total flavonoid content of cowpea sprouts with NaCl elicitation was significantly higher than that of cowpea sprouts without elicitation (0mM NaCl). ROS accumulation in the plant's cell due to osmotic pressure disruption and toxic ion from salt during elicitation lead the phenolic and flavonoid overproduction [22]. Increasing of total flavonoid content in cowpea sprouts prepared with NaCl elicitation was caused by salinity stress that triggers osmotic stress tolerance through osmolyte formation, such as proline. Higher NaCl concentration leads to increasing of proline accumulation [23]. Shetty [24] stated that proline synthesis is related with overproduction of erythrose-4-phosphate that plays a role in shikimate and phenylpropanoid pathway which induce phenolic compounds synthesis. Moreover, Swieca [12] reported that increasing of NaCl concentration cause enhancement of phenolic compound in lentil sprouts. One of the major phenolic compounds in cowpea sprout is p-coumaric acids [10,25,26]. P-coumaric acids might be involved in flavonoids biosynthesis via Coumaryl Coa [10,27–29].

Flavonoid and phenolic compounds exhibit several health benefits related to the antioxidant capacity [30]. Based on its mechanism, an antioxidant is divided into two types, those are a primary and secondary antioxidant. The primary antioxidant has the ability to be free radical scavenger by electron or hydrogen donating, while secondary antioxidant has the ability to reduce another component in order to inhibit free radical generation indirectly [31].

NaCl concentrations not only affects total flavonoid content but also significantly ($p<0.05$) affects the radical scavenging activity of cowpea sprouts. The higher NaCl concentration showed significantly higher ($p<0.05$) radical scavenging activity. This result in agreement with Swieca [12] who reported that lentil sprouts with elicitation using 300mM exhibited higher radical scavenging activity than that of 100mM NaCl ($p<0.05$). The radical scavenging activity enhancement observed in this study could be attributed to the increase in total flavonoid content.

Besides radical scavenging activity, increasing NaCl concentration as elicitor also significantly ($p<0.05$) affects the reducing power of cowpea sprouts. It was interesting that cowpea sprouts without elicitation exhibited the reducing power that did not significantly different ($p>0.05$) from those elicited using 50 and 100mM NaCl, although it showed significantly lower total flavonoid content. This phenomenon is possible because of the reducing power did not only attributed to the flavonoid compounds but also other reductants such as ascorbic acid. Guo et al. [32] reported that the increase in NaCl concentrations decreased ascorbic acid content in the sprouts of three broccoli cultivars. It was could be explained by two mechanisms, i.e NaCl stress inhibits the activity of ascorbic acid generating enzyme as well as ascorbic acid utilization to protect osmotic stress caused by salinity. Reduction of ascorbic acid level in cowpea sprouts elicited using various NaCl concentrations need further investigations.

Table 2. Correlation coefficient between flavonoid total and antioxidant activity of NaCl elicited cowpea sprout.

Antioxidant activity	Correlation coefficient (r)
$ABTS^{\bullet+}$ radical scavenging	0.962**
Reducing power	0.137

** Correlation analysis using Pearson correlation; $p<0.01$

Table 2 showed that the total flavonoid content was significantly correlated ($r=0.962$, $p<0.01$) with radical scavenging activity but not correlated with reducing power ($p>0.05$). It might be caused by major flavonoid contents in cowpea, such as flavonols and flavan-3-ols which have OH group in C3 [33,34]. OH group location in flavonoid compounds will affect energy needed to donate an electron. Flavonols show higher radical scavenging activity due to low energy needed to donate an electron. Moreover, flavan-3-ols also have high radical scavenging activity due to the effect of OH group in C3 and strengthened by the presence of 3'-4'-catechol [35,36]. Several studies showed that reducing power might be more attributed by other antioxidant compounds, such as ascorbic acid than phenolic and flavonoid compound [37,38].

4. Conclusions

The result of this study showed that increasing of the total flavonoid content, radical scavenging activity, and reducing power are affected by NaCl concentration used in elicitation. The higher NaCl concentration produces cowpea sprouts with higher antioxidant capacity. Although the radical scavenging activity of cowpea sprouts has a significantly high correlation to the total flavonoid content, the reducing power did not correlate to the total flavonoid content. The reducing power of cowpea sprouts may be more attributed by other reducing agents such as ascorbic acid rather than flavonoid compounds. Elicitation using 150mM NaCl is a promising strategy to increase the antioxidant capacity of cowpea sprouts thus more potential to be developed as an antioxidant-based functional food.

References

[1] Gulcin I 2012 *Arch Toxicol.* **86** 345–91.
[2] Birben E, Sahiner U M, Sackesen C, Erzurum S and Kalayci O 2012 *World Allergy Organ J.* **5** 9–19.
[3] Rahal A, Kumar A, Singh V, Yadav B, Tiwari R, Chakraborty S and Dhama K 2014 *Biomed Res Int.* **2014** 1–19.

[4] Sies H 2015 *Redox Biol.* **4** 180–3.
[5] Cai R, Hettiarachchy N S and Jalaluddin M 2003 *J Agric Food Chem.* **51** 1623–7.
[6] Ojwang L O, Dykes L and Awika J M 2012 *J Agric Food Chem.* **60** 3735–44.
[7] Tang D, Dong Y, Ren H, Li L and He C 2014 *Chem Cent J.* **8** 1–9.
[8] Tarzi B G, Gharachorloo M, Baharinia M and Mortazavi A 2012 *Iran J Pharm Res.* **11** 1137–43.
[9] Saleh H M, Hassan A A, Mansour E H, Fahmy H A and El-Bedawey A E F A 2018 *J Saudi Soc Agric Sci.* **2017** 1–8.
[10] Khang D, Dung T, Elzaawely A and Xuan T 2016 *Foods.* **5** 27–36.
[11] Aguilera Y, Díaz M F, Jiménez T, Benítez V, Herrera T, Cuadrado C, Martín-Pedrosa M and Martín-Cabrejas M A 2013 *J Agric Food Chem.* **61** 8120–5.
[12] Świeca M 2015 *Saudi J Biol Sci.* **22** 409–16.
[13] Dziki D, Gawlik-Dziki U, Kordowska-Wiater M and Domań-Pytka M 2015 *J Chem.* **2015** 1–8.
[14] Ramakrishna A and Ravishankar G A 2011 *Plant Signal Behav.* **6** 1720–31.
[15] Gorelick J and Nirit B 2014 *Adv Agron.* **124** 201–30.
[16] Azooz M M and Ahmad P 2015 *Legumes under Environmental Stress: Yield, Improvement and Adaptations* (Chichester: Wiley Blackwell).
[17] Saha P, Chatterjee P and Biswas A K 2010 *Indian J Exp Biol.* **48** 593–600.
[18] Taïbi K, Taïbi F, Ait Abderrahim L, Ennajah A, Belkhodja M and Mulet J M 2016 *South African J Bot.* **105** 306–12.
[19] Pękal A and Pyrzynska K 2014 *Food Anal Methods.* **7** 1776–82.
[20] Re R, Pellegrini N, Proteggente A, Pannala A, Yang M and Rice-Evans C 1999 *Free Radic Biol Med.* **26** 1231–7.
[21] Berker K I, Güçlü K, Tor I and Apak R 2007 *Talanta.* **72** 1157–65.
[22] Abogadallah G M 2010 *Plant Signal Behav.* **5** 369–74.
[23] Thakur M and Sharma A D 2005 *J Arid Environ.* **62** 517–23.
[24] Shetty K 2004 *Process Biochem.* **39** 789–803.
[25] Gutiérrez-Uribe J A, Romo-Lopez I and Serna-Saldívar S O 2011 *J Funct Foods.* **3** 290–7.
[26] Singh B, Singh J P, Kaur A and Singh N 2017 *Food Res Int.* **101** 1–16.
[27] Gan R, Lui W, Wang M and Sui Z 2016 *Functional Foods in Health and Disesase* **6** 519–35.
[28] Chemler J A, Yan Y and Koffas M A G 2006 *BioMed Central* **9** 1–9.
[29] Ngameni B 2013 *Medical Plant Research in Africa* (New York: Elsevier).
[30] Chen J, Lu J, Zhao H, Fan W, Dong J, Kong W, Sun J, Cao Y and Cai G 2008 *J Agric Food Chem.* **55** 10994–1001.
[31] Lobo V, Patil A, Phatak A and Chandra N 2010 *Pharmacogn Rev.* **4** 118–26.
[32] Guo L, Yang R, Wang Z, Guo Q and Gu Z 2014 *Int J Food Sci Nutr.* **65** 476–81.
[33] Apea-Bah F B, Minnaar A, Bester M J and Duodu K G 2014 *Food Chem.* **157** 157–66.
[34] Nassourou M A, Njintang Y N, Noubissiie T J B, Nguimbou R M and Bell J M 2016 *Crop J.* **4** 391–7.
[35] Fidrianny I, Elviana D and Ruslan K 2016 *Int J Pharmacogn Phytochem Res.* **8** 470–6.
[36] Heim K E, Tagliaferro A R and Bobilya D J 2002 *J Nutr Biochem.* **13** 572–84.
[37] Chai M Y, Azlan A, Al-Sheraji S H, Hassan F A and Nagendra Prasad K 2014 *Pakistan J Nutr.* **12** 1036–41.
[38] Ramesh C K, Rehman A, Prabhakar B T, Vijay Avin B R and Aditya Rao S J 2011 *J Appl Pharm Sci.* **1** 99–103.

IOP Conf. Series: Materials Science and Engineering **633** (2019) 012035 doi:10.1088/1757-899X/633/1/012035

Effect of different solvents and extraction conditions on fiber contents of local corns silk powders

Haslina[1,2], D Praseptiangga[3], V P Bintoro[4], and B Pujiasmanto[5]

[1] Doctoral Program of Agricultural Sciences, Graduate School of Sebelas Maret University (UNS), Jl. Ir. Sutami 36 A, Kentingan, Surakarta 57126, Central Java, Indonesia
[2] Faculty of Agriculture Technology, Semarang University, Jl. Soekarno Hatta, Semarang 50196, Central Java, Indonesia
[3] Department of Food Science and Technology, Faculty of Agriculture, Sebelas Maret University (UNS), Jl. Ir. Sutami 36 A Kentingan, Surakarta 57126, Central Java, Indonesia
[4] Department of Agriculture, Faculty of Animal and Agricultural Sciences, Diponegoro Univesity, Jl. Prof. Soedarto, SH, Tembalang, Semarang 50275, Central Java, Indonesia
[5] Department of Agrotechnology, Faculty of Agriculture, Sebelas Maret University (UNS), Jl. Ir. Sutami 36 A Kentingan, Surakarta 57126, Central Java, Indonesia

E-mail: chana_panca@yahoo.com

Abstract. The aim of this study was to optimize the different solvents and extraction conditions on fiber contents of local corn silk powders. Each sample of corn silk powder extract was analyzed by enzymatic method for fiber content including soluble fiber, insoluble fiber and total fiber. Design using Factorial Randomized Block Design (RBD) with 4 factors are local varieties of cornsilk (Bisma, Arjuna and Srikandi Putih), different solvents (water, methanol, ethanol and ethyl acetate), different concentration of solvents (methanol (60%, 70% and 80%), ethanol (60%, 70% and 80%), ethyl acetate (60%, 70% and 80%) and water), and ratio of material and solvent (1:5 (w/v), 1:10 (w/v), 1:15 (w/v) and 1:20 (w/v)). Response Surface Methodology (RSM) technique for optimum condition determination was used. Results showed that the difference in experimental conditions affects the content of soluble fiber, insoluble fiber, and total fiber produced by optimization analysis using the DX11.0® Program with RSM-Central Composite Design (CCD). In conclusion, Bisma variety, solvent of methanol, concentration of 80% solvent and ratio of (1:20) was selected as the optimum condition of which resulted in soluble fiber content of 0.85%, insoluble fiber of 0.60% and total fiber of 1.45%.

1. Introduction

The average consumption of dietary fiber Indonesia's population is 10.5 g/day. This number shows that people of Indonesia only meet the needs of fiber around a third of the ideal requirement of 30 g per day [1]. Fiber consumption is not related to where the population lives, but rather to the issue of economic status and knowledge. Another factor that influences is the availability of fibrous food and eating patterns and habits [2]. Important component in food ingredients is fiber which functions

IOP Conf. Series: Materials Science and Engineering **633** (2019) 012035 doi:10.1088/1757-899X/633/1/012035

primarily in maintaining the health and balance of the function of the digestive system. Dietary fiber has an important health value in reducing the accumulation of cholesterol in the blood, improving glucose absorption for people with diabetes mellitus, preventing colon cancer, and helping to lose weight [3, 4]. In Indonesia, currently dietary fiber is still dominated by material from land plants because it is relatively cheap and easily obtained, while the utilization of materials derived from aquatic plants is still limited. Therefore, dietary fiber is not only derived from vegetable ingredients such as cellulose, hemi cellulose, pectin and lignin, but also includes polysaccharides of animal and microbes (aminopolysaccharides) such as chitin, chitosan, keratin sulfate, hyaluronic acid, chondroitin, xanthan gum, dextran and synthetic carbohydrates such as polydecstrose, methyl cellulose and hydroxypropylmethyl cellulose [5,6].

Corn silk is waste that is quite rich in dietary fiber with the characteristic form of set of smooth, soft stigma, it looks like thread or yellowish hair. Corn silk comes from female flowers from corn plants [7]. The chemical content of corn silk such as protein, carbohydrates, fiber of several vitamins such as vitamin B, vitamin C, vitamin K and essential oils. Mineral salts such as Na, Fe, Si, Zn, K, Ca, Mg and P. Phytochemical compounds such as sitosterol and stigmasterol, hasperidine derivatives and quercetin [8,9], contain phenols, terpenoids and glycosides [10]. Besides that, corn silk also contains maysin, β carotene, beta sitosterol, geraniol, hordenin, limonen, menthol and viteskin [11]. Corn silk is rich in phenolic compounds, especially flavonoids [12]. Local varieties of corn silk Bisma Indonesia have water content of 11.58%, fat content of 0.30%, ash content of 3.29%, protein content of 17.70% and carbohydrate content of 67.13%, and total phenol 8262.93 μgGAE/g, total flavonoids 236.03 μgGAEg, beta sitosterol 1343.93 ppm and 75% antioxidant activity. This result is different from previous researcher [13].

The extraction process of corn silk powder is influenced by the extraction technique, extraction time, temperature, type of solvent, solvent concentration and the ratio of the solvent. Comparison of solvent materials, the number of extractants involved in displacement determines the degree of concentration difference that is very important in the diffusion process which will affect the content of the compound.

2. Experimental

The local corn silk varieties used in this study (Bisma, Arjuna and Srikandi Putih) each have different characteristics, obtained from Dukuh Lawangan, Lempuyang Village, Candiroto Districts, Temanggung District. The chemicals used were water, ethanol, methanol and ethyl acetate. Food fiber testing was conducted at the Laboratory of Food Chemistry and Biochemistry, Faculty of Agricultural Technology, Gadjah Mada University, Yogyakarta. Briefly, 3 g of fresh corn silk was washed with distilled water (aquadest) then it was dried in an oven at 60^0C for 24 hours [14] until the water content was 10-11%. Further, it was pounded into powder using a vacuum grinder, packed and stored under -20°C until analysis. Corn silk was extracted using a modification method [15]. 3 g of corn silk powder were mixed with 30 ml of 80% methanol, then they were extracted at 70°C with a water bath shaker. The 1.5 hours later, they were filtered out the pulp using Whatman No. 1 paper. Solvent separation with rotary flash evaporator. Chemical analyses (soluble fiber, insoluble fiber and total fiber) were carried out using enzymatic methods [16]. The experimental design was factorial block randomized design with 4 factors and 2 replications. Local corn silk with different varieties (Bisma, Arjuna and Srikandi Putih) was extracted with different solvents (water, methanol, ethanol and ethyl acetate) with a solvent concentration of water, methanol (60%, 70% and 80%), ethanol (60 %, 70% and 80%) and ethyl acetate (60%, 70% and 80%) with a ratio (material:solvent) (1:5 (w/v), 1:10 (w/v), 1:15 (w/v) and 1:20 (w/v) .The results were processed using Modde software 5 with Response Surface Methodology (RSM) program to determine the optimum concentration of corn silk powder extract.

IOP Conf. Series: Materials Science and Engineering **633** (2019) 012035 doi:10.1088/1757-899X/633/1/012035

3. Results and discussion

3.1. Soluble fiber

The content of soluble fiber in extracts of corn silk powder can be seen at the graph contour plot of Figure 1.

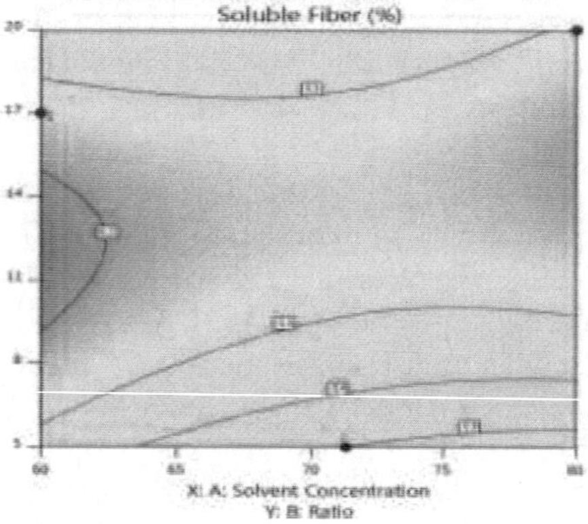

Figure 1. Soluble fiber countour plot

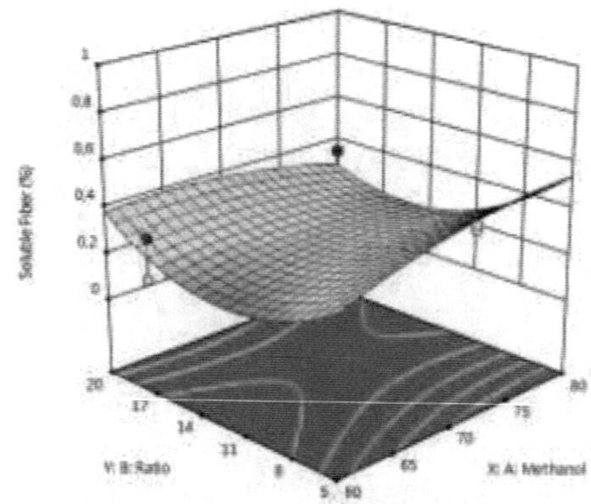

Figure 2. Soluble fiber 3d surface

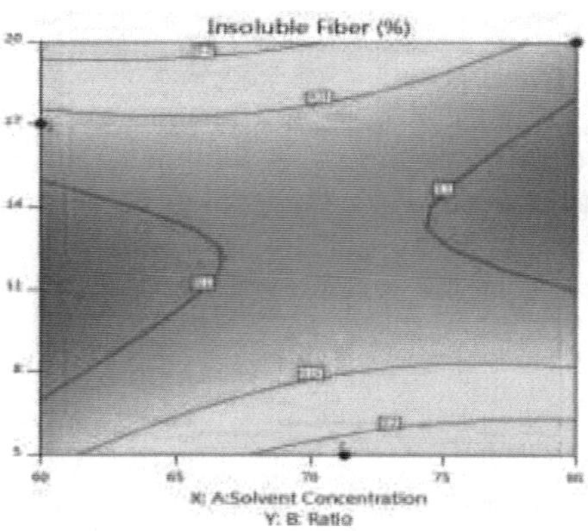

Figure 3. Insoluble fiber countour plot

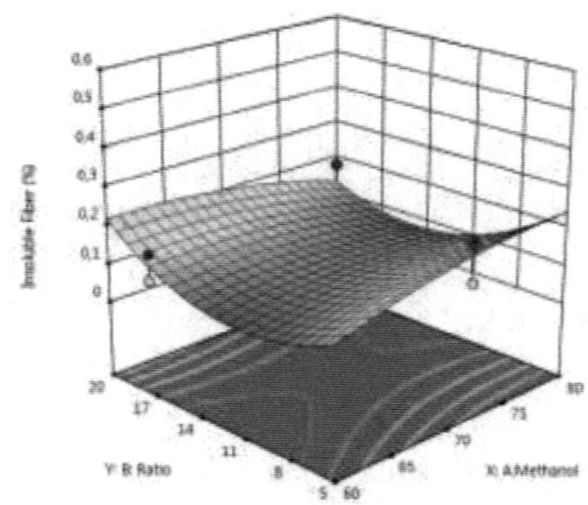

Figure 4. Insoluble fiber 3d surface

Table 1 shows that, the lowest mean soluble fiber was found in a combination of White Srikandi treatment, water solvent, 60% solvent concentration and with a ratio (1:5) is 0.06%, while the highest average soluble fiber was found in Bisma treatment combination, methanol solvent, 80% solvent concentration and with a ratio (1:20) is 0.85%. Design Expert 11.0 program optimization results at figure 1 in blue shows the lowest soluble fiber response value, which is 0.2%. The green color shows the highest soluble fiber response, which is 0.5%. This is presumably because the component of soluble fiber in corn silk is hydrolyzed, causing the release of pectin and oligosaccharides contained in the cell wall of corn silk [17].

3.2. Insoluble fiber

Insoluble fiber (insoluble in water) consists of carbohydrates that contain cellulose, hemicellulose and non carbohydrates that contain lignin. The results in table 1 shows that, the highest average insoluble fiber was found in Bisma treatment combination, methanol solvent, 80% solvent concentration with a ratio (1:20) is 0.60%. Figure 2 shows how the combination between components influences the response value of the insoluble fiber. The dark blue color shows the lowest insoluble fiber response

IOP Conf. Series: Materials Science and Engineering **633** (2019) 012035 doi:10.1088/1757-899X/633/1/012035

value, which is 0.1%. The light blue color shows the highest insoluble fiber response, which is 0.2%. Lines consisting of dots on the contour plot graph show combination of the four different components that produce the same insoluble fiber response. Corn plants are known to contain cellulose, the main hemicellulose in the first cell wall of vegetables are xyloglucan and glukuronoxilan, whereas in the second cell wall is glukuronoxilan [18]. The branch chain in xyloglycans is easier to hydrolyze than the main chain, while glucuronoxilane can be hydrolyzed in an acid medium in the glycosidic bonds between xylose. In addition, hemicellulose has a great ability to bind water molecules [19]. The insoluble fiber content is lower than soluble fiber, it is thought that because the cellulose cell wall in corn silk is degraded, the cellulose cell wall will be hydrolyzed to monosaccharide [20].

Table 1. Description of the chemical content of the fiber

Treatment[a]	component		
	Soluble fiber (%)	Insoluble fiber (%)	Total fiber (%)
$A_1B_1C_1D_1$	0.46	0.15	0.61
$A_1B_1C_2D_2$	0.56	0.24	0.80
$A_1B_1C_3D_3$	0.67	0.33	0.99
$A_1B_1C_3D_4$	0.71	0.42	1.13
$A_1B_2C_1D_1$	0.56	0.24	0.79
$A_1B_2C_2D_2$	0.67	0.41	1.08
$A_1B_2C_3D_3$	0.77	0.49	1.25
$A_1B_2C_3D_4$	0.85	0.60	1.45
$A_1B_3C_1D_1$	0.36	0.11	0.47
$A_1B_3C_2D_2$	0.43	0.17	0.60
$A_1B_3C_3D_3$	0.54	0.24	0.78
$A_1B_3C_3D_4$	0.66	0.32	0.97
$A_1B_4C_1D_1$	0.20	0.08	0.28
$A_1B_4C_2D_2$	0.26	0.12	0.38
$A_1B_4C_3D_3$	0.32	0.18	0.49
$A_1B_4C_3D_4$	0.41	0.25	0.66
$A_2B_1C_1D_1$	0.16	0.08	0.24
$A_2B_1C_2D_2$	0.24	0.12	0.36
$A_2B_1C_3D_3$	0.36	0.21	0.56
$A_2B_1C_3D_4$	0.42	0.27	0.69
$A_2B_2C_1D_1$	0.21	0.14	0.34
$A_2B_2C_2D_2$	0.34	0.17	0.51
$A_2B_2C_3D_3$	0.41	0.20	0.61
$A_2B_2C_3D_4$	0.48	0.25	0.73
$A_2B_3C_1D_1$	0.12	0.05	0.17
$A_2B_3C_2D_2$	0.19	0.08	0.27
$A_2B_3C_3D_3$	0.25	0.14	0.39
$A_2B_3C_3D_4$	0.32	0.20	0.52
$A_2B_4C_1D_1$	0.08	0.02	0.10
$A_2B_4C_2D_2$	0.12	0.06	0.18
$A_2B_4C_3D_3$	0.18	0.10	0.28
$A_2B_4C_3D_4$	0.23	0.14	0.37
$A_3B_1C_1D_1$	0.12	0.05	0.17
$A_3B_1C_2D_2$	0.18	0.10	0.28
$A_3B_1C_3D_3$	0.25	0.15	0.40
$A_3B_1C_3D_4$	0.33	0.20	0.53
$A_3B_2C_1D_1$	0.18	0.10	0.28

Treatment[a]	component		
	Soluble fiber (%)	Insoluble fiber (%)	Total fiber (%)
$A_3B_2C_2D_2$	0.25	0.14	0.39
$A_3B_2C_3D_3$	0.35	0.18	0.52
$A_3B_2C_3D_4$	0.41	0.22	0.63
$A_3B_3C_1D_1$	0.10	0.03	0.13
$A_3B_3C_2D_2$	0.13	0.08	0.21
$A_3B_3C_3D_3$	0.20	0.12	0.32
$A_3B_3C_3D_4$	0.26	0.16	0.42
$A_3B_4C_1D_1$	0.06	0.01	0.07
$A_3B_4C_2D_2$	0.10	0.07	0.17
$A_3B_4C_3D_3$	0.16	0.10	0.26
$A_3B_4C_3D_4$	0.21	0.12	0.33

[a]Description :
Varieties: A_1: Bisma, A_2: Arjuna, A_3: Srikandi Putih
Solvent: B_1: Ethanol, B_2: Methanol, B_3: Ethyl acetate, B_4: Water
Solvent Concentration: C_1: 60%, C_2: 70%, C_3: 80%
Ratio (material:solvent): D_1: (1:5), D_2: (1:10), D_3: (1:15), D_4: (1:20)

3.3.Total fiber

The highest average total fiber in table 1 was found in Bisma treatment combination, methanol solvent, 80% solvent concentration with a ratio (1:20) is 1.45%. The dark blue color shows the lowest total fiber response value, which is 0.3%. The green color shows the highest total fiber response, which is 0.7% (Figure 3). This is presumably due to the total fiber content in the high corn silk extract powder. Soluble fiber content, insoluble fiber content and total fiber results of this study were lower than those carried out by [21] which reported levels of soluble food fiber 2.06% db, insoluble food fiber 15.83% db and total dietary fiber 17.89% db. Phenol components in cereal seeds (including rice) are strongly bound to wall-walled materials such as arabinoxilan and lignin which are components of dietary fiber.There are different locations and environments [22].

Regression coefficients in each response have no significant differences with p=0.05. The results of the quadratic equation show no significant difference for all factors with Lack of Fit values >0.05 (soluble fiber 0.30; insoluble fiber 0.83 and total fiber 0.05). The value of the coefficient of determination (R^2) for all factors in the equation model>0.05.

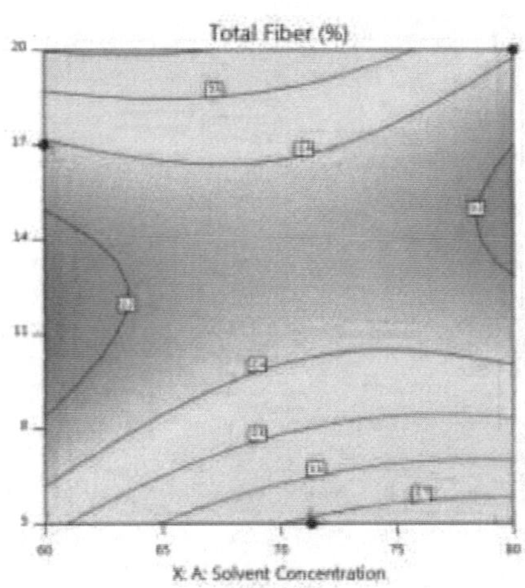

Figure 5. Total fiber countour plot

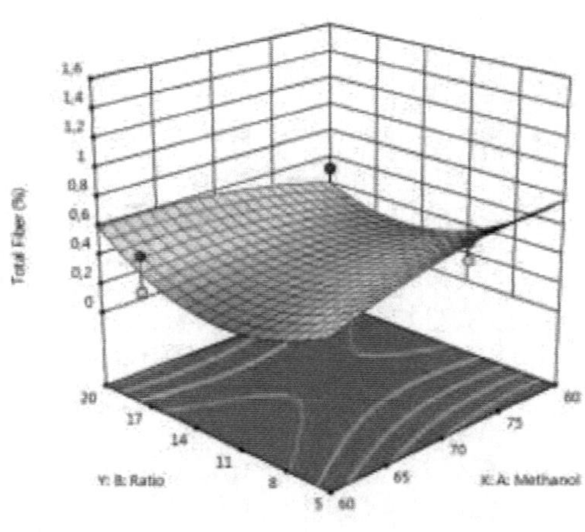

Figure 6. Total fiber 3d surface

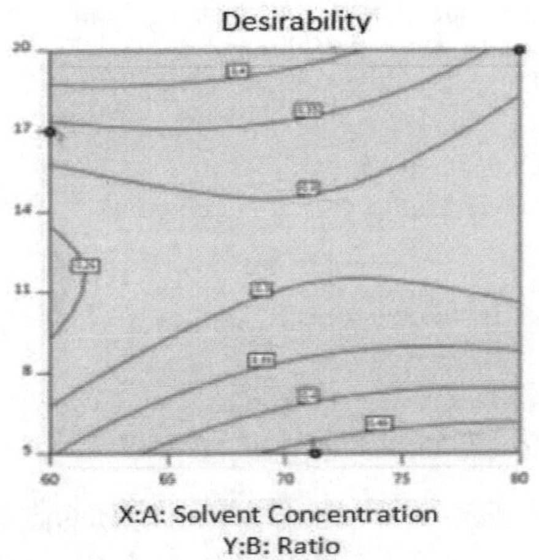

Figure 7. Corn silk powder optimum extract countour plot

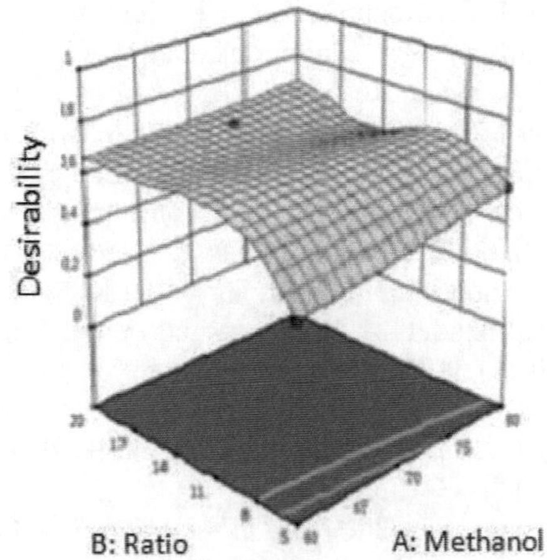

Figure 8. Corn silk powder optimum extract 3d surface

Table 2. Components of optimized responses, targets, limits and interests in the formula optimization stage

Response Name	Goal	Lower Limit	Upper Limit	Lower Weight	Upper Weight	Impor tance
A: Solvent concentration	is in range	60	80	1	1	3
B: Varieties	is in range	Bisma	White Srikandi	1	1	3
C: Solvent	is in range	Etanol	Air	1	1	3
D: Ratio	is in range	5	20	1	1	3
Soluble fiber	Maximize	0.06	0.85	1	1	5
Insoluble fiber	Maximize	0.01	0.6	1	1	5
Total fiber	Maximize	0.07	1.45	1	1	5

Table 2 shows the optimized components, targets, minimum and maximum limits, and the level of importance in the formula optimization stage. The response of soluble fiber, insoluble and total fiber is an optimized component with a maximum target with a level of importance of 5 (+++++). Soluble fiber, insoluble and total fiber is a response that greatly determines the efficiency of a business process, its relationship with economic value that will provide benefits. Whereas varieties, solvents, solvent concentrations and ratios were optimized with a range with a level of importance 3 (+++). Varieties, solvents, solvent concentrations and ratios will affect the quality of corn silk produced.

4. Conclusion

There was an interaction between varieties, solvents, solvent concentration, and the ratio to fiber content with a desirability value 0.717 which produced an optimal experimental formula on bisma varieties, methanol solvent, 80% concentration and ratio (1:20). In this condition, it produced chemical content (soluble fiber is 0.85%, insoluble fiber is 0.60% and total fiber is 1.45%).

References

[1] Astawan M and Wresdiyati T 2004 *Diet Sehat dengan Makanan Berserat* (Surakarta: Tiga Serangkai)

[2] Soerjodibroto 2004 *Asupan Serat Makanan Remaja di Jakarta* vol 54 No 10 Oktober 2004 (Jakarta: Majalah Kedokteran Indonesia) 397-401

[3] Jones P J, Raeini-Sarjaz M, Jenkins D J A, Kendall C W C, Vidgen E, Trautwein E A, Lapsley K G, Marchie A, Cunnane S C, and Connelly P W 2005 *Lipids.* **40(2)**169-174
[4] Wisten A and Messner T 2005 *J.Carring Sci.* **19** 71-76
[5] Furda I 2001 *Chitin and Chitosan, a Special Class of Dietary Fiber (CRC Handbook of Dietary Fiber in Human Nutrition 3rd ed ed Spiller GA CRC Press)* 45-47
[6] Gallaher D D, Gallaher C M, Mahrt G J, Carr T P, Hollingshead C H, Hesslink Jr R H, and Wise J 2002 *J.Am Coll Nut.* **21(5)** 428-433
[7] Bhaigyabati T, Ramya J, and Usha K 2012 *IRJP.* **3(3)**
[8] Ebrahimzadeh M A, Pourmorad F, and Bekhradnia A R 2008 *African Journal of Biotechnology.* **7(18)** 3188-3192
[9] Guo J, Liu T, Han L, and Liu Y 2009 *Journal Nutrition and Metabolism Biomed Central.* **6** 47
[10] Sholihah M A, Nurhanan A R, and Rosli W I W 2012 *International Food Research Journal.* **19 (4)** 1533-1538
[11] Rahmayani A 2007 *Telaah Kandungan Kimia Rambut Jagung* (*Zea mays* L) (Bogor: Institut Pertanian Bogor)
[12] Liu J, Wang C, Wang Z, Zhang C, Lu S, and Liu 2011 *Food Chemistry.* **126** 261-269
[13] Haslina, Praseptiangga D, Bintoro V P and Pujiasmanto B 2017 *International Journal of Advanced Science Engineering Information Technology*. **7** 1957-1963
[14] Hu Q L, Zhang L J, Li Y N, Ding Y J, and Li F L 2010 *International Journal of Physical Sciences.* **5** 321-326
[15] Nurhanan A R, Rosli W I W, and Mohsin S S J 2012 *Sain Malysiana.* **40(2)** 155-161
[16] Association of Official Analytical Chemist (AOAC) 2007 *Official method 980.17 preservatives in ground beef spectrophotometric method* (USA: AOAC International)
[17] Eveline, Antonius H C and Juanita R W 2014 *ISBN* 976-602-99334-3-7
[18] Yuanita L 2006. *Indo J.Chem.* **6(3)** 332-33
[19] Sjostrom E 1993 *Wood Chemistry Fundamentals and Applications Second Edition* (New York: Academic Press)
[20] Yoon K Y, Cha M, Shin S R, and Kim K S 2005 *Journal Food Chemistry.* 9**2** 151-157
[21] Damayanthi E and Listyorini D I 2006 *Jurnal Gizi dan Pangan.* **1(2)** 34-44
[22] Guo W and Beta T 2013 *Food Research International.* **51** 518-525

IOP Conf. Series: Materials Science and Engineering **633** (2019) 012036 doi:10.1088/1757-899X/633/1/012036

Effects of *Kaempferia galanga* L. essential oil incorporation on sensory and physical properties of dark chocolate bar

Y D Handiati, D Praseptiangga, G J Manuhara, L U Khasanah

Department of Food Science and Technology, Universitas Sebelas Maret (UNS), Jl. Ir. Sutami 36 A, Kentingan, Surakarta 57126, Central Java, Indonesia

Email : dpraseptiangga@yahoo.com

Abstract. Chocolate is the world's most popular confectionery product of processed cocoa. The antioxidant content of cocoa flavanols in dark chocolate is higher than other types of chocolate. Widely cultivated in Indonesia, kencur (*Kaempferia galanga* L.) rhizome has many health benefits, including antitussive, expectorant, antioxidant, and hypocholesterolemic. The aims of this research were to evaluate the effects of variation concentration of kencur (*Kaempferia galanga* L.) rhizomes essential oil incorporation on sensory and physical properties of dark chocolate bar. Three formulas of dark chocolate bar with an addition of *Kaempferia galanga* L. rhizomes essential oil (0.1%, 0.2%, and 0.3%) were studied. The results showed that the addition of variation of the essential oil concentration (0.1%, 0.2%, and 0.3%) had no significant effect on the overall sensory properties. However, dark chocolate bar with the addition of 0.3% *Kaempferia galanga* L. essential oil (F3) showed the highest panelist acceptance on aroma attributes. It had very small color difference and the lowest hardness based on the result of physical characterization, suggesting that dark chocolate bar with the addition of 0.3% *Kaempferia galanga* L. essential oil (F3) is expected to be developed as one of chocolate-based functional food products.

1. Introduction

Indonesia is the third world biggest producer of cocoa after Ivory Coast and Ghana in 2016/2017 [1]. Cocoa has contributed Indonesia's fourth crop in area planted and Indonesia's third foreign exchange in the national plantation sub-sector. Indonesia's cocoa area in 2015 was recorded 1,709,284 hectares with 593,331 tons of cocoa production. Total cocoa exports in that year reached 355,321 tons or valued at US$ 1,307,771 [2]. Based on the average data of Indonesian cocoa production over the past five years (2012-2016), there are six cocoa production centers in Indonesia (Central Sulawesi, South Sulawesi, Southeast Sulawesi, West Sulawesi, West Sumatra, Lampung, and North Sumatra). The six provinces contributed 80.19% cumulative contribution [3].

Chocolate is the world's most popular confectionery product of processed cocoa [4]. Chocolate is widely accepted by the public because of its taste, aroma, color [5], and health benefits [6]. There are three types of chocolate based on the composition of the constituent: dark chocolate, milk chocolate, and white chocolate [7,8]. The antioxidant content in dark chocolate is higher than in other types of chocolate [9,10]. Some research have been conducted to enrich the chocolate by using functional material from plant substrate which contains bioactive compound. The addition of dried fruits have been reported in dark and milk chocolates [11].

Spices are potential sources of antioxidants. This ingredient is effective in a number of food products, including confectionery. The main antioxidant compounds in spices are phenolic groups [12]. In some previous researches, spices have been used as additives in chocolate bars, such as cinnamon [8,13,14,15,16], aniseed, and ginger [17]. Kencur (*Kaempferia galanga* L.) as spice is widely known and intensively cultivated in Indonesia [18]. Kencur is fifth rank (6.33%) on the total production of biopharmaca plants in Indonesia [19]. Health benefit of kencur rhizome is antitussive [20], expectorant, antioxidant, and hypocholesterolemic [21]. Dark chocolate bar with the addition of essential oil of kencur rhizomes (*Kaempferia galanga* L.) is expected to be chocolate product that can provide health benefits (functional food product).

2. Experimental

The main materials used in this research were dark chocolate couverture 65% (PT. Tulip, Indonesia) and essential oil of *Kaempferia galanga* L. rhizomes (CV. Orizho Indonesia, Bantul, Indonesia). The process of producing dark chocolate bars included several stages, including material preparation, melting, refining, mixing, dark chocolate block printing, cooling, tempering, moulding, demoulding, and packaging. Dark chocolate was melted using chocolate melter with a temperature of 46°C ± 0.1 for ± 30 minutes. Refining was done by using ball mill with a rotary speed of 90 rpm at 45-46°C. After mixing chocolate and essential oil finished, chocolate paste was printed into chocolate blocks and cooled in a cooling room at 4°C. Tempering process was done by melt down dark chocolate block with a temperature 46°C. ¾ part of the melted chocolate then stirred and slided on marble table for 10-15 minutes until the temperature stabilized at 28°C. ¾ part of the chocolate paste that had been conditioned on the marble table was mixed with ¼ parts. The whole chocolate paste was heated to a temperature of 32°C and printed (moulding). After that, chocolate was put into cold storage 4°C for ± 30 minutes. The demoulding was done by slowly turning the mold on the marble table. Dark chocolate bar which was released from the mold and packed using aluminum foil.

Dark chocolate with addition *Kaempferia galanga* L. rhizomes essential oil were evaluated with Hedonic methods [22]. 30 panelists with varying in age 21-24 years were used in this sensory evaluation. The sensory properties were presented on five point scoring scale (dislike, dislike moderately, neither like nor dislike, like moderately, and like very much) and seven attributes (color, glossiness, aroma, taste, texture, melting, and overall). The physical properties of dark chocolate bar were evaluated with color and hardness measurements. Color measurements were conducted with Chroma Meter CR-400, Konica Minolta Optic, Inc. Before the test was carried out, the test samples were removed from the cold box and conditioned at room temperature (27°C). The results were expressed in the L* (lightness), a*(redness), and b*(yellowness). °Hue value calculated based on the value of a* and b* [23,24]. This color test aimed to determine the effect of variations in the essential oil of kencur rhizomes on the color quality of the sample. Hardness measurements were conducted with Zwick Universal Testing Machine (UTM) series SA/0.5. This measurement was based on the compression method with the ASTM standard. Dark chocolate bar was placed between the two anvils and was given a compressive force with a pressure in the form of a cylindrical acrylic plate until it broke.

Completely Randomized Design with one factor of variation of 0.1% (F1), 0.2% (F2), and 0.3% (F3) concentration of *Kaempferia galanga* L. rhizomes essential oil was used in this study. The data were statistically analyzed using One-Way ANOVA. If there was a significant difference, Duncan's Multiple Range Test (DMRT) was conducted (α=5%).

3. Results and Discussion

3.1 Sensory Properties

Sensory evaluation was used for market needs survey, quality improvement, and product development [5]. Hedonic was defined as the act of choosing based on preference. This scale included a series of statements or panelist statement points that state the level of preference or dislike of the sample [22]. The results of this sensory analysis used 1-5 hedonic scale. The results were statistically processed to obtain conclusions about the preference of dark chocolate bar with the addition of *Kaempferia galanga* L. rhizomes essential oil. Sensory properties results are shown in Table 1.

Table 1. Sensory properties of *Kaempferia galanga* L. dark chocolate bar

Formula	Attributes						
	Color	Glossiness	Aroma	Taste	Texture	Melting	Overall
F1 (0.1%)	4.70^{a}±0.535	4.40^{a}±0.724	4.00^{a}±0.871	4.20^{a}±0.925	4.23^{a}±0.858	4.47^{a}±0.819	4.30^{a}±0.794
F2 (0.2%)	4.77^{a}±0.450	4.37^{a}±0.718	4.17^{ab}±0.699	4.20^{a}±0.664	4.33^{a}±0.758	4.33^{a}±0.758	4.33^{a}±0.547
F3 (0.3%)	4.80^{a}±0.407	4.33^{a}±0.758	4.43^{b}±0.568	4.50^{a}±0.682	4.33^{a}±0.844	4.27^{a}±0.828	4.50^{a}±0.682

[a]Within a row, means the same superscript letters are not significantly different from one another at 5% level of significance based on One-Way ANOVA (α=0.05).

3.1.1 Color Attributes
Color is the first attribute that is valued from product. Color affects the threshold of gustative, sweet perception, preference, and acceptance of panelists [25]. The difference in the composition of dark chocolate bar is only found in variations in the concentration of essential oils of *Kaempferia galanga* L. rhizomes so that the sample did not have significant differences between each other. Table 1 shows that panelist acceptance rates increased insignificant with the increasing concentration of essential oils added. It is suspected because the essential oil of *Kaempferia galanga* L. had clear or colorless so it did not affect the color of dark chocolate bars. The results of the color attribute of the panelist acceptance of this research (4.80 ± 0.407) were relatively better than Dwijatmoko research [8].

3.1.2 Glossiness Attributes
Glossiness interprete as the reflection power of the plane's surface to light. Panelists tend to like glossy chocolate [26]. The decrease in the acceptance of panelists on the glossiness attribute is thought to be related to a fairly high temperature difference between the storage room and the organoleptic test room. It is consistent with Haryadi and Supriyanto [27] which states that chocolate experienced condensation due to extreme temperature changes. The glossiness atribute of dark chocolate bars in this research were not significantly different. The panelists still stated that they liked three samples (F1,F2, and F3) of the dark chocolate bar.

3.1.3 Aroma Attributes
Aroma is formed by the chemical components of cocoa bean compilers such as volatile compounds, namely aldehydes, ketones, and carbonyl. Aroma formation is influenced by precursors of fermentation results, temperature, roasting time, and moisture content of the material. The precursors forming component of chocolate aroma are aliphatic acid, α, β, unsaturated aldehyde, sulfur compounds, and alkylpirazine (Strecker degradation secondary products) [27]. Based on Table 1, aroma attributes had a trend that tend to increase along with the addition of the essential oil of *Kaempferia galanga* L. rhizome. Sample F3 or dark chocolate bar with the addition of 0.3% had the highest level of preference and had a significant difference to sample F1, but not significantly different from F2. It shows that more essential oil of *Kaempferia galanga* L. is added, the panelists increasingly liked the aroma formed.

3.1.4 Taste Attributes
The basic taste of chocolate that consumers want is sour and bitter with a balanced state. Chocolate sweetness is caused by aliphatic inorganic compounds that contain OH groups such as alcohol, some amino acids, aldehydes, and glycerol. The source of sweetness was mainly produced by sugar (sucrose, monosaccharides, or disaccharides). Bitter taste was caused by alcoholoids, such as caffeine, theobromine, quinone, glycosides, and phenolic compounds [27]. Chocolate flavor with the addition of 0.3% (F3) essential oil of kencur (*Kaempferia galanga* L.) gave the highest preference which is not significantly different with the addition of 0.1% (F1) and 0.2% (F2) essential oil of kencur rhizome. The essential oil of rhizomes kencur gave after taste like a strong mint flavor.

3.1.5 Texture Attributes

Chocolate's texture is smooth. It can melt softly in the mouth. Based on the panelists statements that they found dark chocolate with addition of 0.3% essential oil of kencur (*Kaempferia galanga* L.) rhizomes is easier to break down. Therefore, it is suspected that this made the acceptance increased.

3.1.6 Melting Attributes

Cocoa fat in chocolate is composed of stearic acid (34%), palmitic acid (27%), and oleic acid (34%) which functions at room temperature and melts at body temperature giving a smooth texture in the mouth [9]. From the results of hedonic test of melting ability attributes, the difference in the concentration of essential oils did not provide a real value for the resulting dark chocolate bar. Dark chocolate bar with the addition of 0.3% essential oil of kencur rhizome (F3) had a texture that is not harder than 0.1% concentration. Sample F3 had a mouth feel that can be melted in the mouth and had melting point lower than F1. According to research by Limbardo [28], chocolate with a high melting point gave great hardness.

3.1.7 Overall Attributes

The level of acceptance of the three variations of essential oils concentration did not give a significant difference in overall parameters (except aroma parameters) of dark chocolate bars. The acceptance level of panelists in overall attributes had an increasing trend along with the increasing number of essential oils. Overall parameters indicated that the dark chocolate bar with the addition of 0.3% essential oil of kencur (F3) can still be accepted by the panelists.

3.2 Physical Properties

Color is an important indicator or attribute. CIELab is one method in measuring color in food [29]. The values of dark chocolate bars' L*, a*, b*, and °Hue in this study were found in Table 2. Whereas, hardness is the force needed to crush chocolate. Snap is a sound that is heard when the chocolate sample becomes broken. The desired chocolate was chocolate with a hard texture, steady and easily melts in the mouth [9].

3.2.1 Color Measurement

Table 2. Color measurement of *Kaempferia galanga* L. dark chocolate bar

Formula	L*	a*	b*	°Hue
Control (0%)	$28.310^{b} \pm 0.185$	$7.700^{ab} \pm 0.119$	$4.812^{b} \pm 0.141$	$31.996^{b} \pm 0.522$
F1 (0.1%)	$27.877^{ab} \pm 0.387$	$7.712^{ab} \pm 0.231$	$4.677^{ab} \pm 0.274$	$31.212^{b} \pm 0.869$
F2 (0.2%)	$28.087^{ab} \pm 0.474$	$7.888^{b} \pm 0.235$	$4.962^{b} \pm 0.419$	$32.118^{b} \pm 1.559$
F3 (0.3%)	$27.628^{a} \pm 0.446$	$7.617^{a} \pm 0.133$	$4.395^{a} \pm 0.129$	$29.981^{a} \pm 0.336$

[a]Within a row, means the same superscript letters are not significantly different from one another at 5% level of significance based on One-Way ANOVA (α=0.05).

Table 2 shows that the color of three formulas of dark chocolate bar with the addition of essential oils 0.1%, 0.2%, and 0.3% had a positive a* and b* values, the intersection between red and yellow. The addition of the concentration of essential oil of *Kaempferia galanga* L. rhizomes did not significantly affect the °Hue sample of F1 (0.1%) and F2 (0.2%), but significantly affected the sample F3 (0.3%). The three samples (F1, F2, and F3) of the dark chocolate bar were in the red chromatic color (°Hue). The addition of essential oil of *Kaempferia galanga* L. rhizomes gave darker red color to the dark chocolate bar. The effect of the color difference between the three samples were very small (TCD*<1.0), thus it did not affect the level of panelist preference.

3.2.2 Hardness Measurement

The results of the hardness measurement showed a significant difference between one sample and the others. The four samples of the dark chocolate bar had their own subset of a, b, c, and d. The highest hardness of dark chocolate bar was found in the control or sample with no addition of rhizome rhizome essential oil. Meanwhile, the dark chocolate bar that had the lowest hardness is the sample with the addition of the highest concentration of *Kaempferia galanga* L. rhizomes essential oil (0.3%) as shown in Table 3. Based on this research, the higher the concentration of addition of essential oil of *Kaempferia galanga* L. rhizomes reduced the level of hardness of the dark chocolate bar.

Table 3. Hardness measurement of *Kaempferia galanga* L. dark chocolate bar

Formula	Hardness (N)
Control (0%)	89.724[a] ± 2.363
F1 (0.1%)	75.291[b] ± 5.245
F2 (0.2%)	51.346[c] ± 4.273
F3 (0.3%)	22.771[d] ± 2.366

[a]Within a row, means the same superscript letters are not significantly different from one another at 5% level of significance based on One-Way ANOVA (α=0.05).

4. Conclusion

This research evaluated the effects of variation concentration (0.1%, 0.2%, and 0.3%) of kencur (*Kaempferia galanga* L.) rhizome essential oil incorporation on sensory and physical properties of dark chocolate bar. The addition of variation of the essential oil concentration had significant effect on the aroma attribute but had no significant effect on six attributes (color, glossiness, taste, texture, melting, and overall). Three formulas of dark chocolate bar were represented in red of color range. The addition of variation of the essential oil concentration had significant effect on dark chocolate bar' hardness. The higher concentration of kencur rhizome essential oil reduced the hardness of dark chocolate bar. Dark chocolate bar with the addition of 0.3% *Kaempferia galanga* L. essential oil (F3) showed the highest panelist acceptance on sensory properties with insignificant effect and the lowest hardness with very small color difference.

References

[1] International Cocoa Organization (ICCO) 2018 *The World Cocoa Market, Current Cocoa Market and Forecasts, 31 July 2018* (Côte d'Ivoire: ICCO) p 11
[2] Center for Agricultural Data and Information System Ministry of Agriculture Republic of Indonesia 2017 *Agricultural Statistics 2017* (Jakarta: Center for Agricultural Data and Information System Ministry of Agriculture Republic of Indonesia) p 209, 316
[3] Center for Agricultural Data and Information System Ministry of Agriculture Republic of Indonesia 2016 *Outlook Kakao* (Jakarta: Center for Agricultural Data and Information System Ministry of Agriculture Republic of Indonesia) p 12
[4] Nachay K 2010 *Food Tech.* **12** 10
[5] Norhayati H and Ayob M K 2014 *JSM* **43** 9
[6] Scholey A and Owen L 2013 *Nutr.Rev.* **71** 10
[7] Sudibyo A 2012 *JRI* **6** 1
[8] Dwijatmoko M I, Praseptiangga D and Muhammad D R A 2016 *Nusantara Biosci.* **8** 2 301–5
[9] Beckett S T 2008 *The Science of Chocolate 2nd Ed.* (United Kingdom: RSC Publishing) p 205
[10] Ibrić A and Ćavar S 2014 *BCTBH* **42**
[11] Komes, Draženka, Ana B, Svjetlana Š, Aleksandra V and Arijana B 2013 *Food Sci. Tech.* **53**
[12] Santoso U 2016 *Antioksidan Pangan* (Yogyakarta: UGM Press) p 56-60
[13] Ilmi A, Praseptiangga A and Muhammad D R A 2017 *IOP Conf.Series: Materials Science and Engineering* **193** 012031
[14] Muhammad D R A, Praseptiangga D, Van De Walle D and Dewettinck K 2017 *Food Chem.* **231** 356–64
[15] Praseptiangga D, Nabila Y and Muhammad D R A 2018 *Caraka Tani: J. Sustainable Agric.* **33(1)** 78–88

[16] Rasuluntari I N, Muhammad D R A and Praseptiangga D 2016 *Nusantara Biosci.* **8(2)** 297–300
[17] Albak F and Tekin A R 2014 *Academic Food Journal/Akademik GIDA* **12** 2
[18] Pribadi E R 2009 *Prespektif* **8** 1
[19] Direktorat Jenderal Hortikultura 2015 *Statistik Produksi Hortikultura Tahun 2014* (Jakarta: Direktorat Jenderal Hortikultura, Kementerian Pertanian) p 35, 38-40
[20] Jahan Y, Mahmood T, Bagga P, Kumar A, Singh K and Mujahid M 2015 *Int. J. Pharm. Sci. Res.* **6** (9): 1000-09
[21] Ravindran P N and Pillai G S 2006 *Handbook of Herbs and Spices* vol 3 (Cambridge, England: Woodhead Publishing Limited) p 351
[22] Setyaningsih D, Anton A and Maya P 2010 *Analisis Sensori* (Bogor: IPB Press) p 59-60
[23] Misnawi 2011 *CCRJ* **27** 3
[24] Machalkova L, Hrivina L, Nedomova S and Miroslaf J 2014 *MendelNet* **1**
[25] Dias N A A, Lara S B, Miranda L S, Pires I S C, Pires C V and Halboth N V 2012 *Ciênc. Tecnol. Aliment., Campinas* **32** 2
[26] Krasnow M N and Migoya F 2014 *J. Cul. Sci.Tech.* **13** 2
[27] Haryadi and Supriyanto 2017 *Teknologi Cokelat* (Yogyakarta: UGM Press) p 142-44, 149, 154, 237-38, 246-47
[28] Limbardo R P, Santoso H and Witono Judy R 2017 *AIP Conf. Proc.* 1840,060001
[29] Romaniello R, Leone A and Giorgio P 2015 *JAE* **46**

Acknowledgments
This research was supported by the research grant of PNBP 2018 Universitas Sebelas Maret (UNS), Indonesia.

IOP Conf. Series: Materials Science and Engineering **633** (2019) 012037 doi:10.1088/1757-899X/633/1/012037

Fruit identification and effect of starch isolation methods on color attributes of Cilacap bread fruit's starch

N S Rahayu[1,2], D Praseptiangga[3], B Hariyanto[4] and Samanhudi[5]

[1]Graduate School Program of Sebelas Maret University (UNS), Jl. Ir. Sutami 36 A, Kentingan, Surakarta, 57126, Indonesia
[2]Department of Agricultural Products Technology, Faculty of Agricultural Technology, Widya Dharma University, Jl. Ki Hajar Dewantara, North Klaten. Mail Box No.168, Klaten, Central Java, 57438, Indonesia
[3] Department of Food Science and Technology, Faculty of Agriculture, Sebelas Maret University (UNS), Jl. Ir. Sutami 36 A, Kentingan, Surakarta, 57126, Indonesia
[4] Agency for the Assesment and Application of Technology, BPPT Building II 15th Floor, Jl. M.H.Thamrin No.8 Jakarta Pusat, Jakarta Indonesia
[5] Department of Agrotechnology, Faculty of Agriculture, Sebelas Maret University (UNS), Jl. Ir. Sutami 36 A, Kentingan, Surakarta, 57126, Indonesia

E-mail: nunuk@unwidha.ac.id

Abstract. Utilization of Cilacap breadfruit as a food source is very important for supporting local food diversification. The objectives of this study were to identify the Cilacap breadfruit based on morphology identification method and to evaluate the effects of starch isolation methods on color properties of its starch. The results showed that the morphological identification of Cilacap breadfruit was round - slightly oval, and the brownish green white spots-colored was shown on the ripe stage of fruit at the tree. The average weight of Cilacap breadfruit was 1.506 kg, and the yield of starch was 11.08%. The fruit identification was determined by LIPI Bogor Indonesia, of which confirmed that Cilacap breadfruit used in this study was a species of *Artocarpus altilis* (Parkinson ex. F.A. Zorn) Fosberg (Moraceae family). Starch isolation by using sediment separation extraction method after three-time decantation was shown as a better method than the non-separation isolation method. Color attributes of starch by using sediment separation extraction method gave a value of L* = 94.113, a* = 5.253 and b* = -1.506 for the white bottom sediment, while the brown color of the upper sediment gave L* = 87.587, a* = 5.063 and b* = 1.956.

1. Introduction

Breadfruit plants are tropical plants producing quite a lot of breadfruit, an average of 200-300 fruits per tree at harvest time [1,2] with a fruit's weight of about 150-200 kg [3]. This breadfruit is a high carbohydrate-producing fruit [4,5], low in fat [6], a good source of vitamins and minerals [1,5], but its utilization is not yet popular compared to other carbohydrate sources such as cassava, banana or corn.Breadfruit plant distribution in Indonesia is very broad so that many fruit variations are found according to local origin, one of which is Cilacap breadfruit [7]. Breadfruit which becomes the object of this research is ripe-at-tree Cilacap breadfruit. Cilacap breadfruit is one of the breadfruit varieties cultivated in Cilacap district, Central Java, Indonesia since 1980-1995, and has spread to several other regions in Indonesia. The superiority of Cilacap breadfruit is its large, fast-growing fruit, and its ability to grow under extensive ecological conditions, compared to other varieties growing in Java [8]. Cilacap breadfruit is also called bald breadfruit because its peel tends to be smooth and hairless. Since breadfruit has broad distribution and morphology differences due to differences in regional origin, it is significant to identify morphology in Cilacap breadfruit, especially ready-to-harvest breadfruit as used for research, including physical attributes from outside. In order to ensure that Cilacap breadfruit belongs to the *Arthocarpus altilis* species, (Parkinson) Forsberg and obtain scientific clarity that can be accounted for, it is necessary to conduct identification using the identification key by experts. To extend shelf life and expand its utilization, breadfruit can be processed into flour or starch [2,9]. In this case, starch is more beneficial because of its wider use in food and pharmaceutical industries [4,10]. Color is one of the determining factors in starch application in food systems [11]. In food applications, white breadfruit starch without the addition of chemicals is preferred by consumers, but the constraints faced are during processing if there is a lot of air contact which will consequently produce a brownish starch [2,12]. Therefore, it is necessary to modify the starch isolation process with natural extraction methods (without the addition of chemicals) but produce white starch. Subsequently, the starch is tested for color usinga Chromameter CR-400, Konica Minolta Optics, Inc.

2. Experimental

The main materials of this research were ripe-at-tree Cilacap breadfruitsharvested from the people's gardens in Jetis sub-village Klepu village Ceper district Klaten Central Java, unripe fruits, leaves, flower beds, root shoots, distilled water, clean water, ice cubes. Tools include styrofoam box sized 42x31x26 cm, ice box sized 50x40x34 cm, extraction equipment: Knife, 6-liter-volume pan, 2.5-liter-volume pan, Muslim cloth, 30x30x3cm stainless steel pan, cabinet dryer, grinding machine Disk Mill Model FFC-23A , Speed 5800 rpm, serial number 6367, made in Republic of China, Chomameter CR-400, Konica Minolta Optics, Inc.

The outer attribute observation used 9 ripe breadfruits picked from the garden. Outward appearance was observed, including the shape, measurement: length and width, peel color, weight of the fruit with stalks before peeling, weight of edible parts and starch yield after isolation. Starch yield was calculated as described by Akanbi et al [13]. The data obtained is presented in Table.

The fruit identification process is the same as the plant identification process, which is through the search for plant materials employing the identification key, by sending identification materials in the form of root shoots, leaves, flower beds and breadfruit inserted in the styrofoam box, with ice cubes in between the sidelines of the materials and boxes tightly closed. Next, the styrofoam box was inserted in a sealed ice box and sent to LIPI, part of the Bogor Botanical Garden Plant Conservation Center.

The isolation process method Extraction non-Separation (EnS) of sediment is a standard starch isolation procedure that is commonly carried out by many researchers, namely without the separation process after decantation. Starch isolation in this study was carried out through a wet extraction process as described by Nwokocha and Williams [4] with the following modifications: breadfruit starch was isolated from ripe breadfruit. The extraction process was repeated 3 times, each replication consisted of 3 breadfruits. 9 breadfruits were washed, soaked in clean water to avoid browning due to contact with air, sliced into 8 parts, weighed and wet-milled with a grinding machine Disk Mill Model FFC-23A, Speed 5800 rpm, serial number 6367, made in Republic of China. The fruit pulp produced was then weighed, mixed with distilled water with slurry and aquades ratio: 1: 2, and squeezed using Muslim cloth. The filtrate left in the cloth was stored on top while the milk starch passing from the juice was collected in a 6-liter pan. The breadfruit sediment was once again added withaquadest with a volume of 1: 2, filtered and stored in the same panof the first juice of starch. The starch milk was deposited for 4 hours, the supernatant was poured (decanted). The sediment was then added with aquadest by ratio 1: 2, deposited again for 4 hours and the supernatant was poured (decanted). This total process was carried out 3 times. After the third decantation, the starch sediment was stored on a 30x30x3 cm stainlees steelpan and dried in a cabinet dryer at 60° C for 2 days, each for 7 hours.

Starch isolation method Extraction with Separation (EwS) of sediment is a modification of the Cilacap breadfruit extraction process to obtain white starch. The process modification started after 3 decantation processes as the procedure above. After the third decantation, the sediment in the pan was added more aquadest by ratio 1:1, the container in the form of the pan was tilted and allowed to settle for 2 hours. The clear upper liquid was carefully separated, then the sediment was left for another 10 minutes. The Upper Extraction Sediment (UES) in the form of slurry was rocked to separate thebrown upper sediment with white starch sediment below it. The slurry-shaped upper starch sediment was poured in a stainleessteel pan, then dried. The Bottom Extraction Sediment (BES) which was completely precipitated and could not be poured, was taken using a spoon quickly and placed on another stainless steel pan, immediately dried in a cabinet dryer at 60°Cfor two days, respectively 7 hours. After drying, the starch was milled, then sifted with mesh 100. Then, each starch was stored in an airtight container labeled according to its type, then stored in a refrigerator at 4°C.

Starch color analysis useda Chomameter CR-400 (Konica Minolta Optics, Inc.) based on the Hunter scale. The color results were in the form of numbers from the component L*(light = brightness)value ranged from 0 (black) to 100 (white), the value of a*(green-red) ranged from -80 (green) to +100 (red), and the value of b*(blue-yellow) ranged from -80 (blue) to +70 (yellow)[17]which can be read and recorded. Each sample was repeated 3 times.

All data were shown as mean ± SD. The results of color analysis with different extraction methods were analyzed by One Way ANOVA using the SPSS version 20.0 program, with a significance level of $\alpha = 5\%$ ($p \leq 0.05$) was conducted.

3. Result and Discussion

The shape of Cilacap breadfruit is mostly round and slightly oval, with green peel color when still unripe and when it is ripe, it turns brown with white spots. The length of Cilacap breadfruit is around 13.5-17 cm, width 13.23-17.18 cm, weight before peeling 1.62-1.58 kg, weight of the edible part 1.33-1.25 kg, and starch yield 9.9-13.16%. The average value of external attributes is shown in Table 1.

Table 1. Outer appearance (morphology) of Cilacap breadfruit.

Sample	Attribute components	
CilacapBreadfruit	1. Shape	Round and slightly oval
	2. Peel Color	Brownish greenwith white spots
	3. Length	15.36±1.76 cm
	4. Width	14.89±2.05 cm
	5. Weight	1.51±0.07 kg
	6. Weight of edible parts	1.25±0.08 kg
	7. Starch yield	11.08±1.80 %

Breadfruit varies in shape, size and surface texture. Most breadfruit is round or slightly oval 0.25-6 kg in weight, about 9-20 cm in length and 10-20 cm in diameter [6]. Breadfruit is harvested when the tree is ripe which is marked by dark green peel, smooth surface and the presence of latex stains on the peel of the fruit [1]. The most reliable characteristics of ready-to-harvest fruits are the presence of green skin color intensity to brown and the presence of latex stains on the peel surface of the fruit [14]. The fruit used in this study has these characteristics. The presence of latex on the fruit surface is due to the central part of the fruit having a vascular bundle with many latex tubes [6]. Starch yield of each researcher varies, namely 9.60% [15], 14.26-18.5% [13]. The variation in yield of breadfruit starch depends on the level of fruit maturity, variety, differences in climate and agronomic conditions, and the extraction process, namely decantation and purification during starch isolation [13].

The results of analysis undertaken by LIPI, Bogor Botanical Gardens Plant Conservation Center show that baldCilacap breadfruit is confirmed as a species of *Artocarpus altilis* (Parkinson ex. F.A. Zorn) Forsberg, Moraceae family. This is in accordance with the study of Sari et al [15] that breadfruit samples for their research, the results of identification by Bogoriensi Herbarium belong to the same type as those used in this study.

The analysis of starch color attributes with two methods is presented in Table 2. EwS breadfruit starch color having the highest L value is BES sample, which is 94. 11, the second is EnS sample of 88.64, the last isUES sample with L value 87.59. L values range from 0 (black) - 100 (white) [16,17]. The greater the value of L approaches 100, the more white the starch color. This shows that the starch isolated by extraction with separation results in bottom sediment (BES sample) having the whitest color.

Table 2. Color attributes of Cilacap breadfruit starch.

Sample code based on Starch isolation method	HunterScale		
	L*	a*	b*
1. **EnS**	$88.64^{b}\pm0.19$	$5.48^{a}\pm0.05$	$3.27^{a}\pm0.27$
2. **EwS :**			
2.1. UES	$87.59^{b}\pm2.19$	$5.06^{a}\pm0.40$	$1.95^{a}\pm1.05$
2.2. BES	$94.11^{a}\pm0.35$	$5.25^{a}\pm0.39$	$-1.50^{b}\pm0.42$

Description: Numbers followed by different letters show very different ($p \leq 0.05$). EnS = Extraction non-Separation; EwS = Extraction with Separation; UES = Upper Extraction Sedimen; BES = Bottom Extraction Sediment.

This starch is produced without the addition of bleaching chemicals. EnS and UES starch samples are darker (brownish). Starch color is an important factor in acceptance by consumers, since color is one of the important quality components in starch application in various food systems [11,18]. Wheat-based food products, wheat color play an important role in the quality of products, i.e. yellow is the choice of durum wheat pasta, but is not preferred for white bread or noodle products [18]. Nevertheless, the making of breadfruit flour or starch has a problem that is the occurrence of browning reaction during processing, which starts when stripping the skin due to oxidation due to contact with air [2]. To avoid this, in this study, after being peeled, the fruit was immediately soaked in a tub of clean water with the aim of avoiding the peeled fruit in direct contact with air. In addition, breadfruit which is processed into flour or starch quickly turns brown because breadfruit contains polyphenols or phenolic compounds [6,12] and phenolase enzymes causing brown color [12]. Polyphenol compounds reacting with oxygen from the air form polyphenol oxidase which triggers browning of the material in the form of starch. Phenolase enzymes that react with oxygen cause enzymatic browning [2]. Such browning process may also occur in EnS and UES samples, while BES is protected from browning because it does not have direct contact with air.Meanwhile, the use of heating with radio frequency uses 6 kW, 27.12 MHz electromagnetic energy on potato starch is to deactivate the polyphenol oxidase and lipoxygenase enzymes which trigger browning and produce L* 67.72 ± 0.14 [19]. As a comparison, several studies on starch color from various plant sources can be seen in Table 3. To obtain white breadfruit starch color, some researchers added 0.08% NaOH solution and soaked peeled fruit for 45 minutes to produce breadfruit starch with an L* value 74.9 [12], whereas Nwokocha & Williams [4] added 0.3% NaOH to the extraction of breadfruit, and Olatunde et al [17] added a 0.5 g/L citric acid solution to extraction of banana starch with a value of L* 81.23 for Honduran variety banana and 64.74 for plantain, but in this study, extraction without the addition of chemicals produces the highest L* value of 94.11 ± 0.349. In order to standardize, the highest Hunter scale was measured on a standard white tile, L * 97.63; a * 0.78 and b * 0.25 [20]. The higher the L* value means the color of Cilacap breadfruit starch is getting brighter [21]. The value of a* of breadfruit starch in this study for all starch samples is not significantly different, while the value of b* has a difference between the EwS sample and the BES and UES sample with BES. The EnS sample has the highest b* value (3.272in average) followed by UES (1.95in average) as a representation of yellowish color, while the BES sample gives a little blue color because the value of b* is negative (-1.50in average). Cilacap breadfruit starch color in this study has better color attributes compared to the breadfruit starch color of the research result Lubis et al [12], cassava starch [22,23], sweet potato starch, taro starch [23], sago starch [23], and starch potatoes [19].

From Table 3, it is known that in addition to enzyme inactivation and NaOH addition, differences in color measuring instruments are likely to cause differences in results for the L*, a* and b* scales.This is evident in the study [24], i.e. several samples of wheat flour using Chromameter CR-210Minolta and Chromameter II reflectance Minolta, produces different values of L*, a* and b*.In this study, breadfruit starch without NaOH addition and measured with chromameter CR-400 Minolta produces a value of L* 94.11.

Table 3. Starch color from various plant sources

Type of Starch	Hunter Scale			Researcher	Measuring Instrument
	L*	a*	b*		
Breadfruit Starch	73.30-74.9	-	-	Lubis et al [12]	Chromameter
a. Casava starch	82.22±1.74	2.37±0.14	6.64±0.45	Fiorda et al [22]	Digital using microsoft photo editor 3.01 program, converted RGB value.
b. Casava starch	68.95	4.21	-2.64	Polnaya et al [23]	Chromameter Minolta CR-300
Sweet Potato starch	65.91	3.96	-0.69	Polnaya et al [23]	Chromameter Minolta CR-300
Taro starch	67.20	4.32	-2.23	Polnaya et al [23]	Chromameter Minolta CR-300
a. Sago starch	65.01	4.49	-0.22	Polnaya et al [23]	Chromameter Minolta CR-300.
b. Sago starch	86.33	1.66	7.68	Sede et al [21]	Hunterlab Colorflex EZ
Potato starch	67.72±0.14	5.75±0.53	39.68±1.13	Zhang et al [19]	Computer vision system with digital camera, a standardized lighting system with two CIE source D65 lamps
Banana starch					
a. Honduras variety	81.23	3.37	-3.59	Olatunde et al [17]	Chromameter Minolta CR-14
b. Plantain	64.74	3.29	-7.00	Olatunde et al [17]	

4. Conclusion

Cilacap breadfruit has been confirmed as a species of *Artocarpus altilis* (Parkinson ex. FA Zorn) Forsberg, Moraceae family, with a mostly round and slightly oval shape, the brownish green white spots-colored was shown on the ripe stage of fruit at the tree, ±15.36cm in length, ±14.89cm in width, ±1.51 kg in weight, edible parts ±1.25 kg and starch yield ± 11.08%. To obtain white starch breadfruit, Extraction with Separation (EwS) after three-time decantation is better method than Extraction non Separation (EnS), and without the addition of chemicals.Starch isolation extraction method with separation after three-time decantation is a modification process which can produce white bottom starch sediment (BES sample) (L* 94.11) and brownish upper sediment (UES sample) (L* 87. 59).

References

[1] Famurewa J A V, Pele G I, Esan Y O and Jeremiah B P 2016 *Br. Biotechnol. J. Port.* **16** 1–9
[2] Widowati S 2009 *PANGAN* **XVIII** 67–75
[3] Oulaï S F, Gonnety J T, Faulet B M, Djè K M, Ahi A P and Kouamé L P 2013 *Pakistan J. Nutr.* **12** 768–774
[4] Nwokocha L M and Williams P A 2011 *Carbohydr. Polym.* **85** 294–302
[5] Ragone D 2006 *www.traditionaltree.org* **ver.2.1** 1–879
[6] Sikarwar M S, Hui B J, Subramaniam K, Valeisamy B D, Yean L K, and Balaji K. 2014 *J. Appl. Pharm. Sci.* **4** 91–97
[7] Adinugraha H A and Kartikowati N 2012 *Wana Benih* **13** 99–106
[8] Achmad B, Kuswantoro D P and Mulya S 2016. Balai Penelitian Kehutanan Ciamis. *Kajian Potensi Usaha Sukun* In http://www.pohoninvestasi.com/2016/04/kajian- potensi-usaha-sukun.html
[9] Adebowale K O, Olu-Owolabi B I, Olawumi E K and Lawal O S 2005 *Ind. Crops Prod.* **21** 343–351
[10] Beyer R 2007 Breadfruit as a Candidate for Processing. *Acta Hortic.* **(ISHS)** 209–214
[11] Adebowale K O and Lawal O S 2002 *Nahrung/Food* **46** 311–316
[12] Lubis Y M, Kumalaningsih S and Susanto T 2006 *J. Teknol. Pertan.* **7** 173–183
[13] Akanbi T O, Nazamid S and Adebowale A R 2009 *Int Food Res. J.* **16** 151–157
[14] Deivanai S and Bhore S J 2010 *Middle - East J. Sci. Res.* **6** 418–428
[15] Sari L V, Yanti Eff A R and Boedijono E P 2017 *digilib.esaunggul.ac.id/ UEU J* 1–8
[16] Song X, Zhu W, Pei Y, Ai Z and Chen J 2013 *J. Cereal Sci.* **58** 400–407
[17] Olatunde G O, Arogundade L K and Orija O I2017 *Cogent Food Agric.* **3** 1–12
[18] Hidalgo A, Fongaro L and Brandolini 2017 *J. Cereal Sci.* **77** 9–16
[19] Zhang Z, Wang J, Zhang X, Shi Q, Xin L, Fu H and Wang Y 2018 *Food Chem.* **248** 173–182
[20] Alvarez-Ramirez J, Vernon-Carter E J, Carrillo-Navas H and Meraz M 2018 *LWT-Food Sci. Technol.* **91** 203–212
[21] Sede V J, Mamuaja C Fand Djarkasi G S 2015 *J. Ilmu dan Teknol. Pangan* **3** 24–35
[22] Fiorda F A, Soares M S, da Silva F A, Grosmann M V E and Souto L R F 2013 *LWT - Food Sci. Technol.* **54** 132–138
[23] Polnaya F J, Breemer R, Augustyn G H and Tuhumury H C D 2015 *Agrinimal* **5** 37–42
[24] Hidalgo A, Fongaro L and Brandolini A 2014 *Food Res. Int.* **64,** 363–370

Acknowledgement

This research has been supported by research project Widya Dharma University based on Number: 083/F.02.83/PUSLIT/XI/20l6

Utilization of carrageenan and chitosan as coating material in phycocyanin encapsulation

Hadiyanto[1,2], H Sutanto[3], M Suzery[4], N P Adetya[1*], A M Nilamsari[1], A Yunanda[1]

[1]Department of Chemical Engineering Faculty of Engineering, Diponegoro University, Jl. Prof Soedarto, SH Tembalang, Semarang, Central Java, Indonesia
[2]Master Program of Environmental Science, School of Postgraduate Studies, Diponegoro University, Jl. Imam Bardjo No 3-5, Semarang, Central Java, Indonesia
[3]Physic Department, Faculty of Science and Mathematic, Diponegoro University. Jl. Prof. Soedarto, SH-Tembalang Semarang
[4]Departement of Chemistry, Diponegoro University, Jl. Prof.Soedharto SH, Tembalang, Semarang
Email: naispinta26@gmail.com

Abstract. Phycocyanin is one of the pigments found in *Spirulina platensis.* Phycocyanin has potential as natural blue dye and antioxidant. The application of phycocyanin as food ingredient is still limited due to its vulnerability to environment. Microencapsulation can keep phycocyanin from environmental influences such as temperature, light and pH. This research aimed to determine the antioxidant activity of phycocyanin microencapsulation with carrageenan and chitosan coating material caused by the effect of thermal degradation overtime and its influence on the physical characteristics of the microencapsulation. The highest antioxidant activity of 61.48% was found in chitosan coating material that occurred at temperature of 40°C with 30 minutes heat treatment.Phycocyanin microcapsule with chitosan has more vibrant blue color compared to carrageenan.Phycocyanin microencapsulation either by using carrageenan or chitosan have functional groups such as alkane, alcohol, ether and ester.

1. Introduction

Phycocyanin is a natural pigment which has high economic value, *Spirulina platensis* may contain this compound up to 20% of its dry mass [1]. The blue-green pigment can be used as natural coloring for foods. This substance can substitute the synthetic coloring because it can produce a bright blue color [2]. Besides, it can also be developed as additive, functional nutrition, and non-toxic. Phycocyanin has ability as antioxidant which inhibits peroxidation of lipid in rats [3].

The use of natural antioxidant has been increasingly demanded by society because it is considered saver than synthetic antioxidant. One of the chemical compounds in *Spirulina*sp is phycocyanin which is known to have the ability to boost immune system, contains antioxidant, anti-inflammation, also neuroprotective [4, 2]. However, phycocyanin is very susceptible to degradation when it is exposed to light and high temperature thus it has limited storage life [5]. One of the alternatives to overcome the problem is microencapsulation. Microencapsulation can be literally translated as a process of coating of a core particle with another material in order to obtain the expected physical and chemical characteristics. Microencapsulation is a process which transform solid or liquid substance into micro (0.2 – 5000 μm) capsule form. This process is intended to protect and preserve the active component from environmental influence [6].

The coating material used in the process of microencapsulation should have emulsifier property so that it can be added into food ingredient. The commonly used materials for coating are carrageenan and chitosan. Besides used as chemicals in various industries, chitosan also have antibacterial property [7]. The use of carrageenan in food products is able to improve emulsion stability [8]. Therefore, carrageenan and chitosan can be used as edible coating material which inhibit the growth of bacteria on food product so that the storage life can be extended [7]. This research is aimed to determine the antioxidant activity of phycocyanin microcapsules with carrageenan and chitosan coating materials caused by thermal degradation overtime. Moreover it also aimed to find out the influence of the use of coating materials on phycocyanin microcapsules' physical characteristics.

2. Materials and methods

2.1. Materials

Phycocyanin powder from Spirulina platensis microalgae was used in this study. Carrageenan and chitosan are used as coating materials. It also used other materials in the process of microencapsulation and measurement of antioxidant activity such as acetic acid, NaCl, methanol, DPPH, STTP, KCl and aquades.

2.2. Phycocyanin microcapsule with carrageenan coating material

Carrageenan is dissolved in NaCl 1% solvent and heated while being stirred at 40°-50^{0} C. The stirring process is stopped after 100ml of 2% w/v carrageenan is obtained. Phycocyanin is dissolved in distilled water (1:10) and then combined with carrageenan and stirred. The solution is put into a syringe and injected into KCl 0.5 M solution. The formed granules are then washed and dried by using freeze drying method.

2.3. Phycocyanin microcapsule with chitosan coating material

2 gram of chitosan is dissolved with 100ml of acetic acid 1% solution and then stirred for 4 hours. The stirring process is stopped after 100ml of 2% w/v chitosan is obtained. Meanwhile, phycoctanin is dissolved in distilled water with 1:10 ratio. The solution then mixed with chitosan solution with 1:1 ratio and then stirred. After it turns homogenous, the solution is put into syringe and injected into STTP 0.5 M. The formed granules are washed with NaCl 1% and then dried with freeze drying at 10°C [9].

2.4. Antioxidant activity, color stability test and FTIR test

The chemical for testing antioxidant activity by using DPPH method is 1.1-diphenyl-2-picrylhydrazyl (DPPH) and methanol. The variables used were heating temperature (40°C, 50°C, and 60°C) and heating time (0, 30, 60, and 90 minutes). 10 mg of microcapsules were dissolved into 10 ml of distilled water, and then heated according to the temperature variable. After that 1 ml sample was taken based on heating time variable. 1 ml of DPPH 0,1mM was added into 1ml of pre-heated phycocyanin microcapsules and 5ml of methanol, then incubated for 1 hour. The absorbency measurement was conducted in 515-520nm wavelength [10]. Antioxidant activity is determined by the following equation (1):

$$\text{Antioxidant activity (100\%)} = \left\{1 - \left(\frac{A_{\text{sample}}}{A_{\text{blank}}}\right)\right\} \times 100\% \quad (1)$$

Color stability test was conducted by using digital colorimeter. 5 ml sample of each variables was dripped right under the lens of the camera which was connected to Digital Colorimeter device. Then the camera is directed on the part of which the color was analyzed to obtain the value of L, a and b on the device's screen. While FTIR test is used to determine the vibration of the functional groups of the compound. From the FTIR spectrum will be known by the function of functional vibrations in the

Phycocyanin microencapsulation. Shimadzu FTIR 8400 Japan was used with a spectral resolution of 1 $cm^{-1}$and the spectral profiles were collected in region between 4,000 $cm^{-1}$and 400 cm^{-1}.

3. Results and discussion

3.1. The effect of coating material on antioxidant activity

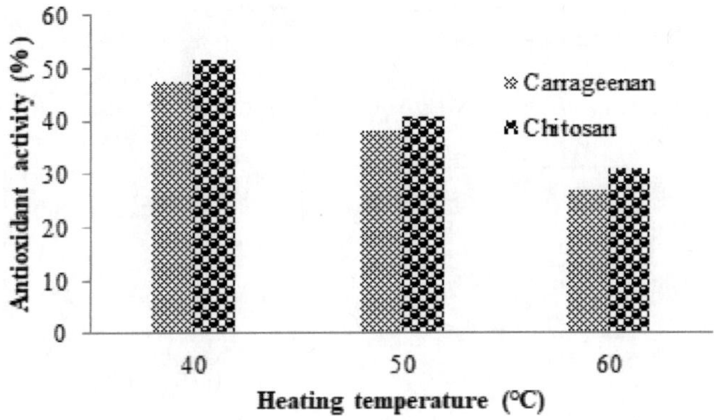

Figure 1. Effect of heating temperature to antioxidant activity at 60 minutes heating time

Based on Figure 1, it shows that the longer the heat treatment, the graphic tends to decrease. This phenomenon happened to both phycocyanin microcapsule with carrageenan and chitosan coating material. It is because phycocyanin is biliprotein. Biliproteinis just like globular protein in general can have structural damage caused by heat [11]. This structural damage is called denaturation, which is the systematic change of organic structure become random three dimensional configuration [12]. Therefore the longer the heat treatment the more the number of damaged phycocyanin. The decrease in the number of active phycocyanin the decrease in antioxidant activity.

Besides, based on the test that had been conducted, antioxidant activity with chitosan coating material is higher than carrageenan coating material. The highest antioxidant activity is 61.48% which is found in chitosan coating material at 40°C temperature and 30 minutes heat treatment. This happened because the process of phycocyanin microencapsulation was conducted at temperature of 40-50°C which make some phycocyanin become damaged because phycocyanin is very vulnerable to heat [5].

3.2. The color stability test comparison between carrageenan and chitosan coating material

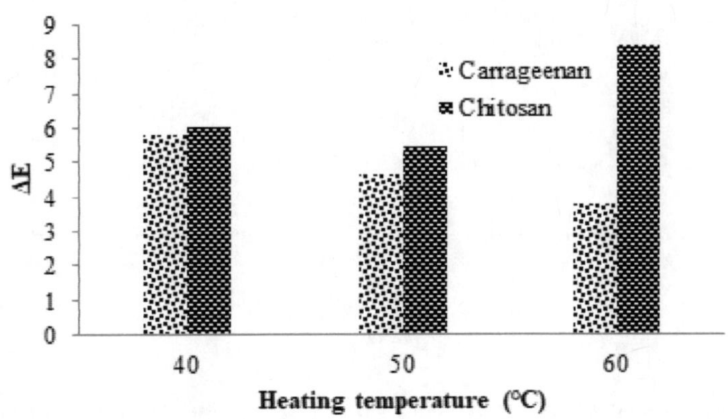

Figure 2. The color stability of carrageenan and chitosan coating agent

Phycocyanin microencapsulation with both chitosan and carrageenan coating material have L(-) color which indicate that the color of phycocyanin solution is getting darker. The value of a(-) indicates greenish color and the value of b(-) indicates bluish color, while ΔE (Figure 2) indicate how big the color change that happens [13].

Phycocyanin microcapsule with chitosan has more vibrant blue color compared to phycocyanin microcapsule with carrageenan. The blue color indicates abundant of phycocyanin content. Phycocyanin is a natural blue pigment which can be found in Spirulina sp. which has antioxidant property. However, phycocyanin is vulnerable to temperature [2, 14]. If treated with high temperature, then the phycocyanin extract will be damaged.

3.3. FTIR test of phycocyanin microencapsulation with carrageenan and chitosan

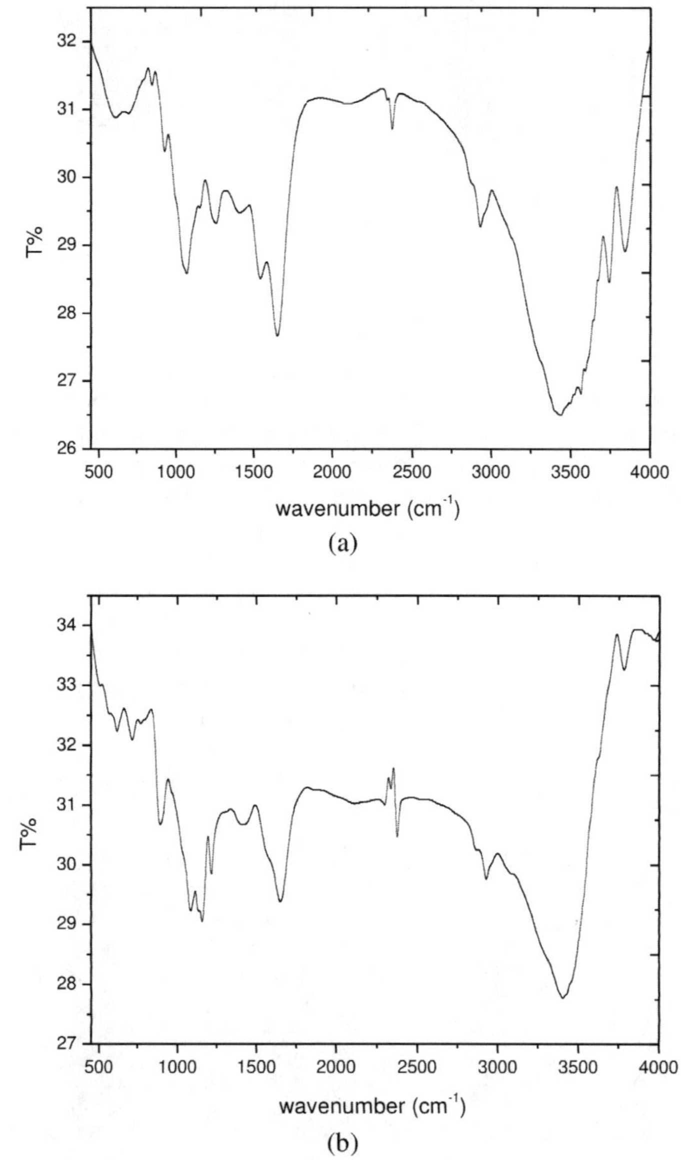

Figure 3. FTIR test result of (a) phycocyanin+carrageenan; (b) phycocyanin+chitosan

IOP Conf. Series: Materials Science and Engineering **633** (2019) 012038 doi:10.1088/1757-899X/633/1/012038

Figure 3a shows the result of phycocyanin+carrageenan FTIR test have functional group namely alcohol, alkane, ester and ether. The absorption band on a surface area around 3439 cm^{-1} shows –OH linkage (alcohol), while on the 2930 cm^{-1} area is C-H linkage (alkane). The absorption band on 1649 cm^{-1} indicates ester linkage and 1068 cm^{-1} absorption is ether linkage [15]. This is in accordance with the reference that algae biomass absorb metal ion through complex formation interaction between metal ion and functional group such as -COOH the main constituent of polysaccharide and peptide group (-CO, $-NH_2$ and $-CONH_2$) as the constituent of pectin and protein which act as electron pair donor [16].

Figure 3b shows the result of phycocyanin+chitosanhave functional group namely alkane, ester, ether and alcohol. Absorption band on around 3404 cm^{-1} area shows O-H linkage (alcohol) while on 2928 cm^{-1} is C-H linkage (alkane). The absorption band on 1650 cm^{-1} C=O linkage (ester). The absorption band on 1157 $cm^{-1}$and 1215 cm^{-1}indicates ether linkage [17].

4. Conclusion

The highest antioxidant activity is 61.48% which is found in chitosan coating material at 40°C temperature and 30 minutes heat treatment. Phycocyanin microencapsulation with both chitosan and carrageenan coating material have L(-) color which indicate that the color of phycocyanin solution is getting darker. The value of a(-) indicates greenish color and the value of b(-) indicates bluish color. Phycocyanin microcapsule with chitosan has more vibrant blue color compared to phycocyanin microcapsule with carrageenan. Phycocyanin microencapsulation either by using carrageenan or chitosan have functional groups such as alkane, alcohol, ether and ester.

References

[1] Antelo F S, Costa J A V, and Kalil S J 2008 Thermal degradation kinetics of the phycocyaninfrom Spriluna platensis *Biochemical Engineering Journal* **41** 43-47

[2] Chaiklahan R, Chirasuwan N and Bunnag B 2012 Stability of phycocyanin extracted from Spirulina sp. : Influence of temperature, pH and preservatives. *Process Biochemistry* **47** 4 659-664

[3] Bertolin T E, Farias D, Guarienti C, Petry F T S, Colla L M and Costa J A V 2011 Antioxidant Effect of Phycocyanin on Oxidative Stress Induced with Monosodium Glutamate in Rats. *Brazilian Archives of Biology and Technology* **54** 733-738

[4] Hadiyanto, Suttrisnorhadi, Susanto H and Suzery M 2016 Phycocyanin extraction from microalgae Spirulina platensis assisted by ultrasound irradiation: effect of time and temperature *Songklanakarin J. Sci. Technol.* **38** 391-398

[5] Suzery M, Hadiyanto, Majid D. Setyawan D, and Susanto H 2016 Improvement of stability and antioxidant activities by using phycocyanin – chitosan encapsulation technique *IOP Conf. Series: Earth and Environmental Science* **55** 1-7

[6] Purmayati L, Dewi E N and Kurniasih R A 2016 Physical characteristics of Spirulina phycocyaninmicroencapsules using different concentration of coating materials *JurnalTeknologiHasilPertanian* **9** 1-8

[7] Bourtoom A 2008 Chitosan film mechanical and permeation properties as affected by acid, plasticizers, and storage *Journal of Food Science* **63** 1049–52

[8] Rojas-Nery E, Garcia-Martinez I and Totosaus A 2015 Effect of Emulsified Soy Oil with Different Carrageenans in Rennet-Coagulated Milk Gels *International Food Research Journal* **22** 606-612

[9] Nilamsari A M, Yunanda A and Hadiyanto 2017 Thermal degradation kinetics of phycocyanin encapsulation as an antioxidant agent *IOP Conf. Series: Earth and Environmental Science* ***102*** 1-6

[10] Molyneux P 2004 The Use Of the stable Free Radical Diphenylpicryl-Hrdrazyl (DPPH) For Estimating Antioxidant Activity *Songklanakarin Journal of Science Technology* **26** 211-9

[11] Woon YC and Hyeon YL 2018Effect of Ultrasonic Extraction on Production and Structural Changes of C-Phycocyanin from Marine Spirulina maxima International *Journal of Molecular Sciences* **19** 1-11

[12] Moosavi-Movahedi, Saboury, Hosseinkhani, Lohrasbi-Nejad, Habibi-Rezaei, Maghami, Atri and Fotouhi 2016 *Biomacromol J* **2** 8-20

[13] Woo SP, Hye-Jin K, Min L, Dong HL, Jungmin K, Sang-Soo K, Chang-Min K, Mario GF and Mi-Jeong A2018 Two Classes of Pigments, Carotenoids and C-Phycocyanin, in Spirulina Powder and Their Antioxidant Activities*Molecules* **23** 1-11

[14] Yan M, Liu B, Jiao X and Qin S 2014 Preparation of phycocyanin microcapsules and it's properties *J. Food and Bioproducts Processing***92** 89-97

[15] Bajpai and Daheriya P 2014 Kappa-Carrageenan/PVA Filmswith Antibacterial Properties: Part 1. Optimization of Preparation Conditions and Preliminary Drug Release Studies*Journal of Macromolecular Science, Part A:Pure and Applied Chemistry* **51** 286–295

[16] Gupta VK and Rastogi A 2008 Biosorption of lead from aqueous solution by green algae Spirogyra species : Kinetics and equilibrium studies*Journal of Hazardous Materials* **152**407-414

[17] Yasmeen S, Kabiraz M K, Saha B, Qadir M R, Gafur M A, Masum S M 2015 Chromium (VI) Ions Removal from Tannery Effluent using Chitosan-Microcrystalline Cellulose Composite as Adsorbent *International Research Journal of Pure &Applied Chemistry***10** 1-14

IOP Conf. Series: Materials Science and Engineering **633** (2019) 012039 doi:10.1088/1757-899X/633/1/012039

Antioxidant capacity of pigeon pea (*Cajanus cajan* L.) sprouts elicited using NaCl with various elicitation time

L S Kristiani, F M Rajendra and S Ariviani

Departement of Food Science and Technology, Faculty of Agriculture, Universitas Sebelas Maret, Jl. Ir. Sutami 36 A, Kentingan, Surakarta 57126, Indonesia

Email: setya_ariviani@yahoo.com; setyaningrum_ariviani@staff.uns.ac.id

Abstract. Previous research showed that pigeon pea has potential as an anti-diabetic functional drink related to its antioxidant capacity. Antioxidant capacity in legumes can be increased by elicitation. Improvement of antioxidant capacity in legumes elicited by NaCl was significantly higher than that of other abiotic elicitation. The objective of this study was to determine the antioxidant capacity of pigeon pea (*Cajanus cajan* L.) sprouts which were elicited using NaCl (50mM) with various elicitation times (8 hours, 12 hours, 16 hours) and germinated for 48 hours. The results showed that elicitation time did not have a significant effect on the total flavonoids content, but significantly increased antioxidant activity (DPPH radical scavenging, and reducing power) of pigeon pea sprouts along with increasing elicitation time. This study proved that elicitation using NaCl 50mM for 16 hours increases the total flavonoid content (69,36%), DPPH radical scavenging activity (134,18%), and reducing power (24,54%) of pigeon pea sprouts. Elicitation using 50mM NaCl with 16 hours elicitation time and 48 hours germination time can be considered as a technique to enhance antioxidant capacity in legumes sprouts.

1. Introduction

ROS naturally occurring as a result of energy synthesize reaction in mitochondria. Several conditions such as infection, respiratory burst, ischemia, and strenuous exercise lead to excessive ROS production. Air pollutants, tobacco smoke, radiation were further increased ROS accumulation in the body [1,2,3,4]. An Imbalance between ROS and antioxidant compounds in a human body leads to oxidative stress which further triggers various degenerative diseases [5]. Therefore, it is important to keep the balance of ROS and human antioxidant status through an intake of food which were rich in antioxidant compounds as one of the simple strategy.

Leguminous known as a natural source of antioxidant [6,7]. Previous researches had reported antioxidant capacity both in pigeon pea seed [8,9,10] and pigeon pea sprouts [11]. Previous studies reported that pigeon pea seed contains several antioxidant compounds, such as phenolic and flavonoids content [8,9,10], and has potential as anti-diabetic functional drink related to its antioxidant capacity [12].

Elicitation is one of the effective technique to improve antioxidant capacity in plants. Elicitation can cause stress so that the plants activated a series of defense system such as enzymatic and non-enzymatic antioxidant production as a response to the stress [13,14]. There are two major groups of elicitor used for elicitation, namely biotic and abiotic elicitor. Abiotic elicitor available as a chemical

IOP Conf. Series: Materials Science and Engineering **633** (2019) 012039 doi:10.1088/1757-899X/633/1/012039

compound that is safe, cheap, easy to be applied, and effective to increase phenolic compounds [14,15,16].

The use of NaCl as an abiotic elicitor in lentil proved to produce sprouts with a higher antioxidant capacity than other abiotic elicitors [14]. Several studies showed the antioxidant capacity improvement of mungbean [17] and common bean [18] by NaCl elicitation. Elicitation using various concentration of NaCl to enhance antioxidant capacity of legumes sprouts also have been reported [14,19,20,21]. The higher NaCl concentration produces legumes sprouts with higher antioxidant capacity. Na+ and Cl- are the most easily salt ions entered the plant cells, therefore induce environmental stress to the plants which further induce ROS production [22]. Swieka [14] stated that plants have developed an antioxidant defense system to reduce the ROS excess. Thus, the higher NaCl concentration can trigger higher stress so that generate a higher antioxidant defense through higher antioxidant compounds production. To the best of our knowledge, the study about the effect of NaCl elicitation time on the antioxidant capacity enhancement of pigeon pea sprouts hasn't been reported. The authors hypothesized increasing NaCl elicitation time are going to increase NaCl accumulation thus provide higher environmental stress which further induces a higher antioxidant compounds production. This study aims to determine the antioxidant capacity of pigeon pea (*Cajanus cajan* L.) sprouts which were elicited using NaCl (50mM) with various elicitation times (8 hours, 12 hours, 16 hours) and germinated for 48 hours. In this study, the potential of NaCl elicitation in increasing the antioxidant capacity of pigeon pea sprouts was also examined. It was determined by comparing the antioxidant capacity of pigeon pea sprouts prepared using NaCl elicitation at the best elicitation time with that of control which was prepared using 0mM NaCl (without elicitation).

2. Methods

2.1 Preparation of pigeon pea sprouts

Pigeon pea (*Cajanus cajan* L.) seeds were purchased from local market in Surakarta, Indonesia. The selected seeds were rinsed thrice with distilled water and soaked in 50mM NaCl solution (1:3 w/v) with different elicitation time (8, 12, 16 hours). Elicited seeds were germinated at room temperature for 48 hours on bamboo weaved (15x15 cm). Distilled water was sprayed every 12 hours.

2.2 Extract preparation

2 grams of dehulled pigeon pea sprouts were extracted with 20 ml methanol (methanol: distilled water of 80:20 v/v) for 2 hours at 50°C and 200 rpm using water bath shaker (HAAKE SWB 20, Fisher Scientific Haake, Germany). The mixture was centrifugated using a tabletop centrifuge (PLC-05 Centrifuge, Gemmy Industrial Corp, Taiwan) at 10,000 rpm for 15 minutes. The clear supernatant was collected in an amber bottle and stored at 10°C until further analysis.

2.3 Total flavonoid content analysis

Total flavonoid content was determined using a method previously described by Pekal and Pyrznska [23] and expressed as μM Quercetin Equivalent (QE)/g sample dry weight.

2.4 Radical scavenging activity analysis

Radical scavenging activity was analyzed using the DPPH method [24] and expressed as μM Gallic Acid Equivalent Activity (GAEA)/g sample dry weight.

2.5 Reducing power analysis

The reducing power was measured following the method described by Berker et al [25] and expressed as μM Ascorbic Acid Equivalent Activity (AAEA)/g sample dry weight.

2.6 Statistical analysis

The Total flavonoid content, reducing power, as well as DPPH radical scavenging activity data of pigeon pea sprouts were analyzed using the program IBM SPSS Statistics 22 (SPSS Inc., Chicago, USA) by analysis of variance (ANOVA) followed by Duncan Multiple Range test to evaluate the significant differences between mean ($p<0.05$). The t-test was used to evaluate the potential of NaCl elicitation for enhancing total flavonoid content and antioxidant activity of pigeon pea sprouts compared with control.

3. Results and Discussion

3.1 The effect of NaCl elicitation time on the antioxidant capacity enhancement of pigeon pea sprouts

Flavonoids are secondary metabolites that have the capability to reduce the production of and quench ROS. Flavonoids have low redox potential and can undergo single electron oxidation so it is capable to reduce strong free radicals. Flavonoids is also can directly quench free radical due to proton transfer from the A and or B ring [26]. Elicitation increases total flavonoid content and antioxidant activity of plants [13,14]. Total flavonoid content and antioxidant activity of pigeon pea sprouts elicited using 50mM NaCl with various elicitation time were presented in Table 1.

Table 1. Total flavonoid content and antioxidant activity of NaCl elicited pigeon pea sprouts with different elicitation time.

Elicitation time (hours)	Total flavonoid content (µM QE/ g sprouts dry weight)	Radical scavenging activity (µM GAEA/ g sprouts dry weight)	Reducing power (µM AAEA/ g sprouts dry weight)
8	395.19 ± 6,48^{a}	464.01 ± 16,63^{a}	7942.46 ± 105.30^{a}
12	408.37 ± 11,14^{a}	521.60 ± 10.31^{b}	8375.56 ± 226.84^{b}
16	413.38 ± 6,91^{a}	551.85 ± 9.93^{c}	9259.24 ± 272.28^{c}

Different superscript within a similar column means significant differences ($p<0.05$)

Although elicitation time did not have an effect on the total flavonoid content ($p>0.05$), however, the higher elicitation time produces significantly higher both radical scavenging activity and the reducing power of elicited pigeon pea sprouts. This results indicated that might be antioxidant compounds other than flavonoids were produced as a response to the NaCl stress. So even though the total flavonoid contents of elicited pigeon pea sprouts were not significantly different but it showed significantly different in both radical scavenging activity and the reducing power. Accumulation of Na^{+} and Cl^{-} ions lead to ionic and osmotic stresses as well as water deficit [22]. The plants activated a series of defense system as a response to stress [13] among others through increasing both enzymatic and non-enzymatic antioxidant production [14].

Phenolic acid, flavonoid, and tannin were naturally synthesized to protect plants from stress during the growth [27]. The antioxidant compounds such as phenolic and flavonoids have a high correlation to the antioxidant activities [7,14,28]. Table 1 showed that pigeon pea sprouts prepared using 50mM NaCl elicitation for 16 hours have the highest antioxidant capacity in term of radical scavenging activity and the reducing power. Based on this result, the potential of NaCl elicitation in enhancing the antioxidant capacity of pigeon pea sprouts was conducted by comparing it with the control which was prepared without elicitation (soaked using 0mM NaCl) for 16 hours. The data were presented in Table 2.

3.2 The potential of elicitation using NaCl in increasing antioxidant capacity on pigeon pea sprout

Table 2. Antioxidant capacity of NaCl elicited pigeon pea sprout compared to control.

Pigeon pea sprout	Total flavonoid content (μM QE/g sprouts dry weight)	Radical scavenging activity (μM GAEA/g sprouts dry weight)	Reducing power (μM AAEA/g sprouts dry weight)
Control	244.09 ± 6.99[a]	235.66 ± 7.10[a]	7434.60 ± 276.85[a]
Elicited	413.38 ± 6.91[b]	551.85 ± 9.93[b]	9259.24 ± 272.28[b]

Different superscript within a similar column means significant differences ($p<0.05$)

The results presented in Table 2 showed that total flavonoid content and antioxidant activity (radical scavenging activity and reducing power) of pigeon pea sprouts elicited by 50mM NaCl were significantly higher than that of pigeon pea sprouts prepared without elicitation. This result was relevant to previous research [14] which reported that total flavonoid, antiradical activity, and the reducing power of the NaCl elicited lentil sprouts were higher than that prepared without elicitation (0mM NaCl) [14]. Total flavonoid content, radical scavenging activity, and the reducing power of NaCl elicited pigeon pea sprouts were 69.36%, 134.18%, and 24.54% higher compared to pigeon pea sprout without elicitation. This study proved that germination technique using NaCl elicitation significantly increase the antioxidant capacity of legumes sprouts.

4. Conclusion

Elicitation time had no significant effect on the total flavonoid content of NaCl elicited pigeon pea sprouts but significantly affect the radical scavenging activity and reducing power. The radical scavenging activity and reducing power of elicited pigeon pea sprouts were increased along with increasing of elicitation time. The research result exhibited that elicitation using 50mM NaCl for 16 hours improves the total flavonoid content, radical scavenging activity and the reducing power of pigeon pea sprouts reached to 69.36%, 134.18%, and 24.54% respectively, compared to control which was elicited using 0mM NaCl for 16 hours.

References

[1] Bhattacharyya A, Chattopadhyay R, Mitra S and Crowe S E 2014 *Physiol Rev* **94** (2) 329-54
[2] Forman H J and Torres M 2002 *Am J Respir Crit Care Med* **166** (122) S4-8
[3] He F, Li J, Liu Z, Chuang C C, Yang W and Zuo L 2016 *Front Physiol* **7**(486) 1-10
[4] Yang W and Omaye S T 2009 *Mutation Research* **674** (1-2) 45-54
[5] Huy L A P, He H and Huy C P 2008 *Int J Biomed Sci* **4** (2) 89-96
[6] Wang S, Melyk J P, Tsao R and Marcone M F 2011 *Food Research International* **44** (1) 14-22
[7] Lin P Y and Lai H M 2016 *J Agric Food Chem* **54** (11) 3807-14
[8] Rani S, Poswal G, Yadav R and Deen M K 2014 *Int J of Pharmaceutical Sciences Review and Research* **28** (2) 90-4
[9] Devi R R, Premalatha R and Kayathri R 2016 *International Journal of Current Microbiology and Applied Sciences* **5** (10) 688-97
[10] Marathe S A, Rajalakshmi V, Jamdar S N and Sharma A 2011 *Food Chem Toxicol* **49** (9) 2005-12
[11] Unchegbu N N and Ishiwu C N 2016 *Food Science and Nutrition* **4** (5) 772-7
[12] Ariviani S, Affandi D R, Listiyaningsih E and Handajani S 2016 *Earth and Environmental Science* **102** 1-9
[13] Baenas N, Viguera C G and Moreno D A 2014 *Molecules* **19** (9) 13541-63
[14] Świeca M 2015 *Saudi J Biol Sci* **22** (4) 409-16
[15] Fu T J, Gummer S and Wayne R C 2012 *Plant Cell and Tissue Culture for the Production of Food Ingredients* (Berlin: Springer Science and Bussiness Media)

[16] Naik P M and Al-Khayri J M 2016 *J Adv Res Biotech* **1**(2) 1-7
[17] Yan X, Wang P, Zhou T, Gu Z and Yang R 2017 *Emir J Food Agric* **29** (2) 123-30
[18] Taïbi K, Taïbi F, Abderrahim L A, Ennajah A, Belkhodja M and Mulet J M 2016 *South African Journal of Botany* **105** 306-12
[19] Ahmad F, Singh A and Kamal A 2017 *Biosci Biotech Re. Comm* **10** (3) 481-9
[20] Roychoudhury A and Ghosh S 2013 *UJPB* **01** (03) 11-21
[21] Saraf N 2013 *Asian Journal of Biomedical and Pharmaceutical Sciences* **3** (17) 6-8
[22] Azooz M M and Ahmad P 2015 *Legumes under Environmental Stress: Yield, Improvement and Adaptations* (Chichester: Wiley Blackwell)
[23] Pękal A and Pyrzynska K 2014 *Food Anal Methods* **7** (9) 1776-82
[24] Sharma O P and Bhat T K 2009 *Food Chemistry* **113** (4) 1202-05
[25] Berker K I, Güçlü K, Tor I and Apak R 2007 *Talanta* **72** (3) 1157-65
[26] Mierziak J, Kostyn K and Kulma A 2014 *Molecules* **19** (10) 16240-65
[27] Khang D T, Dung T N, Elzaawely A A and Xuan T D 2016 *Foods* **5** (2) 27-37
[28] Xue Z, Wang C, Zhai L, Yu W, Chang H, Kou X and Zhou F 2016 *Czech J Food Sci* **34** (1) 68-78

IOP Conf. Series: Materials Science and Engineering **633** (2019) 012040 doi:10.1088/1757-899X/633/1/012040

The effect of reduced acetic acid concentration on nano-chitosan formulation as fish preservative

I J Winayu[1], N Ekantari[1], I D Puspita[1], Ustadi[1], WBudhijanto[2] and P S Nugraheni[1*]

[1]Department of Fisheries, Faculty of Agriculture, Gadjah Mada University, Bulaksumur Yogyakarta 55281, Indonesia
[2]Department of Chemical Engineering, Gadjah Mada University, Bulaksumur Yogyakarta 55281, Indonesia

E-mail: nugrahenips@ugm.ac.id

Abstract. Nano-chitosan has antibacterial activity and can be used as fish preservative. Acetic acid concentration of 1% is commonly used to produce nano-chitosan but it affected consumer acceptance because of its acidity. This study aimed to reduce the concentration of acetic acid and observed its effects on particle size, pH, and bacterial growth inhibition activity. Besides, it was also observed the effect of nano-chitosan with acetic acid concentration reduction on total bacteria and sensory value when applied on tilapia fillets. The results showed that the higher concentration of acetic acid resulted smaller particle size of chitosan and higher bacterial growth inhibition. The similar trend was also shown by acetic acid solution in same concentration. However, the bacterial inhibition zone of nano-chitosan solution was higher than acetic acid solution for each variation of acetic acid concentrations. This was evidenced by the observation of the clear zone formation of Gram positive and negative bacteria culture grown on nutrient agar. The optimum acetic acid concentration in nano-chitosan production was 0.43% because it was able to inhibit the growth of bacteria on the tilapia fillet and resulted in good consumer acceptance even when tilapia fillet was soaked up to 25 minutes in the nano-chitosan solution.

1Introduction

Chitosan is a natural polysaccharide obtained from chitin through deacetylation process [1]. Size reductionof chitosan particle into nano-chitosan increasesits antibacterial activity [2]. Nano-chitosan is more effective in inhibiting bacterial growth than chitosan itself and it is capable of damaging the bacterial cell membrane such as in *Salmonella choleraesuis*[3],*Staphylococcus aureus*, *Escherichia coli, Bacillus licheniformis* and *Vibrio parahaemolyticus*[4].

The most widely used method for nano-chitosan production is ionic gelation method because the process is easy and inexpensive. The ionic gelation method is prepared by dissolving chitosan into organic acid and then added with tripolyphosphate (TPP) as crosslinker[5]. The chitosan molecules in organic acid solution become soluble and swollen due to the protonation of chitosan's amine group with H^+ groups of acids. The swelling process of chitosan's positive charge causes a large size of chitosan; thus, reduce the reactivity of the molecules to bind the bacterial cell membrane. Tripolyphosphate addition plays an important role in the formation of chitosan particle size into nano-chitosan due to the bond constructed between chitosan's positive groups that have been protonated by acid with TPP'snegative groups[6]. It makes the positive charge of nano-chitosan being more concentrated and increasing the effectivity in inhibiting bacterial growth. The common organic acid used to dissolve chitosan is acetic acid [7]. The concentration of acetic acid commonly used for nano-chitosanmanufacture is ranged from 1 to 2% (v/v)[7,8,9,10]. The use of 1%acetic acid (v/v) already influenced the taste and smell of the preserved products with nano-chitosan[8], so it is necessary to optimize the amount of acetic acid in nano-chitosan in order to reduce bacterial growth but still acceptable for consumers when it is applied to a fishery product.

2 Experimental Method

2.1 Preparation of Nano-chitosan by Ionic Gelation Method

Chitosan powder (Bio Chitosan Indonesia) were weighed for 0.6 g and then dissolved in acetic acid with varied concentration of 0.2%; 0.4%; 0.6%; 0.8%; and 1% (v/v). The dissolution of chitosan powder into acetic acid solution was sequentially in a small amount to preventa clot. It was homogenized using a magnetic stirrer for 2 hours. Then sodium tripolyphosphate solution (0.84 g/L) was added in ratio 5:2 of chitosan: TPP (v/v) then stirred again for an hour to produce nano-chitosan with final acetic acid concentrations of 0.14%; 0.29%; 0.43%; 0.57%; and 0.71% (v/v), respectively. This reduction of acetic acid was done gradually until obtained the minimum concentration with high performance (antibacterial and sensory effect) of nano-chitosan.

2.2 Characterization of Nano-chitosan

Nano-chitosan was analyzed for its particle size, pH, and bacterial growth inhibition activity. Particle size analysis was conducted using HORIBA SZ 100 PSA with Dynamic Light Scattering (DLS) technique. The pH measurement was performed to determine the acidity level of the nano-chitosan solution.

Bacteria used in this study were *Bacillus licheniformis* (collection culture of Quality and Fisheries Product Lab, UGM), *Staphylococcus aureus* (FNCC 0047) (Gram-positive), *Escherichiacoli*(FNCC 0091), and *Vibrio parahaemolyticus* (JCM 2147) (Gram-negative). These four types of bacteria are often found in fishery products because of the possibility of contamination during the process. So, by using this bacterial strain on the antibacterial test will give early information the probability of nano-chitosan for preserving fish. Each bacterium was derived from a pure culture that was inoculated on TSB medium (Tryptone Soya Broth) and incubated at 37°C for ± 24 hours. The activity of bacterial growth inhibition was performed using paperdisk method [11]. The bacterial isolates were planted on nutrient agar(NA) in a petri dish, then flattened using drigalsky. Paper disc with diameter of 5 mmwas moistened with 20 μl of nano-chitosan solution, placed on the agar media, then incubated at 37°C for 24 hours. Inhibition activity on bacterial growth was proven by clear zone formation around the paper disk. Inhibition zone was calculated by subtracting clear zone diameter with diameter of paper disk.

2.3 Nano-chitosan Application on Tilapia Fillet

Tilapia fillet was soaked in nano-chitosan solution with acetic acid concentrations of 0.14%; 0.29%; 0.43%; 0.57%; 0.71% (v/v). The soaking procedure was carried out by dipping the tilapia fillet into solution with the ratio of tilapia filet: solution = 1:1 for 3 minutes. Total plate count of bacteria was measured on those samples. As comparison, the sample treated with only acetic acid solution (similar concentration used in nano-chitosan treatment) was also analyzed.

The sensory analysis was performed by the trained panelists that were selected by a series of triangle test. The samples of tilapia fillet were prepared previously by soaking in nano-chitosan solutions for 3 minutes or 25 minutes then cooked in the microwave at 40°C for 8 minutes. Panelists were asked to describe whether the presence of acid in the product was detectable. The threshold test was also performed to reinforce the explanation in the description test [12]. The test was carried out using an acetic acid solution in varied concentration to get the smallest concentration of acetic acid recognized by the panelists.

3 Results and Discussion

3.1 Characterization of Nano-chitosan

Table 1. Particle size and pH of nano-chitosan with variation of acetic acid concentration

Acetic acid concentration in nano-chitosan	pH	Particle size (nm)
0.14%	3.9	162.5±0.00
0.29%	3.7	157.7±0.00
0.43%	3.5	153.4±42.63
0.57%	3.4	149.2±0.70
0.71%	3.3	134.3±32.38

Reduction of acetic acid concentration affects the particle size of nano-chitosan (Table 1). Acetic acid is used to protonate amine groups of chitosan. The sufficient amount of acetic acid causes protonation to take place so that many positive charges is available to crosslink with the negative charge of TPP and form a smaller particle size. Reduction of acetic acid concentration up to 0.14% only slightly changes the size of the nano-chitosan particles. The smaller pH value of nano-chitosan compare to the pKa value of chitosan (6.5) is advantageous for the stability aspect of the solution since the high pH value triggers agglomeration.

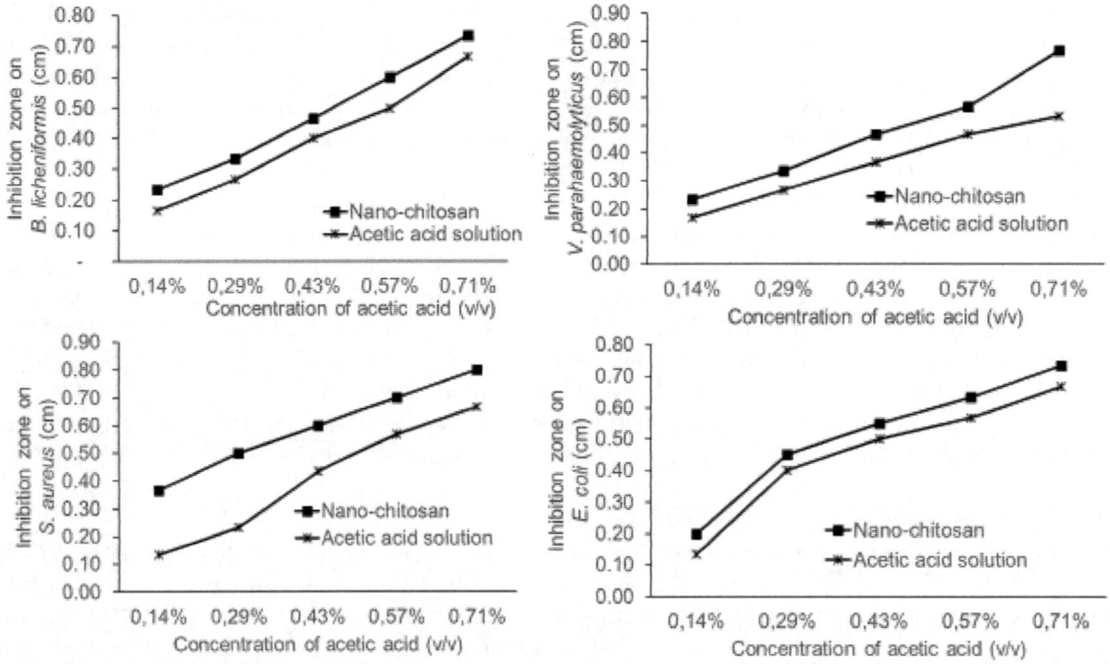

Fig 1. Effect of reduced acetic acid concentration in nano-chitosanon its antibacterial activity measured by inhibition zone on Gram-positive bacteria (*Bacillus licheniformis* and *S. aureus*) and Gram-negative bacteria (*E. coli* and *V. parahaemolyticus*) culture grown on NA.

The effectivity of nano-chitosan in inhibiting the bacterial growth is shown in Fig 1. Nano-chitosan and acetic acid solution have a similar trend for bacterial inhibition but the activity of nano-chitosan was higher than the acetic acid solution in every concentration for all types of bacteria. The presence of nano-chitosan disrupts the permeability of membrane cell of bacteria so that cells experience nutrient deficiencies to grow and cause cells death [13]. The bacterial cell wall of the Gram-positive and Gram-negative has a different structure that affects the reaction process between bacteria and nano-chitosan. Gram-negative bacteria are more sensitive to the presence of chitosan because the outer part of the cell membrane consists of lipopolysaccharide, lipoprotein, and phospholipids, which easily interact with the protonated amino group of chitosan due to the presence of anionic carboxyl groups and phosphate groups on the membrane cell layer. Whereas in Gram-positive bacteria, the outer part of the cell membrane consists mainly of peptidoglycan that leads to the lower ability of chitosan to adhere to the cell membrane [14,15].However, in Gram-positive bacteria, amine groups from protonated chitosan are still possible to interact with negatively charged from teichoic acid located in the outer part of the cell membrane[16].

3.2Effect of reduction of acetic acid concentration on tilapia fillet

The effect of the concentration acetic acid reduction on tilapia fillet was observed immediately after the sample was soaked in nano-chitosan or acetic acid solution. The higher number of bacteria were suppressed with the use of higher concentrations of acetic acid. The initial total amount of bacteria in tilapia fillets before being treated with nano-chitosan or acetic acid solution was 2.30 log cfu/mL. Even in the lowest concentration, acetic acid (0.14%) was able to reduce the number of bacteria (Fig 2). The statistical analysis shows that the number of bacteria logs in tilapia after immersion in nano-chitosan solution was significantly different ($P < 0.05$) with acetic acid solution, indicating a higher antibacterial activity of nano-chitosan compare to that of acetic acid solution

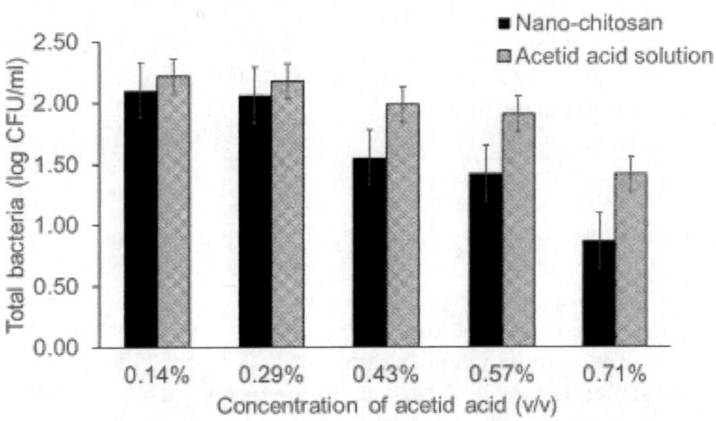

Fig 2.The total amount of bacteria in tilapia fillets after soaking in nano-chitosan prepared in varied concentration of acetic acid and acetic acid solution

The acetic acid also has an antibacterial property[7] so it is expected that the optimal results in inhibiting bacterial growth were obtained from the combination of acetic acid and nano-chitosan. The effectiveness of acetic acid in reducing the growth of L. monocytogenes was reported previously. The results showed that the acetic acid with the concentration of 1% [8] provide a higher bacterial inhibitory effect than 0.3% [7].This result is in line with the data of the total bacteria count and inhibition zones that was conducted in this study. Acids are very influential on the antibacterial activity of nano-chitosan. The H+ group of acetic acid is capable of protonating amine groups from chitosan, which causes the amine group to be more reactive in interacting with bacterial cells that have a negative charge so that it can cause lysis and inhibit growth.

Sensory evaluation of tilapia fillet treated by soaking in nano-chitosan for3 minutes showed no effect on taste and flavor of the product due to the low absorption of nano-chitosan into tilapia fillet.

Meanwhile, the addition of acetic acid concentration greatly influences consumer acceptance in term of taste and flavor for the sample soaked in nano-chitosan for 25 minutes. The taste and flavor of the acid were not detectable until the concentration of acetic acid reached 0.43%. The remaining acidic taste was detected (Table 2). Threshold values of acetic acid that can be detected by human organs is very low. The threshold value was also evaluated in order to obtain the information of minimum concentration detected by the human senses that will facilitate in formulating of the product. Panelists started to detect the difference between solution sample with minerals water starting at 0.004% acetic acid concentration while not yet explaining the taste (absolute threshold). The introduction of flavor types and the different impression of flavor that started to be detected were at a concentration of 0.0075% (recognition threshold and difference threshold). The terminal threshold that can be detected was at 0.04% acetic acid concentration so it is indicating that all panellists were able to sense the difference between sample and mineral water.

Table 2. Sensory evaluation of the taste and flavor of tilapia fillet acid with nano-chitosan treatment

Concentration of acetic acid in nano-chitosan	Taste of acid on fillet after soaking for		Flavor of acid on fillet after soaking for	
	3 minutes	25 minutes	3 minutes	25 minutes
0.14%	undetected	undetected	undetected	undetected
0.29%	undetected	undetected	undetected	undetected
0.43%	undetected	undetected	undetected	undetected
0.57%	undetected	start detected	undetected	Detected
0.71%	undetected	detected	undetected	Strong detected

Based on the data of total bacterial and sensory evaluation on tilapia fillets, it was found that the concentration of 0.43% acetic acid to produce nano-chitosan showed good bacterial inhibition results without changing the taste and aroma of tilapia fillets so that it gives chance to be utilized further on fish preservation.

4 Conclusion

The acetic acid concentration used for formulating nano-chitosan affects the consumer acceptance and the effectiveness in inhibiting bacterial growth. The antibacterial activity of nano-chitosan with various concentrations of acetic acid was higher than acetic acid itself. Nano-chitosan prepared by using 0.14-0.43% acetic acid to preserve the tilapia fillet were accepted by the consumers, even when it soaked up to 25 minutes. The optimum concentration of acetic acid in terms of its ability to reduce bacterial growth and consumer acceptance was at 0.43% acetic acid concentration. Furthermore, research on the effect of nano-chitosan on fish shelf life can be carried out, with reduced acetic acid concentration.

References

[1] Rhazi M, Desbrieres J, Tolaimate A, Alagui A and Vottero P 2000*Polym. Int.* **49**(4):337–44
[2] Ramezani Z, Zarei M and Raminnejad N 2015 *Food Control***51**:43–8
[3] Qi L, Xu Z, Jiang X, Hu C and Zou X 2004 *Carbohydr. Res.* **339**(16):2693–700
[4] ShofiyatiDA, Nugraheni P S, Murwantoko, Puspita I D, Kusumastuti Y and Budhijanto W 2017 *Proceedings of the XIV Annual National Seminar on Fisheries and Marine Research*
[5] Primaningtyas A, Budhijanto W, Fahrurrozi M and Kusumastuti Y 2017*AIP Conference Proceedings***1840**, 080004; doi: 10.1063/1.4982302
[6] Chattopadhyay DP and Inamdar MS 2012 *Res. J. Engineering Sci.* **1**(4)**,** 9-15
[7] Li M, Wang W, Fang W and Li Y 2013 *J Food Prot.***76**(8):1377–83
[8] Beverlya RL, Janes ME, Prinyawiwatkula W and No HK 2008 *Food Microbiol.* **25**(3):534–7

[9] Fernandez-Saiz P, Soler C, Lagaron JM and Ocio MJ 2010 *Int J Food Microbiol.* **137**(2–3):287–94
[10] RamezaniZ, Zarei M and Raminnejad N 2015 *Food Control* **51**:43-48
[11] Ericsson BH, Tunevall G and Wickman K1960*Scand J Clin Lab Invest.* **12**(4):414–22
[12] Wise PM and Breslin PAS 2013*Chem Senses***38**(4):333–42
[13] Helander I, Nurmiaho-Lassila EL, Ahvenainen R, Rhoades J and Roller S 2001*Int. J. Food Microbiol.***71**(2–3):235–44
[14] Sabaa MW, Mohamed NA, Mohamed RR, Khalil NM and Abd El Latif SM 2010 *Carbohydr. Polym.***79**(4):998–1005
[15] Kouchak M, Ameri A, Naseri B and Boldaji SK 2014 *Iran J. Basic Med. Sci.***17**(1):7
[16] Pasquina LW, Santa Maria JP and Walker S 2013 *Curr Opin. Microbiol.* **16**(5):531–7

Acknowledgment
This research was funded from Doctoral Dissertation Research Grant 2017, awarded by The Ministry of Technology Research and Higher Education. Some of the research material was supported by Research Grant of Agriculture Faculty in 2016.

IOP Conf. Series: Materials Science and Engineering **633** (2019) 012041 doi:10.1088/1757-899X/633/1/012041

Sensory, chemical, and nutritional characteristic of coffee-chocolate instant drink (chocomix-fee) from Nglanggeran, Gunungkidul, Yogyakarta

E R N Herawati*, R Nurhayati, M Angwar
Research Unit for Natural Product Technology, Indonesian Institute of Sciences, Jl. Jogja-Wonosari km 31,5 Gading, Playen, Gunungkidul, Yogyakarta- 55861, Indonesia

E-mail : ervika.lipi@gmail.com

Abstract. Cocoa (*Theobroma cocoa*) is one of the potential commodities in Indonesia. There are many cocoa fields in Yogyakarta, and one of cocoa field is Nglanggeran, Gunungkidul. Diversification product must be done to increase the economy value of local chocolate from Nglanggeran. The aim of this research is to determine sensory, chemical, and nutritional characteristic of Chocomix-Fee. 'Chocomix-Fee" is instant drink made from chocolate powder and coffee powder. Chocomix-Fee made with four different coffee powder addition (22%,20%,18%,16%). Sensory test results indicated that the most preferred product was Chocomix-Fee that made from 16% coffee powder addition. The selected Chocomix-Fee contain water content about 2.11%, ash content about 1.41%, fat content 6.83%, protein content about 7.29%, and carbohydrate content about 82.37%. Chocomix-Fee contain polyphenol about 49.55 mg GAE/g sample and antioxidant activity about 56.35%. Chocomix-fee is packaged using metallized package with a weight about 25 grams per sachet. The energy obtained by consuming chocomix-fee drinks is 105 kcal / 25 g product

1. Introduction

Cocoa (*Theobroma cocoa*) is an important product and highly consumed in several countries. Cocoa is one potential commodities in Indonesia because it has high economic value and can high contribute to economic sector. Indonesia is the third largest cocoa producing country in the world after Ivory Coast and Ghana [1, 2]. The main producing area of cocoa is Sulawesi island and the remaining Indonesian cocoa production takes place in North Sumatra, West Java, and Papua, with some small production areas in Bali, Flores, and other islands. As one of the cocoa-producing countries, Indonesia must be able to compete with other cocoa-producing countries by increasing the quality of cocoa beans and cocoa products. Cocoa has many benefit in health because it contain high polyphenols and antioxidant activity [3,4]. Some of the beneficial effects of polyphenols are such as anti-carcinogenic, anti-atherogenic, anti-ulcer, anti-thrombotic, anti-inflammatory, anti-microbial, immune modulating, analgesic effects. One of the cocoa producing areas in Gunungkidul district is Nglanggeran, Patuk. Nglanggeran also known for tourism village because ancient mountain and Nglanggeran Ponds were located in Nglanggeran. Post-harvest processing technology of cocoa beans in Nglanggeran is still limited. Processing of cocoa into intermediate product (chocolate powder, chocolate fat or chocolate paste) or end product (chocolate instant drink, chocolate bar) in Indonesia is mostly controlled by large companies, while cocoa processing by farmers is still minimal [5]. One of cocoa product diversification that potential to developed is instant chocolate drinks which blend with coffee powder (Chocomix-Fee). Coffee is one of the most popular beverages in the world. Coffee also contain polyphenol and antioxidant capacity [6]. Some researcher reported about blending coffee with cocoa powder to make an instant drink [7,8]. 'Chocomix-Fee" is instant drink made from Nglanggeran chocolate powder and coffee powder. The aim of this research is to determine the sensory, chemical, and nutritional characteristic of Chocomix-Fee.

2. Materials and Methods

2.1. Materials

The materials used in the research were cocoa powder obtained from Nglanggeran, Gunungkidul, coffee powder, skim milk, and sugar obtained from local market. The tools used for Chocomix-Fee were blending machine, cabinet dryer, tray, digital scales, and metallize packaging.

2.2. Methods

Research methodologies consist of two stages : Chocomix-Fee making and characterization the sensory, chemical, and nutritional composition of the product. Chocomix-Fee was made with blending coffee powder, cocoa powder, skim milk, and sugar with composition as described in Table 1.

Table 1. Formulation of chocomix-fee

Sample	Coffee Powder (%)	Cocoa powder, Skim milk, sugar (%)
A	22	78
B	20	80
C	18	82
D	16	84

Chocomix-Fee was made with four different addition of coffee powder, i.e. 22%, 20%, 18%, and 16%. Addition higher and/or lower addition of coffee (> 22% and/or < 16%) had not acceptable from previous research. Cocoa powder was mixed with coffee powder, cocoa powder, skim milk, and sugar using blending machine during 15 minute, then sieved (1 mm sieve size). Chocomix-Fee then packed in metallize pack. Characterization of the product has been carried out including sensory, chemical, and nutritional analysis. Sensory properties using hedonic test were determined with attributes: color, taste, odor, and overall acceptance [9]. Selected product from sensory analysis then tested for chemical analysis consist of water content [10], protein content [10], fat content [10], ash content [6], carbohydrate by difference, polyphenol and antioxidant activity [11,12]. Nutritional analysis was done with measured the calorie value. The design of the experiment used was a completely randomized design (CRD). Statistical data were analyzed using the software of SPSS version 16 with One Way Anova method with a significant level of 5% in comparison of means using Duncan method.

3. Results and Discussion

3.1 Sensory Analysis

Sensory properties are some of the most important factors on consumer liking and preference; thus, it is very important to determine factors affecting the product attributes, acceptance, and preference, especially for foods and drinks [13]. Sensory analysis of four Chocomix-Fee formulation with attributes i.e. color, odor, taste, and overall acceptance were done as shown in Table 2. The same superscript symbol in the same column indicate that sample are not significantly different at a significance level of 95%. Sensory analysis was conducted with 22 panelist using hedonic scale 1-5, which 1 = very dislike; 2 = dislike; 3 = a little like; 4 = like; 5 = very like. Panelist were random and untrained who were interested in participating in research, were not disease and allergies to chocolate and coffee drink.

Table 2. Sensory analysis of chocomix-fee

Sample	Sensory Parameter			
	Color	Odor	Taste	Overall Acceptance
A	3.91±0.68^{a}	3.32±1.04^{ab}	2.64±0.95^{a}	2.95±0.95^{a}
B	3.59±0.90^{a}	3.00±0.75^{a}	2.73±0.93^{a}	2.86±0.88^{a}
C	3.68±0.89^{a}	3.18±0.85^{ab}	3.00±0.97^{a}	3.05±0.84^{ab}
D	4.00±0.75^{a}	3.59±0.79^{b}	3.23±1.19^{a}	3.59±0.95^{b}

Values are expressed as mean ± standard deviation. The same superscript symbols in the same column indicate that sample are not significantly different at a significance level of 95%

As seen in Table 2, all sensory parameter indicated that the highest score was sample D, with lowest addition of coffee powder. For color and taste parameter, addition of coffee powder with different concentration did not change the product color and product taste significantly. The odor of sample D was significantly different with sample B. The highest overall acceptance score was sample D that significantly different with sample A and B. From sensory analysis, it can be concluded that the selected product was sample D, with addition of 16% coffee powder. The selected product is then analyzed including chemical and nutritional properties.

3.2. Chemical Characteristic

Chemical characteristic of selected Chocomix-Fee (D) can be shown in Table 3.

Table 3. Chemical characteristic of selected chocomix-fee

Chemical characteristic	Content (% wet basis)
Water content	2.11±0.10
Ash content	1.41±0.03
Fat	6.83±0.40
Protein	7.29±0.39
Carbohydrate	82.37±0.53

Values are expressed as mean ± standard deviation

The chemical characteristic of selected Chocomix-Fee product based on wet basis was 2.11% water content, 1.41% ash content, 6.83% fat content, 7.29% protein content, and 82.37% carbohydrate content. Based on dry basis, the chemical characteristic was 1.44% ash content, 6.97% fat content, 7.44% protein content, and 84.14% carbohydrate content. The results showed that the carbohydrate content of Chocomix-Fee was 82.37% in wet basis and 84.14% in dry basis. The high content of carbohydrates in instant chocolate-coffee beverage products due to the contribution of other constituent ingredients, that was sugar. Formulation of Chocomix-fee include 15-20% chocolate powder, 15-20% coffee powder, 5-10% skim milk, and 55-65% granulated sugar. Chocomix-Fee contain fat about 6.83% in wet weight and 6.97% in dry weight. The fat content of Chocomix-Fee was lower than instant chocolate drinks with the addition of sweeteners which fat content about 16-18% [14]. The difference in fat content due to differences material and formulation used. About fatty acid composition, Riberio et al [8] reported that coffee which blend 3% cocoa powder contain 44% saturated fatty acids, 11% monounsaturated fatty acids, and 45% polyunsaturated fatty acids.

Functional analysis including polyphenol and antioxidant activity of Chocomix-Fee, coffee powder, and cocoa powder were done as shown in **Table 4**. As shown in Table 4, Chocomix-Fee product contain polyphenol about 49.55 mg GAE/g and antioxidant activity about 56.35%. The higher polyphenol and antioxidant activity of Chocomix-Fee related to the basic ingredient : cocoa powder and coffee powder. Higher antioxidant activity due to polyphenol content [15]. From Table 4, it shown that polyphenol content and antioxidant activity of cocoa powder was higher than coffee powder. Polyphenol content of cocoa powder was 49.28 mg GAE/g sample. Hii et al [5] reported that cocoa powder contain polyphenols about 45.30-60.20 mg GAE/g. Awe et al (2013) reported that total phenolic content of the cocoa, Hibiscus flower extract, and ginger beverage blends ranges from 11–21 mg GAE/g [16]. Cocoa and its derived products (cocoa powder, cocoa liquor, and chocolates) contain varied polyphenols contents and has antiradical capacity [17]. Three main groups of cocoa polyphenols can be distinguished namely the catechins (37%), anthocyanins (4%), and proanthocyanidins (58%) [5]. Phenolic compounds are major plant secondary metabolites. They are found in rather large quantities in plant based foods and beverages and considered beneficial for human health. The key role of phenolic compounds as scavengers of free radicals is emphasized in several reports [18,19]. High antioxidant activity of Chocomix-Fee can increase antioxidant intake of humans, and may protect from diseases as oxidative events in vivo may play a role in the pathogenesis of many diseases which icludes cancer, cardiovascular diseases and arthritis [20,21]. Beside content polyphenol and antioxidant activity, cocoa and coffee beverages also contain caffeine [22].

Table 4. Polyphenol content and antioxidant activity

Sample	Polyphenol Content (mg GAE/g)	Antioxidant Activity (%)
Chocomix-Fee	49.55 ± 0.05^{c}	56.35 ± 0.07^{b}
Cocoa powder	49.28 ± 0.08^{b}	57.03 ± 0.06^{c}
Coffee powder	41.36 ± 0.05^{a}	39.04 ± 0.01^{a}

Values are expressed as mean ± standard deviation. The same superscript symbols in the same column indicate that sample are not significantly different at a significance level of 95%

3.3. Nutritional value

Nutritional value of the product can described by measure the calorie value and nutritional adequacy that shown in **Figure 1.**

Serving size : 25 g
Total energy : 105 kcal
Energy from fat : 15 kcal

% Nutritional adequacy based on 2000 kcal
Fat : 4%
Protein : 2%
Carbohydrate : 6%

Figure 1. Chocomix-fee nutritional value

Chocomix-fee is packaged using metallized package with a weight about 25 grams/sachet. The serving size of chocomix-fee was 105 kcal. Queiroz et al [23] reported serving size of sorgum drink was 126.5-128.3 kcal.

4. Conclusion

The most preferred Chocomix-Fee from sensory analysis is chocomix with addition of 16% coffee powder. The selected Chocomix-Fee contain water content about 2.11%, ash content about 1.41%, fat content 6.83%, protein content about 7.29%, and carbohydrate content about 82.37%. Chocomix-Fee contain polyphenol about 49.55 mg GAE/g sample and antioxidant activity about 56.35%. Chocomix-fee is packaged using metallized package with a weight about 25 grams per sachet. The energy obtained by consuming chocomix-fee drinks is 105 kcal / 25 g product.

Acknowledgement
This project is a part of collaboration program between Bank Indonesia KPw Yogyakarta, Research Unit for Natural Product Technology - LIPI, and Agriculture Division of Gunungkidul District.

References

[1] FAO 2015 Top 10 Commodities of Cocoa Beans in 2013 Food and agricultural organization of united nation faostat3.fao.org.
[2] Neilson J 2013 *Bulletin of Indonesian Economic Studies* **43**(2) p 227-250
[3] Lee K, Kim Y, Lee H and Lee C 2003 *Journal of Agriculture and Food Chemistry* **51**(25) p 7292-7295
[4] Hii C, Law C, Suzannah S, Misnawi and Cloke M 2009 *Asian Journal of Food and Agro-Industry* **2**(04) p 702-722
[5] Nurhayati R, Herawati E R N, Angwar M 2015 Sensory and Chemical Charasteristic of Chocolate Dodol with Fat Source Variation. Prosiding of Seminar Nasional Inovasi Teknologi untuk Memperkuat Peran Industri Menuju Akselerasi Pemenuhan Pangan Nasional, Univ. Katolik Soegijapranata, Semarang. ISBN : 978-602-6865-06-9
[6] Niseteo T, Komes D, Cvitanovic A, Horzic D and Budec M 2012 *Food Chemistry* **134** (4) p 1870-1877
[7] Riberio V S, Leitao A E, Ramalho J C and Lidon F C 2014 *Food Research International* **61** p 39-47
[8] Riberio V S, Lidon F C, Campos P S, Pais I, Leitao A E and Ramalho J C 2016 *Emirates Journal of Food and Agriculture* **28**(10) p 676-682.
[9] Kartika B 1992 Petunjuk Evaluasi Sensori Hasil Industri Produk Pangan (Sensory evaluation methods for Food Product). Yogyakarta: Departemen Pangan dan Gizi.
[10] AOAC 1984 Official Methods of Analysis. Association of Official Analytical Chemists, Washington
[11] Brand-Williams W, Cuvelier M E and Berset C 1995 *Lebensmittel-Wissenschaft und Technologie* **29**: 2530
[12] Belscak A, Komes D, Horzic D, Ganic K and Karlovic D 2009 *Journal Food Research International* **42** p 707-716
[13] Dos A, Ayhan Z and Sumnu G 2005 *J Sens Stud.* **20**(3), p 228–42
[14] Cvitanovic B, Benkovic M, Komes D, Bauman I, Horzic D, Dujmic F and Matijasec M 2010 *J Agric Food Chem.* **58**(12) p 7187-95
[15] Ali F, Ranneh Y, Ismail A and Esa N M 2015 *Journal of Food Science and Technology* **52**(4) p 2103-2111
[16] Awe F B, Fagbemi T N, Ifesan B and Badejo A 2013 *Food Research International* **52**(2) p 490–495
[17] Genovese M I and Lannes S 2009 *Cienc Tecnol Aliment Campinas* **29**(4) p 810-814
[18] Duenas M, Hernandez T and Estrella I 2006 *Food Chemistry* **98**(1) p 95–103
[19] Savikin K, Zdunic G, Jankovic T, Tasic S, Menkovic N and Stevic T 2009 *Foods for Human Nutrition*, **64**(3) p 212–217
[20] Oboh H and Omorogie I 2011 *Nigerian Journal of Basic and Applied Science* **19**(1) p 68-75
[21] Ghanim H, Sia C L, Abuaysheh S, Korzeniewski K, Patnaik P, Marumganti A, Chaudhuri A and Dandona P 2010 *J. Clin. Endocrinol Metab* **95**(9) p 1-8
[22] Ogah C O and Obebe O T 2012 *Journal of Innovative Research in Engineering and Sciences* **3**(1) p 404-411
[23] Queiroz, V, Aguiar A, de Menezes C B, de Carvalho C, Paiva, C L, Fonseca P C and da Conceicao R 2018 *Journal of Cereal Science* **79** p 43–49

IOP Conf. Series: Materials Science and Engineering **633** (2019) 012042 doi:10.1088/1757-899X/633/1/012042

Influence of withering time and leaf condition on physical and chemical characteristics of fig leaf tea

B S Amanto, F N Laily, A Nursiwi

Department of Food Science and Technology, Universitas Sebelas Maret, Indonesia

E-mail: asrinursiwi@gmail.com

Abstract. The fig leaf (*Ficus carica*) is a good source of bioactive compounds, especially phenolic. Fig leaf can be used for producing herbal tea. The processing of tea includes withering, rolling, oxidation, and drying. Withering is one of the major steps for tea quality. The objectives of the research were to study the effect of withering time and leaf condition on physical (color) and chemical (total phenol content, tannin content, antioxidant activity) characteristics of fig leaf tea. The fig leaves were withered in four diferrent times (0 hours, 8 hours, 13 hours, and 18 hours). The leaf conditions were normal and rusty leaves. The results showed that withering time significantly affects the physical and chemical characteristics of fig leaf tea. The longer withering time the darker tea color, and the lower total phenol content, tannin content, and antioxidant activity of fig leaf tea infussions. Leaf condition significantly affects the tea color (L, a, b, hue). Normal leaves produce the tea with higher total phenol content, tannin content, and antioxidant activity than the rusty leaves.

1. Introduction

Fig plants (*Ficus carica*) is one of the plants from the Mediterranean that has high functional value in the roots, fruit, and leaves [1]. Leaves are one part of the plants that can be used easily. Leaves are an excellent source of bioactive compounds, especially phenolic compounds. The fig leaves contain phenolic components which effectively scavenge free radical DPPH. Phenolic compounds can act as antioxidants that can prevent cancer, heart disease, cataracts, asthma, and can improve immune function [2].

Today fig plants are widely cultivated in Indonesia. Fig plants grows easily in Indonesia, but because of the origin of these plants from the hot and dry Mediterranean, there are often problems with fig plants in Indonesia. Indonesia has higher humidity and rainfall compared to the Mediterranean so that it causes the leaf diseases such as leaf rust. The leaf rust is caused by the fungi, the *Cerotelium fici*. The characteristics of fig rusty leaves are yellow to reddish brown on the leaves [3].

In Indonesia, fig leaf is used as a material for making tea. Herbal tea is tea made not from from *Camellia sinensis* plants but from flowers, skin, seeds, leaves, and roots of various plants. Herbal tea with various ingredients are widely developed in Indonesia, including rosella tea [4], coffee leaf tea [5], lotus flower tea [6], soursop leaf tea [7], cocoa leaf tea [8], gambir leaf tea [9], and melinjo skin tea [10].

Black tea is produced from tea leaves through several steps, namely withering, rolling, fermentation, drying, and sorting. Withering time is one of the important factors in the tea processing because it can affect the composition of the tea. The withering process is carried out as an initial step to reduce the moisture content in the leaves so that subsequent processing takes place properly [11]. Flavor, color, and flavor-forming compounds are formed due to the polyphenol oxidation during the withering process [12]. The withering process causes biochemical reactions. The main change during withering is the breakdown of proteins into amino acids, amino acids along with carbohydrates and catechins form aromatic compounds [13].

Withering time has a significant effect on tannin content and flavor of soursop leaf tea [11]. In the processing of lotus tea, withering time affected the decrease of moisture content, antioxidant activity, tannin content, and sensory [6]. Therefore, the objectives of the reasearch were to study the effect of withering time and leaf condition on physical (color) and chemical (total phenol content, tannin content, antioxidant activity) characteristics of fig leaf tea.

2. Material and Methods

2.1. Material

Fig leaves (*Ficus carica*) Green Jordan variety was obtained from fig farm in Yogyakarta. The leaves picked were the 2nd to 5th leaves with the condition of normal and rusty leaves.

2.2. Fig leaf tea processing

Fig leaves were sorted and separated based on the leaves condition (normal and rusty leaves). Then the leaves were washed and drained. Next, the leaves were withered at 28-30° C and using a fan for 0,8,13, and 18 hours. After the withering process was complete, the withered leaves were weighed. Then the leaves were rolled mannualy and cut to 0.5 cm in size. Next, the leaves were oxidized at 27° C for 2 hours and dried using a cabinet dryer at 65° C for 4 hours.

2.3. Physical and chemical analysis

The analysis of color was determined by the Hunter CIE Lab system method [14]. The analysis of total phenol content and tannin content was determined by folin ciaocalteu method [15], [16], and antioxidant activity was determined by the DPPH method [17].

3. Result and Discussion

3.1. Physical Characteristics

3.1.1. Color

3.1.1.1. L value*

Table 1. L* value of fig leaf tea infussions under the different withering time and leaf condition

L* value			
Withering time (hours)	Leaf condition		Average
	Normal	Rusty	
0	56.117 ± 0.200	55.125 ± 0.400	55.621^{d} ± 0.600
8	54.972 ± 0.150	54.762 ± 0.219	54.867^{c} ± 0.210
13	54.352 ± 0.186	53.228 ± 0.327	53.790^{b} ± 0.639
18	52.175 ± 0.107	50.895 ± 0.347	51.535^{a} ± 0.712
Average	54.404^{B} ± 1.473	53.503^{A} ± 1.729	

Note: Different notations in the same row and column show a significant difference at 0.05.

According to Table 1, leaf condition significantly affected the L* value of fig leaf tea infussions. The L* value rusty leaves were lower than normal leaves. The L* value were 54.404 and 53.503 for normal and rusty leaves respectively. It means rusty leaves produce a darker color on fig leaf tea infussions. Rusty leaves have brownish spots, are more withered, and look dull. Rusty leaves cause infection in the leaves. The structure of the damaged leaves causes chlorophyll in the leaves degraded easily [18]. Chlorophyll in leaves is degraded into blackish feofitin compounds so that rusty leaves look darker.

The L* value under the different withering time was various, beetween 51.535 to 55.621 (Table 1). Withering time significantly affected the L* value of fig leaf tea infussions. The longer withering time, the lower L* value. The same result in lotus flower tea [6] and in kakao leaf tea [8] that showing the longer withering time the lower L* value. During withering, the leaves contact with oxygen. In black tea processing, the longer withering time, the lower tea's lightness [19]. Tannin compounds in the leaf tea changes into it derivates, namely theaflavins and thearubigin. During the process, the color of the leaves changes from green to light brown, then dark brown. The longer withering time, the more phenol oxidized and forming brown pigmen and the lightness was reduced [20].

3.1.1.2. a value and b* value.* Notation a is the chromatic mixture of red and green while notation b is the chromatic mixture of blue yellow [10].

Table 2. a* vaue of fig leaf tea infussions under the different withering time and leaf condition

a* value			
Withering time (hours)	Leaf condition		Average
	Normal	Rusty	
0	1.230 ± 0.044	0.827 ± 0.027	1.028^{a} ± 0.213
8	1.323 ± 0.039	1.213 ± 0.062	1.268^{b} ± 0.076
13	1.685 ± 0.019	1.918 ± 0.052	1.802^{c} ± 0.127
18	2.462 ± 0.019	2.420 ± 0.051	2.441^{d} ± 0.043
Average	1.675^{B} ± 0.496	1.595^{A} ± 0.632	

Note: Different notations in the same row and column show a significant difference at 0.05.

Table 3. b* value of fig leaf tea infussions under the different withering time and leaf condition

b* value			
Withering time (hours)	Leaf condition		Average
	Normal	Rusty	
0	15.043 ± 0.055	11.877 ± 0.100	13.460^{a} ± 1.655
8	16.093 ± 0.085	15.650 ± 0.099	15.872^{b} ± 0.248
13	22.478 ± 0.077	16.833 ± 0.145	19.656^{c} ± 2.950
18	24.165 ± 0.172	21.882 ± 0.118	23.023^{d} ± 1.201
Average	19.445^{B} ± 4.026	16.560^{A} ± 3.655	

Note: Different notations in the same row and column show a significant difference at 0.05

According to Table 2 and Table 3, the leaves condition significantly affected the value of a* and b* fig leaf tea infussions. The a* value were 1.675 and 1.595 for normal and rusty leaves respectively, while the b* value were 19.445 and 16.560 for normal and rusty leaves respectively. Rusty leaves had lower a* value than normal leaves, it means rusty leaves produce a lower intensity of red on fig leaf tea infussions. Rusty leaves also have a lower b* value than normal leaves. It means rusty leaves produce a lower yellow intensity in the fig leaf tea infussions. In the rusty leaves, photosynthesis process is disrupted by spots that cover the leaf surface from the entry of sunlight, thereby reducing the ability to synthesize organic matter [18]. One of the photosynthetic organic compounds is a phenol. Tannin is one of the phenol components. Tannin oxidation in the leaves tea by polyphenoloxidase enzyme allows the change of catechins to theaflavins and thearubigin. Theaflavins play a yellow color, while thearubigin plays a brownish red color [13]. Low tannins in rusty leaves form a small amount of thearubigin and also theaflavin, so it has a lower a * value and b* value.

The a* value and b* value under the different withering time was various, beetween 1.028 to 2.441 (Table 2) for a* value and beetween 13.460 to 23.023 (Table 3) for b* value. Withering time significantly affected the a* value and b* value of fig leaf tea infussions. The longer withering time, the higher a* value. It means the infussions color was more redness. The longer withering time, the higher b * value. It means the infussions color was more yellowness. Catechin oxidation in the leaves tea by polyphenoloxidase enzymes allows the change of catechins to theaflavins and thearubigin. The longer catechin oxidation in the withering process causes the color changes to more brownish. Thearubigin plays a brownish red color [13]. A high a* value is associated with high levels of thearubigin in the infussions [21]. Theaflavins play a yellow color [13]. High b* values are associated with high levels of theaflavins in the infussions [21].

3.1.1.3. °Hue

Table 4. °Hue of fig leaf tea infussions under the different withering time and leaf condition

Withering time (hours)	°Hue Leaf condition Normal	Rusty	Average
0	85.326 ± 0.171	86.018 ± 0.121	85.672^{d} ± 0.388
8	85.300 ± 0.143	85.568 ± 0.203	85.433^{c} ± 0.218
13	85.713 ± 0.056	83.497 ± 0.220	84.605^{b} ± 1.167
18	84.183 ± 0.068	83.589 ± 0.151	83.936^{a} ± 0.281
Average	85.130^{B} ± 0.594	84.693^{A} ± 1.150	

Note: Different notations in the same row and column show a significant difference at 0.05

According to Table 4, leaf condition significantly affected the value of the hue of fig leaf tea infussions. The °Hue value were 85.130 and 84.693 for normal and rusty leaves respectively. Tea infussions with a lower °Hue has a greater intensity of red color and a smaller intensity of yellow. Rusty leaves have a lower hue value than normal leaves because chlorophyll in rusty leaves is more easily degraded due to damaged leaf structure. In the processing of black tea, the enzymatic oxidation process causes the green chlorophyll to decompose into black feofitin [13].

Withering time significantly affected the °Hue value. The longer the withering time the lower °Hue value. A high °Hue value indicates the intensity of red is getting smaller and the intensity of yellow is getting bigger. The °Hue value under the different withering time was various, beetween 85.672 to 83.936 (Table 4). In this range of values, the color of the fig leaf tea infussinos is yellow red [22]. The same result in lotus flower tea [6] that showed the longer withering time the lower ° Hue value.

3.2. Chemical Characteristics

3.2.1. Total phenol content

Table 5. Total Phenol Content of fig leaf tea infussions under the different withering time and leaf condition

Withering time (hours)	Total phenol content (%) Leaf condition Normal	Rusty	Average
0	1.494 ± 0.023	1.076 ± 0.009	1.285^{c} ± 0.219
8	1.486 ± 0.018	1.070 ± 0.018	1.278^{c} ± 0.218
13	1.315 ± 0.014	1.042 ± 0.020	1.179^{b} ± 0.143
18	1.241 ± 0.012	0.709 ± 0.012	0.975^{a} ± 0.278
Average	1.384^{B} ± 0.113	0.975^{A} ± 0.158	

Note: Different notations in the same row and column show a significant difference at 0.05

According to Table 5, the leaves condition significantly affected the total phenol content of fig leaf tea infussions. The total phenol content was 1.384% and 0.975% for normal and rusty leaves respectively. Rusty leaves have a lower total phenol content compared to normal leaves. It is due to the *Cerotelium fici* which attack the leaves results in the appearance of spots that can cover part of the leaf surface so that the sunlight cannot enter. Thus the photosynthesis process on the leaf surfaces is disrupted, thereby reducing the ability to synthesize organic materials [18]. One of the photosynthetic organic material is a phenol.

The total phenol content under the different withering time was various, beetween 0.975% to 1.285% (Table 5). Withering time significantly affected the total phenol content of fig leaf tea infussions. The longer withering time the lower total phenol content. The longer withering time, the more compounds in the leaves undergo biochemical changes [10] so that the total phenol content decreased. Phenols have volatile properties, are easily oxidized, sensitive to light and oxygen. The longer withering time, the more phenol oxidized and forming brown pigmen and the lightness was reduced [20].

3.2.2. Tannin content

Table 6. Tannin content of fig leaf tea infussions under the different withering time and leaf condition

Withering time (hours)	Tannin cotent (%) Leaf condition Normal	Rusty	Average
0	0.208 ± 0.001	0.201 ± 0.001	0.205^d ± 0.004
8	0.200 ± 0.001	0.197 ± 0.001	0.198^c ± 0.002
13	0.198 ± 0.001	0.173 ± 0.002	0.186^b ± 0.013
18	0.186 ± 0.002	0.165 ± 0.001	0.175^a ± 0.011
Average	0.198^B ± 0.008	0.184^A ± 0.016	

Note: Different notations in the same row and column show a significant difference at 0.05

According to Table 6, the leaves condition significantly affected the tannin content of fig leaf tea infussions. The tannin content was 0.198% and 0.184% for normal and rusty leaves respectively. Same with total phenol content, in the rusty leaves the tannin content also lower than in the normal leaves. It is due to the *Cerotelium fici* which attack the leaves results in the appearance of spots that can cover part of the leaf surface so that the sunlight cannot enter. Thus the photosynthesis process on the leaf surfaces is disrupted, thereby reducing the ability to synthesize organic materials [18]. Tannins are one of the phenol components [13].

The tannin content under the different withering time was various, between 0.175% to 0.205% (Table 6). Withering time significantly affected the tannin content of fig leaf tea infusions. The longer withering time the lower tannin content. The same result with tannin content in the lotus flower tea, tannin content decresed because the tannin slightly oxidized during the withering process[6]. The content of catechins decreases during withering in the tea processing. The catechins are converted into theaflavins and thearubigin to produce a distinctive flavor [13].

3.2.3. Antioxidant activity

Table 7. Antioxidant activity of fig leaf tea infussions under the different withering time and leaf condition

Witering time (hours)	Antioxidant activity (%) Leaf condition Normal	Rusty	Average
0	89.066 ± 0.198	86.048 ± 0.146	87.575^d ± 1.566
8	87.930 ± 0.109	83.547 ± 0.284	85.738^c ± 2.298
13	87.256 ± 0.145	75.364 ± 0.335	81.310^b ± 6.215
18	68.016 ± 0.239	62.513 ± 0.191	65.264^a ± 2.881
Average	83.067^B ± 8.903	76.877^A ± 9.391	

Note: Different notations in the same row and column show a significant difference at 0.05.

According to Table 7, the leaves condition significantly affected the antioxidant activity of fig leaf tea infussions. The antioxidant activity was 83.067% and 76.877% for normal and rusty leaves respectively. Rusty leaves have lower antioxidant activity when compared to normal leaves. Antioxidant activity in the fig leaves is strongly influenced by phenol content [2]. Antioxidant properties are caused by phenolic compounds. Phenolics compounds have function as antioxidants because of their ability to stabilize free radicals by giving hydrogen atoms [23]. Besides, tannins has activity as antioxidants [24], so that phenol and tannin content affect the antioxidant activity. Rusty leaves have a lower phenol and tannin content than normal leaves. So the lower the phenol and tannin content, the lower the antioxidant activity.

The antioxidant activity under the different withering time was ranged beetween 65.264% to 87.575% (Table 7). Withering time significantly affected the antioxidant activity of fig leaf tea infussions. The longer withering time the lower antioxidant activity. It is assosiate with total phenol content and tannin content. The longer withering time, the lower total phenol content and tannin content (Table 5 and Table 6).

IOP Conf. Series: Materials Science and Engineering **633** (2019) 012042 doi:10.1088/1757-899X/633/1/012042

4. Conclusions

Withering time and leaf condition significantly affect the physical and chemical characteristics of fig leaf tea. The longer withering time the darker of fig leaf tea infussions color and the lower total phenol content, tannin content, and antioxidant activity of fig leaf tea infussions. Leaf condition significantly affects the tea color (L*, a*, b*, hue) of fig leaf tea infussions. Rusty leaves produce fig leaf tea infussions with a dark/ old/brownish color. Normal leaves produce fig leaf tea infussions with higher total phenolic content, tannin content, and antioxidant activity than the rusty leaves. From these parameters, normal leaves that are withered for 8 hours produce the best fig leaf tea.

References

[1] Barolo MI, Mostacero NR and López SN 2014 *Food Chemistry* **164** 119-27
[2] Mahmoudi S, Khali M, Benkhaled A, Benamirouche K and Baiti I. 2016 *Asian Pac. J. Trop. Biomed.* **6** 239-245
[3] Verga A. and Nelson S 2014 *Plant Disease* **100** 1-5
[4] Ekanto B and Sugiarto 2011 *Jurnal Media Ilmu Keolahragaan Indonesia* **1** 171-180
[5] Siringoringo FHT, Lubis Z and Nainggolan RJ 2012 *Jurnal Rekasaya Pangan dan Pertanian* **1** 1-5
[6] Kusumaningrum R, Supriadi A and Hanggita RJS 2013 *Jurnal FishTech* **2** 9-21
[7] Adri D and Hersoelistyorini W 2013 *Jurnal Pangan dan Gizi* **4** 1-12
[8] Supriyanto, Darmadji P and Susanti I 2014 *Jurnal Agritech* **34** 422-9
[9] Dinata A 2015 *Jurnal Nasional Ecopedon* **3** 93-7
[10] Ardiyansyah and Apriliyanti M 2016 *Jurnal Ilmiah Inovasi* **1** 89-92
[11] Manik DMT, Rusmarilin H and Limbong LN 2015 *Jurnal Rekayasa Pangan dan Pertanian* **3** 14-8
[12] Andrianis Y 2009 *Pengolahan Teh Hitam* (Jakarta: Penerbit Gagas Media)
[13] Towaha J 2013 *Warta Penelitian dan Pengembangan Tanaman Industri* **19** 12-6
[14] Manera FJ, Martinez JMB, Conesa A and Porras I 2012 *Aust. J. Crop Sci.* **6** 1051-58
[15] Senter SD, Robertson JA and Meredith FI 1989 *J Food Sci* **54** 1259-1268
[16] AOAC 1999 *Official Methods of Analysis of The Association Analytical Chemistry* (Washington: AOAC)
[17] Subagio A and Morita N 2001 *Food Research International* **34** 315-20
[18] Kumar N and Khurana SMP 2015 *Int.J.Curr.Microbiol.App.Sci.* **4** 749-54
[19] Turkmen N, Sari F and Velioglu YS 2009 *Akademik Gida*, **7** 29-40
[20] Wiranata G, Yuwono SS and Purwantiningrum I 2016 *Jurnal Pangan dan Agroindustri* **4** 449-57
[21] Laddi A, Sharma S, Kumar S and Prakash NR 2011 *International J. of Computer Applications* **34** 7-9
[22] Hutching J B 1999 *Food Colour and Appearance* (Maryland: Aspen Publisher)
[23] Suryatno H, Basito and Widowati E 2012 *Jurnal Teknosains Pangan* **1** 118-125
[24] Uysal S, Zengin G and Aktü msek A 2015 *J. Med. Herbs Ethnomed.* **1** 97-102

IOP Conf. Series: Materials Science and Engineering **633** (2019) 012043 doi:10.1088/1757-899X/633/1/012043

Chemical characteristics of composite flour based on white corn and okara

D Ishartani, R Sagita and D Praseptiangga

Department of Food Science and Technology, Universitas Sebelas Maret, Surakarta, Indonesia

E-mail: dwiishartani@staff.uns.ac.id

Abstract. Composite flour is a flour based from two or more types of flours. In this research white corn and okara were used as composite flour raw materials. This study aimed to determine the ratio of white corn and okara flour effect on chemical characteristics of the composite flour. The research design used in this study was Completely Randomized Design with 1 factor which was the difference ratio of white corn flour and okara flour (100%:0%, 95%:5%, 90%:10%, 85%:15% and 80%:20%). The data obtained was analysed statistically using one-way ANOVA and Duncan's Multiple Range Test at a significance level of α = 0,05. The results showed the ratio of white corn and okara flour significantly affect the chemical characteristics of the composite flour except moisture content. Higher ratio of okara flour could increase ash, lipid, protein, resistant starch, total dietary fiber, insoluble dietary fiber and soluble dietary fiber content of the composite flour, but decrease carbohydrates, starch, amylose and amylopectin content of the flour.Moisture content of the composite flour were 9.53-9.68%, ash 0.31-0.76%, lipid 1.03-2.62%, protein 7.19-8.73%, carbohydrates 87.89-91.48%, starch 51.00-61.95%, amylose 12.69-13.84%, amylopectin 36.76-48.15%, resistant starch 0.76-1.41%, dietary fiber 10.40-20.15%, insoluble dietary fiber 8.98-17.72%, and soluble dietary fiber 0.83-1.37%.

1. Introduction

Flour is a semi-finished food products that can be used as raw materials in the manufacture of various types of products. There are several kinds of flour that have been used. However, the characteristics of flour sometimes did not meet the expectations. Efforts in the manufacturing of flour with desirable characteristics may be carried out by composite flour manufacture. Composite flour is a flourbased from two or more types of flours.

A raw material containing most carbohydrates was needed in the manufacture of flour, because basically the flour was used by its starch content. Otherwise the other content was also needed such as protein, lipid, and minerals. Materials containing functional components such as resistant starch and dietary fiber can also be added to enrich the chemical content of the composite flour that can be helpful for body health. One potential fiber and linoleic acid source that could be explored is okara, besides pea pod and broad bean pod [1]. However, study on composite flour based on okara has not been reported yet.

In this research white corn and okara was used as the composite flour materials.White corn was used as a source of carbohydrates whether okara was used as a source of protein and dietary fiber. White corn has high content of dietary fiber (12-15%) and low glycemic index (28.66-41.37) [2,3]. Okara contains protein 25.4-28.4% and dietary fiber 52.8-58.1% [4]. So this research was conducted by manufacturing composite flour based on white corn and okara flour on various ratio. The aim of this research was investigating the effect of white corn and okara flour ratio on the composite flour chemical characteristics.

2. Research methods

2.1. Materials
The ingredients for producing composite flour were white corn flour obtained from Technopark Grobogan Factory managed by Indonesian Agency for The Assessment and Application of Technology (BPPT) and okara obtained from SMEs of Tofu "Mr.Gundul" in Purwogondo, Kartasura, Sukoharjo, Central Java. All chemicals being used for analysis were pro-analysis grade.

2.2. Tools
Hydraulic press, steamer, pan, gas stove,cabinet dryer, dry mill blender and sieving device were used to produce okara flour, sealed plastic container to mix white corn flour and okara flour, spectrophotometer Thermo Scientific Genesys 20 and glassware to analyse chemical characteristic.

2.3. Research Stages

2.3.1. Okara flour making.

The process of okara flour making was begun by pressing wet okara to separate water. Pressed okara then was steamed for 30 minutes.At the first half time the steamer lid was opened and at the last half time the steamer lid was closed and the temperature was maintained at 90±5°C. The process was continued by pressing the okara and then roasting with continuous stirring for 15 minutes, and drying the okara using cabinet dryer at 60±5°Cfor 24 hours. The dried okara further was milled using drymill blenderand sieved 80 mesh to obtain okara flour [5].

2.3.2. Composite flour making.

Composite flour was made by mixing white corn flour and okara flour with ratio 100%:0% (F1=control), 95%:5% (F2), 90%:10% (F3), 85%:15% (F4), 80 %:20% (F5). Mixing was done manually in a sealed plastic container.

2.3.3. Chemical characterization.

Chemical characteristics being analysed were moisture [6], protein [7], lipid [7], ash [7], carbohydrate by difference [7], starch [8], amylose-amylopectin [8], resistant starch [8], and dietary fiber [8].

2.3.4. Data analysis.

The data obtained was analysed using One Way Analysis of Varian. If there was a significant different between samples so further data analysis was conducted using Duncan's Multiple Range Test at $\alpha = 0.05$.

3. Results and discussion

3.1. Proximate content
The proximate content of composite flour based white corn and okara can be found in **Table 1.**It showed significantly differences in ash, lipid, protein and carbohydrate content as the percentage of okara flour was increasing. While the moisture content did not show any significantly difference. Okara flour addition did increase the protein,lipid and ash, yet did decrease the carbohydrate content of the composite flour.

The protein content raised up to 21.42%. Previous research reported that protein content of okara flour was ranging from 25.40%-34.70% [4,9,10] which was higher compared to protein content of white corn flour (10.00%-10.44%)[11,12].White corn flour used in this research itself contained 6.5% protein.

Table 1. Proximatescontent of composite flour based on white corn and okara

Formula[a]	Moisture content (%wb)[b]	Ash content (%db)[b]	Lipid content (%db)[b]	Protein content (%db)[b]	Carbohydrate content (%db)[b]
F1	9.53 ± 0.82[a]	0.31 ± 0.03[a]	1.03 ± 0.10[a]	7.19 ± 0.78[a]	91.48 ± 0.86[d]
F2	9.59 ± 076[a]	0.43 ± 0.05[b]	1.79 ± 0.19[b]	7.84 ± 0.36[ab]	89.94 ± 0.51[cd]
F3	9.60 ± 0.73[a]	0.51 ± 0.05[c]	1.84 ± 0.13[b]	8.24 ± 0.66[bc]	89.41 ± 0.65[bc]
F4	9.67 ± 0.63[a]	0.72 ± 0.02[d]	2.23 ± 0.23[c]	8.45 ± 0.65[bc]	88.60 ± 0.80[b]
F5	9.68 ± 0.65[a]	0.76 ± 0.04[d]	2.62 ± 0.29[d]	8.73 ± 0.44[c]	87.89 ± 0.66[a]

[a] Ratio of white corn flour: okara flour = F1 (100%:0%, control), F2 (95%:5%), F3 (90%:10%), F4 (85%:15%), F5 (80 %:20%)

[b] Different letter in the same column showsa significant difference at $\alpha = 0.05$.

3.2. Starch content

Starch contentof the composite flour will affect the characteristics of the product in the cooking process. During thermal processing starch will allow gelatinization which causes starch granule expand. Starch is a glucose polymer consisting of two components, namely amylose and amylopectin [12]. Amylose content in composite flour affect the difference of peak, setback and final viscosity during pasta formation. While amylopectin affect the characteristics of granule swelling, viscosity, peaktemperature, peak viscosity, pasta formation, and gel strength during storage [13]. The ratio between amylose and amylopectin will affect the texture of the product.

Starch,amylose and amylopectin content of the composite flour were shown in **Table 2**. It showed significantly differences in all parameters with decreasing tendency as the okara flour increased. The decrease in these contents was probably influenced by lower levels of starch in the okara flour compared to white corn flour. The composite flour based on white corn and okara contained low amylose which was ranging from 12.69%-13.84%.

Table 2. Starch content of composite flour based on white corn and okara

Formula[a]	Starchcontent (%db)[b]	Amylose content (%db)[b]	Amylopectincontent (%db)[b]
F1	51.00 ± 1.82[a]	13.84 ± 0.29[b]	36.76 ± 0.58[a]
F2	61.95 ± 2.86[d]	13.80 ± 0.19[b]	48.15 ± 2.97[d]
F3	58.12 ± 0.22[c]	12.69± 0.15[a]	45.07 ± 0.31[c]
F4	56.43 ± 0.60[bc]	12.89 ± 0.34[a]	43.55 ± 0.52[bc]
F5	55.24 ± 0.54[b]	12.96 ± 0.30[a]	42.28 ± 0.51[b]

[a] Ratio of white corn flour: okara flour = F1 (100%:0%, control), F2 (95%:5%), F3 (90%:10%), F4 (85%:15%), F5 (80 %:20%)

[b] Different letter in the same column showsa significant difference at $\alpha = 0.05$.

3.3. Dietary fiber content

Resistant starch and dietary fiber were functional compounds that have potential benefit to human health. Both resistant starch and dietary fiber cannot be digested by digestive enzymes and resistant to stomach acid. They were important component that have the necessary characteristics in functional food formulation [14]. Dietary fiber consist of soluble dietary fiber and insoluble dietary fiber. Insoluble dietary fiber affected the decrease of glycemic response [15]. Insoluble fiber would form a gel-like texture in the digestive tract that can slow the digestive rate in the intestine and slow the appearance of blood glucose, and provide a longer satiety. Blood sugar levels become more stable because of the need for insulin in transferring glucose into bodycells and converted into less energy [16]. Meanwhile insoluble dietary fiber serves to prevent the occurrence of some diseases,especially those associated with the digestive tract [17].

Table 3 showed significantly difference in theresistant starch and dietary fiber content of composite flour which tends to increase along by the increasing ratio of okara flour. Resistant starch content on the food can be improved by addingnuts and the derived and by products [18].

Table 3. Dietary Fiber Content of Composite Flour Based On White Corn and Okara

Formula*	Resistant Starch Content(%db)**	TotalDietary Fiber Content (%db)**	Insoluble Dietary Fiber Content(%db)**	Soluble DietaryFiber Content (%db)**
F1	0.76 ± 0.03^{a}	14.71 ± 2.10^{c}	12.41 ± 1.33^{b}	1.37 ± 0.12^{d}
F2	0.89 ± 0.04^{b}	10.40 ± 0.65^{a}	8.98 ± 1.00^{a}	0.83 ± 0.07^{a}
F3	1.12 ± 0.09^{c}	12.92 ± 0.40^{b}	11.15 ± 0.84^{b}	1.02 ± 0.05^{b}
F4	1.23 ± 0.13^{c}	16.04 ± 0.28^{d}	13.92 ± 0.86^{c}	1.16 ± 0.12^{c}
F5	1.41 ± 0.14^{d}	20.15 ± 0.64^{e}	17.72 ± 1.44^{d}	1.23 ± 0.09^{c}

* Ratio of white corn flour: okara flour = F1 (100%:0%, control), F2 (95%:5%), F3 (90%:10%), F4 (85%:15%), F5 (80 %:20%)

** Different letter in the same column showsa significant difference at $\alpha = 0.05$.

4. Conclusions

White corn and okara flour ratio did not affect the moisture content of the composite flour. Higher ratio of okara flour did result higher ash, lipid, protein, resistant starch, total dietary fiber, insoluble dietary fiber and soluble dietary fiber content of the composite flour. However, higher ratio of okara flour tended to decrease carbohydrates, starch, amylose and amylopectin content of the flour. A further research is needed to find out the physical characteristics and gelatinization profile of the composite flour based on white corn and okara as well as the application of the composite flour in various food products.

References

[1] Mateos-Aparicio I, Redondo-Cuenca A, Villanueva-Suárez M, Zapata-Revilla M andTenorio-SanzM. 2010. *LWT - Food Sci. Tech.* **43** p. 1467-1470
[2] Richana N,Ratnaningsih, Arif A B and Hayuningtyas M. 2013. *Proceeding of International Maize Conference 2012*p. 178 – 183
[3] Bressani R. 1990. *Food Rev. Int.***6** p. 225-264
[4] Van der Riet W B, Wight A W, Cilliers J J L and Datel J M. 1989. *Food Chem.* **34** p. 193-202
[5] Yuwono, Setyo S and Zulfiah A A. 2015. *Jurnal Pangan dan Agroindustri* **3** p.1465-1472
[6] AOAC. 1984. *Official Methods of Analysis.* Associationof Official Analytical Chemists, Arlington.
[7] AOAC. 2000. *Official Methods of Analysis.* Associationof Official Analytical Chemists International, Gaithersburg.
[8] Vishwanathan KH, Singh V and Subramanian R. 2011. *J. Food Eng.***102**p. 240-246
[9] Ostermann-Porcell MV, Quiroga-Panelo N, Rinaldoni A.N and Campderros ME. 2017. *J. Food Qual.***2017** p.1-8
[10] Aini N and Purwiyatno. 2010. *Int.Food Res. J.***17** p. 961-968
[11] Noviasari S, Kusnandar F and Budijanto S. 2013.*Jurnal Teknologi dan Industri Pangan***24**

p.194-200
[12] Thomas DJ and Atwell W A. 1999. *Starches*. Eagan Press Handbook Series. Minnesota. USA.
[13] White PJ and Tziotis A. 2004. *New Corn Starches*. Wood Head Publishing, Cambridge.
[14] Cuenca A R, Suarez M J Vand Aparicio I M. 2008. *Food Chem.* ***8***p. 1099-1105
[15] ElleuchM., Bedigian D, Roiseux O, Blecker C and Attia H. 2011. *Food Chem.* **124** p. 411-421
[16] Alsaffar A A. 2011. *Int. J. Food Sci. Tech.***46** p. 455-462
[17] Eckel RH. 2003. *Am. J. Clin. Nutr.***78** p.671-672
[18] Rahman S, Bird A, Regina A, Li Z, Ral JP, McMaugh S, Topping D and Morell M. 2007. *J. Cereal Sci.***46** p. 251-260

Preparation and preliminary characterization of sago flour and semi refined kappa carrageenan-based biocomposite film incorporated with coconut crabs chitosan nanoparticles

H Rasulu[1,2], D Praseptiangga[3], I M Joni[4,5] and A H Ramelan[6]

[1] Department of Agricultural Science, Graduate School of Sebelas Maret University (UNS), Jl. Ir. Sutami 36A, Kentingan, Surakarta 57126, Central Java, Indonesia
[2]Department of Food Technology, Faculty of Agriculture, Khairun University Jl. Yusuf Abdurahman Kampus Gambesi Ternate 97719, North Maluku, Indonesia
[3]Department of Food Science and Technology, Faculty of Agriculture, SebelasMaret University (UNS), Jl. Ir. Sutami 36A,Kentingan, Surakarta57126, Central Java, Indonesia
[4]Department of Physics, Faculty of Mathematics and Natural Sciences, UniversitasPadjadjaran, Bandung, West Java, Indonesia
[5]Nanotechnology and Graphene Research Center, UniversitasPadjadjaran, Jl. Raya Bandung-Sumedang KM 21, Jatinangor, West Java 45363, Indonesia
[6] Department of Physics, Faculty of Mathematics and Natural Sciences, SebelasMaret University (UNS), Jl. Ir. Sutami 36A,Kentingan, Surakarta57126, Central Java, Indonesia

Email: hamidinrasulu@yahoo.com

Abstract. The use of nano-sized chitosan natural polymers served as biopolymers containing antimicrobial substances is promising to be utilized as preservatives that could maintain quality and extend the shelf-life of foods.The objective of this research was to evaluate the effect of chitosan coconut crabs nanoparticles (CsCC NPs) incorporation to the sago flour (SF) and semi-refined kappa carrageenan (SRKC) based biocomposite film on their mechanical and barrier properties. The CsCC NPs suspension was prepared by the beads-milling method. The CCCsNPs was incorporated with various concentration (0, 1, 2, 3%) and introduced into 2 g of SF and SRKC (2:1) with the addition of Carboxymethyl Cellulose(CMC) and glycerol (2: 1) as a plasticizer. The results showed that the brightness value (L*) of prepared biocomposite film was at the range of 59.21-60.44. The thickness, tensile strength, and elongation at break of the films were at the range of 0.082-0.172mm, 3.22-8.09 MPa, 18.49-24.25%, respectively, while the water vapor permeability (WVP) of the films were at a range of 3.61-7.99 $gm/m^2h^{-1}kPa^{-1}$. It is indicated that the incorporation of CsCC NPs on the investigated biocomposite film could improve the films' properties and have enormous potential to be developed as biodegradable active packaging.

1. Introduction

Nonbiodegradable polymers come from plastic-based synthetic polymers that are widely used in food packaging because the production costs are relatively cheap, durable, water resistant and can be consumed directly. The use of nonbiodegradable plastic packaging will cause problems with environmental damage, so as to encourage the development of biodegradable and environmentally friendly materials, especially food packaging [1, 2]. Other applications of materials that have a nanoparticle size function as antimicrobials mainly prevent bacterial growth [3]. The use of nanotechnology is currently attracting attention and has a major influence on the food packaging industry for the development of food packaging in ensuring product quality and food safety [4].

The potential of chitosan to act as a food preservative that comes from nature. Chitosan is also an excellent film-forming material [5]. Chitosan film has a permeability to gas properties [6]. Polymer nanotechnology is an area that dynamically develops materials loaded with special particulates on the nanoscale [7]. Nano-bio-hybrid biopolymers are extensively studied earlier for food packaging applications [8]. Chitosan film loaded with cellulose nano-crystals improves the permeation properties of films. Nano-biocomposite is a newer biopolymer matrix applying physical-chemical interactions [9]. The use of chitosan particles as fillers in the manufacture of biocomposite films can improve mechanical properties and barrier in films [10, 11]. The use of chitosan as a packaging material has the advantage of being an antibacterial activity [12]. The addition of chitosan nanoparticles as a filler in the manufacture of films using carrageenan kappa and tapioca starch showed the ability of accessibility and antimicrobial properties [13]. In filmmaking research using filler nanoparticles to improve the film's ability to significantly resist water vapor permeability [14].

Edible films have emerged as alternative packaging for synthetic plastic packaging in food applications. The development of seaweed-based biopolymers for edible films[15].Characterization of novel composite semi-refined iota carrageenan based edible film [16]. Mechanical and barrier properties of semi-refined kappa carrageenan-based composite edible film and its application [17]. Edibility and biodegradability are the most beneficial properties of edible films and edible coating that can be eaten [18]. Edible nano-biocomposite film is a new application for food packaging. The layer of the adenine polysaccharide nanoparticles serves as an active molecular inhibitor [19]. The use of nano-sized natural polymer chitosan serves as a biopolymer which has antimicrobial properties so it can be used as a preservative that extends the shelf life of foodstuffs. Therefore this study aims to determine the mechanical properties and barrier properties of sago flour (SF) films substituted by kappa carrageenan (KCRG) with filler chitosan coconut crabs nanoparticle (CsCC NPs) as a result of the beads-milling method. The novelty of this is research was to develop biocomposite films sago flour-kappa carrageenan semi-refined as matrix and then CsCC NPs suspension as fillers.

2. Experimental

2.1. Materials

Chitosan coconut crabs shell is the result of chitin deacetylation process, which is then processed using the beads-milling method to produce chitosan nanoparticles, which are used as fillers in films (data not shown) conducted at the Laboratory of Nanotechnology and Graphene Research Center, Padjadjaran University. Sago flour is obtained from the traditional market of Ternate City, while semi-fine kappa-carrageenan is supplied by Galic Artabahari, Co., Ltd. (West Cikarang, Indonesia), these two materials are used as the main components of the film matrix. Carboxymethyl cellulose (CMC) and glycerol are used as plasticizers and purchased from Brata Kimia (Surakarta, Indonesia). Aquadest is used for all sample preparations. As well as chemicals to be used as the analytical material.

2.2. Methods

2.2.1. Preparation of Bio-nanocomposite Film

The edible film made in this study is a film made from chitosan nanoparticle products as the filler with the main raw materials, namely sago flour (SF) and kappa carrageenan (KCRG). The solvents used were distilled water, the use of the main ingredients was the concentration of sago flour and carrageenan, while the chitosan coconut crabs nanoparticles (CsCC NPs) with a certain concentration only as a filler with plasticizers used glycerol and carboxyl methyl cellulose (CMC) (2: 1) [20]. The procedure for making films was modified from Harris's (1999) method. While the formulation of the material used refers to research [25]. Briefly the procedure for making edible films is as follows, the main ingredients of sago flour and carrageenan kappa (2: 1 b/v) are dissolved in 100 ml of distilled water and add glycerol and CMC (2: 1 v/v), then add 0% CCCsNPs concentration, 1%, 2%, 3% (v/v) results from the beads-milling (BM) method. The process of mixing the solution using a heating magnetic stirrer at 1500 rpm to a temperature of 65 ^{0}C is maintained for 10 minutes. Furthermore, the temperature is lowered to 45 ^{0}C, then the printing process on the plastic plate size (W x L x H: 15 × 23 × 2 cm) is evenly distributed and cooling is done at room temperature for 20 minutes, and to get the film followed by drying using cabinet drying for 12 hours at a temperature of 60 ^{0}C with relative room humidity. After cooling to room temperature, each dry film is carefully peeled from the casting surface into an aluminum foil packaging to avoid hygroscopicity and storage in a plastic box containing silica gel (0% RH) at 28 ± 2 ^{0}C for 24 hours before analysis [21].

2.2.2. Color Measurements
Color values (L*, a*, b*) of the film were measured with CR 310 Minolta Chroma Meter (Minolta Camera Co., Ltd.,). The film is placed on a standard white plate (calibration plate) and the Hunter Lab color scale is used to measure color [22]. Each film sample is measured in four different reading positions. The total color difference (ΔE) is calculated as:

$$\Delta E = (\Delta L2 + \Delta a2 + \Delta b2)0.5 \qquad (1)$$

2.2.3. Film Thickness (FT)
Measurement of film thickness with an accuracy rate of 0.001 mm using a digital micrometer 0-25 mm (Krisbow, Indonesia). The sample is measured randomly at four different points, then the average thickness is calculated.

2.2.4. Tensile Strength (TS) and Elongation at Break (E)
Tensile strength (TS) and elongation (E) are measured by Universal Testing Machine (Zwick I Model Z0.5, UK) with the standard method ASTM D882-00 (ASTM, 2000) [23]. The Edible Film (EF) samples are cut in rectangular strips with a length of 100 mm and a width of 25 mm using a printing device, measured by the thickness of the film with a micrometer, then mounted on the clamp on the tool. The tool is set with a pre-load value of 2 N/mm^2, a test speed of 10 mm/minute, and a clamping distance of 50 mm. The start button is pressed 2 times where the 1st pressure activates the tool and the 2nd pressure will operate the tool so that the test takes place.

2.2.5. Water Vapour Transmission Rate
Testing of water vapor transmission rate(WVTR) levels in film samples was carried out gravimetrically following the ASTM E96-00 (ASTM, 2000) method with modifications [24]. 10 grams of silica gel is placed into the Petri dish. The film sample was cut round and glued to the top surface of a 6 x 6 cm 2 Petri dish using plasticine. Every 1 hour for 8 hours of testing, weight gain from the Petri dish was measured. The weight change of silica gel (g) in the Petri dish is recorded as a function of time and the slope is calculated using linear regression (change in weight vs. time). Slope (g/hour) divided by the area or surface area of the petri dish (m^2) is the rate of water vapor transmission (WVTR). The film thickness is measured by a digital micrometer (Digital Outside Micrometer, Inside, Germany) with an accuracy of 0.0001 mm. Then the Water Vapor Permeability (VWP) value with units of g.mm/m^2.KPa.jam can be calculated using the following formula:

$$VWP = \frac{WVTR.x}{\Delta P} \qquad (2)$$

in which x (mm) is the film thickness and ΔP (Kpa) is the air pressure.

2.3. Statistical Analysis
The data were analyzed statistically by one-way variance analysis (ANOVA) at a significance level of 0.05 and the difference in mean values was determined by Duncan's Multiple Range Test (DMRT) ($p < 0.05$) to determine the difference between concentration treatments. For statistical analysis using the SPSS 16 program.

3. Results and Discussion

The characterization of the bio-nanocomposite edible film propertieswith the chitosan coconut crabs (CsCC NPs) nanoparticles suspension by the beads-milling method with an average diameter (D10) of ±75 – 528 nm that applicable as nano reinforcement for biocomposite films.is presented in **Table 1**. Each parameter showed a certain tendency as the concentration between refined kappa semi-carrageenan sago flour and chitosan nanoparticle concentration as filler. The effect of using chitosan nanoparticles for each edible film is discussed separately.

Table 1. Characterization of mechanical and barrier of nano-biocomposite edible film properties of semi refined kappa sago-carrageenan flour with coconut crabs chitosan nanoparticle filler.

Film	CsCC NPs (% v/v)	Color (L*)	Thickness (mm)	Tensile Strength (MPa)	Elongation (%)	WVP ($g.mm/m^2h^{-1}kPa^{-1}$)
SF : SRKC (2:1)	0	59.61 ^{ab}±0.13	0.14 ^{b} ±0.02	5.11 ^{a} ±0.38	24.25 ^{b} ±2.25	3.61^{a} ±1. 07
	1	59.21 ^{a} ±0.09	0.08 ^{a} ±0.01	3.22 ^{a} ±0.31	23.17 ^{b} ±0.24	3.98^{a} ±0.55
	2	59.54 ^{ab}±0.44	0.12 ^{b} ±0.01	7.57 ^{a} ±0.08	18.49 ^{a} ±0.64	7.99^{b} ±1.71
	3	60.44 ^{c} ±0.58	0.17 ^{c} ±0.02	8.10 ^{a} ±0.09	24.13 ^{b} ±0.28	2.61^{a} ±1.07

Note: Value in the same line followed by a different letter was significantly different ($p < 0.05$).

3.1. Color Measurements

Color values in edible films are the most important factor in the appearance of food products and overall are an assessment of consumer acceptance of the product [27]. The color properties of the SF-KCRG nano-biocomposite film with different concentrations of CsCC NPs according to the system of color values (L*) are shown in Table 1. Brightness color values (L*) with 3% CsCC NPs concentrations show the brightness level is higher than the lower concentration ($p <0.05$). Color values are indicated that the film has good brightness transparency in addition to being influenced by different CsCC NPs concentrations, also because of the use of SF-KCRG material as a matrix that leads to a transparent formation. the effect of wrapping treatment by composite SRKC film in maintaining the color [17]. Transparency of biocomposite films with nano chitosan as filler is indicated by the value of the sample [28]. On the other hand, the addition of plasticizers such as glycerol in film composites using a number of concentrations can produce better films [29]. In low concentrations, fillers as CSNPs suspension such as carrageenan-tapioca with suspension do not affect the transparency of composite films [30].

3.2. Thickness

The film thickness is affected during the printing process on plastic molds. The uniformity of film thickness is also influenced by the SF-KCRG material which is easily gelatinized due to heat and easily retrogressed in room temperature conditions. Film thickness with CsCC NPs filler concentrations varied between 0.08-0.17 mm as shown in Table 1. From the results of the observations showed that the CsCC NPs concentration was low, then the thickness was obtained 0.08 mm in the film.

The use of concentrations of 0.5% sorbitol plasticizer in semi-refined carrageenan materials showed that the film thickness level was significantly influenced by sorbitol concentration [18]. The thickness of the film is influenced by the wide mold, the volume of the solution, and the total dissolved solids [31]. Drying conditions also affect the thickness of edible films. Higher temperatures and longer drying times will evaporate more water, allowing edible films to make thinner thicknesses [32].

3.3. Tensile Strength and Elongation at Break

The results in Table 1 shows that the tensile strength of films with various concentrations of NPs CsCC ranges from 3.22 to 10.10 MPa. The addition of CsCC NPs 3% (v/v) shows the highest tensile strength compared to other concentrations. However, from the results of the statistical test using the DMRT test ($\alpha = 0.05$), it was shown that the CsCC NPs concentration used was not significantly different. Elongation at break of edible films with various NPs CsCC concentrations were 18.49-24.25%, respectively (Table 1.). The results show that the higher the concentration of CsCC NPs, the lower the elongation. Elongation of edible films is inversely proportional to tensile strength. The difference in 1% CsCC NPs concentration interval is too low so it does not significantly affect elongation.

The higher concentration of plasticizer sorbitol affects the film thickness level [26]. The plasticizer is usually added to the polymeric matrix to overcome the fragility of the film, because it can reduce this strength, thereby increasing film flexibility and elongation. The mechanical properties with higher tensile strength, as well as percentage elongation, will be preferable for food packaging films [21]. The more compact the edible film matrix, the more force is needed to break the film. This leads to a decrease in the film's ability to change shape when it is subjected to certain strengths so that the film tends to break or lose elasticity [26]. These mechanical and solubility properties suggest that bio-nanocomposite film of semi-refined kappa carrageenan and nanoparticle ZnO can be effectively used as food packaging material [32].

3.4. Water Vapor Transmission Rate (WVTR)

The results in Table 1 shows that the water vapor transmission rate (WVTR) of the SF-KCRG edible film with concentrations of CsCC NPs filler (0, 1, 2, 3%) is around 3.61-7.99 $g.mm/m^2h^{-1}kPa^{-1}$. The results showed that WVTR films with an addition of 3% (v/v) NPs CsCC concentration (3.61 $g.mm/m^2h^{-1}kPa^{-1}$) were the lowest values compared to other CsCC NPs concentrations. However, the results of the statistical test using the DMRT test ($\alpha = 0.05$) showed that the various CsCC NPs concentrations used were not significantly different. The lowest WVTR shows that edible film has the best properties. That means water molecules, both from the surrounding environment or from packaged products, do not easily penetrate the film layer, thus, the quality of food products, such as weight, crispness, or freshness, is not affected. Water vapor transmission rates depend on the ratio of hydrophilic and hydrophobic materials in film formulas. WVTR is often used to study the moisture transport through the film and is of interest in almost all research into the characterization of biopolymers because it is a key factor in food packaging to support food against water adsorption and desorption [33].

4. Conclusion

Mechanical at barrier properties of SF-KCRG edible film with concentration of CsCC NPs filler (0, 1, 2, 3%) showed that the brightness color value (L*) with 3% CsCC NPs concentration showed higher brightness level, as well as film thickness and tensile strength, and elongation increased with the addition of 3% CsCC NPs concentration, but the water vapor transmission rate (WVTR) decreased with increasing CsCC NPs concentration. Thus, it can be concluded that the best properties of SF-KCRG edible film with concentrations of 3% CsCC NPs filler for further production in advanced research applications.

References

[1] Spotti ML, CecchiniJP, Spotti M J andCarrara C R 2016 *J. Food Science and Technology* ***68*** 127-134
[2] Sukhija S, Singh S and Rear C S 2016 *J. Food Hydrocolloids* ***60*** 128-137
[3] Pathankot K, Manubolu M and Hwang H M 2017 *J. of Food and Drug Analysis* ***25*(2)** 245-253
[4] Sudibyo A and T F Hutajulu 2013 *J. Kimia Kemasan* **35(1)** 6-19

[5] Rhim JW, Hong SI, Park HM and Ng P KW 2006 *J. of Agricultural and Food Chemistry* **54** 5814–22
[6] Yen MT, Yang J H and Mau J L 2009 *J. Carbohydrate Polymers* ***75*(1)** 15-21
[7] Rhim J W, Park HM and Ha C S 2013 *J. Progress in Polymer Science **38***1629-52
[8] Kochumalayil J J, Zhou Q and Berglund L A 2013 *J. Carbohydrate Polymers**93*** 466-472
[9] Konwar A, Gogoi N, Majumdar G andChowdhury D 2015 *J.Carbohydrate Polymers **115*** 238-245
[10] de Moura M R, Avena-Bustillos R J, McHugh T H, Krochta J M and Mattoso L H C 2008 *J. of Food Science **73*(7)** N31-N37
[11] de MouraMárcia R, Fauze A Aouada, Roberto J. Avena-Bustillos, Tara H McHugh, John M Krochta and Luiz HC Mattoso 2009 *J. of Food Engineering* **92** 448-453
[12] DuttaPK, Tripathi, Mehrotra G K and Dutta J 2009 *J.Food Chemistry**114***(**4**) 1173-82
[13] Rochima E, Utami S, Hamdani H, Azhary S Y, Praseptiangga D, Joni I M and Panatarani C 2018 *AIP Conf Proceedings* Vol. 1927 (1) 030-032
[14] Dogan N and McHugh T M 2007 *J. Food Science* **72** (1) 16–22
[15] Praseptiangga D 2017 In *IOP Conf. Series: Materials Science and Engineering* Vol. 193 (1) 012003
[16] Praseptiangga D, Giovani S, Manuhara G J and Muhammad DRA 2017 In *AIP Conf. Proceedings* Vol. 1884 (1) 030006
[17] Praseptiangga D, Maimuni B H, Manuhara G J and Muhammad D R A 2018 In *IOP Conf. Series: Materials Science and Engineering* Vol. 333 (1) 012086
[18] Praseptiangga D, Fatmala N, Manuhara G J, Utami R and KhasanahLU 2016 In *AIP Conf. Proceedings* Vol. 1746 (1) 020036
[19] Basu A, Kundu S, Sana S, Halder A, Abdullah M F, Datta S and Mukherjee A 2017 *J.Food Packaging and Shelf Life**11***98-105
[20] Ghasemlou M, Khodaiyan Fand A Oromiehie 2011 *J.Carbohydrate Polymers* **84** 477–483
[21] Aji A I, Praseptiangga D, Rochima E, Joni I M and Panatarani C 2018 In ***AIP** Conf. Proceedings* Vol. 1927 (1) 030039
[22] BitencourtCM, Fávaro-TrindadeCS, SobralP JA andCarvalhoRA 2014 *J.Food Hydrocolloid* 40 145–152
[23] Al-HassanAA and NorziahM H 2012 *J.Food Hydrocolloids* 26 108-117
[24] OzdemirM and FlorosJ D 2008*J. of Food Engineering* **86** 215–224
[25] Sofia I, Murdiningsih H and Yanti N 2016*J. BahanAlamTerbarukan* **5**(**2**) 54-60
[26] Manuhara G J, Praseptiangga D, Muhammad D R A and Maimuni B H 2016 In *AIP Conf. Proceedings* Vol. 1710 (1) 030043
[27] Balti R, Mansour M B, Sayari N, Yacoubi L, Rabaoui L, Brodu N, andMassé A 2017 *International J. of Biological Macromolecules **105*** 1464-72
[28] RhimJ W, HongS I, ParkH M and K W N G Perry 2006 *J. Agric. Food Chem***54** 5814–22
[29] FarahnakyA,SaberiB and MajzoobijM 2013 *J. Texture Stud***44** 176–186
[30] Rochima E, Fiyanih E, Afrianto E, Subhan U, Praseptiangga D, Panatarani C, and Joni I M 2018 In *AIP Conf. Proceedings* Vol. 1927 (1) 030042
[31] Herliany, N Ervina, J Santoso and E Salamah 2013 *J. Aquatica* **4** 1
[32] Saputri A E, Praseptiangga D, Rochima E, Panatarani C, and Joni I M 2018 In *AIP Conf. Proceedings* Vol. 1927 (1) 030040
[33] KhoirunnisaAR, Joni I M, Panatarani C, Rochima E and Praseptiangga D 2018 In *AIP Conf. Proceedings* Vol. 1927 (1) 030041

Acknowledgments
This research is fully supported by using the scholarship funds of the Government of the Republic of Indonesia through the Indonesian Institute of Education Scholarship Fund Management Institution (LPDP BUDI-DN) Indonesia. Number: PRJ-4952/LPDP.3/2016.

IOP Conf. Series: Materials Science and Engineering **633** (2019) 012045 doi:10.1088/1757-899X/633/1/012045

The drying rate of mushroom on tray dryer and effect of mushroom powder on organoleptic properties of Batagor

J Waluyo[1], Y Prasetyaningsih[2], MW Sari[2], N Ekawandani[2]

[1]Department of Chemical Engineering, Universitas Sebelas Maret, Surakarta, Indonesia
[2]Department of Chemical Engineering, Politeknik TEDC Bandung, Bandung, Indonesia

E-mail: jokowaluyo@staff.uns.ac.id, yusi.prasetyaningsih@poltektedc.ac.id

Abstract. Oyster mushrooms and straw mushrooms are currently great demand in the food industry. Mushrooms have the characteristics that are easily damaged and decomposed. The effort of preserving mushrooms is by drying using hot air at a certain temperature and flow rate. The purpose of this study was to determine the drying rate of oyster mushrooms and straw mushroom in variations temperature and flow rate of hot air. In this study, the drying of mushrooms was carried out with a tray dryer that has a stacking rack with dimensions of 30 cm x 25 cm. The temperature variations used were 30, 40, 50°C and the dryer air flow rate was 1, 2, 3 m/s. The dried mushroom with a moisture content of less than 10% is mashed to a powder then added to the food (Batagor) for an organoleptic test. The best drying result is obtained at temperature 50°C and flow rate of 3 m/s with a drying rate of 1.52 kg water/m^2.h. Organoleptic test on Batagor produces the most preferred flavour by using type mushroom of *Volvariella volvacea*, and for aroma and texture preferably with *Pleurotus ostreatus.*

1. Introduction

Oyster mushroom (*Pleurotus ostreatus*) and straw mushroom (*Volvariella volvacea*) are edible fungi that are easily obtained and cultivated. Mushrooms are foods that contain protein, vitamins, and minerals. Protein content in oyster mushrooms is 20-25%, fiber 37-48%, fat 4-5%, carbohydrates 37-48%, minerals 37-48% and moisture content 86-87%. Oyster mushrooms also contain 19 types of amino acids such as cysteine, glutamic acid, aspartic acid, and lysine. Vitamin C and folic acid are also found in oyster mushrooms [1]. The high potassium to sodium ratio makes this fungus ideal for treating hypertension and heart disease. Straw mushroom (*Volvariella volvacea*) have been tested and added to Cantonese sausages [2]. With the addition of edible mushroom powder, it can enhance the flavor and nutrition to Cantonese sausages. Mushrooms are cultivated to be used as food ingredients and can be used as medicine. Mushrooms can be used to reduce cholesterol levels, high blood pressure, obesity, inflammation, and cancer [3]. From the research results of oyster mushroom (*Pleurotus ostreatus*) which has heteropoly saccharide characteristics can be used to maintain health including antidiabetic activity [4], anti-hyperlipidemia [5] and anti-tumor [6].

Mushrooms have a short shelf life, are quickly damaged after post-harvest because the mushroom has a high moisture content. To extend the shelf life of mushroom can be done by drying, canning, and pickling [7]. Dry mushrooms are usually used as an ingredient in various foods such as instant soup, snacks etc. Drying can maintain taste, color, and aroma.

IOP Conf. Series: Materials Science and Engineering **633** (2019) 012045 doi:10.1088/1757-899X/633/1/012045

Drying is the most inexpensive method for storing mushrooms, if stored in an airtight condition, the fungus can be stored for more than 1 year [8]. Various conditions for drying mushroom operations will affect the quality of the mushroom powder produced. These conditions are drying flow rate and drying air temperature. Various ways of drying can be done to make mushroom powder. Drying using tray dryer is an inexpensive and effective tool in drying mushrooms because mushrooms are low density. To dry the mushroom from 7.5% moisture to 2% the time required for the type of cabinet/tray will be faster than the fluidized bed or microwave oven. But it is slower than using vacuum drying. However, the tray dryer has the advantage of being easy to operate rather than vacuum drying.

The purpose of this study was to determine the drying rate of drying mushrooms and investigate the effect of adding mushroom powder on texture, color and sensory properties on snack Batagor (tofu with meatballs frying), as an alternative to improve the quality and nutrition of Batagor. Batagor is a snack made from tofu and meatballs, which are fried with a mixture of flour dough.

2. Material and method

2.1 Experimental procedures

The Oyster mushrooms and straw mushrooms were got from Bandung, West Java, Indonesia. Early preparation was done by washing the mushrooms and cut to a size of 4 mm. Mushrooms were arranged on the top of the tray and put into the tray dryer. The dryer used in this study is presented in Figure 1. The mushrooms were dried at various temperature (30°C, 40°C, 50°C) and air flow rate (1m/s, 2m/s, 3 m/s). During the drying process, temperature, weight, and humidity were observed. The drying was terminated when the sample had reached a constant weight. Furthermore, dried mushrooms were mashed and used as additional ingredients in batagor.

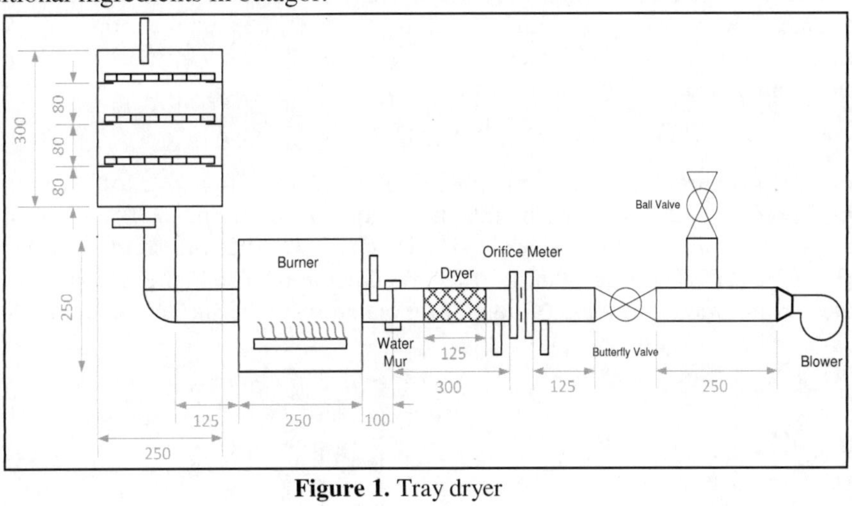

Figure 1. Tray dryer

2.2 Data Analysis

Moisture content was defined as

$$Xt = \frac{(W - Ws)}{Ws} \quad (1)$$

$$X = Xt - X^* \quad (2)$$

For drying rate, could be determine as

$$R = -\frac{Ls}{A}\frac{dX}{dt} \quad (3)$$

for

Xt	=	moisture content at the time (kg water/kg dry solid)
W	=	weight of the wet solid (kg)
Ws	=	weight of the dry solid (kg)

X	= moisture content (kg free water/kg dry solid)
X*	= equilibrium moisture content (kg equilibrium moisture/kg dry solid)
R	= drying rate (kg water/h.m^2)
Ls	= weight of dry solid used (kg)
A	= surface area for drying (m^2)

2.3 Organoleptic test

Organoleptic tests conducted in this study used three samples with code A (sample without mushroom powder); B (sample with oyster mushroom powder); C (sample with straw mushroom powder). The parameters tested by panelists include savory flavors, aroma, color, and texture. The assessment conducted by the panelists is by observing the sample and filling out the questionnaire with a numerical assessment in each parameter. Organoleptic test on batagor samples with mushroom-based flavorings was carried out on 20 random panelists aged 18 – 45 years.

The organoleptic test method used in this study is the preference test. The quality determining factor in this method is the feeling of liking or dislike of the panelists for the product given. The level of the panelist's preference is expressed on a hedonic scale that distinguishes some product offered.

3. Result and discussion

3.1 Effect of temperature on drying time

The drying operating conditions determine the results of the products got. The operating conditions are temperature and drying air flow rate. The temperature used in the food dryer ranges from 30 - 60°C [9]. If the temperatures used under 30°C will require a longer drying time compared to the use of a higher temperature. Whereas if drying is carried out at temperatures above 60°C it will damage the micronutrients in the material. The higher drying temperature results in more starch or carbohydrates being converted into sugar. When carbohydrates are heated, the sugar in the carbohydrates will turn brown [10].

Effect of drying temperature of Oyster mushrooms is presented in Figure 2. At a Temperature of 50°C drying time is faster than 30°C which is half the time required for drying at 30°C. The phenomenon also occurs in drying straw mushroom (Figure 2 (b)) which shows that the shortest time required is 390 minutes. Drying time is faster because of the higher temperature. The higher the heat energy carried out by air, the greater the amount of mass of liquid evaporated from the surface of dried material.

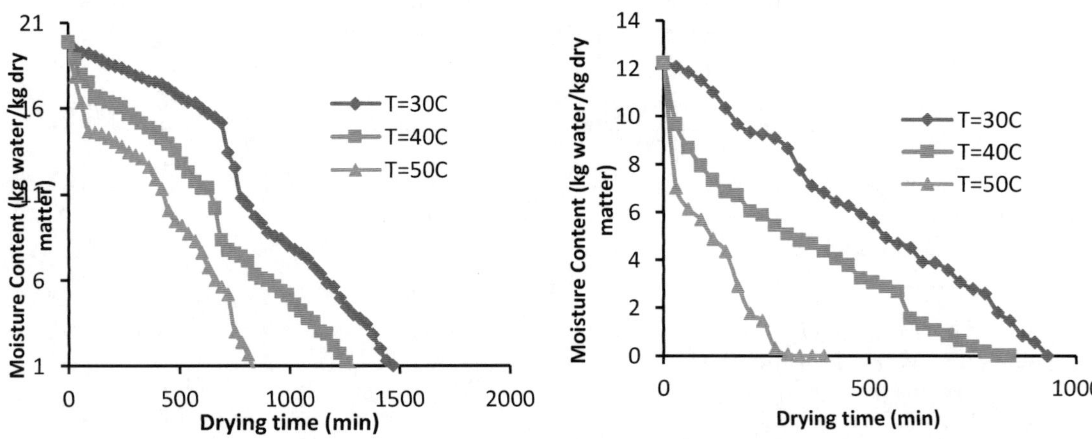

(a) (b)

Figure 2. Effect temperature at air flow rate of 1.5 m/s (a) oyster mushrooms (b) straw mushroom

IOP Conf. Series: Materials Science and Engineering **633** (2019) 012045 doi:10.1088/1757-899X/633/1/012045

3.2 Effect of air flowrate on drying time

As the drying air flow rate increases, it will also increase the diffusion of hot air into each sample. So that the water content in the sample can be reduced faster and shorten the drying time. In Figure 3, the fastest drying time is achieved at the highest air flow rate of 3 m/s for both oyster mushrooms and straw mushrooms. At the end of the drying process, the moisture content of solid materials with varying flow rates experience almost the same value. This happens because the relative humidity the drying air used has the same value. The relative humidity is obtained from the use of the same drying air temperature of 40°C.

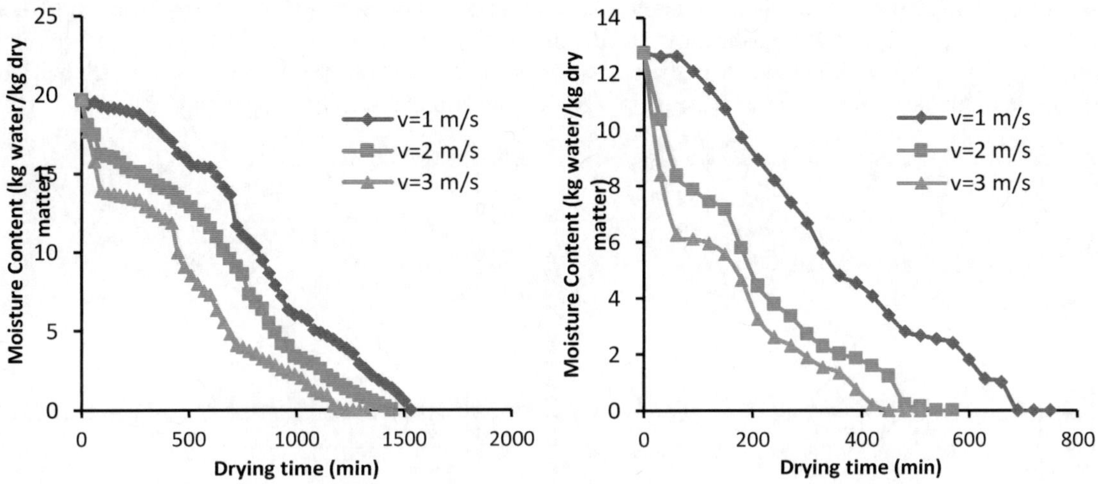

(a) (b)

Figure 3. Effect air flow rate at 40°C (a) oyster mushrooms (b) straw mushroom

3.3 Effect of temperature on drying rate

The effect of temperature on the relationship between moisture content in solids and drying rate is shown in Figure 4. At the beginning of the drying process, the water contained in the solids is still large enough so that the drying rate is faster because a lot of water is evaporated. However, after some time, drying reaches a constant condition so the existing water vapor is surface water vapor. This constant period is visible at 50°C. After that, the curve has decreased which proves that the evaporated water is bound to water vapor resulting in a decrease in the drying rate.

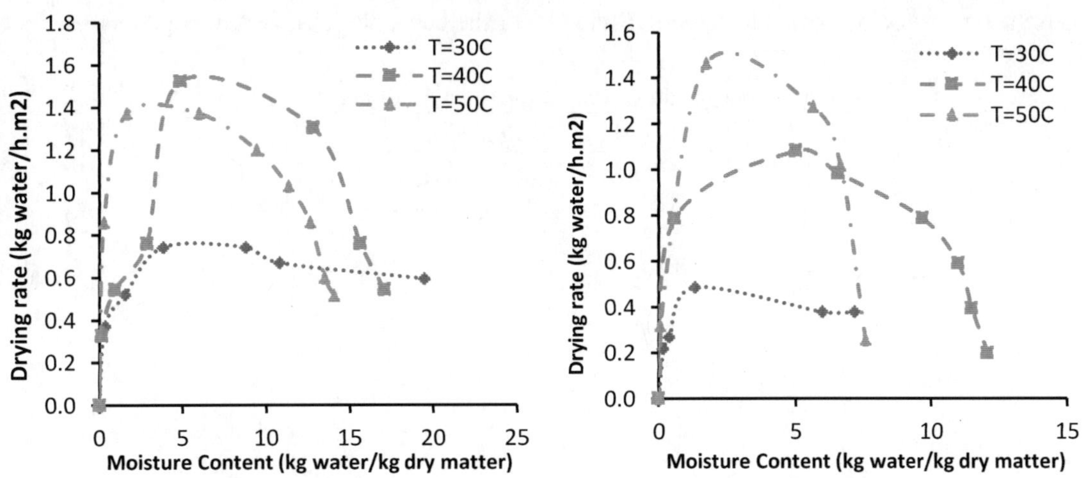

(a) (b)

Figure 4. Effect temperature on drying rate at 3 m/s (a) oyster mushrooms (b) straw mushroom

Increasing temperatures results in increased drying rate so that the water content in the material decreases. This is due to the diffusivity of water on the surface and inside the solid increases with increasing temperature. This phenomenon occurs as shown in Figure 4, where for the type of mushroom, drying at a temperature of 50°C gives a higher drying rate than the temperature of 30°C and 40°C which is 1.52 kg of water/h.m^2.

3.4 Effect of air flow rate on drying rate

The air flow rate affects the drying rate of the mushrooms. Figure 5 shows the fluctuating drying rate of mushrooms because the air flow is not evenly distributed throughout the surface of mushrooms. Effect of air flow rate on drying straw mushrooms has the opposite phenomenon of drying oyster mushrooms. The highest drying rate is obtained from the air flow rate of 2 m/s. Tray dryer has disadvantages, dried mushrooms in the bottom tray get heat and a larger air flow compared to the upper tray.

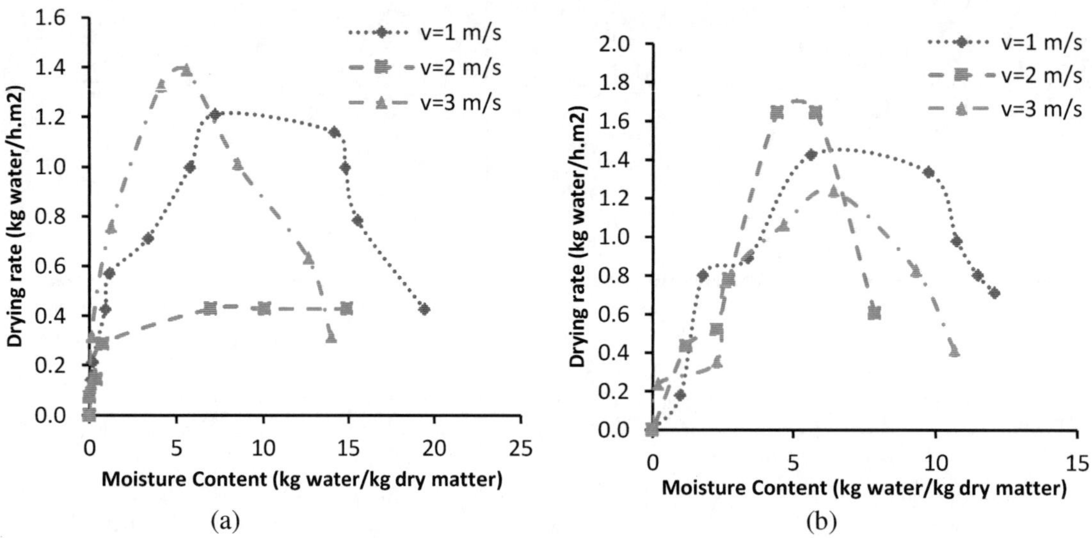

Figure 5. Effect air flow rate on drying rate at 40°C (a) oyster mushrooms (b) straw mushroom

3.5 Organoleptic test

Organoleptic test results are presented in Table 1. The results of the assessment of savory flavors in a various sample given, sample C obtained the highest score of 4.7. For aroma and texture, sample B obtained the best value compared to sample A and C. For the best color assessment is given to sample A, which gets a perfect score (very preferred).

Table 1. Organoleptic test of mushroom powder in Batagor

Sample	Parameter			
	savory taste	**aroma**	**color**	**texture**
A	4.65	4.30	5.00	4.65
B	4.50	4.50	4.50	4.85
C	4.70	4.45	3.70	4.30

Figure 6. Batagor with a variety of mushroom powder: (A) Original Batagor without mushroom powder, (B) Batagor with oyster mushroom powder, (C) Batagor with straw mushroom powder

The color of the sample is influenced by several factors, such as frying time, the temperature of cooking oil, and the raw material color. Sample with straw mushrooms is browner than other varieties. This is because the straw mushroom in the initial conditions has been brown and cannot be removed before processing into flavoring powder. Based on the result of organoleptic tests, it can be concluded that the use of straw mushroom powder as an alternative to MSG is very possible to use. It is proven to get a higher taste for savory sample compared to other variations.

4. Conclusion

Based on this study, the most effective temperature for drying mushrooms is 50°C. The air flow rate of 3 m/s during drying of oyster mushrooms provides a drying rate of 1.52 kg of water/h.m^2. The organoleptic results tested using mushrooms powder on food (Batagor) give results: straw mushrooms are preferred for savory flavours, while oyster mushrooms for aroma and texture.

5. References

[1] Patil SS, Ahmed SA, Telang SM and Baig MMV 2010 *Innov. Romanian Food Biotech.* **7** 66-76
[2] Wang X, Zhou P, Cheng J, Chen Z and Liu X 2018 *Meat Sci.* **146** 18-25
[3] Ma G, Yang W, Zhao L, Pei F, Fang D and Hu Q 2018 *Food Sci. Human Wellness* **7** 125-33
[4] Zhang Y, Hu T, Zhou H, Zhang Y, Jin G and Yang Y 2016 *Int. J. Biol. Macromol.* **83** 126-32
[5] Zhang Y, Wang Z, Jin G, Yang X and Zhou H 2017 *Int. J. Biol. Macromol.* **101** 107-16
[6] Cao X-Y, Liu J-L, Yang W, Hou X and Li Q-J 2015 *Mol. Med. Rep.* **12** 2383-9
[7] Darvishi H, Azadbakht M and Noralahi B 2018 *Renew. Energ.* **120** 201-8
[8] Walde SG, Velu V, Jyothirmayi T and Math RG 2006 *J. Food Eng.* **74** 108-15
[9] Mewa EA, Okoth MW, Kunyanga CN and Rugiri MN 2018 *LWT* **92** 484-9
[10] Pornpraipech P, Khusakul M, Singklin R, Sarabhorn P and Areeprasert C 2017 *Agr. Nat. Res.* **51** 402-9

IOP Conf. Series: Materials Science and Engineering **633** (2019) 012046 doi:10.1088/1757-899X/633/1/012046

Effect of various heat treatment on physical and chemical characteristics of red rice bran (*Oryza nivara* L.) Rojolele

Siswanti, R B K Anandito, E Nurhartadi, and B D Iskandar

Department of Food Science and Technology, Universitas Sebelas Maret (UNS), Jl. Ir. Sutami 36A Kentingan Surakarta 57126, Central Java, Indonesia

Email: siswanti@staff.uns.ac.id

Abstract. This research aims to study the effect of various heat treatment on physical and chemical characteristics of red rice bran (*Oryza nivara* L.) Rojolele. This study used a Completely Randomized Design (CRD) with one-factor variations of heat treatment applied on the bran. The various heat treatments use Oven an hour (80ºC and 100ºC), Microwave maximum temperature (2 and 5 minutes), and Autoclave 150^{0}C (5 and 10 minutes). Results showed that all heat treatment methods decrease L value. The highest moisture level is on heating by microwave for 2 min. There were no significant difference in the level of ash, fat, and antioxidant activity. The FFA content was decreased by all heating treatment. The fenol level with heating with Oven 80ºC an hour was the highest. Among the treatments, Oven an hour at 80ºC is the best method. This method produces red rice bran with characteristics of lightness, water content, ash, fat, protein, total phenol, antioxidant activity, crude fiber, and free fatty acid content respectively 49.78; 5.51%; 8.37%; 27.11%; 10.50%; 0.25%; 75.88%; 9.30%; and 13.48%. Red rice bran Rojolele also contains anthocyanin and thiamin obtained from spectroscopy FTIR.

1. Introduction

Red rice (*Oryza nivara* L.) is a type of rice, that color of red. The red color of rice is caused by the presence of anthocyanins on the outside of the rice seeds. Red rice contains many higher nutritional compound compared to white rice. Red rice has soluble fiber and rich in vitamin B [1]. Nutritional compounds in red rice are 14.38% water, 9.16% protein and 2.5% fat [2].

Rice bran, a natural by-product obtained from the milling process of rice, is the outer layer of rice grains and includes a small portion of starch endosperms [3]. Rice bran generally consists of several pericarp tissues, seed coat, nuselus, aleurone layer, embryo (germ), a little chaff and a little endosperm. The process of rice milling or silting produces 8-18% rice bran [4]. Red rice bran has a high nutritional content there are carbohydrate content of 48.3-50.7%, crude protein 15.7–17.2%, fat 23.3–24.9%, ash 9.2–11.3 %, and moisture content 9.61–14.74% [5]. The high-fat content in rice bran caused it to be able to experience both oxidative and hydrolytic damage [6]. Rancidity in rice bran is caused by enzymes such as lipases, that hydrolyzing fat into fatty acids and glycerol. Free fatty acids are oxidized to peroxide, ketones, and aldehydes causing rancidity [7]. Hydrolytic damage is caused by the interaction of fat with water [8]. Rice bran damage during the storage process due to the formation of free fatty acids from enzymatic reactions. The three free fatty acids that are formed are palmitic acid, oleic acid, and linoleic acid. The free fatty acids formed an oxidative reaction which causes a rice bran rancidity [9]. One way to increase the stability of rice bran is by using heating treatment.

Rice bran stabilization method can be done by wet heating, dry heating, and microwaves. Wet heating is one of the effective methods for the rice bran stabilization process. Pressure is needed to maintain moisture loss during the heating process. The combination of wet and autoclave pressurization methods was carried out at 1050C for 5 and 10 minutes [4]. The stabilization method with dry heating using an oven is heated at a temperature of 800C and 1000C for 1 hour. Stabilization with the oven causes increased the shelf life of white rice bran and prevents damage from oxidative damage [10].

Another method that can be used for stabilization is a microwave. The use of a microwave to reduce the water content of rice bran so that the water in the material does not interact with fat causes rancidity [11]. Microwave heating aims to activate the lipases so that the enzyme does not interact with the surface of water and oil so that it does not cause damage to the rice bran [9]. Research conducted by Kim et al. (2014) shows that the stabilization of microwave white rice bran can be done with a maximum temperature of 2 and 5 min. This research aims to study the effect of various heat treatment on physical and chemical characteristics of red rice bran (Oryza nivara L.) Rojolele.

2. Materials and Methods

2.1 Materials

The red rice bran (Oryza nivaraL.) Rojolele was provided by Abdi Bumi Lestari Farmers Group, Karangpandan Village, Karanganyar Regency. The materials used for chemical analysis were purchased from Merck (Germany) and Sigma-Aldrich (USA). The tools used for chemical analysis are weigh bottles (Pyrex), porcelain crucibles, ovens (Memert UNB 400), desiccators (AS ONE), ignition furnaces (Neycraft), clamps, analytical balance sheets (Ohaus), kjeldahl flasks, measuring cups (Pyrex), volume pipette, erlenmeyer, distillation apparatus, electric stove, drop pipette, burette, soxhlet flask, measuring cup, water bath (Memert), vortex, test tube, rotary evaporator (KNF RC 600), UV-Vis spectrophotometer (Spectronic 20 D +), spectrophotometer IR (Shimadzu IRPrestige 21), and Chromameter (Konika Minolta CR-400).

2.2 Methods

2.2.1. Preliminary Treatment of Red Rice Bran

Red rice bran before analysis was screened through a 60-mesh sieve to have a uniform particle size.

2.2.2. Stabilization by Various Heat Treatment (Oven, Microwave, and Autoclave)

Stabilization by heat treatment using an oven refers to the research of Kim et al., (2014) [10]. Red rice bran was transferred into shallow pans and spread uniformly in a layer of about 0.5 cm thickness. The pans were then placed in an oven at 80 and 100°C for 1 h. Heat treatment using a microwave begins with the addition of distilled water to a moisture content of 20% [10]. Before heating the rice bran, the microwave is heated at maximum temperature for 3 min to get a stable temperature and effective heating. Rice bran was transferred into shallow pans and spread uniformly in a layer of about 1.5 cm thickness and covered with plastic wrap then heated using a maximum temperature (120°C) for 2 min and 5 min [9]. Autoclave heating was carried out with 1 cm rice bran thickness wrapped in a filter cloth then heated at a 105°C pressure of 1 atm for 5 min and 10 min. The sample is then dried with a cabinet dryer at 50°C for 1 hour. All heat treated samples were sealed polyethylene bags and then stored in a freezer at -20°C until analysis.

2.2.3 Method of Analysis

Chemical analysis was conducted on water content by thermogravimetric method [12], ash content by dry ignition method [12], fat content by the Soxhlet extraction method [12], protein content by micro-Kjeldahl method [12], carbohydrate content by difference method [12], crude fiber content [12], antioxidant activity by DPPH immersion method [13], total phenol by the folin-ciocalteu method [13], free fatty acids by gravimetric method [10] and structure FTIR functional group [14]. The physical analysis was conducted on the form of color values were measures using a chromameter [15].

3. Results and Discussion

3.1 Effects of Various Heat Treatment on Color of Red Rice Bran

The color analysis conducted using chromameter to get the value of L *, a *, and b *. Then the °Hue was calculated with the formula °Hue = arc tan (b * / a *). The color values of various heat treatment samples are shown in Table 1.

Table 1. Lightness (L*), Redness (a*), Yellowness (b*), and °Hue) of Red Rice Bran (*Oryza nivara*L.) Rojolele by Various Heat Treatment

Treatment	(L*)	(a*)	(b*)	($h^{0)}$
Control	$50.45{\pm}0.06^{f}$	$13.85{\pm}0.06^{d}$	$15.50{\pm}0.12^{f}$	$46.30{\pm}0.00^{e}$
F_1	$49.78{\pm}0.15^{e}$	$13.18{\pm}0.10^{bc}$	$14.35{\pm}0.17^{e}$	$46.13{\pm}0.17^{e}$
F_2	$49.38{\pm}0.29^{d}$	$13.23{\pm}0.10^{c}$	$14.13{\pm}0.13^{e}$	$45.83{\pm}0.25^{d}$
F_3	$46.38{\pm}0.19^{c}$	$14.60{\pm}0.12^{e}$	$13.75{\pm}0.13^{d}$	$42.75{\pm}0.17^{c}$
F_4	$46.20{\pm}0.14^{c}$	$14.53{\pm}0.05^{e}$	$13.30{\pm}0.20^{c}$	$42.73{\pm}0.13^{c}$
F_5	$43.48{\pm}0.13^{b}$	$12.73{\pm}0.15^{a}$	$10.03{\pm}0.24^{a}$	$41.15{\pm}0.17^{b}$
F_6	$43.15{\pm}0.13^{a}$	$13.05{\pm}0.06^{b}$	$10.30{\pm}0.12^{b}$	$40.75{\pm}0.13^{a}$

Values with different superscripts within the same column are significantly different (p<0,05). F1 = 80°C oven for 1 h; F2 = 100°C oven for 1 h; F3 = microwave 2 min maximum temperature; F4 = microwave 5 min maximum temperature; F5 = 5 min of autoclave temperature 105°C; F6 = 10 min of autoclave temperature 105°C.

The redness of red rice brand after microwave heat treatment (2 min and 5 min) increases but decreased by heating the oven (80°C and 100°C) and autoclave (5 min and 10 min). Yellowness decreased in all samples. The values of a* and b* in the sample, were significantly different from the control. The lightness (L*) of red rice bran after all various heat treatment decreased. Similarly, the decreased in lightness (L*) of white rice bran with increased in heat temperature has been reported by Aliva and Silva, 1999 [16] and Kim et all, 2004 [10]. The heat treatment induce formation some products from the Maillard reaction [10].

3.2 Effects of Various Heat TreatmentonChemical Properties of Red Rice Bran

The proximate characteristics of the heat-treatment red rice bran sample are shown in Table 2.

Table 2. Proximate characteristics (water, ash, fat, protein, and carbohydrates) of Red Rice Bran (*Oryza nivara* L.) Rojolele by Various Heat Treatment

Treatment	Parameter				
	Water (%wb)	Ash (% db)	Fat (% db)	Protein(% db)	Carbohydrate (% db)
Control	$10.29{\pm}0.32^{e}$	$7.90{\pm}0.65^{a}$	$28.55{\pm}0.34^{c}$	$9.75{\pm}0.17^{b}$	$42.34{\pm}0.87^{c}$
F_1	$5.51{\pm}0.23^{b}$	$8.37{\pm}0.66^{ab}$	$27.11{\pm}0.27^{b}$	$10.50{\pm}0.08^{d}$	$48.19{\pm}0.96^{f}$
F_2	$4.13{\pm}0.22^{a}$	$8.87{\pm}0.24^{b}$	$26.24{\pm}0.15^{a}$	$10.38{\pm}0.13^{d}$	$50.23{\pm}0.45^{g}$
F_3	$12.51{\pm}0.17^{g}$	$7.90{\pm}0.32^{a}$	$32.39{\pm}0.35^{d}$	$9.48{\pm}0.10^{a}$	$35.97{\pm}0.44^{a}$
F_4	$8.60{\pm}0.36^{d}$	$8.70{\pm}0.24^{b}$	$27.40{\pm}0.38^{b}$	$10.10{\pm}0.34^{c}$	$44.40{\pm}0.79^{d}$
F_5	$11.36{\pm}0.20^{f}$	$7.89{\pm}0.31^{a}$	$28.80{\pm}0.66^{c}$	$9.35{\pm}0.17^{a}$	$41.16{\pm}0.27^{b}$
F_6	$8.05{\pm}0.31^{c}$	$7.79{\pm}0.42^{a}$	$27.22{\pm}0.18^{b}$	$10.10{\pm}0.08^{c}$	$46.13{\pm}0.44^{e}$

Values with different superscripts within the same column are significantly different (p<0,05). F1 = 80°C oven for 1 h; F2 = 100°C oven for 1 h; F3 = microwave 2 min maximum temperature; F4 = microwave 5 min maximum temperature; F5 = 5 minof autoclave temperature 105°C; F6 = 10 minof autoclave temperature 105°C.

Table 2 shows that the differences in heat treatment methods an effect on the water content of red rice bran. The highest water content was found in F3 treatment (12.51%). This is due to the raw red rice bran was first adjusted to 21% by adding deionized water. Water added to the rotation process and absorb microwave energy cannot be completely evaporated. The maximum rice bran water content is 12% [17]. The differences in heat treatment methods also effect on the ash contentof red rice bran. The stabilized red rice bran has an ash content of 7.79% - 8.87%. The maximum ash content of rice bran flour is 10% [17]. The ash content obtained was not significantly different from the control. High-temperature heating cannot remove the mineral content contained in bran [4].

The differences in heat treatment methods also effect on the protein content of red rice bran. The protein content of rice bran flour is at least 8% [17]. Decreased protein content caused by a decrease in water content of rice bran. When the water content is decreased, it will reduce the possibility of chemical reactions such as oxidation, deamination, denaturation, and aggregation. In high water content, the protein content in the ingredients becomes unstable and degraded. Increased chemical reactions due to increased water content cause conformation of the flexibility of the protein molecule. The lower of water content in a material, the possibility of a chemical reaction can decrease. Protein damage reactions can be minimized by reducing the moisture content of the ingredients [18]. According to SNI 01-4439-1998, the fat content of rice bran flour is at least 3%. Non-stabilized control rice bran and stabilized rice bran have fat content greater than 3%. The highest fat content was found in F3. The increase in fat content is caused by a disturbance in the fat complex and protein or fat and carbohydrates so that the opening of the bond is due to heating in a short time. Decrease in fat content in microwave heating for 15 minutes due to triglyceride decomposition by releasing free fatty acids and volatile components that provide aroma, at a longer duration of heating [19].

The level of bran carbohydrate of red rice brand Rojolele ranged between 35.97% - 50.23%. The highest carbohydrate levels were found in F2. The process of heating and drying causes an increase in the carbohydrate of red rice bran. Carbohydrate increase is caused by a decrease in other components in the ingredients.

Table 3. Phenolic Compound, Antioxidant Activity, Crude Fiber dan FFA of Red Rice Bran (*Oryza nivara* L.) Rojolele by Various Heat Treatment

Treatment	Content of			
	Phenolic Compound (%)	AntioxidantActivity(%)	Crude Fiber (%)	Free Fatty Acid (FFA) (%)
Kontrol	$0.22{\pm}0.002^{e}$	$73.90{\pm}0.11^{ab}$	$10.63{\pm}0.19^{d}$	$17.89{\pm}0.52^{e}$
F_1	$0.25{\pm}0.005^{g}$	$75.88{\pm}0.21^{b}$	$9.30{\pm}0.14^{c}$	$13.48{\pm}0.00^{c}$
F_2	$0.24{\pm}0.004^{f}$	$73.93{\pm}1.02^{ab}$	$9.40{\pm}0.27^{c}$	$11.80{\pm}0.12^{b}$
F_3	$0.16{\pm}0.008^{d}$	$74.98{\pm}2.23^{ab}$	$7.83{\pm}0.25^{b}$	$16.41{\pm}0.35^{d}$
F_4	$0.13{\pm}0.003^{c}$	$75.93{\pm}1.02^{b}$	$6.45{\pm}0.24^{a}$	$8.04{\pm}0.17^{a}$
F_5	$0.05{\pm}0.002^{b}$	$75.10{\pm}1.80^{ab}$	$7.85{\pm}0.13^{b}$	$13.47{\pm}0.12^{c}$
F_6	$0.03{\pm}0.002^{a}$	$72.85{\pm}2.02^{a}$	$6.50{\pm}0.16^{a}$	$12.24{\pm}0.69^{b}$

Values with different superscripts within the same column are significantly different (p<0,05). F1 = 80^{0}C oven for 1 h; F2 = 100^{0}C oven for 1 h; F3 = microwave 2 min maximum temperature; F4 = microwave 5 min maximum temperature; F5 = 5 minof autoclave temperature 105^{0}C; F6 = 10 minof autoclave temperature 105^{0}C.

Rice bran contains many unique bioactive compounds, such as tocopherols, tocotrienols phytosterols, ferulic acids and gammaoryzanol, Loypimai et al., 2009 [20]; Sharma et al., 2004 [21] that have various health-beneficial effects on glucose metabolism, lipid profiles, and blood pressure Cicero and Derosa, 2005 [22]; Shirakawa et al., 2006 [23]. The highest content of phenol red bran rice was found in F1 (0.25%). This value is higher than control. The content of phenol compounds is very sensitive and more susceptible to degradation. The most important phenol degradators are temperature, oxygen, and light. Increasing temperature will cause damage to most phenolic compounds. Increased temperature and heating time can decrease the total phenol content in the material [24]. The treatment of stabilization using oven at 800C that showed the greatest phenol yield. Similarly, the increased in tocol of white rice bran in dry heating (800C) has been reported by Kim et all, 2004 [10].

The antioxidant activity of red rice bran Rojolele ranged from 72.85% -75.93%. The antioxidant activity of all samples was not significantly different from control. The stabilized white rice bran then extracted, also obtained the results of antioxidant activity which was not significantly different from the control [25]. The level of crude fiber ranged from 10.63% - 6.45%. The stabilization process causes a decrease in crude fiber content. Decreased levels of crude fiber caused by softening of the fiber structure, so that in testing the addition of acid-base can dissolve several parts of the fiber [26]. The highest free fatty acid content is found in the control sample. The stabilization process using an oven, microwave and autoclave can inhibit enzymatic activity. The content of oleic fatty acids in white rice bran was higher than that of stabilized white rice bran using microwave and stored for 4 months with storage treatment using plastic and in cold temperatures. White rice bran-free fatty acids control about 20% and white rice bran stabilized using microwave less than 10% [9]. The increase in free fatty acids for the control sample was much higher than sample with various heat treatment, indicating that the various heat treatments were effective in retarding the enzyme degradation. Similarly result also has been reported by Kim et all (2004) in stabilization of white rice bran.

In this study, the FTIR method was used to determine the bioactive compounds contained in brown rice bran. FTIR results showed that the red rice bran Rojolele contained anthocyanin and thiamin (data not shown). The treatment of stabilizing red rice bran using various heat treatment does not eliminate the anthocyanin and thiamin content in the rice bran. Stabilized rice bran with various heating methods has anthocyanin and thiamin content which is almost the same with control. The best stabilization process is determined by Sullivan et al. method [27]. The results showed that Oven an hour at 80°C is the best method, with a weight of 0.809 (data not shown). A heating method using an Oven temperature of 800C for 1 hour is the best method because it show a high value of quality the red rice bran Rojolele.

4. Conclusion

The results showed that all heat treatment methods decrease L value. The highest moisture level is on heating by microwave for 2 min. There were no significant difference in the level of ash, fat, and antioxidant activity. The FFA content was decreased by all heating treatment. The fenol level with heating with Oven 80°C an hour was the highest. Among the treatments, Oven an hour at 80°C is the best method. This method produces red rice bran with characteristics of lightness, water content, ash, fat, protein, total phenol, antioxidant activity, crude fiber, and free fatty acid content respectively49.78; 5.51%; 8.37%; 27.11%; 10.50%; 0.25%; 75.88%; 9.30%; and 13.48%. Red rice bran Rojolele also contains anthocyanin and thiamin obtained from spectroscopy FTIR.

References

[1] Mulyani M dan Sukesi 2011 Analisis proksimat beras merah (*Oryza sativa*) varietas slegreg dan aek sibundon. Prosiding Tugas Akhir Semester Genap 2010/2011 Institut Teknologi Sepuluh November.Surabaya.

[2] Kristamtini dan Heni P 2009 Potensi pengembangan beras merah sebagai plasma nutfah yogyakarta. *J. Litbang Pertanian* **3** 88-95

[3] Hadipernata M 2007 Mengolah dedak menjadi minyak (rice bran oil). *Artikel warta penelitian dan pengembangan pertanian* 8-10

[4] Orthoefer F T 2001 Rice bran and oil. Dalam. Juliano B O 2016 Rice: Overview *Encyclopedia Of Food Grain* **1**125-129

[5] Wulandari P., Suter I., Putra N K., dan Widarta I. 2013 Bekatul beras merah sebagai salah satu alternatif sumber antioksidan *J. Universitas Udayana.* Bali.

[6] Anggraini E dan Dwi K 2015 Pengaruh substitusi bekatul (rice bran) terhadap sifat organoleptik donat. *E-Jurnal Boga* **8** 63-70

[7] Ketaren S 1986 Pengantar teknologi minyak dan lemak pangan UIPress. Jakarta

[8] Champagne E, Horn R. J, dan Abraham G 1992 Stabilizing brown rice to lipolytic hydrolysis by ethanol vapors *J. Cereal Chem* **69** 152-156

[9] Ramezanzadeh, Fatemeeh M., Ramu M. R., Marlene W., Witoon P., Richard T., and Wayne E. M. 1999 Prevention of hydrolytic rancidity in rice bran during storage. *J.Agricultural Food Chemistry* 47 (**8**) 3050-3052

[10] Kim, Sung-Min., Hyun-Jung Chung dan Seung-Taik Lim. 2014 Effect of various heat treatment on rancidity and some bioactive compounds of rice bran. *J.of Cereal Science* **60**: 243-248
[11] Hadipernata, M., Supartono dan Falah 2012 Proses stabilisasi dedak padi (Oryza sativa L) menggunakan radiasi far infra red (fir) sebagai bahan baku minyak pangan. *J.Teknologi Pangan* 1 (**4**) 103-107
[12] AOAC 2000 Official method of analysis. Association of Official Analytical Chemists. Washington D.C
[13] Sugiat D 2010 Penetapan kadar fenol total dan aktivitas antioksidan ekstrak methanol dedak beberapa varietas padi (Oryza sativa L.) *Skripsi* Fakultas Matematika dan Ilmu Pengetahuan Alam, Universitas Indonesia. Depok
[14] Sari M 2011 Identifikasi protein menggunakan Fourier Transform Infrared (FTIR). *Skripsi* Jurusan Teknik Kimia. Fakultas Teknik Universitas Indonesia. Jakarta
[15] Andarwulan N, Feri K, dan Dian H 2011 Analisis pangan. *Dian Rakyat*. Jakarta.
[16] Aliva, I., Silva, C.L.M., 1999. Modeling kinetics of thermal degradation of colour in peach puree. *J. Food Eng*. 39, 161e166.
[17] Badan Standarisasi Nasional 1998 SNI 01-4439-1998 tentang Bekatul. *BSN*. Jakarta
[18] Towns J K 1995 Moisture content in proteins: its effect and measurement *J.of Chromatography A* **705**: 115 – 127
[19] Garcia, M., Belassi M., and Junior M 2012 Physicochemical and sensory profile of rice bran roasted in microwave *Ciênc. Tecnol. Aliment., Campinas* 32 **4** 754-761
[20] Loypimai, P., Moonggarm, A., Chottanom, P., 2009. Effects of ohmic heating on lipase activity, bioactive compounds and antioxidant activity of rice bran. Aust. *J. Basic Appl. Sci*. 3, 3642e3652.
[21] Sharma, H.R., Chauhan, G.S., Agrawal, K., 2004. Physico-chemical characteristics of rice bran processed by dry heating and extrusion cooking. Int. *J. Food Prop*. 7, 603e614
[22] Cicero, A.F.G., Derosa, G., 2005. Rice bran and its main components: potential role in the management of coronary risk factors. *Curr. Top. Nutraceutical Res*. 3, 29e46.
[23] Shirakawa, H., Koseki, T., Ohinata, K., Hazhizume, K., Komai, M., 2006. Rice bran fractions improve blood pressure, lipid profile, and glucose metabolism in stroke-prone spontaneously hypertensive rats. *J. Agric. Food Chem*. 54, 1914e1920
[24] Li, W. L., Pickard, and Beta 2007 Effect of ohmic heating on lipase activity bioactive compounds and antioxidant activity of rice bran. *Australia Journal of Basic and Applied Science* **3** 3642-3652
[25] Ulfa, S 2016. Identifikasi dan uji aktivitas antioksidan dalam bekatul dengan menggunakan variasi pelarut. *Skripsi*. Fakultas Sains dan Teknologi, UIN Malang
[26] Aunillah, A 2009 Stabilisasi tepung bekatul dengan metode pemanasan bertekanan dan pengeringan rak serta pendugaan umur simpannya. *Skripsi*. Fakultas Teknologi Pertanian, Institut Pertanian Bogor. Bogor
[27] Sullivan, W.G., Wicks, E.M., Koeling, C.P. 2015 Engineering Economy 16th Edition. Pearson Education Ltd.London

IOP Conf. Series: Materials Science and Engineering **633** (2019) 012047 doi:10.1088/1757-899X/633/1/012047

Chemical and sensory characteristics of white sweet potato (*Ipomoea batatas* L.), rice (*Oryza sativa* L.), and tapioca (*Manihot esculenta*) flours - based seasoning composite flour

N N A Alfani, D Ishartani, C Anam, D Praseptiangga, G J Manuhara

Department of Food Science and Technology, Universitas Sebelas Maret (UNS), Jl. Ir. Sutami 36A Kentingan Surakarta 57126, Central Java, Indonesia

E-mail: dwiishartani@staff.uns.ac.id

Abstract. Seasoning flour was a food commodity that has a good market trend but currently the raw materials were still dominated by wheat flour that has been known as the triggering agent of some allergic disease. So in order to reduce the domination of wheat flour as raw material of the seasoning flour, an innovation of seasoning composite flour made from non-wheat flour that have suitable characteristics is needed. The objective of this study were to determine the chemical characteristics of seasoning composite flour that made from white sweet potato, rice and tapioca flours, and to determine and characterize the sensory profile of the best seasoning composite flour which was chosen using weighting test. The seasoning composite flour made using 9 formulas contain 13.261 - 14.089% water; 4.467 – 5.729% ash; 1.238 – 2.018% fat; 4.324 – 5.838% protein; 74.371 – 74.863% carbohydrate; 72,781 – 75.613% starch; 21.904 – 25.760% amylose and 0.086 – 0.499% crude fiber. The best formula of seasoning composite flour has 30% tapioca flour, 30% rice flour and 40% white sweet potato flour ratio. The best one contains 13.261% water; 5.728% ash; 1.194% fat; 4.646% protein; 74.427% carbohydrate; 72.781%starch; 21.904% amylose and 0.499% crude fiber. Its sensory characteristic is significantly different from the commercial seasoning flour and overall get hedonic score 3.76 which means neutral.

1. Introduction

Along with the growing trend of fried foods, seasoning flour has experienced a good market increase. Seasoning flour makes the preparation of crispy fried foods easier and faster [1]. Rice flour and tapioca flour are Indonesian major flour commonly applied in seasoning flour recipes. Rice flour and tapioca flour can improve and maintain the crispiness of the fried foods. In addition, rice flour also gives bright effect on fried foods. Tapioca starch consists of 83% amylopectin and 17% amylose, rice starch consists of 20-25% amylose and 75-80% amylopectin, while white sweet potato flour consists of 24.37% amylose and 75.63% amylopectin [2].

Tapioca flour, rice flour, and white sweet potato flour are gluten-free commodities that intended safe for gluten intolerance, such as autism. Gluten causes brain dysfunction in autistic people because autistic people have digestive problems in absorbing abnormal gluten proteins so that proteins turn into peptides that enter the blood and poison the brain. Impaired brain function by this peptide is similar to the effect of morphine which can cause dependence [3]. Gluten has also been known as an initiator of celiac disease. Other facts also mention that accumulation of gluten can also cause other diseases such as ataxia, skin rashes (eczema), decreased muscle strength (hypotonia), stunting, learning difficulties, depression, migraines and headaches. The number of cases of gluten intolerance is an opportunity for seasoning flour producers to provide gluten-free seasoning flour.

Several studies related to composite flour as a substitute for flour in several types of food have been carried out including research on formulas of taro flour and sweet potato as raw material for dried noodles [4], composite flour formulations made from rice, sweet potatoes, potatoes, soybeans, and xanthan gum [5], cake formulations based on red bean, soybean, and corn composite flour [6], and a hypoglycemic effect of composite flour made from purple sweet potato in streptozotocin induced diabetic rats [7]. Yet, a gluten-free seasoning flour that formulated from composite flour has not been studied.

Based on physical characteristic, composite flour made from tapioca flour, rice flour, and white sweet potato flour has the potential to be a raw material for gluten-free seasoning flour [4]. However, the chemical and sensory characteristic of seasoning flour made from the composite flour is not well understood yet. Thus, it is necessary to determine chemical and sensory characteristic of composite seasoning flour based on tapioca flour, rice flour, and white sweet potato flour as an alternative for making gluten-free seasoning flour.

2. Experimental

Seasoning composite flour formulation and manufacture was based on previous research [4]. The materials, tapioca (*Rose Brand*), rice flour (*Rose Brand*) and white sweet potato flour (CV. KUSUKA UBIKU), were sieved 100 mesh and manually mixed into composite flour. The composite flour was further mixed with blended spices consisting refined salt, pepper powder, onion, garlic, and candlenut. Mixing was done using a hand mixer at medium speed for 20 minutes. The nine formulas of seasoning composite flour are shown in **Table 1**.

Table 1. Seasoning composite flour formula

Tapioca Flour (%)	Rice Flour (%)	White Sweet Potato Flour (%)
30	70	0
30	65	5
30	60	10
30	55	15
30	50	20
30	45	25
30	40	30
30	35	35
30	30	40

2.1. Chemical analysis methods

Water content was analyzed using thermo-gravimetric method [5], ash content using dry method [5], fat content using Soxhlet extraction method [5], protein content using micro Kjeldahl method [5], carbohydrate content using by difference method [2], starch content using acid hydrolysis method [6], amylose content using spectrophotometric methods [7], and crude fiber content using acid-base hydrolysis [8].

2.2. Determination of the best formula

The best formula was determined using weighting test method with a modification [9]. The parameters being tested were physical [4] and chemical characteristic.

2.3. Sensory analysis methods

Sensory analysis was performed using difference test (i.e., triangle test) and also hedonic test [10].

IOP Conf. Series: Materials Science and Engineering **633** (2019) 012047 doi:10.1088/1757-899X/633/1/012047

3. Results and discussion

3.1. Chemical characteristics

Chemical characterization was carried out on nine formulas of composite seasoning flour and two control seasoning flour. Both control were commercial seasoning flour. The chemical characteristics of composite seasoning flour are shown in **Table 2.**

Table 2. Chemical characteristics of composite flour based on rice flour, tapioca flour and white sweet potato flour

Tapioca Flour : Rice Flour : White Sweet Potato Flour	Water content (%bb)	Ash content (%bb)	Fat content (%bb)	Protein content (%bb)	Carbohydrate content (%bb)	Amilose content (%)	Starch content (%)	Crude fiber (%)
30 : 70 : 0	14.09^{d} ± 0.04	4.49^{a} ± 0.03	1.24^{b} ± 0.12	5.84^{e} ± 0.01	74.35^{bc} ± 0.00	25.76^{c} ± 0.01	75.61^{e} ± 0.00	0.09^{a} ± 0.07
30 : 65 : 5	14.14^{d} ± 0.03	4.47^{a} ± 0.03	1.24^{b} ± 0.04	5.24^{d} ± 0.03	74.86^{c} ± 0.01	24.88^{d} ± 0.01	74.82^{e} ± 0.00	0.29^{b} ± 0.04
30 : 60 : 10	14.10^{d} ± 0.00	4.77^{a} ± 0.05	1.29^{b} ± 0.09	5.25^{d} ± 0.07	74.60^{bc} ± 0.01	24.36^{de} ± 0.01	74.59^{cd} ± 0.00	0.30^{bc} ± 0.02
30 : 55 : 15	13.95^{cd} ± 0.00	5.07^{b} ± 0.06	1.59^{c} ± 0.07	4.76^{bc} ± 0.13	74.59^{bc} ± 0.00	24.35^{f} ± 0.02	74.69^{d} ± 0.01	0.30^{bc} ± 0.03
30 : 50 : 20	13.69^{bcd} ± 0.01	5.37^{bc} ± 0.03	1.63^{c} ± 0.03	4.94^{cd} ± 0.55	74.37^{bc} ± 0.00	24.04^{ab} ± 0.01	74.33^{cd} ± 0.01	0.34^{cd} ± 0.03
30 : 45 : 25	13.57^{bc} ± 0.05	5.46^{cd} ± 0.05	1.77^{d} ± 0.01	4.73^{bc} ± 0.44	74.47^{bc} ± 0.00	23.76^{a} ± 0.00	74.09^{cd} ± 0.00	0.37^{d} ± 0.07
30 : 40 : 30	13.54^{bc} ± 0.01	5.65^{cd} ± 0.03	1.19^{de} ± 0.03	4.46^{abc} ± 0.05	74.45^{bc} ± 0.00	22.49^{ef} ± 0.01	74.00^{c} ± 0.00	0.46^{e} ± 0.09
30 : 35 : 35	13.51^{bc} ± 0.00	5.73^{ed} ± 0.04	2.02^{e} ± 0.05	4.32^{ab} ± 0.27	74.42^{bc} ± 0.00	22.28^{bc} ± 0.01	73.20^{b} ± 0.00	0.499^{f} ± 0.09
30 : 30 : 40	13.26^{b} ± 0.01	5.73^{e} ± 0.02	1.19^{e} ± 0.03	$4,65^{a}$ ± 0.51	74.04^{b} ± 0.05	21.90^{a} ± 0.01	72.78^{b} ± 0.00	0.19^{f} ± 0.02
Sajiku Brand	11.27^{a} ± 0.01	6.42^{f} ± 0.13	0.58^{a} ± 0.12	7.48^{e} ± 0.20	74.25^{bc} ± 0.04	24.38^{ed} ± 0.02	71.44^{a} ± 0.06	0.27^{a} ± 0.02
Rizqiyya Brand	13.53^{bc} ± 0.02	5.47^{cd} ± 0.02	0.46^{a} ± 0.12	4.32^{f} ± 0.27	74.42^{a} ± 0.04	22.49^{c} ± 0.02	73.20^{b} ± 0.04	0.22^{a} ± 0.01

Same notation in the same column is not significantly different at α: 0.05

The increasing ratio of white sweet potato flour in seasoning composite flour has a significant effect on the decreasing moisture content of composite seasoning flour. White sweet potato flour itself has lower water content than tapioca and rice flour. The moisture content of composite seasoning flour is higher than the moisture content of Sajiku Brand but is not significantly different compared to Rizkiyya Brand. It caused by the different form of the spice being added into the seasoning flour. The composite seasoning flour and Rizkiyya Brand were using wet blended spices while Sajiku Brand was using dry blended spices. The water content of composite seasoning flour is not too different from the

IOP Conf. Series: Materials Science and Engineering **633** (2019) 012047 doi:10.1088/1757-899X/633/1/012047

moisture content of composite flour rice flour, sweet potatoes, potatoes, soybeans and xanthan gum [5].

Ratio of the raw material flour has a significant effect on the ash, lipid, and protein content of seasoning composite flour. The greater ratio of white sweet potato flour, the higher ash and lipid content, yet the lower protein content of composite seasoning flour. Compared to Sajiku Brand, ash and lipid content of composite seasoning flour is lower. Composite seasoning flour with 20-35% white sweet potato contain ash content similar with Rizkiyya Brand. Level of seasoning composite flour's ash was higher than previous composite flour ash content [5]. This is due to the addition of seasonings and sodium salts to into seasoning composite flour. However, fat content of seasoning composite flour was lower than that previous composite flour [5] due to the form difference of the soy source being added. Previous research [5] using soy flour containing more than 28.26% fat while composite seasoning flour did not use high-fat base flour ingredients.

Ratio of white sweet potato flour also has a significant effect on the protein content of composite seasoning flour. It also has a real effect on carbohydrate content of composite seasoning flour. The greater the ratio of white sweet potato flour, the lower the carbohydrate content of composite seasoning flour. This is because the carbohydrate content of white sweet potato flour is lower than the raw material for tapioca flour and rice flour. But this carbohydrate level is still higher than commercial seasoning flour. The carbohydrate content of seasoning composite flour manufactured in this research is lower than the previous reported composite flour [4].

Amylose and starch content tend to decrease as the white sweet potato flour ratio increase, especially above 20%. Starch content in composite seasoning flour would affect its functional properties, namely water absorption and flour viscosity. The greater the starch content in the composite seasoning flour, the greater the water absorption and flour viscosity [8]. Amylose content of the composite seasoning flour, which ranged from 21.90 - 25.76%, was higher than the amylose content of previous composite flour [15], which was 15.77%.Higher ratio of white sweet potato flour tends to ascend the level of crude fiber in seasoning composite flour. The level of crude fiber of composite seasoning flour is higher than the commercial ones.

3.2. Determination of the best formula

The best formula was determined using result value (RV) calculation. The best one is formula which got highest RV. Thus the best composite seasoning flour was made from 30% tapioca, 30% rice flour and 40% white sweet potato flour. The best formula composite flour contains 13,261% water; 5,728% ash; 1,194% fat; 4,646% protein; 74.427% carbohydrate; 72,781% starch; 21,904% amylose and 0.499% crude fiber.

3.3. Sensory characteristics

Triangle test was conducted to determine whether there are any differences or not in sensory properties between best composite seasoning flour and commercial control seasoning flour. The result of the triangle test for overall parameter is shown in **Table 3.** The best formula composite seasoning flour is significantly different either from Sajiku Brand and Rizkiyya Brand.

Table 3. Triangle test result of seasoning composite flour

Control	Compared to Sajiku Brand	Compared to Sajiku Brand
The number of panels who stated different	25	24
The smallest amount to be significantly different at α 5%	18	18

The best formula composite seasoning flour, however, get lower hedonic score compare to Sajiku and Rizkiyya Brand for appearance, crispiness, and overall. It has darker color and less crispy compared to both control flour. Crispness has a correlation with the parameters of starch and amylose content [16]. Overall, it gain hedonic score 3.76 which means neutral.

Table 4. Hedonic score of seasoning composite flour

Seasoning Flour	Hedonic Score		
	Appearance	Crispiness	Overall
Sajiku Brand	5.68[c]	5.64[b]	5.76[c]
Rizkiyya Brand	4.92[b]	4.68[a]	4.80[b]
Best Formula Composite	2.72[a]	4.28[a]	3.76[a]

Score 1 = very dislike, 2 = dislike, 3 = rather dislike, 4 = neutral, 5 = rather like, 6 = likes, 7 = really like

4. Conclusion

The best formula of composite seasoning is 30% tapioca, 30% rice flour, and 40% white sweet potato flour which contains 13.26% water; 5.73% ash; 1.19% fat; 4.65% protein; 74.43% carbohydrate; 72.78% starch; 21.90% amylose and 0.50% crude fiber. Sensory characteristics of composite seasoning flour are significantly different from commercial seasoning flour.

References

[1] Bagja JS, Yuwono SS and Widyaningtyas D. 2015. *J. Food and Agroindustry* **3** (4), p. 1627 – 1636
[2] Winarno FG. 2008. *Food Chemistry and Nutrition*. (Bogor - Embrio Press)
[3] Dewanti HW, and Machfud S. 2014. JKKI **6** (2), p. 67 – 74
[4] Aurum FS and Elisabeth DAA. 2015. *J. Assessment and Development of Agricultural Technology* 18 (3)
[5] Julianti E, Rusmarilin H, Ridwansyah and Yusraini E. 2017. *J. Saudi Soc. Agric. Sci.* **16** p. 171–177
[6] Astuti SD, Andarwulan N, Hariyadi P and Agustia FC. 2014. *J. Food Technology's Aplication* **3** (2)
[7] Djunaidi CS, Affandi DR and Praseptiangga D. 2014. *J. Indonesia Clinical Nutrition* **10** (3)
[8] Alfani NNA, Ishartani D, Anam, C, Praseptiangga D, and Manuhara GJ. 2017. *IOP Conference Series: Materials Science and Engineering* **193** (1), p. 1 – 6
[9] AOAC. 2005. *Official Methods of Analysis 18th Edition*. (Washington DC - Association of Official Analytical Chemist).
[10] AOAC. 1984. *Official Methods of Analysis of Association of Chemist*. (Washington DC - Association of Official Analytical Chemist).
[11] Fardiaz D, Puspitasari NL and Palupi H. 1986. *Food Analysis Practicum Guide*. (Bogor - Faculty of Agricultural Technology, Bogor Agricultural University).
[12] Apriyantono A, Fardiaz D, Puspitasari NL, Sedarnawati and Budianto S. 1992. *Food Analysis Laboratory Instructions*. (Bogor – Bogor Agricultural University).
[13] Indrianti N, Rima K, Riyanti E and Doddy A D. 2013 *AGRITECH* **33**.
[14] Setyaningsih D, Apriyanto A, and Puspitasari M. 2010. *Sensory Analysis for Food and Agro Industries*. (Bogor – IPB Press).
[15] Zaqie F, Rusmarilin H, and Nurmimah M. 2018. *J. Food Engineering and Agriculture* **6** (2)
[16] Asgar A and Rahayu ST. 2014. *J. Biology News* **13** (3), p. 283 – 293

IOP Conf. Series: Materials Science and Engineering **633** (2019) 012048 doi:10.1088/1757-899X/633/1/012048

The influence inter-day colorimetric method in quantification of chicken meat's total cholesterol with lipid extraction - saponification pretreatment

U Laila[1] and T J Putra[2]

[1]Research Unit for Natural Product Technology, Indonesian Institute of Sciences, Jl Jogja-Wonosari km 31.5, Desa Gading, Kecamatan Playen, Kabupaten Gunung Kidul, Yogyakarta 55861, Indonesia
[2]Mathematical and Science Program, Indonesia Islamic University, Yogyakarta, Indonesia

E-mail: umilaila38@gmail.com

Abstract. Unskinned chicken meat's cholesterol consisting of thigh, breast, and wing meat were determined inter-daily by ferric chloride reagent spectrophotometric methods. The pretreatments before cholesterol analysis were conducted which involved with lipid extraction, saponification, and subsequent extraction of unsaponified matter. Afterwards, spectrophotometry was carried out. For understanding precision of quantification, the total cholesterol was measured inter-daily in which the difference of measurement time were one day for each same part of chicken meat. The research showed unskinned thigh meat had significantly higher total cholesterol 186.19 mg/100 g as compared to breast, and wing meat (156.05 mg/100g and 135.38 mg/100 g; $p<0.01$ for both). Inter-daily spectrophotometry analysis insignificantly affected on thigh, breast, and wing meat's cholesterol content. This method can become solution for analysis time limitation problem in cholesterol spectrophotometric analysis.

1. Introduction

Cholesterol is fat-like substance of mammalian tissue containing 27 atom carbon [1]. Their combination of steroid and alcohol make the substance usually bounded with lipoprotein and can be esterified, besides free-form cholesterol is also found [2]. The cholesterol plays role in composing cellular membrane and becoming a precursor of bile acid, vitamin D, and essential hormones, such as cortisol, progesterone, testosterone, etc. During growth stage of human life, cholesterol is essential substance for composing myelin, thus it aids in developing nervous system [3]. Cholesterol is commonly synthesized by body itself, but it can be served by dietary intake such as egg, meat, fish, etc. Cholesterol in optimum amount give advantages for human. However, its exceed level is correlated to arteriosclerosis and cardiovascular disease, so that the cholesterol dietary intake to our body should be controlled [4].

To estimate cholesterol dietary intake, the precise determination is essentially needed. The spectrophotometric or colorimetric method is often used in cholesterol calculation because of low cost and simplicity, so that it can be applied in all vary laboratory [5]. However, the pretreatments before spectrophotometer analysis including lipid extraction and saponification spend more time and they were complicated. To overcome it, many researchers proposed saponification without lipid extraction pretreatment (direct saponification in other word). This pretreatment is effective and efficient because it doesn't consume time, labor, and apparatus [6]. It is commonly known that saponification has role to change esterified cholesterol to free cholesterol, so that total cholesterol is resulted from the analysis [7]. However, based on the fact that direct saponification is unavailable for sample extract whose the color disturb the turbidity, the lipid extraction coupled with saponification technique is still remain to be applied in cholesterol analysis.

This study was aimed to assess the quantification of total cholesterol by lipid extraction coupled with saponification technique. The broiler chicken meat was used as analyzed sample because it was consumed worldwide, moreover its cholesterol content is rather high based on some literatures [8]. The extract sample was analyzed for its cholesterol by $FeCl_3$ reagent - spectrophotometric methods in triplicate repetition but at different day. The aim was to prove that total cholesterol quantification by lipid extraction coupled with saponification technique resumed by inter-day colorimetric method is available or not.

2. Material and Methods

2.1. Material

Broiler chicken meat was purchased from local market in Jogjakarta. The number of broiler was single with consideration that the cholesterol will be differ among the individual chickens, moreover the research focused in inter-day testing. Part of meat analyzed were thigh, breast, and wing. Standard cholesterol $C_{27}H_{45}OH$ was purchased from Sigma Aldrich, USA. Chloroform, KOH, ethanol, sulfuric acid, $FeCl_3$. $6H_2O$, acetic acid, petroleum ether were purchased from Merck.

2.2. Methods

2.2.1. Cholesterol standard calibration curve preparation. Ferric chloride reagent ($FeCl_3$ reagent) was initially prepared by dissolving $FeCl_3.6H_2O$ by glacial acetic acid at portion 10% (w/v), then followed by diluting $FeCl_3$ solution in H_2SO_4 with dilution factor of 10. Meanwhile, cholesterol stock solution was prepared by dissolving cholesterol standard by glacial acetic acid to get final concentration 0.1 mg/ml. Then serial dilution was carried out to stock solution by glacial acetic acid for getting cholesterol concentration of 0.075; 0.050; 0.025; 0.0125; and 0.0050 mg/ml. The cholesterol stock and its diluted solution were reacted by $FeCl_3$ reagent in volume ratio (1:1). Absorbance of solutions were measured by UV-Vis spectrophotometer (Dynamica HALO RB-10 Spectrophotometer) at wavelength 515 nm. Cholesterol standard curve depicted the relationship of cholesterol concentration and its absorbance value. The linearity of standard curve was high with R square value was 0.9997.

2.2.2. Pretreatment sample before colorimetric analysis. Chicken meats comprising thigh, breast, and wing were unskinned and then dried in cabinet dyer at temperature 105°C for overnight. The pretreatment sample before colorimetric analysis followed method by Ismail et al. [9] with slight modification. The dried sample was ground for obtaining fine powder. The meat powdered was weighed for 2 gr and was conducted for fat extraction by soxhlet method with petroleum ether 40 ml for 8 hr. Afterwards, the fat extract was diluted by chloroform 10 ml. A portion of 2 ml solution was saponified by reflux boiling with alcoholic-KOH 10 ml for 1 h in order to convert esterified cholesterol to free cholesterol. It was cooled, then water was used for rinsing it. The unsaponified matter was separated by extraction with petroleum benzene for three times. Then, cholesterol extract was separated from the solvent by evaporation at 60°C. The extract sample was ready to be analysed. For samples would analysed at different day, samples were stored in desiccator at ambient temperature.

2.2.3. Spectrophotometric analysis of sample. Cholesterol extract sample was diluted with glacial acetic acid 4.5 ml. One ml portion of sample solution diluted again with glacial acetic acid until the

IOP Conf. Series: Materials Science and Engineering **633** (2019) 012048 doi:10.1088/1757-899X/633/1/012048

volume reached 10 ml. The next step was similar to that of standard cholesterol. Cholesterol sample solution was also reacted by reacted by $FeCl_3$ reagent with volume ratio (1:1) and then absorbance was measured at 515 nm. Spectrophotometric analysis for a cholesterol sample extract was in triplicate but repetition was conducted at different day. Total cholesterol was calculated as following formula:

$$\text{Cholesterol (mg/100g)} = (Cs \times DF \times Vs \times F)/(m \times 100)$$

Where: Cs is cholesterol concentration based on standard calibration curve, DF is dilution factor (in this case, the DF value is 10), F is factor because of one fifth of fat extract that was saponified (F=5), m is dried chicken meat weight. Vs is total volume of cholesterol extract, in this case Vs was 4.5 ml.

2.2.4. Statistical analysis. Collected data were statistical analysed by SPSS 12. One way Analysis of Variance (one way ANOVA) was used for compared mean of thigh, breast, and wing meat cholesterol with significance level of 0.01. Furthermore, effect of inter-daily analysis on cholesterol content was studied by Pearson correlation. Pearson correlation was considered to be chosen because it is used for parametric test.

3. Results and Discussion

It was considered that chicken skin contained high cholesterol, about 71 mg / raw chicken skin [10]. Before chicken meats were analysed, they were unskinned for avoiding ambiguous result. The analytical data for cholesterol content of thigh, breast, and wing chicken meat can be seen in **Table 1** and **Table 2**.

Table 1. The result of inter daily cholesterol analysis of chicken meat by spectrophotometric method

Part of meat	Day of analysis after extraction	Cholesterol content (mg / 100 g meat)*
Thigh	1st	188.94
	2nd	186.18
	3rd	183.45
Breast	1st	155.90
	2nd	157.71
	3rd	154.53
Wing	1st	141.25
	2nd	130.25
	3rd	134.63

*All measurement were conducted as dry weight basis

Table 2. The result of correlation test inter day cholesterol analysis spectrophotometric method on cholesterol content

Part of meat	Cholesterol content (mg / 100 g meat)	Effect of differ day analysis on cholesterol content
Thigh	186.19 ± 2.75^a	insignificant
Breast	156.05 ± 1.60^b	insignificant
Wing	135.35 ± 5.54^c	insignificant

^{a,b} Means within a column with no common superscript differ significantly (P< 0.01)
All measurement were conducted as dry weight basis, values are means ± standard deviation for 3 meats per group.

From **Table 1** and **Table 2**, it can be shown that thigh chicken contained more cholesterol than breast and wing. It agreed with results obtained by Anandhi et al.[11] that thigh chicken cholesterol was higher than that of breast. Salma et al.[12]found that thigh cholesterol (194.2 mg /100 g) was higher than breast (93.6 mg /100 g). The cholesterol determination was carried out by enzymatic method with direct saponification pretreatment and the result gave cholesterol content was almost similar to the range of cholesterol content in this research.

The value total cholesterol from this research was also similar to range of by Ponte et al. [13] research in which the breast cholesterol was 185.04 mg /100 kg dried meat. Ponte's cholesterol quantification was conducted by HPLC method with direct saponification pretreatment. Dinhet al. [14] also investigated that chicken thigh had more cholesterol than breast.

Meanwhile, cholesterol spectrophotometric analysis conducted in different day (1st, 2nd, 3rd day) for every sample did not affect cholesterol content significantly. It proved that inter-day colorimetric method was precise for cholesterol quantification. It means that the method achieve acceptable repeatability. As we known, that pretreatment before spectrophotometric analysis spend more time, so that the analysis at the day after is available. The reproducibility of inter-day also obtained acceptable repeatability for cholesterol analysis by gas chromatography – tandem mass spectrometry [15].

4. Conclusions

Analysis of cholesterol content by $FeCl_3$ colorimetric method revealed that thigh meat had significantly higher total cholesterol 186.19 mg/100 g as compared to breast, and wing meat (156.05 mg/100g and 135.38 mg/100 g; $p<0.01$ for both). Inter-day spectrophotometry analysis achieve acceptable repeatability because the different-day analysis insignificantly affected on thigh, breast, and wing meat's cholesterol content. The value total cholesterol from this research was in range of other research in which different method were used, i.e. HPLC method and enzymatic method with direct saponification pretreatment.

References

[1] Saini Y 2007 *Standardization of a Method for Cholesterol Estimation in Milk Fat Using O-Phthaldehyde*, Dairy Chemistry Division, National Dairy Research Institute, (Deemed University), India.
[2] Guler Z and Park Y W 2009 *Handbook of Dairy Foods Analysis*(Boca Raton : CRC Press) pp 233-261.
[3] Álvarez-SalaA, Garcia-Llatas G, Barberá Rand Lagarda M J 2015 *Nutricion Hospitalari* **32** 1535-1540.
[4] Osman H and Chin Y K 2006 *The Malaysian Journal of Analytical Sciences* **10(2)**205-210.
[5] Viturro E, Meyer H H, Gissel E and Kaske M.2010 *Journal of Dairy Research* **77** 85–89.
[6] Dinh TTN, Blanton JR, Brooks JC, MillerMF and Thompson LD 2008 *Journal of Food Composition and Analysis* **21** 306-314.
[7] Agullo E and Gelos BS 1996 *Food Research Internationa l* **29(1)**77-80.
[8] Sudarman A, Sumiati and Solikhah SH 2011 *Media Peternakan* 64-68.
[9] Ismail I, Huda N, Ariffin F and Ismail N 2010 *International Journal of Poultry Science* **9(6)**556-561.
[10] van de Bovenkamp Pand Katan MB 1981 *Journal of Food Science* **46 (1)**
[11] Anandhi M, Poorani K and Mathivanan 2017 *International Journal of Science, Environment and Technology* **6(6)**3350 – 3353.
[12] Salma U, Miah AG, Maki T, Nishimura M and Tsujii H 2007 *Poultry Science* **86** 1920–1926.
[13] Ponte P IP, MendesI, Quaresma M, Aguiar MNM, Lemos JPC, Ferreira LMA, Soares MAC, Alfaia CM, Prates JAM and Fontes CMGA 2004 *Poultry Science* **83** 810–814.
[14] Dinh TTN, Thompson L D, Galyean ML, Brooks JC, Patterson KY and Boylan LM 2011*Comprehensive Reviews in Food Science and Food Safety***10**269-289.
[15] ChenY, Kao S, Jian H, YuY, Li J, WangW and Tsai C 2015 *Journal of Food and Drug Analysis* **23** 636-644

Acknowledgements
This research was fully supported by Research Unit for Natural Product Technology, Indonesian Institute of Sciences. The authors also thank to Mr Andri Suwanto for assistance.

IOP Conf. Series: Materials Science and Engineering **633** (2019) 012049 doi:10.1088/1757-899X/633/1/012049

Physical and hedonic properties of cow milk yogurt containing different levels of avocado pulp (*Perseaamericana, Mill*)

Atmanaji[1], B S Hertanto[2] and A Pramono[3]
[1]UndergraduatedStudent of Animal Science Department, Faculty of Agriculture, Universitas Sebelas Maret
[2]Laboratory of Industry and Animal Product Processing, Department of Animal Science, Faculty of Agriculture, Universitas Sebelas Maret
[3]Laboratory of Animal Production, Department of Animal Science, Faculty of Agriculture, Universitas Sebelas Maret

E-mail: ahmad_pram@staff.uns.ac.id

Abstract. The research aimed to analyze physical and hedonic properties of cow milk yogurt containing different levels of avocado pulp (*Perseaamericana, Mill*) and determine the best level of avocado pulp on cow's milk yogurt. The material used were fresh milk of Holstein Friesian crossbred dairy cow, avocado pulp and commercial starter plain yogurt. The research design used a completely randomized design (CRD) of direct pattern with five treatments includingP0 = 100 ml yogurt + 0% avocado pulp, P1 = 97.5 ml yogurt + 2.5% avocado pulp, P2 = 95 ml yogurt + 5.0% avocado pulp, P3 = 92.5 ml yogurt + 7.5% avocado pulp, and each treatment consisted of five replications. The data were analyzed by variance analysis,and difference treatment was continued with Duncan's New Multiple Range Test (DMRT). The results showed that different levels of avocado pulp affected on viscosity adhesiveness and cohesiveness of yoghurt ($P<0.05$). However, the treatment did not affect hardness ($P>0.05$). Hedonic test showed that adding avocado pulp into yogurt did affect ($P<0.05$) color and flavor, and also taste of yogurt was high significanly effected ($P<0.01$). Overall, it can be seen that the different levels of avocado pulp in cow milk yogurt can increase physical properties of yogurt including viscosity and hedonic quality. The best level of avocado pulp added to cow milk yogurt was 7.5%.

1. Introduction

The dairy products have a special target for the functional food market due to effiecient carriers of probiotic bacteria[1,2]. Yoghurt is common fermented milk product and widely consumed because of its functional value and good vehicle to deliver probiotics toconsumers[3]. Numerous studies have been reported on utilization of probiotics to achieve a product with satisfactory quality characteristics and good acceptance by consumers [4].

The texture is a fundamental characteristic of yogurt. It can be improved by the use of gelling, thickening, or stabilizing agents[5].Textural properties of yogurt, such as viscosity[6], smoothness and thickness [7], and structural resistance to stress [8], are important attributes to determine its consumer acceptance. Hardness is the most important parameter for evaluation of yogurt texture. It is regarded as the force required to attain a certain deformation and is considered as a measure of firmness of the yogurt[9]. Many methods have been used to improve the quality of the yogurt, such as increasing the solids in milk (adding fat, proteins, or sugars such as sucrose and fructose), addition of stabilizers (pectin, starch, alginate, and gelatin)[10,11]. However, these approaches did not satisfy the consumers demand for products with as fewfood additivesaspossible.

Yogurt has various types included set yogurt, stirred yogurt, sweet drinking yogurt, fruit yogurt, yogurt cheese, frozen yogurt dan dried yogurt[12]. Several yogurt are marketed with the addition of either fruit or vegetables rich in bioactive food [13]. The addition of flavors using essences, fruit and/or fruit extracts and honey may be a better option than artificial flavorings for use in the development of new dairy products [14]. The product is accepted by consumers for its flavor, (mainly acetaldehyde) and pleasant texture[15].

Some studies on avocado processing involved the incorporation of additives to improve thesensory quality and avoid syneresis[16]. The processed avocado pulp is an alternative to utilize fruits, which can be used in various value-added food products[17]. Based on all yogurt categories, 74% is yogurt mixed with fruit (fruit yogurt). Additional fruit flavor, pure fruit of fruit extract can improve taste, color and texture flexibility of yogurt to consumers [18]. Avocado in the immature state that has been freshly harvested is characterized by having an extremely firm texture[19]. The composition differences of the fruit also influence its physical properties such as texture and color[20]. In this case, the higher avocado fat and water content of harvest can be another factor that influences the progressive changes in fruit texture [21].The physical attributes of yogurts, including the lack of visual whey separation and perceived viscosity, are crucial aspects of the quality and overall sensory consumer acceptance of yogurts [11].

Based on the description, it is necessary to study cow's milk yogurt incorporated by fruit pulp. Hence, the aims of this research were to analyze physical and hedonic properties of cow milk yogurt containing different levels of avocado pulp (*Perseaamericana, Mill*) and determine the best level of avocado pulp on cow's milk yogurt.

2. Materials and Methods

The materials of this research werefresh milk of Friesian Holstein crossbred, avocado pulp brought at local market, and commercial starter plain yogurt.

2.1. Sample preparation

The treatment of this research wasconcentration ratio between yogurt and avocado pulp (v/v): P0 = 100 ml:0 ml; A2 = 97.5 ml:2.5 ml; A3 = 95 ml:5 ml; A4 = 92.5 ml:7.5 ml.Yogurt manufacturing according to [22]. Milk from local dairy farmer was pasteurized at 85^0C for 15 minutes, and it was cooled around40 – 45^0C. Then, addition of bacteria to milk using plain starter with the *Biokul trademark*, and it was incubated for 24 hours at 43^oC.

The procedure for making avocado pulp followed the method of research conducted by [23]. First,all avocados with same charateristics (age and cultivar) was sorted from dirt and splited in the middle and discarded. The flesh of avocados was cut into pieces and then blended. Second, mixing avocado pulp into yogurt by putting 100 ml of yogurt into each glass and then adding avocado pulp (P0 = 0%; P1 = 2.5%; P2 = 5.0%; P3 = 7.5%, P4 = 10.0%). And the last, mixture of avocado pulp and cow's milk yogurt was evenly mixed.

2.2. Measurement of physical properties

The samples were mixed and analyzed in duplicatefor pH, viscosity, and hardness. Brookfield Viscometer was used to determine viscosity of yogurt. The viscosity was expressed in centipoise (cP)). Hardness was measured by TA.XT plus, *Texture Analyzer, Stable Micro Systems Ltd.*, England[24], and The pH was measured with a pH meter(Hanna Instruments 8521).

2.3. Measurement of hedonic properties

Hedonic testof yogurt was carried out using semi-trained panelists with a age range of 20-24 years. The measurement of hedonic qualities included color, aroma, and taste using a 9-point hedonic scale[25].

2.4. Statistical analysis

The experimental design of thisresearch used a completely randomized design method. Data of this study included physical test (viscosity and texture) and hedonic test (color, aroma, taste and overall acceptance)were analyzed using ANOVA with significance differences at α=0.05, and differences between treatment means were further analyzed using Duncan's New Multiple Range Test. However, pH was analyzed using descriptive statistical by comparing to means of the treatments.

3. Result and Discussion

Physical properties

The resultof descriptive analysis showed that the addition of avocado pulpwas still within the normal range of a pH of cow's milk yogurt.The pH value in this study was 4.0–4.1 (Table 1.). Fruit-flavored yogurt has pH between 3.93 –4.00[26]. Yogurt containing acid is the spesific product resulted from fermentation process using *Streptococcus thermophillus* and *Lactobacillus bulgaricus* cultures[24]. Change of the pH values from milk inoculated with bacterial cultures to be yogurt decreased from 6.70 to 4.34[27]. The increase solids content in yogurt milk as a result of fortification, and also creates buffering that requires additional acid development by the starter cultures to achieve a similar pH target [28].

Table 1. Evaluation of physical properties of cow's milk yogurt containing avocado pulp

Variables	Treatment (*mean*±SD)					*p-value*
	P0	P1	P2	P3	P4	
pH	4.10±0.00	4.00±0.00	4.10±0.00	4.10±0.00	4.10±0.00	-
Viscosity (cP)	173.50^{a}±1.48	205.37^{ab}±7.25	228.23^{b}±1.30	230.87^{b}±0.45	234.57^{b}±0.89	0,042*
Hardness(g)	28.80±1.37	26.60±1.29	28.80±0.97	28.80±0.97	27.40±1.11	0,767^{ns}

^{a,b}Different superscript in the same line showed significantly different(P<0,05).
- = variance was not analyzed, * = different significantly, ns = not significantlydifferent.

The addition of avocado pulp showed a significant difference (P <0.05) to viscosity of cow's milk yogurt. Addition of avocado pulp increases the viscosity of cow's milk yogurt due to the large concentration of total solids added.The amount of solids determines the viscosity of the yogurt liquid, therefore viscosity can also be used as an index of the solid amount contained in the liquid[29]. Addition of avocado pulp up to level5% increases the quality of yogurt, but viscosity of yogurt at levels 5 to 10% did not increase. The viscosity produced by addition of avocado pulp into cow's milk yogurt was high viscosity with ranging from 173.5–234.5 cP. The addition of honey was able to increase total solids content and increase the consistency of the product. In accordance to [30], the difference of viscosity level of yogurt was caused by the total solids found in each product and also the difference in acid and pH value.

The results of this research showed that addition of avocado pulp had no effect (P> 0.05) on the texture (hardness) of cow milk yogurt. Addition of avocado pulp does not decrease the texture of yogurt. Hardness of yogurt remains almost same and then decreased with increase in partially hydrolyzed guar gum concentration[9]. Similar effects when yogurt samples were enriched with oat-maltodextrin[31], and also the increase in yogurt softness with addition of mango pulp[32].

Hedonic properties

The yogurt samples containing different levels of avocado pulp showed significanly different to color, taste and flavor. The addition of avocado pulp gave a significant difference (P <0.05) in increasing yogurt color. Greenish color of yogurt can increase panelists' hedonic for yogurt which are generally white.The green and violet colors of most vegetables are contributed by chlorophyll and anthocyanins[33], therefore the degradation of color in avocado fruit spreads can be most likely to be related to the degradation of chlorophyll [34]. Significant differences were shown in the addition of avocado pulp at levels 7.5 and 10% to 0%, while the level of P4 showed a significant difference also to P1 at the level of 2.5%. Adding avocado pulp to level 7.5 on yogurt can increase the preference for color in the product. Color and appearance decrease may be due to mango pulp particles settling at the bottom in set yoghurt at higher mango pulp concentrations[32].

Table 2. Evaluation of hedonic properties of cow's milk yogurt containing avocado pulp

Variables	Treatment (*mean*±SD)					*P-value*
	P0	P1	P2	P3	P4	
Color	$5{,}83^{a}$±1,37	$6{,}00^{ab}$±1,29	$6{,}23^{abc}$±0,97	$6{,}53^{bc}$±0,97	$6{,}73^{c}$±1,11	0,018*
Flavor	$5{,}93^{ab}$±1,57	$5{,}93^{ab}$±1,46	$5{,}67^{a}$±1,45	$6{,}63^{b}$±1,03	$6{,}37^{ab}$±1,03	0,042*
Taste	$4{,}47^{a}$±1,41	$5{,}17^{ab}$±1,32	$4{,}57^{a}$±1,52	$5{,}97^{c}$±1,43	$5{,}70^{bc}$±1,49	0,0001**

^{abc}Different superscript in the same line do not differ significantly (p> 0.05).
* = different significantly, ** = highsignificantlydifferent.

The results of variance analysis showed that the addition of avocado pulphad a significant difference (P <0.05) in increasing flavorof yogurt. The typical flavor of yogurt is obtained by the formation of lactic acid, acetaldehyde, acetic acid and diacetyl [35]. The avocado's distinctive flavorwas able to eliminate the distinctive flavor of yogurt so that there was an increase of panelists hedonic of yogurt products. The difference was shown by addition of 5% avocado fruit pulp (P2) to an addition level of 7.5% (P3). All stirred fruit yogurts were scored higher on average by the panelists than plain yogurt in terms flavor, possibly reflecting a combined contribution from flavor compounds in fruit juice and higher viability of *L. acidophilus* LA-5, which may also produce flavor compounds. Acetaldehyde for example is recognised as a major flavor component in yogurt and the presence of lactobacilli in the starter culture can influence the total content of acetaldehyde in final product [36].

The results of this study indicated that there was an increase in taste score compare to control treatment. Taste produced by cow's milk yogurt with addition of avocado pulpwas able to reduce sour taste due to a typical avocado flavor. A previous study found that the addition of fruit juice positively influenced the acceptance of goat yogurt, improving its taste and suggesting a positive influence of sugars naturally found in fruit on these sensory attributes [37]. Consumers prefer yogurt which has a less acidic taste and still tastes sweet. The acidity of yogurt therefore also affects the taste preference in cow milk yogurt. The typical avocado flavor and mixed acid produced can increase panelists' preference for the product[38]. A very significant difference was shown by the addition of avocado pulp level P3 and P4 at level 7.5 and 10% to levels P0 and P2 at levels 0 and 5%, while at the level of P3 it was significantly different from P1. Adding avocado juice to level 7.5 in yogurt can increase the taste for the product. Flavored yogurts including fruit (apple, apricot, black cherry, black currant, blue berry, lemon, mandarin, raspberry, strawberry, peach), cereal, vegetables, chocolate, vanilla, caramel, ginger) gave a wide array of tastes, but also increases sweetness of the product[39].

4. Conclusion

Addition of avocado pulpinto cow milk yogurt can improve the physical quality of yogurt including viscosity (thickness) and hedonic level of yogurt, namely color, taste,and flavor. Thebest level of adding avocado pulpinto cow milk yogurt was 7.5%.

References

[1] Amirdivani S and Baba A S 2011 *LWT -Food Sci. Technol.***44** 1458–68
[2] do Espírito Santo A P, Perego P, Converti A and Oliveira M N 2012 *LWT - Food Sci. Technol.***47** 393–99
[3] Reid G, Sanders M E, Gaskins H R, Gibson G R, Mercenier A, Rastall R, Roberfroid M, Rowland I, Cherbut C and Klaenhammer T R 2003 *J. Clin. Gastroenterol.***37** 105–18
[4] Chapman C M C, Gibson G R and Rowland I 2011 *Eur. J. Nutr.***50** 1–17
[5] Lucey J A, Munro P A and Singh H 1998 *J. Texture Stud.***29** 413–26

[6] Marshall V M and Rawson H L 1999 Effects of exopolysaccharide-producing strains of thermophilic lactic acid bacteria on the texture of stirred yoghurt *Int. J. Food Sci. Technol.***34** 137–43
[7] Jaworska D, Waszkiewicz-Robak B, Kolanowski W and Swiderski F 2005 *Int. J. Dairy Technol.***58** 39–46
[8] Skriver A, Holstborg J and Qvist K B 1999 *J. Dairy Res.***66** 609–18
[9] Mudgil D, Barak S and Khatkar B S 2017 *J. Food Sci. Technol.***54** 3810–7
[10] Duboc P and Mollet B 2001 *Int. J. Dairy***11** 75 –68
[11] Lee W J and Lucey J A 2010 *Asian-AustralasianJ. Anim. Sci.***23** 1127–36
[12] Yildiz F 2010 *Overview of Yogurt and Other Fermented Dairy Product. In: Development and Manufacture of Yogurt and Other Fucntional Dairy Product* (New York: CRC Press)
[13] Konstantinos B. Petrotos, Fani K. Karkanta, Paschalis E. Gkoutsidis, Ioannis Giavasis, Konstantinos N. Papatheodorou A C N 2012 *World Acad. Sci. Eng. Technol.***6** 170–5
[14] Borba K K S, Silva F A, Madruga M S, de Cássia Ramos do Egypto Queiroga R, de Souza E L and Magnani M 2014 *Int. J. Food Sci. Technol.***49** 1279–86
[15] Farahat A M and El-Batawy O I 2013 *World J. Dairy Food Sci.***8** 38–44
[16] Akua O 2012 *Formulation and Shelf Life Evaluation of Avocado (Persea Americana) Fruit Spread* (Ghana: Kwame Nkrumah University of Science And Technology) p 4
[17] Duarte P F, Chaves M A, Borges C D and Mendonça C R B 2017 Cienc. Rural **46** 747–54
[18] Čakmakči S, Četin B, Turgut T, Gürses M and Erdoğan A 2012 *Turkish J. Vet. Anim. Sci.***36** 231–7
[19] Li X, Xu C, Korban S S and Chen K 2010 *Crit. Rev. Plant Sci.***29** 222–43
[20] Bustos M C, Mazzobre M F and Buera M P 2015 *LWT - Food Sci. Technol.***61** 89–97
[21] Blakey R J, Tesfay S Z, Bertling I and Bower J P 2014 *J. Hortic. Sci. Biotechnol.***89** 655–62
[22] Tamime A Y and Robinson R K 1999 *Yoghurt: Science and Technology* (Boca Raton, FL: CRC Press)
[23] Purnamayati L 2008 *Kajian substitusi krim dengan daging buah alpukat (Persea americana, Mill) terhadap sifat es krim* (Surakarta: Universitas Sebelas Maret)
[24] Han X, Yang Z, Jing X, Yu P, Zhang Y, Yi H and Zhang L 2016 *Biomed Res. Int.***2016** 1–6
[25] Lawless T H and Heymann H 2010 *Sensory Evaluation of Food: Principles and Practices*(New York: Springer)
[26] Tarakçi Z, Küçüköner E, Yıl Y, Ziraat Ü, Gıda F, Bölümü M and Türkiye V 2003 *YYÜ Vet Fak Derg***14** 10–4
[27] Cais-Sokolinska D, Michalski M and Pikul J 2004 *Bull. Vet. Inst. Pulawy***48** 437–41
[28] Lee W J and Lucey J A 2010 *Asian-Australasian J. Anim. Sci.***23** 1127–36
[29] Setianto Y C, Pramono Y B and Mulyani S 2014 *J. Apl. Teknol. Pangan***3** 110–3
[30] Mahdian E and Tehrani M M 2007 *American-Eurasian J. Agric. & Environ. Sci.*2 587–92
[31] Domagała J, Sady M, Grega T and Bonczar G 2006 *Int. J. Food Prop.***9** 1–11
[32] Balasubramanyam B V and Kulkarni S 1991 *J Food Sci Technol.***28** 389–90
[33] Lazcano C ., Yoo K . and Pike L M 2001 *Sci. Hortic.***90** 321−4
[34] Taoukis P ., Bournakis A and Giannakourou M . 2001 *Shelf life study of bottled tomato ketchup as a function of packaging and storage conditions IFT Annual Meeting*(New Orleans, Louisiana) pp 7–9
[35] Chandan R . and Shahani K . 1993 *Yogurt ed Hui YH* (New York: VCH Publishers)
[36] Güler-Akin M B and Akin M S 2007 *Food Chem.***100** 788–93
[37] Senaka Ranadheera C, Evans C A, Adams M C and Baines S K *Food Chem.* **135** 1411–8
[38] Gad A S, Kholif A M and Sayed A F 2010 *Am. J. Food Technol.***5** 250–9
[39] Modhu A L 2016 *Development of good quality of Yogurt in Terms of Texture, Appearance, Color, Taste and Determination of Fat Percentage in Milk and Yogurt*(Dhaka: Brac University)

IOP Conf. Series: Materials Science and Engineering **633** (2019) 012050 doi:10.1088/1757-899X/633/1/012050

Textural and sensory properties of little tuna fish balls (*Euthynnus affinis*) arrowroot flour substitutions (*Maranta arundinacea* Linn.) added with sodium tripolyphosphate

R Y Kurniasari, D R Affandi, B Yudhistira and A P Sanjaya

Department of Food Science and Technology, Faculty of Agriculture, Universitas Sebelas Maret, Surakarta, Indonesia

Email: adhit.sanjaya@staff.uns.ac.id

Abstract. Physical properties (water holding capacity, cohesiveness, chewiness, springiness) and sensory characteristics (cohesiveness, chewiness, springiness and overall) of little tuna fish balls that substituted with arrowroot flour and added with sodium tripolyphosphate (STPP) were investigated. Substituted little tuna fish ball was added with 0%; 0.1%; 0.2%; 0.3% of sodium tripolyphosphate. Results of this study showed that the addition of sodium tripolyphosphate increased water holding capacity, cohesiveness, chewiness, and springiness parameters with respect to better textural properties of substituted little tuna fish balls ($p<0.05$). Sensory evaluation test showed that overall, substituted little tuna fish balls with 0.1% addition of STPP were the most favored by panelists. Water holding capacity, cohesiveness, chewiness, and springiness parameters of substituted little tuna fish balls added with 0.1% of STPP were increase up to 4.7%, 9.1%, 8.6%, and 2.0%, respectively. The multiple comparison tests showed that color, odor, flavor parameter of substituted little tuna fish balls added with sodium tripolyphosphate were relatively the same level.

1. Introduction

Little tuna fish is one of the marines catches with a high production amount in Indonesia. Little tuna has a high protein content of 26.2% per 100 g and rich in omega-3 fatty acids [1]. However, the utilization and food processing of little tuna as the main ingredient in food products is still limited. Little tuna fish are usually marketed in the local fish market and most people prefer to consume little tuna fish in the original form, such as fried fish, steamed fish, etc. Little tuna fish has good potential as a major ingredient in a food product and it can diversify little tuna fish into various processed products such as fish balls, nuggets, and shredded fish to increase the consumption number of little tuna. In this study, little tuna fish was used as the main ingredient for fish balls production. Presence study reported that arrowroot flour can be used as a filler for fish balls and was used to substitute tapioca flour because arrowroot flour has a high starch content of 98.10% with the carbohydrate content of 85.2% [2]. Dietary fiber in food product could affect health and able to cure many chronic diseases. Harmayani et al [3] reported that arrowroot flour contained 13.7% of dietary fiber (8.7% insoluble dietary fiber and 5.0% soluble dietary fiber). The good quality parameter of fish balls is influenced by appearance, textural properties and taste, and nutritional value [4]. Surimi has advantages, which can be processed into various products such as *kamaboko*, fish balls, and nuggets [5]. Surimi has been used as the main material for little tuna fish balls production and aims to provide good results for its textural characteristic. The utilization of surimi as a major ingredient in the fish product, e.g. fish balls, can provide benefits that are a good gel formation [6].

Surimi is minced fish meat or fish paste that has been washed from sarcoplasmic proteins, lipids, and other fish waste so that it is called a wet concentrate protein [5]. In this study also used sodium tripolyphosphate (STPP) as a gelling agent. STPP is commonly used in the restructured meat product. STPP has the ability to increase the water holding capacity. Glorieux et al state that STPP affects ionic strength by forming polyelectrolytes in water, causing electrostatic repulsion between the meat proteins, which allows more space for binding water and hence, increased WHC [7]. On the other hand, the maximum amount of the application of STPP in a food product which was described by Codex Alimentarius Abriged Version is 0.3% of the weight of used meat in processed meat product [8]. Due to the textural characteristic of surimi and gelling formation by STPP in the processed meat product, the textural and sensory characteristics of little tuna fish balls with arrowroot flour substitution with STPP addition were investigated to study consumer acceptance to little tuna fish balls product.

2. Materials and Methods

2.1. Materials

Little tuna was used as the main ingredient for the production process of little tuna fish balls. Little tuna was obtained from the local fish market, Solo (Indonesia). Arrowroot flour as supporting material was obtained from CV. Kusuka Ubiku, Yogyakarta (Indonesia). STPP food grade was kindly provided by a local shop in Surakarta (Indonesia). The other supporting ingredients such as tapioca flour, salt, sugar, egg, and pepper were purchased from Pasar Gede Solo (Indonesia).

2.2. Surimi production

The little tuna surimi processing method was previously described by the Center for Testing the Application of Fisheries [9]. The little tuna fish were cleaned and filleted to obtain only the white meat. The fish was minced and added with ice block to keep the temperature of the minced fish not more than 22°C. The minced fish were mixed, stirred and washed twice with cold water (at 5-10°C) with fish meat and water ratio was 1:4 (v/w). At the second washing step, 0.3 % of salt was added into the mixture and then stirred. After that, the washed minced fish meat was pressed with a hand press technique in order to reduce the remaining water. The minced fish meat was added with 2% of sugar. After that, the surimi was stored at -20°C until fish balls processing time.

2.3. Fish ball production

The little tuna fish balls were prepared from surimi and other ingredients (garlic, salt, sugar, and pepper), eggs, tapioca flour, and arrowroot flour. Surimi and other ingredients were weighed according to the specified composition before processing carried out and then, these mixtures were mixed the fish balls dough until well mixed and form a thick dough. STPP were added into the fish balls dough and shaped manually into balls with a diameter of the fish balls up to 2 cm. The uncooked fish balls were boiled in boiling water at a temperature of 100°C. In this study, the arrowroot flour substitutions ratio was 22.5% of total surimi weigh. The concentration of STPP which was added to the fish balls dough was 0%; 0.1%; 0.2% and 0.3%. The percentage of STPP concentration on the fish balls dough was obtained from a preliminary study.

2.4. Analytical methods of little tuna fish balls

The physical, chemical and sensory properties of little tuna fish balls were determined. Textural properties (cohesiveness, chewiness, and springiness) were performed using the Texture Profile Analyzer [10] and water holding capacity (WHC) method described by Muchtadi [11]. Chemical properties i.e. water content was performed using the thermogravmetry method and ash content was analyzed using oven method [12]. All experiments were performed in triplicate. Determination of sensory evaluation was investigated using scoring and multiple comparison tests [13,14]. In this study, twenty five semi-trained sensory panels have been used to evaluate the sensory characteristics of little tuna fish balls. The panelists should be passed a preliminary screening program, and then the panelist was selected and trained to discriminate differences and communicate their reactions on sensory characteristics of little tuna fish balls.

3. Results and Discussion

3.1. Chemical and Physical Properties of Fish balls

Little tuna surimi as the raw material was investigated before being processed into fish balls. Little tuna surimi contains 73.75% water content, the ash content of 1.02% and 23.27% protein content. The results of the cemical properties of little tuna fish balls showed in **Table 1**. The highest water content and ash content of little tuna fish balls were achieved by the addition of 0.3% STPP, which was 66,60%. At higher STPP concentration, the water content of little tuna fish balls was an increase. Effect of STPP addition in little tuna fish balls could increase the ability of proteins to absorb the water. Phosphate groups in STPP could bond with NaCl increased the solubility of the protein. This fact was in line with Aulawi and Retty [15] that the addition of STPP can increase the water content and water holding capacity of meat balls. The water content of little tuna fish balls with the addition of STPP has been in accordance with the water content in fish balls product which reported in SNI 01-3819-1995, which was a maximum of 70% (wet basis). The ash content of fish balls shows that the increasing concentration of the STPP addition, the ash content of fish balls tends to increase. Dewanti said this because STPP can bind nutrients which were dissolved in salt solutions such as vitamins, minerals, and proteins so that it can increase the ash content of fish balls [16].

Table 1 Chemical and physical properties of little tuna fish balls substituted with arrowroot flour with the addition of STPP

STPP (%)	Water Content (%)	Ash (%)	WHC (%)	Cohesiveness (%)	Chewiness (%)	Springiness (%)
0.0	63,53^{a}	2,59^{a}	61,74^{a}	0,44^{a}	4197,47^{a}	8,51^{a}
0.1	64,20^{a}	2,66^{b}	64,66^{b}	0,48^{b}	4556,49^{ab}	8,68^{ab}
0.2	65,43^{b}	2,77^{c}	68,63^{c}	0,48^{b}	5128,79^{ab}	8,76^{ab}
0.3	66,60^{c}	2,88^{d}	72,50^{d}	0,49^{b}	5836,66^{b}	8,83^{b}

*Notation different letters in the same column indicate significant difference at a significance level of 5%

The textural properties showed in **Table 1**. The water holding capacity (WHC) of little tuna fish balls was increased significantly different ($p<0.05$) by the addition of STPP. Water holding capacity of little tuna fish balls will increase as the concentration of STPP increases. Long et al [17] reported that STPP is one of phosphate salt which could improve the water holding capacity, stabilize pH and reduce weight loss when meat products were cooked. Water holding capacity is the ability of protein gels to maintain water. Water holding capacity was strongly influenced by water content, protein and salt use. The higher protein content also could increase the WHC value [18]. Salt gives a large effect on ionic strength and it is able to extract myosin from the structure of meat myofibril. Salt can increase the swelling of the protein structure but does not dissolve protein much [19]. Cohesiveness level of little tuna fish balls with STPP addition is higher than that of the little tuna fish balls without STPP addition. In this study, the addition of STPP could increase significantly ($p<0.05$) the cohesiveness of little tuna fish balls with STPP addition. Increasing of STPP concentration addition There was no significant difference ($p>0.05$) cohesiveness value of little tuna fish balls with different concentration of STPP addition. Yuanita reported that the increasing of STPP concentration on tilapia fish balls can increase cohesiveness properties of tilapia fish balls [20]. In addition, the compact texture was also influenced by the type of flour which was used as a filler, especially those starch content [21].

The previous study reported that STPP which was added as a gelling agent could increase the hardness and cohesiveness properties of meatball [22]. The higher the STPP concentration which was added to little tuna fish balls affected and increased the water holding capacity, hence little tuna fish balls have a chewy texture and more stable [23]. Bonded water in fish balls structure affects the texture of fish balls to become chewy and was not easily destroyed [23]. Elasticity properties of little tuna fish balls were increased with the addition of STPP. STPP has the ability to extract meat protein which could give affect to increase the water holding capacity. Increasing the water holding capacity is caused by the binding power of the water. The elasticity of the fish ball produced was related to the strength of the gel formed during the heating process [24].

An increasing the water content could affect springiness, cohesiveness and hardness properties of little tuna fish balls [25]. Pandisurya reported that the elasticity properties of the fish ball are related to the ability of starch molecules to form gels or elastic three-dimensional networks [26]. The texture of little tuna fish balls with the addition of STPP was also influenced by the use of surimi as raw materials. At the second washing process on surimi processing could eliminate sarcoplasmic protein which can inhibit gel formation and dissolve myofibril protein to form sol actomyosin so it could increase the strength of the gel.

3.2 Sensory Properties of Fish ball

Little tuna fish balls were assessed for sensory attributes using an acceptance test (**Figure 1**) and the differentiation test after production (**Figure 2**). Based on the acceptance test for cohesiveness, chewiness and springiness parameters, panelists most like little tuna fish balls with the addition of 0.3% STPP. The results of the acceptance test are consistent with the results of the STPP addition of the tuna fish balls test found in **Figure 2** which was presented that fish balls with the addition of 0.3% STPP have a stronger cohesiveness than little tuna fish balls without the addition of arrowroot flour and STPP (reference).

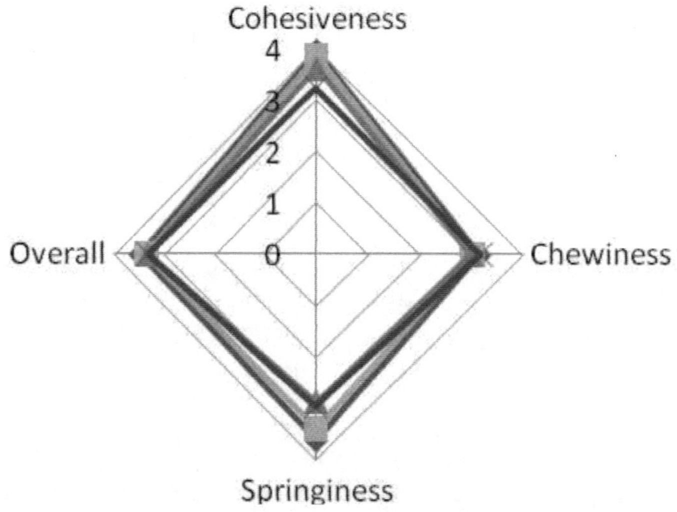

Figure 1. Sensory properties (scoring method) of little tuna fish balls with 0% (♦), 0.1% (■), 0.2% (▲), or 0.3% (●) STPP concentration.

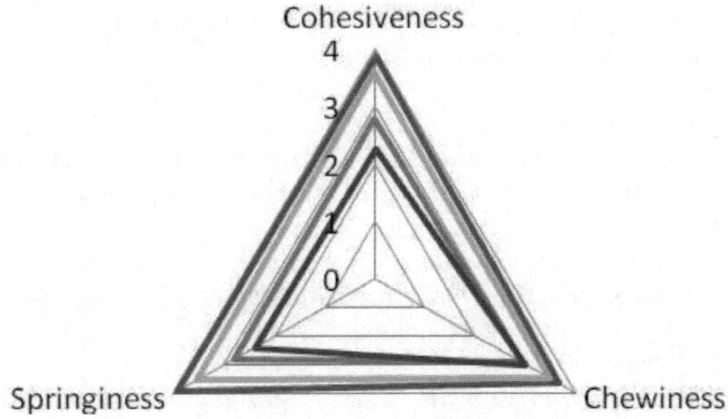

Figure 2. Intensity differentiation's score of sensory properties of little tuna fish balls arrowroot flour substitutions with with 0% (♦), 0.1% (■), 0.2% (▲), or 0.3% (●) of STPP addition compared to little tuna fish balls without addition of arrowroot flour and STPP (as reference)

Based on the results of the acceptance test and differentiation test, the result showed that the addition of STPP, the cohesiveness of little tuna fish balls was increased. STPP addition could increase texture due to the incorporation of polymer chains, so as to form a continuous mesh that could catch water in it and form a strong, compact and rigid structure [20]. According to [27], the springiness of little tuna fish balls was influenced by water holding capacity. A high water holding capacity indicates that the protein contained in a product was also high, so fish ball becomes more elastic [27]. The overall score of little tuna fish balls with the addition of 0%; 0.1%; 0.2% and 0.3% STPP showed in **Figure 1**. In general, these results showed the good acceptance of the little tuna fish balls. Overall, little tuna fish balls with 0.1% addition were favored by panelists.

4. Conclusions

Water content, ash content, water holding capacity, and texture (cohesiveness, chewiness, springiness) of little tuna fish balls arrowroot substitution were increased with the addition of STPP. Based on the acceptance test and differentiation test of litlle tuna fishball substitution with arrowroot flour and of 0.1% addition of STPP was most preferred by panelists.

Acknowledgment

The authors wish to thank PNBP UNS project number 543/UN27.21/PP/2018 for providing financial support for this work.

References

[1] Sanger G 2010 *Warta WIPTEK* **35** 1-2
[2] HakimU N, Djalal R, and Aris S W 2013 *Jurnal Ilmu dan Teknologi Hasil Ternak* **8** 9-22
[3] Harmayani E, Ika D K and Marsono 2011 *International Research Journal of Microbiology (IRJM)* **2** 278-284
[4] Uju N R and Ibrahim B 2004 *BuletinTeknologi Hasil Perikanan* **8** 1-10
[5] Okada M 1992 *History of Surimi Technology in Japan* (New York: Marcel Dekker)
[6] Oktaviani, D 2012 *Karakteristik Fisika Kimia Gel dan Bakso dari Surimi Ikan Layaran (Istiophorus Sp.) Frekuensi Pencucian Dua Kali* (Bogor: Skripsi)
[7] Glorieux S, Olivier G, Liselot S and Ilse F 2017 *Technol Biotechnol* **55**
[8] Codex Alimentarius Abridged Version 1990 *Food Standarts Programs. Food Aditive no. Codex 452 a.* FAO

[9] Balai Besar Pengembangan dan Pengendalian Hasil Perikanan 2006 *Teknologi Pengolahan Surimi dan Produk Fish Jelly*. (Jakarta: BBPPHP)
[10] Bourne M C 2002 *Food Texture and Viscosity: Concept and Measurement* (London. Elsevier Science and Technology Books)
[11] Muchtadi T R 1988 *Teknologi Pemasakon Ekstrusi* (Bogor: PAU Pangan dan Gizi Press)
[12] AOAC 2002 *Official Methods of Analysis of Association of Official Analytical Chemist* (Washington DC)
[13] Stone H and Joel L S 2004 Sensory Evaluation Practices (Elsevier Academic Press)
[14] Kemp S E, Tracey H and Joanne H 2009 *Sensory Evaluation a Practical Handbook* (Singapore: Macmillan Publishing Solutions Singapore)
[15] Aulawi T dan Retty Ninsix 2009 *Jurnal Peternakan* **6** 44-52
[16] Dewanti T 2009 *Sodium TripoliPhosfat (STPP) Sebagai Pengganti Garam Bleng pada Krupuk Puli* (Malang: Skripsi)
[17] Long N H and Gál R Buňka 2011 African *Journal of Biotechnology* **10** 19874-19882
[18] Kramlich J E 1971 *Sausage Product Technology 2nd edition* (San Fransisco: W H Freeman Company)
[19] Shu Q X U, Guang H Z, Zeng Q P, Li Yan Z and Rui Y A O 2009 *Journal of Muscle Foods* **20** 101-116
[20] Yuanita L 2008 *Berk Penel Hayati* **13** 179-186
[21] Triatmojo S 1992 *Buletin Peternakan* **16** 63-71
[22] Sunarlim R 1997 *Seminar Nasional Peternakan dan Veteriner*
[23] Keton J T 2001 *Formed and Emulsion Product* (Boca Raton: CRC Press)
[24] Indarmono T P 1987. *Pengaruh Lamanya Pelayuan dan Jenis Daging Karkas serta Jumlah Es yang Ditambahkan ke dalam Adonan terhadap Sifat Fisik Kimia Bakso Sapi* (Bogor: Skripsi)
[25] Indiarto R, Bambang N and Edy S 2012 *Jurnal Teknologi Hasil Pertanian* **5** 106-114
[26] Pandisurya C 1983 *Pengaruh Jenis Daging dan Penaruhan Tepung Terhadap Mutu Bakso* (Bogor: Skripsi)
[27] Poernomo D, Sugeng H S and Bayu P S 2013 *Jurnal Pengolahan Hasil Perikanan Indonesia* **16** 1

IOP Conf. Series: Materials Science and Engineering **633** (2019) 012051 doi:10.1088/1757-899X/633/1/012051

Chemical, textural, and sensory properties of eastern little tuna fish ball (*Euthynnus affinis*) with rice bran flour (*Oryza sativa*) substitution

D R Affandi, E Purnama, B Yudhistira and A P Sanjaya

Department of Food Science and Technology, Faculty of Agriculture, Sebelas Maret University, Surakarta, Indonesia

Email: adhit.sanjaya@staff.uns.ac.id

Abstract. The aim of this study was to investigated the characteristic of chemical, physical, functional, and sensory properties of the fish balls with substitution rice bran flour. Substitution of rice bran flour and tapioca as fish ball fillers used in this research were 25:75, 40:60, 55:45, and 100:0 as a control. The result showed that the substitution of rice bran affected the chemical, functional, physical and sensory characteristics of eastern little tuna fish balls. The substitution of tapioca flour with rice bran flour as fillers in eastern little tuna fish ball could increase moisture, ash, fat, and protein content. In this study, antioxidant activity, vitamin E, and WHC of eastern little tuna fishballs were also increased. Texture properties (cohesiveness, chewiness, springiness) of eastern little tuna fish ball affected by the substitution of rice bran flour. Substitution of 25% rice bran and 75% tapioca flour as fillers in eastern little tuna fish ball was more acceptable than another formula. The moisture, ash, fat, protein, and total starch content of the eastern little tuna fish ball substituted 25% rice bran and 75% tapioca flour were 68,07%; 6,53%; 1,47%; 38,12%; and 41,51%, respectively. Sensory analysis of the eastern little tuna fish ball substituted 25% rice bran and 75% tapioca flour formula showed that it had a brownish color, not-strong-fishy aroma, a slightly weaker fishy flavor, a rather strong rice bran flavor, and slightly weak texture.

1. Introduction

Eastern little tuna (*Euthynnus affinis*) has a high-productivity sea catch and also has a high market value of fish. Ministry of Marine Affairs and Fisheries Indonesia in 2015 reported that an average production of 552.410 tons/year. This a huge number of production was not in line with the number of consumption of Eastern little tuna which still limited. In daily consumption, eastern little tuna usually is consumed as a protein source food. According to Intarasirisawat et al [1], the nutritional composition in eastern little tuna was 73.03% of moisture content, 18.16% of protein content, 4.26% of fat content, 1.79% of ash content, and 2.76% of carbohydrate. Hernandez et al [2] also mentioned that eastern little tuna contains omega 3 (1.5 g / 100 g) and omega 6 (1.8 g / 100 g) which can prevent heart disease, stroke or high blood pressure. Accordingly, eastern little tuna has good potential as a major ingredient in a food product.

There are several processed fish products that can be produced to increase the consumption number of eastern little tuna, such as nuggets, fish balls, shredded fish, and fish chips. In this study, eastern little tuna was prepared as the main ingredient to produce fish balls product. Fish balls were can be consumed by a varied age range. Holley and Patel [3] described that fish ball is an emulsion of fish meat that is mashed and then added flour as filler, salt, and spices. According to Huda [4], fish balls have a brighter color than beef meatballs, fish balls tend to have a white color and elastic properties. The elasticity of fish ball is one of the parameters which are used to determine the textural quality of fish balls. The elasticity of fish balls can be obtained by the addition of texture-forming food additives and fillers. In general, tapioca was used as a filler in fish ball production. In this study, the filler component of the eastern little tuna fish ball was substituted with rice bran flour in order to provide functional value that has a potentially positive effect on health. Rice bran has a higher protein content and antioxidant activity compare to that of tapioca. Moongngarm et al. [5] reported that rice bran contains 13.66% of protein content, 40.63% of carbohydrate content, 18.80% of fat content, and high antioxidant activity of tocopherol as 40.94 mg/g. Due to the potential health of the rice bran flour, eastern little tuna fish balls were prepared using rice bran flour as filler in order to substituted tapioca. The nutritional component, functional characteristics and physical properties of the eastern little tuna fish balls which substituted with rice bran flour were investigated.

2. Materials and Methods

2.1. Materials

Eastern little tuna was obtained from the local fish market, Surakarta (Indonesia). Rice bran flour as filler material was obtained from CV. Kusuka Ubiku, Yogyakarta (Indonesia). Rice bran flour was sifted using an 80 mesh sieved before used and then roasted for 15 minutes. The roasting process of rice bran was aimed to reduce the bad odor of rice bran flour. Chitosan and sodium tripolyphosphate (STPP) food grade was kindly provided by a local shop in Surakarta (Indonesia). The other ingredients such as salt, tapioca flour, sugar, egg, and pepper were purchased from Pasar Gede Solo (Indonesia).

2.2. Preparation of Eastern Little Tuna Fish Balls

Fish balls were produced according to the following traditional recipe. In this study, fresh eastern little tuna fish was prepared as the main ingredient on fish balls production. The first step, the internal organs and another part of fish by product of fresh eastern little tuna fish was removed and separated from the meat part. After that, eastern little tuna meat was washed and then fish meat was minced at cold temperature. Ice cube was added during the mincing process in order to maintain the temperature at 4°C. The minced fish was weighted and added STPP, chitosan, spices (salt, sugar, pepper, garlic) and flour. The fish meat and other ingredients were kneaded for 30 min, by hand, to obtain a homogeneous dough. The fish balls dough was shaped into ball form with 2 cm diameter. Fish balls dough was cooked for 15 minutes at 100 °C. In this study, the fish balls formula was prepared in four ratios (w/w) of tapioca : rice bran flour which was F0 (100: 0) as a control, F1 (75:25), F2 (60:40), F3 (45:55). A preliminary study was done in order to determine the basic formula of the fish balls dough and then to obtain ratios of tapioca : rice bran flour (w/w) on the fish balls dough. All experiments were performed in triplicate.

2.3. Analytical Methods

The moisture and ash content were determined according to the methods no. 971.28 and no. 940.26 of AOAC by drying the samples using a convection drying oven at 105°C to achieve a constant weight, and burning the samples at 575°C [6]. Fat content was analyzed using soxhlet extraction method and protein content was determined according to the Micro-Kjedahl method [6]. Antioxidant activity was determined using DPPH method, described by Gow-Chin [7]. Water holding capacity (WHC) analysis was performed using a centrifuge to remove the water content in the sample [8]. Texture profile analysis was determine according to the method described by Bourne [9] using Analyzer merk Brookfield model LFRA 4500. Determination of sensory evaluation was investigated using scoring [10,11]. In this study, thirty untrained sensory panels have been used to evaluate the aceptance test of little tuna fish balls product. The selected of the untrained panel has an age range between 19 and 21 years and has the interest to consume the fish product. Consumer evaluation was performed according to an hedonic scoring from 1 (like very much) to 7 (dislike very much) for cohesiveness, chewiness, springiness, and overall parameters. Samples were caried out under similar conditions (temperature and quantity) for all panelists. Analysis of Variance (ANOVA) that followed by the Duncan Multiple Range Test (DMRT) with a significance level of 0.05 (P<0.05) was used to analyze the significance of the means differences of each sample. This test was performed using the Statistical Package for the Social Sciences version 19 (SPSS).

3. Results and Discussion

3.1. Chemical and Functional Properties of Eastern Little Tuna Fish balls

Eastern little tuna fish balls were analyzed chemical composition in order to determine the chemical characteristic of each rice bran substitution treatment. Chemical properties of eastern little tuna fish balls are shown in **Table 1**. The moisture content of eastern little tuna fish balls increased significantly different (P<0.05) due to the difference concentration of rice bran substitution. The highest moisture content of eastern little tuna fish balls was achieved by eastern little tuna fish balls with 55% rice bran substitution. The corresponding value was 69.65%. The moisture content of eastern little tuna fish balls can be affected by the chemical composition of eastern little tuna, rice bran flour, tapioca flour, and other ingredients. The increasing of eastern little tuna fish balls moisture content was caused by the increasing WHC of eastern little tuna fish balls (**Table 2**). The moisture content of eastern little tuna fish balls can be affected by starch gelatinization which is content in tapioca and rice bran flour. During the boiling process, the water is gradually absorbed in the amorphous space of starch, which leads to a swelling phenomenon [12]. Standard Nation of Indonesia (SNI) 01-3819-1995 reported that the water content level of fish balls has maximal 80%. Previous study reported that the moisture content of fish ball which was made from alburnus fish was 71.3% [13]. Poernomo et.al. [14] also reported that fish ball from the indo-Pacific sailfish has 79.11% of moisture content.

Table 1. Chemical and Functional Properties of Eastern Little Tuna Fish balls

Rice Bran Substitution	Moisture (%wb)	Ash (%db)	Fat (%db)	Protein (%db)	Antioxidant (%)	Vitamin E (mg/100g)
0% (control)	68.31^{a}	5.70^{a}	0.67^{a}	38.98^{a}	2.37^{a}	0.85^{a}
25%	68.07^{a}	6.53^{b}	1.47^{b}	38.12^{a}	10.88^{b}	0.91^{a}
40%	68.26^{a}	7.13^{c}	1.57^{c}	41.37^{b}	15.51^{c}	1.06^{b}
55%	69.65^{b}	8.52^{d}	2.02^{d}	47.43^{c}	16.45^{d}	1.15^{c}

*Different notations in the same column show a significant difference at $\alpha = 0.05$.
Formula of rice bran substitution based on the percentage of the total fish portion in fish balls dough formula. Control was defined as 100% tapioca flour used in fish balls dough formula.

In **Table 1**, the ash content of eastern little tuna fish balls increased significantly different (P<0.05). In this study, the ash content of rice bran flour used was higher than that of tapioca ash which were 8.01% and 0.13%, respectively (data not shown). The substitution of rice bran could increase the ash content of eastern little tuna fish balls. The highest of ash content of eastern little tuna fish balls was achieved by the 55% rice bran substitution which was 8.52%. The 55% rice bran substitution eastern little tuna fish balls were significantly higher than that of the other rice bran substitution treatment. Bhosale [15] reported that the mineral content contained in rice bran as calcium of 52.10 mg, phosphor 1185.20 mg, zinc at 6.02 mg, and iron of 28.10 mg. In this study, the ash content of eastern little tuna fish balls with rice bran substitution was significantly higher (P<0.05) than that of the control which was made from 100% tapioca as filler. Arifin [16] mentioned that the ash content of threadfin bream fish balls made with various flour has various ash content depending on the type of flour. The ash content value of the threadfin bream fish balls which made from wheat flour, banana flour, or ubi badak flour were 2.19%, 2.78%, and 2.45%, respectively.

Fat content of eastern little tuna fish balls increased significantly different (P<0.05). The highest fat content was obtained by the 55% rice bran substitution which was 2.02% (**Table 1**). The fat content of eastern little tuna was influenced by the composition of the fish balls. Moongngram et al. [5] rice bran flour have a high-fat content which was 2.77%. The rice bran substitution resulted in an increasing of

IOP Conf. Series: Materials Science and Engineering **633** (2019) 012051 doi:10.1088/1757-899X/633/1/012051

the fat content of eastern little tuna fish balls. The rice bran substitution of eastern little tuna fish balls also showed an increasing protein content. The highest protein content was achieved by the 55% rice bran substitution, which was 47.43%. The rice bran substitution could increase significantly the protein content of eastern little tuna fish balls.

Antioxidants in food can be defined as a substance that has the capability to prevent the oxidation process in foods. Antioxidants can prevent the formation of free radicals in food that that can impair the taste of food [17]. The antioxidant activity of eastern little tuna fish balls with rice bran substitution was presented in **Table 1**. The highest antioxidant activity of eastern little tuna fish balls with rice bran substitution was obtained by the 55% rice bran substitution and the lowest antioxidant activity was achieved by control. The increasing of antioxidant activity of eastern little tuna fish balls followed by the amount of rice bran substitution on eastern little tuna fish balls. Vitamin E is one of the powerful antioxidants preventing free radicals [18]. Rice bran contains a high level of vitamin E (tocopherol). The result showed that vitamin E of eastern little tuna fish balls was increased significantly ($P<0.05$). As an antioxidant in rice bran, the increasing of vitamin E level in eastern little tuna fish balls also followed by the amount of rice bran substitution on eastern little tuna fish balls. The highest vitamin E was achieved by the 55% rice bran substitution.

3.2. Physical Characteristic of Eastern Little Tuna Fish balls

Eastern little tuna fish balls were analyzed water holding capacity (WHC), cohesiveness, chewiness, and springiness in order to determine the physical characteristic of each rice bran substitution treatment. Physical characteristic of eastern little tuna fish balls is presented in **Table 2**. Water holding capacity is one of a physical parameter of food that describes as an interaction of the protein with water in the food [19]. Jideani [20] mentioned that WHC is influenced by several factors namely, amino acid profile, pH, temperature, ionic bond, a hydrophobic group, and protein concentration. Water binding strength increases with the addition of rice bran composition. The highest WHC of eastern little tuna fish balls was obtained by the 55% rice bran substitution. The corresponding value was 8.53%. The eastern little tuna fish balls with 55% rice bran substitution also had the highest protein content, which was 47.43%. In this result, the increasing of WHC properties of eastern little tuna fish balls was caused by the protein content of eastern little tuna fish balls.

Table 2. Physical Characteristic of Eastern Little Tuna Fish balls

Rice Bran Substitution	WHC (%)	Cohesiveness	Chewiness (gf)	Springiness (mm)
0% (control)	6.08^a	0.61^b	19338.27^b	7.96^b
25%	6.04^a	0.54^a	15904.81^a	7.56^a
40%	6.46^a	0.54^a	15416.21^a	7.54^a
55%	8.53^b	0.54^a	15374.19^a	7.52^a

*Different notations in the same column show a significant difference at $\alpha = 0.05$.
Formula of rice bran substitution based on the percentage of total fish meat portion in fish balls dough formula.
Control was defined as 100% tapioca flour used in fish balls dough formula.

Texture Profile Analysis (TPA) is used to determine the bite size and shape of the food samples placed on the baseplate and pressed twice with the platter. This method was also used to investigate the bite force, i.e. the action of biting and chewing by the teeth, preferably given high pressure [9]. In general, the texture profile analysis of the fish balls is influenced by the elasticity (springiness), chewiness (chewiness), and cohesiveness (cohesiveness). Texture profile analysis of eastern little tuna fish balls with rice bran substitution presented in **Table 2**. Cohesiveness profile was defined as the ability of fish balls can steady after it is being given a force or pressure by the teeth. Fish balls which can steady after pressure giving process mean that cohesiveness level of fish balls is greater [21]. The increasing level of bran flour substitution leads to decrease the value of cohesiveness. According to

Serdaroglu et al. [22], the gelatinization starch content influences the cohesiveness of fish balls. The total starch content of rice bran flour and tapioca flours are 47.21% and 83.97%, respectively (data not shown). The cohesiveness value of eastern little tuna fish balls with rice bran substitution was higher than that of eastern little tuna fish balls without rice bran substitution (control). This result could be related to the total starch content that was higher in the tapioca than that of the rice bran resulted in the cohesiveness value of tapioca was higher than that of rice bran. Chewiness is described as the energy that is needed to chew fish balls into small sizes [21]. The increasing amount of rice bran substitution on eastern little tuna fish balls caused a decreasing of the chewiness value of fish balls. The higher the starch content in fish balls, the chewiness is greater. The higher the chewiness value was indicated a good texture quality of fish balls. Szczesniak [21] defined springiness as the ability of the fish balls to be able to return to their original shape after the first emphasis until the second emphasis will begin. According to Serdaroglu et al. [22], springiness is influenced by the amount and type of starch used for the produce of fish balls. In **Table 2**, the increasing amount of rice bran substitution could affect to a decreasing of springiness profile. Springiness is influenced by the amount and type of starch used for the fish balls production.

3.3. Sensory Properties of Eastern Little Tuna Fish balls

Sensory properties of eastern little tuna fish balls were determined using an acceptance test (**Figure 1**). In this study, acceptance test of eastern little tuna fish balls was performed using a scoring method. The score in **Figure 1** presented the panelist acceptance and preference level of eastern little tuna fish balls with rice bran substitution compare to control. The scale consists of seven labeled categories which range from 'dislike very much' to 'like very much, and have a neutral category in the middle which the score levels are 1= like very much; 2= moderately like much ; 3= slightly like ; 4= slightly like; 5= slightly dislike; 6= moderately dislike; 7= dislike very much. The acceptance test of cohesiveness, chewiness, springiness, and overall parameters showed that panelists most like eastern little tuna fish balls with the 25% rice bran substitution (**Figure 1**). Sensory analysis of the eastern little tuna fish ball substituted 25% rice bran and 75% tapioca flour formula showed that it had a brownish color, not-strong-fishy aroma, a slightly weaker fishy flavor, a rather strong rice bran flavor, and slightly weak texture.

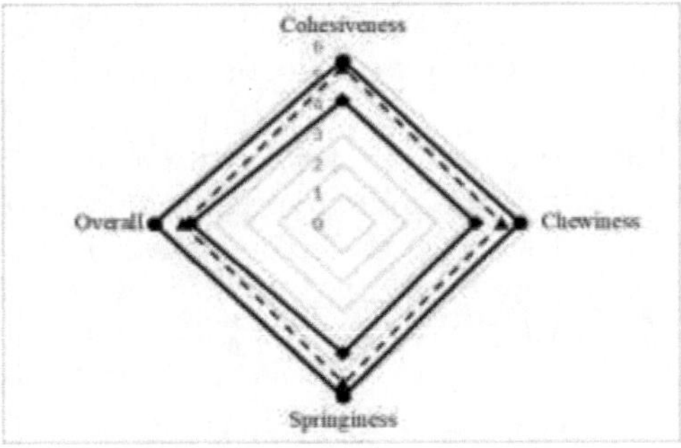

Figure 1. Sensory properties of eastern tuna fish balls with 25% (■), 40% (▲), or 55% (●) rice bran substitution.

3.4. Determination the Best Formula of Eastern Little Tuna Fish Balls

Best formula of rice bran substitution on eastern little tuna fish balls was determined according to the method described by De Garmo and Sullivan [23] using score compensatory method. The method was carried out to determine the score from each parameter on each sample, and then the samples which have the highest score would be chosen as the best formula of rice bran substitution on eastern little tuna fish balls. The eastern tuna fish balls with 25%, 40%, or 55% rice bran substitution had a total score as 0.526, 0.428, and 0.421, respectively. The best formula had been determined, i.e. eastern little tuna fish balls with the 25% rice bran substitution.

4. Conclusions

The chemical composition (moisture, ash, fat, and protein content) of eastern little tuna fish balls were increased by the substitution of tapioca flour with rice bran flour. In this study, antioxidant activity, vitamin E, and WHC of eastern little tuna fish balls were also increased. Texture properties (cohesiveness, chewiness, springiness) of eastern little tuna fish ball affected by the substitution of rice bran flour. Due to the sensory characteristics of the eastern little tuna fish balls with rice bran substitution, the panelists provide high values for eastern little tuna fish balls with 25% rice bran substitution.

References

[1] Intarasirisawat R, Benjakul S and Visessanguan W 2011 *Food Chem.* **124** 1328-1334
[2] Chaparro-Hernández S Ruíz-Cruz S, Márquez-Ríos E Ocaño-Higuera V M Valenzuela-López C C Ornelas-Paz J DE J and Del-Toro-Sánchez C L 2015 *Food Sci. Technol.* **35** (4) 734-741
[3] Holley R A and Patel D 2005 *Food Microbiol.* **22** (4) 273-292
[4] Huda N, Shen Y H, Huey Y L and Dewi R S 2010 *Pakistan J. Nutr.* **9** (12) 1183-1186
[5] Moongngarm A, Daomukda N and Khumpika S 2012 *APCBEE Procedia* **2** 73-79
[6] AOAC 2005 *Official Methods of Analysis of AOAC International* (Washington: Benjamin Franklin Station)
[7] Yen G-C and Chen H-Y 1995 *J. Agric. Food Chem.* **43** 383-386
[8] Lin K W and Huang H Y 2003 *Meat Sci.* **65** (2) 749-755
[9] Bourne M 2002 *Food texture and viscosity: Concept and Measurement* (Bath: Elsivier Science and Technology Books)
[10] Singh-Ackbarali D and Maharaj R 2014 *J. Curric. Teach.* **3** (1) 10-27
[11] Lawless H T and Heymann H 2010 *Sensory Evaluation of Food Principles and Practices* (New York: Springer)
[12] Olkku J and Rha C K 1978 *Food Chem.* **3** (4) 293-317
[13] Duman M and Peksezer B 2016 *J. Appl. Ichthyol.* **33** (3) 285-290.
[14] Poernomo D, Suseno S H and Subekti B P 2013 *Karakteristik Fisika Kimia Bakso dari Daging Lumat Ikan Layaran (Istiophorus orientalis)* (Bogor: Dep. Teknol. Has. Perairan, Fak. Perikan. dan Ilmu Kelautan, IPB)
[15] Bhosale S and Vijayalakshmi D 2015 *Curr. Res. Nutr. Food Sci.* **3** 74–80
[16] Arifin B, Zulkifli N and Zainol Z 2013 *Journal of Food Sciences and Engineering* **3** 131–40
[17] Pokorný J Yanishlieva N and Gordon M 2001 *Antioxidants in food. practical applications* (Cambridge: Woodhead Publishing Limited)
[18] Eitenmiller R R and Landen W O 2010 *Vitamin Analysis for the Health and Food Sciences* (New York: CRC Press)
[19] Traynham T L Myers D J Carriquiry A L and Johnson L A 2007 *JAOCS, J. Am. Oil Chem. Soc.* **84** 151-155
[20] Jideani V A 2011 Functional Properties of Soybean Food Ingredients in Food Systems *Soybean* (Rijeka: IntechOpen)
[21] Szczesniak A S 2002 *Food Qual. Prefer.* **13** 215–225
[22] Serdaroğlu M, Yildiz-Turp G and Abrodímov K 2005 *Meat Sci.* 70 (1) 99-105
[23] De Garmo E and Sullivan J 1984 *Engineering Economy The 7th Edition* (New York: Macmillan Publishing Comp)

Acknowledgment

The authors wish to thank PNBP UNS project number 1075/UN27.21/PP/2017 for providing financial support for this work.

Effects of health consciousness and environmental attitude on intention towards organic food purchase

D Kusumaningsih, H Irianto and E Antriyandarti

Department of Agribusiness, Faculty of Agriculture, Universitas Sebelas Maret, Surakarta, Indonesia 57126

E-mail: dikusuma@student.uns.ac.id

Abstract. The objective of this research is to study the effect of health consciousness and environmental attitude towards intention to purchase organic food. The research is using Theory of Planned Behaviour (TPB). The study variables are health consciousness, environmental attitude, availability, subjective norm, attitude and intention to purchase organic foods. Survey method is applied, with the 150 respondents intending to purchase organic food in Surakarta. Structural equation model (SEM) is applied to analyze data. The result shows that health consciousness and environmental attitude are the determinants of an individual's positive attitude to purchase organic food. In addition, the attitude and subjective norm are also being the determinants of intention to purchase organic food. The study suggests that increasing organic food consumption can be achieved by increasing the knowledge about the benefit of organic food on health and environmental maintenance of potential consumers.

1. Introduction

Environmental damage due to conventional agriculture and food safety issues are factors that make people switch to organic products. Health consciousness and Environmental attitude make people assess that organic food is healthier and environmentally friendly. One of the most consumed organic products is organic vegetables which are a source of minerals and vitamins that are important for human health. Communities with healthy lifestyles emphasize activities to maintain health such as eating organic food, caring for health, and the existence of a balance of life. They will tend to have a positive attitude towards organic food. Consumers have a perception that organic labeled foods are healthier than conventional foods [1].

Availability will affect the purchase intention of organic food. The availability of products in the market and the diversity of types of vegetables will influence consumers in making purchases. Availability of shopping places is also a consideration, considering that not all shopping places sell organic food that consumers need [2]. This research was conducted in Indonesia which is one of the agrarian countries who has potention as producers and consumers of organic food. Therefore, this study aims to test the variables affecting the consumers attitude and intention to purchase organic food by using the Theory of Planned Behavior (TPB). This theory is appropriate approach to determine the attitude and intention to buy organic food since the theory was intended to explain all behaviours over which people have the ability to exert self-control with specific time and place [1, 2, 3]. The study of health consciousness has not been explored well in Indonesia. There is only one study about health consciousness and environmental attitude on intention towards organic food purchase in Indonesia [3]. However, this study differs with the previous [3]. In this study, variable of availability organic food is

IOP Conf. Series: Materials Science and Engineering **633** (2019) 012052 doi:10.1088/1757-899X/633/1/012052

added as one of important explanatory variable. This study uses variables of health consciousness, environmental attitude, availability, subjective norms, attitudes, and intentions to buy organic food.

2. Development of research hypotheses

The structure of this research design is shown in figure 1. In total 6 hypotheses are drawn from 6 constructs. There are health consciousness, environmental attitude, availability, subjective norm, attitude and purchase intention.

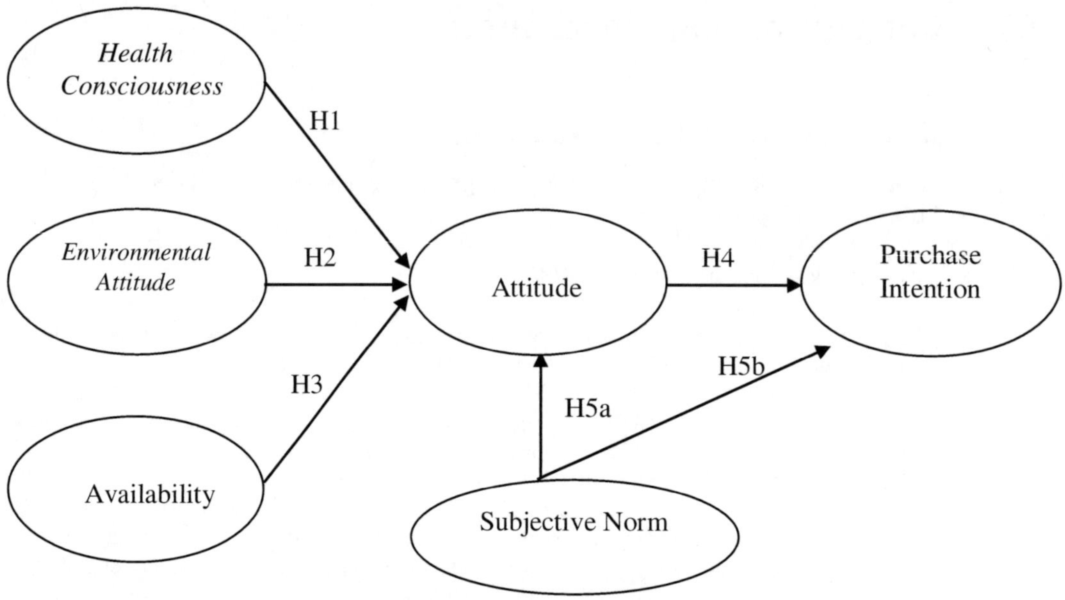

Figure 1. The research framework

2.1. Health Consciousness

Health consciousness is a person's tendency to focus on health. Previous research shows that this variable is a strong factor that influences attitudes and buying intentions of organic food [1,2,3,4]. If individual health awareness increases, the attitude and intention to purchase organic food will increase positively [4]. Based on these facts, the first hypothesis is determined as H1.
H1: Health Consciousness influences the attitude of organic food purchase

2.2. Environmental Attitude

Organic food is considered more environmentally friendly [5] and is considered less dangerous and more beneficial to the environment [6], which has been accepted as one of the factors that encourage consumers to buy organic food. Eating organic foods is strongly recommended, compatible with ecology and sustainability [7]. This shows that the attitude towards organic food is influenced by environmental issues, consumers who have more knowledge about environmental problems will be more likely to consume organic food. Based on the results of previous studies [5, 6, 7], the second hypothesis is defined as below.
H2: Environmental Attitude affects the attitude of organic food purchase

2.3. Availability

One of factors that influence consumers' purchasing intention is product availability. Previous research identifying the lack of availability of organic food in stores is considered as one obstacle for consumers to buy organic products [7,8]. The availability feature also includes product range. Some stores only sell in limited quantities and variety of organic products so that consumer choices are often

limited [9]. Therefore, product availability is important to encourage organic food purchases. Surakarta City has a number of Supermarkets that provide organic food, so that people can easily get organic food for consumption. Based on the above explanation, the third hypothesis can be arranged as follows.
H3: availability affects the attitude of organic food purchase

2.4. Attitude to Purchase Organic Food
Attitude describes the results of a person's evaluation of an entity (object or deed), whether that person likes or dislikes. Attitude towards behaviour refers to the extent to which a person evaluates or assesses the behaviour [10]. Attitude serves as a determinant of behavioural intention. A positive attitude towards organic food is a starting point to stimulate the consumption of organic food [11]. Previous research also showed positive results of the influence of attitudes towards the purchase intention of organic food [2,3,11]. Based on the above facts, the fourth hypothesis can be constructed as below.
H4: Attitude affects the intention of organic food purchase

2.5. Subjective norm
According to the Theory of Planned Behaviour [10], subjective norm has been identified as important determinant of the impact of social influence on behavioural intentions. Previous research also identified significantly positive relationships between consumers' subjective norms and intention to buy organic food [2,3,11]. If the consumers have the positive attitudes about organic food, they tend to have positive intentions to buy organic food. Trends in society have been shifted to healthier diets and increased individual responsibility for their health, this trend has greatly influenced consumer decisions to buy organic food. Therefore, the fifth hypothesis is defined as follows.
H5a: Subjective norm affect the attitude of organic food purchase
H5b: Subjective norm affect the intention of organic food purchase

3. Methods

3.1. Sample and Data Collection
The location of the research was determined purposively in Surakarta, Central Java of Indonesia. The Surakarta city is the one of the major cities in Central Java, known as trading city in the traffic of Jakarta-Surabaya. Surakarta area is the potential one for organic food market because it is surrounded by the organic food producers and also the per capita income of the population increases every year [15]. Primary data obtained by survey method by using standard questionnaires. Non probability sampling applied with the 150 respondents intending to purchase organic food. Structural Equation Modeling (SEM) applied to analyze and test the hypotheses.

3.2. Instrument testing
The instrument is tested by validity, reliability and normality tests. The validity test conducted based on the factor loading, according Hair et al [12], the factor loading should be greater than 0.50. The reliability tests were conducted by using Cronbach' alpha coefficient and applying the reliability margin 0.70. The results of validity and reliability tests on 150 samples showed that all variables were valid and reliable. The result of normality test showed that there were 10 outlier observations to be eliminated from the analysis model. Then, the data were analyzed by SEM analysis.

4. Results and Discussion
SEM was conducted by AMOS 22.0 using a maximum likelihood parameter to evaluate the hypotheses. The results of the evaluation of goodness of fit show a chi square value of 396,956 which had met the minimum chi square based on the output of AMOS 22.0. The probability value is 0.000. The value of CMIN/df, RMSEA, TLI and CFI with the results of 1,821, 0.077, 0.901 and 0.915 meet

the reference value so that it can be concluded to be good fit. The value of another fit indices such as RMR, GFI, NFI are 0.052, 0.809, and 0.832 that close to the reference value, so that it marginally fit.

Table 1. Hypotheses results

Hypotheses	Variable			Parameter Estimation	Standard Error	P
H1	AT	<---	HC	.198*	.118	.068
H2	AT	<---	EA	.615***	.080	.000
H3	AT	<---	AV	-.102	.085	.424
H4	PI	<---	AT	.098***	.103	.000
H5a	AT	<---	SN	.613	.099	.441
H5b	PI	<---	SN	.306***	.069	.000

Source: Primary data analysis (2018)
Description: ***= p< 0.01 **= significant p< 0.05 *= significant p< 0.1

Based on the results in Table 1, it is known that health consciousness (HC) positively influences attitude towards organic food purchase (AT) with an estimated value of 0.198 and is significant at a significance level of 90%. Environmental attitude (EA) positively influences attitudes towards purchasing organic food (AT) with an estimated value of 0.615 and is significant at 99% significance level. Environmental attitudes (EA) have a stronger influence on attitudes towards the purchase of organic food compared to health consciousness. Attitude (AT) positively affects the purchase intention of organic food (PI) with an estimated value of 0.613 at a 99% significance level. Subjective norm (SN) positively affect the purchase intention of organic food with an estimated 0.098 at and significant at 99% significance level.

The result shows that Health Consciousness (HC) has a positive effect on the attitude of organic food purchase (AT). This indicates that the increase of health awareness will affect the attitude of organic food purchase. The results of this study contradict the findings of Hossain and Lim [2] that there was no influence between health consciousness and attitude towards organic food. However, this study remains consistent with previous studies [1,3,4,13], which is the study shows a positive relationship between health consciousness and attitude towards organic food purchase. Based on these findings, the people of Surakarta City who have health awareness and aware of the benefits of consuming organic food for health will encourage the emergence of a positive attitude towards organic food as a healthy food product, which will encourage the emergence of buying organic food.

Environmental Attitude (EA) has a positive influence on attitude towards organic food purchase (AT). This means that the higher of someone's caring attitude towards the environment, the more people will have a positive attitude towards organic food. This finding is consistent with previous study [3]. Respondents are the people of Surakarta who have an environmental care attitude, that the environment needs to be maintained and preserved will be positive towards organic food. The respondents aware the negative impact of using synthetic chemicals on conventional agriculture on environmental damage such as soil, water and air pollution due to pesticides and land degradation. Thus, they have a positive attitude towards organic food as environmentally friendly products.

Availability (AV) does not have an influence on the attitude towards the purchase of organic food (AT). This finding contrasts with previous studies by [2,8,14] where the availability of organic food has a positive effect on attitudes towards the purchase of organic food and is an important factor for the intention of organic food purchase. In Surakarta City, there have been some modern/self-service markets that provide organic food and can make it easier for consumers to purchase organic food, but the result shows that there is no influence between availability and attitudes towards purchasing organic food. This is possible due to other factors such as price which affects consumer attitudes towards purchasing organic food. The price of organic food is more expensive than non-organic food, so respondents will rethink buying organic food even though their availability is abundant.

IOP Conf. Series: Materials Science and Engineering **633** (2019) 012052 doi:10.1088/1757-899X/633/1/012052

Attitude (AT) has a positive influence on the intention to buy organic food (BI). This finding is consistent with previous studies [2, 11]. A person's positive attitude towards organic food will encourage them to consume organic food. The results of this study are in accordance with the Theory of Planned Behaviour where the attitude will affect a person's intention to do something.

Subjective Norm (SN) does not have an influence on the attitude towards the purchase of organic food (AT). This finding contrasts with Irianto [3] which shows that subjective norms influence attitudes to the purchase of organic food. This is due to the attitude of people around who are considered to play an important role in the life of a person (family, friends, friends or community leaders) towards organic food may not immediately change someone's attitude towards organic food. In the hypothesis test it is also known that the Subjective Norm (SN) has a positive influence on organic food purchasing Intention (PI). This finding is consistent with Irianto [3] and Pomsanan [14]. The higher subjective norms will affect someone's intention to purchase organic food. Intention to buy organic food is influenced by people who influence the lives of people. If people who are considered influential for the life of a person such as family, friends, friends and community leaders consider good organic food and have already consumed organic food, then it will affect them to intend to buy organic food as well. This findings show that subjective norm does not affect a person's attitude, but affects a person's intention to perform an action, in accordance with the Theory of Planned Behaviour that the subjective norm is one of the things that can affect a person's intention to do something.

5. Conclusion

Health Consciousness (HC) and Environmental Attitude (EA) have positive effect to the attitude of organic food purchase (AT), while Availability (AV) and Subjective Norm (SN) do not influence it. In addition, Attitude (AT) has a positive influence on the intention to buy organic food (BI). These findings highlight several implications that may help developing strategies for organic product purchasing. Increasing consumption of organic food can be done by increasing the knowledge of the community as potential consumers of the benefits of organic food for environmental health and sustainability through advertisements in print and electronic media. Organic farming entrepreneurs can use well-known figures/celebrities endorses strategies to sound the organic food benefit. Intention to buy organic food is influenced by people who influence the life of consumers. Manufacturers and marketers can promote the benefits of organic food to the community, for example safe, healthy, environmentally friendly, and accessible to increase consumer recognition about organic food. Word of mouth strategies can also be used as a marketing strategy that is useful to increase the purchase intention of organic food consumers. This study considers organic food as organic product, so the finding could be different for different products. Future research should test the proposed model for another organic product as organic rice, organic fruits and so on or the general of organic foods. This study regarding the area of Surakarta, so the finding could be different in another city. Future research is expected to be done in another area, as metropolitan city with the dynamic lifestyle and more pollution to know the attitude and intention of purchasing organic food. The future research also expected to expand the variables of research such as price that involve in someone intention of organic food purchase.

References

[1] Chen M F 2009 *British Food J.* **111** 165-178

[2] Michaelidou N, Hassan L M 2008 *Intl. J. Cons. Studies* ***32*** 163-170

[3] Irianto H 2015 *Intl. J. Management, Economics and Soc. Sci.* **4** 17-31

[4] Cabuk S, Tanrikulu C, Gelibolu L 2014 *Intl. J. Cons. Studies* **38** 337-345

[5] Hanzaoui L, Zahaf M 2008 *J. Intl. Business and Economics* **8** 25-39

[6] Gronhoj A, Olander F 2007 *J. Cons. Behaviour* **6** 218-235

[7] Zanoli R, Naspetti S 2002 *British Food J.* **104** 643-653

[8] Beardsworth A, Bryman A, Keil T, Goode J, Haslam C, Lancashire E 2002 *British Food J.* **104** 470-491

IOP Conf. Series: Materials Science and Engineering **633** (2019) 012052 doi:10.1088/1757-899X/633/1/012052

[9] Dettmann R L, Dimitri C 2010 *J. Food Prod. Marketing* **16** 79-91
[10] Ajzen I 2005 *Attitides, Personality, and Behaviour. Second Edition.* (London: Open University Pr/McGraw-Hill Education)
[11] Teng C C, Wang Y M 2015 *British Food J.* **117** 1066-1081
[12] Hair J F, Black W C, Babin B J, Anderson R E 2010 *Multivariate Data Analysis Seventh Edition.* (Pearson: Prentice Hall)
[13] Kriwy P, Mecking R A 2012 *Intl. J. Cons. Studies* **36** 30-37
[14] Pomsanan P, Napompech K, Suwanmaneepong S 2014 *Asian J. Sci. Res.* **7** 434-446
[15] BPS Kota Surakarta 2018 *Surakarta in Figure* (Surakarta:BPS-Statistics of Surakarta)

IOP Conf. Series: Materials Science and Engineering **633** (2019) 012053 doi:10.1088/1757-899X/633/1/012053

Implementation of organic farming system and consumer satisfaction

Suswadi[1], E S Rahayu[2], M Harisudin[2], S Anantanyu[2]
[1]Doctoral Student of Development Extension Program of Sebelas Maret University Surakarta
[2]Post Graduate Lecturer of Development Extension Program of Sebelas Maret University Surakarta

E-mail: suswadi_slo@yahoo.com

Abstract. The research objectives are: (1) to analyse the implementation of organic farming of Boyolali farmers. (2) to find out consumer satisfaction level of organic rice in Surakarta and Boyolali.The research location was determined purposively in Dlingo Village, Boyolali and Surakarta. This research used qualitative method with interactive data analysis model of Miles and Huberman and descriptive analysis. The research resulted in the application of organic farming systems by the farmers in the following ways: (a) the selection of locations not contaminated with chemicals, (b) appropriate local types of rice plant, (c) programming of appropriate crop rotation, (d) processing soil with tools not contaminated with chemicals, (e) intermitten irrigation, (f) use of fermented compost, (g) control of pests with natural enemies and vegetable pesticides, (h) appropriate post-harvest handling, and (i) organic rice marketed with the packaging that attaches the *SNI organik*logo. The results of research on the level of consumer satisfaction with organic rice are: (a) the quality of the organic rice is 90.41% meaning that the consumers are satisfied; (b) the availability of the organic rice is 73% meaning that the consumers are satisfied; and (c) the price of the organic rice is 60% meaning that the consumers are dissatisfied.

1.Introduction

Food is an important and strategic commodity because it is an essential human need that must be fulfilled at all times. Food needs need to be made available in sufficient quantities, adequate quality, safe for consumption and easily obtainable at affordable prices for all levels of society. The awareness of hazards caused by the use of synthetic chemicals in agriculture makes organic farming attract the attention of both the producer and consumer levels. The Organic Standard Operational Procedure (SPO) in Indonesia refers to the best operational methods that can be carried out by the actors of organic rice production in accordance with the Indonesian National Standard (SNI) 01-6729-2016 concerning Organic Food System [1]. This standard is a revision of Indonesian National Standard 6729: 2013. This Indonesian National Standard adapts all material in the CAC / GL 32 - 1999 standard document, Guidelines for the production, processing, labelling and marketing of modified organically produced foods in accordance with Indonesian conditions [2]. For the application of good organic rice cultivation and in accordance with the provisions of the Indonesian National Standard in order to produce quality, safe, and consumable organic food, provisions concerning Good Agriculture Practices (GAP-organic) are provided which explain how to cultivate organic crops, Good Handling Practices

IOP Conf. Series: Materials Science and Engineering **633** (2019) 012053 doi:10.1088/1757-899X/633/1/012053

(GHP -organic) guidelines for good post-harvest handling methods of organic farming and Good Manufacturing Practices (GMP-organic), a good guideline for processing organic agricultural products. The organic market is growing rapidly in both international and domestic markets. The world market for organic products increases by 20% per year. Based on 2015 Statistics and Organic Trends published by the Research Institute of Organic Agriculture (FIBL) and the International Federation of Organic Agriculture Movements (IFOAM) [3], Boyolali Regency is one of the regencies in Central Java which has the second largest organic land which is 318.45 ha. One of them is APPOLI (Boyolali Organic Rice Farmers Alliance) with 15 organic farmer groups that have developed certified organic farming. The sustainability of organic groups must be supported by good practices of organic farming, so that the rice production of farmer groups is able to satisfy consumers.

The optimization of the application of organic farming systems by the farmers can be seen from the level of consumer satisfaction of organic agricultural products, whether consumers feel satisfied or vice versa. Even, this assessment cannot be separated from consumer perceptions. Perception also makes consumers know about what are the advantages, weaknesses, opportunities or threats to the product. It shows that Consumer Perception has a significant influence on the purchase decision of a product / service [4]. Perception not only is important in the information processing stage but also plays a role in post-product consumption that is when consumers evaluate their purchasing decisions. Consumer perceptions about organic products vary greatly. Therefore, organic farmers must pay attention to consumers' needs and tastes for the sustainability of marketing of organic rice because currently market forces are in the hands of buyers. Inevitably, organic farmers must guarantee the quality of their products so that they can meet the demand and needs of the consumers and customer satisfaction. Demands for various attributes such as product quality, product design, and product benefits for consumers must be fulfilled by organic rice producers. The problem of developing organic farming in Boyolali Regency is due to the absence of information between producers and consumers. The information on consumer behaviour and organic farming practices is important to ensure the sustainability of organic farming development. Therefore, this research aimed to: (1) analyse the application of organic farming systems, (2) find out the level of perception of consumers of organic rice.

2. Methods

The research location was determined purposively that is the organic farming implementation system was applied to the farmers of the Boyolali Organic Rice Farmers Alliance (APPOLI) in Boyolali Regency, Central Java. This research used qualitative approach. The data were collected using observation, interviews and documentation. The data were then analysed using "Miles and Huberman Interactive Analysis Models" and descriptive analyses. The interactive analysis model includes: Data collection, data reduction, data display and verification/conclusion drawing [5]. The objectivity test of the research data was carried out using the source triangulation and informant review techniques. The research on the consumer satisfaction in consuming organic rice was carried out on the organic rice consumers in Surakarta and Boyolali with 30 respondents selected purposively. The data were then analysed using descriptive analysis method.

3. Results and Discussion

3.1. Application of organic farming systems.

3.1.1. Selection of land

The location of organic agricultural land according to the provisions of SNI 6729 of 2016 is the land with a slope of less than 45%. The suitability of the land chosen for the development of organic rice from APPOLI farmers has been in accordance with the rules of SNI so that it has good potential for the development of organic rice. The location of the land was in Wates Village, SimoSubdistrict of 9.35 ha with a slope of <10%, in the Catur Village, Sambi District, covering an area of 17.77 ha, a slope of less than 15% and Glonggong Village, NogosariSubdistrict with a slope of <10%.

3.1.2. Land preparation.
To prevent the contamination of chemicals according to the direction of BSN (2016), the farmer group must create a buffer zone with a width of about 2 meters to limit the land of non-organic farmers [1]. From this provision in the three villages, APPOLI's development of organic farming has not met the requirements. The buffer zone that has been made by the farmers is 0.6 meters wide with grass planted. To clear agricultural land from the contamination of chemicals derived from chemical fertilizers and pesticides, all 122 farmers in the research location had converted the land for 2 years before certification in accordance with the rules in BSN in 2016. For land processing in accordance with SOP, the farmers must use special tools used only for organic farming. However, all the farmers have not been able to do it according to SOP. They used cleaned tools with clean water before used.
3.1.3. Watering.
The application of SOP for organic rice irrigation systems is very important to be implemented by APPOLI farmers as a form of their compliance with the rules established by the SOP of APPOLI organic rice irrigation system when the plant is at the age of 8 days and 9 to 10 days after the plant is inundated for 24 hours before the water is discharged. At the age 18, 19 to 20 days, the rice is re-inundated. After that, drying is done until the flowering plants. When the flowering plants are re-inundated and after the rice plants are ripe, they are re-dried until before harvest. In fact, only 36 farmers (29.5%) had followed the rules of SOP. Others always inundated their agricultural land until the rice plants were flowering before they were drained. According to BSN policy, the Irrigation for organic rice plants must be economical. The irrigation system of rice plants showed that with dry wet irrigation method there was a saving of 55.03% [6].
3.1.4. Production Process.
(a). In using organic rice according to the provisions of BSN, farmers may not use the seeds derived from genetically modified products (GMOs). As many as 122 farmers (100%) in the research area did not use GMO seeds. They used the seeds from the selected farmers' own land crops. The selected types of rice were the local species i.e. "Pandan Wangi", "MerahSlegreng" and "HitamCempo". (b) The SOP for planting is that the age of the seeds used is 18-20 days after planting, with a spacing of 25x25 cm. However, only 36 farmers (29.5%) followed the SOP. Others used the seeds at the age of >20 days with planting spacing of <25 cm. (c) The SOP for fertilization on organic farming using manure and fermented agricultural waste is at a dose of 5 tons per hectare, but only 36 farmers (29.5%) followed it. Others used unfermented manure. The application of cow manure *bokashi* fertilizer with doses ranging from 5-7.5 tons ha gives a better influence on the growth and production of rice plants [7]. (d) The SOP for controlling plant disturbing organisms (OPT) prioritize preventive measures by rotating crops to cut the cycle of pests and diseases and using plant-based pesticides.(e) The SOP for weed cleaning and soil loosening was carried out using a traditional tool "gosrok". A total of 122 farmers (100%) farmers have followed the SOP of fertilization and weed control methods.
3.1.5. Post-harvest
The SOPs in the post-harvest are as follows: (a) Harvesting or collection of agricultural products is carried out based on visual appearance if 90% to 95% of the rice grain is yellow or golden yellow. This is in accordance with the research of [8] that the determination of harvest as a critical moment in the harvest process is determined visually. Cutting the rice plants with sickles is done by down cutting because the farmers do the threshing with "thresher pedal". (b). Rice drying by drying method is the process of drying wet grain by utilizing the sun heat until the grain reaches the moisture content of 12-14%. (c) Storage moisture content to obtain premium quality of rice in the milling process grain moisture content ranges from 12% to 14%.

The temperature in the paddy grain storage can be used to improve the quality of milled rice and the yield of milled rice and the percentage of head rice [9]. The storage of organic products using group warehouses is always cleaned to be free of pests and not contaminated with chemicals. In storing organic products, the farmer must do good recording and separate between organic and non-organic rice. (d) Organic paddy grain milling was carried out on a special milling machine which is not used for milling non-organic grain. Milling workers have been trained and worked by prioritizing work safety. 122 organic farmers (100%) have implemented the SOP for post-harvest.

3.1.6. Packaging and labelling.

Packaging is a product container that includes the physical appearance of the container including design, colour, labelling, shape, and materials used [10]. Packaging and labelling were not carried out by each farmer, but were carried out by the trained personnel in special places managed by APPOLI. The SOPs of packaging and labelling are as follows: (a). Prior to packaging, the rice must be sorted first to separate it into various quality fractions based on the characteristics: physical (water content, shape, size, specific gravity, texture, colour and foreign matter / dirt), chemical (material composition, odour and rancidity) and biological (type and amount of damage by insects). (b). Packaging rice used plastic with a thickness of 0.8 mm. (g) Organic rice marketed by farmer groups was packaged and had *Organik Indonesia* logo on it, which is a circular symbol that reads *Organik Indonesia*. Packaging with good quality material tends to be the choice of consumers [11].

3.2. Level of consumer satisfaction with organic rice.

The level of consumer satisfaction shows high or low consumer satisfaction with organic rice. This consumer satisfaction research is reviewed from three aspects, namely (1) quality of rice, (2) availability of rice, and (3) price of rice. To be clearer, the average score achieved can be seen in table 1.

Table 1. Level of Consumer Satisfaction on Organic Rice

No.	Parameters	Scores	Categories
1.	Quality of rice		
	a. Taste	100%	Satisfied
	b. Colours	100%	Satisfied
	c. Aroma	100%	Satisfied
	d. Composition	66,6%	Satisfied
	e. Efficacy	93,3%	Satisfied
	f. Hygiene	80%	Satisfied
	g. Legality	93%	Very satisfied
	Sub Total	90,41%	
2.	Availability of rice	73%	Satisfied
	Sub Total	73%	
3.	Price	60%	Less satisfied
	Sub Total	60%	

The results of the research are as follows: (1) Quality of rice, the consumers interviewed were satisfied with an average score of 90.41%. The aspects assessed include: (a) Taste, 100% of the consumers consider organic rice to be more tasteful and fluffier and not easily spoilt. (b) Colour, 100% of the consumers considered organic rice to be more typical white and bright. (c) Aroma,the consumers considered organic rice to be more fragrant. (d) Nutrient composition contained,66.6% of the consumers was satisfied while 33.4% of them was not. They considered rice from organic rice plant to be easily digested by the body and relatively safe for consumption. (e) Efficacy, 93.3% of the consumers was satisfied and 6.7% was dissatisfied. They considered organic rice to be beneficial to the body. (f) Hygiene, 80% of the consumers was satisfied and 20% was dissatisfied. They considered that the hygiene was still lacking because sometimes there was still residual rice husk. (g) Legality, 93% of the consumers was satisfied and 7% was dissatisfied. They considered that the existence of the *SNI organik*logo attached to the packaging shows the guaranteed quality of the organic rice. The image of the product has a positive and significant effect on perceived quality and performance expectation at the evaluation stage. In addition, the purchase satisfaction is influenced by the image of the product [12].

(2). Availability of rice, 73% of consumers were satisfied and 27% were dissatisfied. They considered that the availability of rice was still not satisfactory because sometimes it is not available when consumers need it. (3) Price, 60% of consumers thought that the price of organic rice was still high even though they did not mind it.The consumers perceived that there is a positive relationship between price and quality of a product. Therefore, they will compare one product with another. After that, they make the decision to buy a particular product [13].

4. Conclusion

The results of this research indicate the application of organic farming systems by the farmers in the following ways: (a) the selection of locations not contaminated with chemicals, (b) appropriate local types of rice plant, (c) programming of appropriate crop rotation, (d) processing soil with tools not contaminated with chemicals, (e) intermitten irrigation, (f) use of fermented compost, (g) control of pests with natural enemies and vegetable pesticides, (h) appropriate post-harvest handling, and (i) organic rice marketed with the packaging that attaches the *SNI organic* logo. The results of research on the level of consumer satisfaction with organic rice are: (a) the quality of the organic rice is 90.41% meaning that the consumers are satisfied; (b) the availability of the organic rice is 73% meaning that the consumers are satisfied; and (c) the price of the organic rice is 60% meaning that the consumers are dissatisfied.

References

[1] National Standardization Agency (BSN). 2016. Indonesian National Standard (SNI) 01-6729-2016 about Organic Food System. Jakarta

[2] National Standardization Agency (BSN). 2010. Indonesian National Standard (SNI) 01-67292010. Organic Food System. Jakarta.

[3] AOI. 2015. Indonesian Organic Farming Statistic (SPOI). Bogor: Indonesian Organization Alliance (AOI).

[4] Taroreh, O., R.J. Jorie., R. Wenas. 2015. Pengaruh Persepsi Konsumen Dan Kepercayaan Terhadap Penggunaan Jasa Asuransi Pada Asuransi Jasindo Manado (The Effect of Consumer Perception and Trust in the Use of Insurance Services at Jasindo Manado Insurance).*Jurnal Emba*. Vol. **3** No. 3 Sept. 2015, Pg. 312-321

[5] Milles, M.B. and Huberman, M.A. 1984. Qualitative Data Analysis. London: Sage Publication

[6] Adriati Y. 2008. Kajian beberapa metode system pemberian air irigasi padi sawah (A study of several methods of providing irrigation systems for paddy rice.). Pekan Baru. Universitas Islam Riau

[7] Tufaila M., Yusrina, Alam S. 2014. Pengaruh Pupuk Bokashi Kotoran SapiTerhadap Pertumbuhandan Produksi Padi Sawah pada Ultisol Puosu Jaya Kecamatan Konda, Konawe Selatan (The Effect of Cow Manure Bokashi Fertilizer on Paddy Rice Growth and Production in UltisolPuosu Jaya Konda District, South Konawe). *Jurnal Agroteknos*, March 2014. Vol. **4** No. 1. pg 18-25

[8] Kobarsih, M. and N.Siswanto. 2015. Penanganan Susut Panen dan Pasca Panen Padi Kaitannya dengan Anomali Iklim di Wilayah Daerah Istimewa Yogyakarta (Coping with Harvest and Post-Harvest Rice Shrinkage Related to Climate Anomalies in the Special Region of Yogyakarta). PlantaTropika *Journal of Agro Science* Vol. **3** No. 2. pg. 101-106

[9] Millati, T. , Y. Pranoto, N. Bintoro, T. Utami, 2017. Pengaruh Suhu Penyimpanan pada Gabah Basah yang Baru Dipanen terhadap Perubahan Mutu Fisik Beras Giling (The Effect of Storage

Temperature on Newly Harvested Wet Paddy Grain to Changes in Physical Quality of Milled Rice), *AGRITECH*, Vol. **37**, No. 4, November 2017, pg. 477-485

[10] Arens. 1996. Contemporary advertising. United States of America: Irwin.

[11] Danger, E.P. 1987. Selecting colour for packaging. England: Gower Technical Press.

[12] Yu, H. J. &Kincade, D. H. 2001. Effects of product image at three stages of the consumer decision process for apparel products: alternative evaluation, purchase and post-purchase. *Journal of Fashion Marketing and Management*, Vol. **5** Iss: 1 pg. 29 – 43.

[13] Akshay, R.K. 1989. The Effect of Price, Brand Name, and Store Name on Buyers Perception of Product Quality: An Integrative Review. *Journal of Marketing Research*. Vol. **26**, pg. 351-357.

IOP Conf. Series: Materials Science and Engineering **633** (2019) 012054 doi:10.1088/1757-899X/633/1/012054

Analysis of social economic factors affecting maize farming income in Marga Tiga district of East Lampung Regency

A R Wardani, E S Rahayu and Agustono

Department of Agribusiness, Faculty of Agriculture, Universitas Sebelas Maret
Jl. Ir. Sutami No.36A Kentingan Surakarta 57126 Telp/Fax (0271) 637457

E-mail: annisarizkywardani@gmail.com, endangsiti@staff.uns.ac.id, agustono0108@gmail.com

Abstract. This study aims to determine the cost and income of maize farming in Marga Tiga district of East Lampung regency and to discover the influence of any social and economic factors that affect the income of maize farming in Marga Tiga district. The basic method used in this research is descriptive method. The location of research in Marga Tiga district, then it is selected a village which represents as research location by considering the highest production in Marga Tiga district. Sampling is undertaken by using simple random sampling method. The results of this study indicate that the income of farmer in Marga Tiga Distric is profitable. Factors the area of land, wage labor outside the family, the price of phonska, and land ownership significantly influence the income of maize farmers. Variable dummy land ownership that significantly indicate a difference between the ownership of the owned land own and land rent. The results of the test VPM, on land rent variable land area and the price of phonska significantly affect the income of corn farmers, while in their own land variable area of land, wage labor, and the price of phonska significantly affect the income of maize farming.

1. Introduction

Agricultural development in Indonesia is increasingly rapidly. The development of the industry in the field of agriculture is an overview of the increasing public awareness to build another farm in Indonesia. One of agricultural commodity in Indonesia, which is the amount of production is increasing is the food crop maize. In an effort to increase farmers incomes, maize will continue to play an important role in farming activity. That is, success in increasing the income of corn farmers can't be separated from the maize itself. There are two factors that affect costs and revenues, namely internal factors and external factors. Both of these factors will together affect the costs and revenues. External factors in the form of output and input. Whereas, internal factors in the form of socio-economic factors such as age, education, knowledge, experience, skills, number of family labor, land and capital [1].

Internal factors can't be released from the farming activity because it is one subsystem that determines the amount of output issued. The amount of output issued subsequently can affect how big is the total acceptance and the revenue to be received by corn farmers. Therefore, the need for further analysis related to the socio-economic factors affecting the farm income of maize crops in the District of Marga Tiga, East lampung Regency. This research is based on the theory of farming to determine the income of corn farmers. Researchers using the variable price of seed, land area, wage labor, experience, number of family dependants, the price of urea fertilizer, price of fertilizer phonska, education, age, and dummy variable of ownership of land own and rent.

IOP Conf. Series: Materials Science and Engineering **633** (2019) 012054 doi:10.1088/1757-899X/633/1/012054

2. Hypotheses

Socio-economic factors which includes the price of seed, land area, price of the wages of labor, experience, number of family dependants, the price of urea fertilizer, price of fertilizer phonska, education, age and the variable dummy which consist of the ownership of the owned land own and land rent affect the income of maize farming in the Marga Tiga district of East Lampung Regency.

3. Methodology

3.1. Sample and Data Collection

The location of the research was determined purposively in Marga Tiga Distrct, East Lampung Regency, Lampung Province of Indonesia. The district of Marga Tiga is one of the district which has a production maize is quite high [2]. The primary data obtained by interview method using a questionnaires 70 maize farmers selected as respondents at random simple. The types of data collected include primary data and secondary data. SPSS is used to analyze and test the hypothesis.

3.2. Instrument Testing

The Instrument was tested with test of determination (R^2), test of simultaneous (F test), the test of significance of regression coefficient (t test) and classic assumption test consist of normality test, multicollinearity, autocorrelation test, and heteroscedasticity test. Concurrent test (F test) and test of significance of regression coefficient (t test) are said to be influential or significant as some value of alpha with a reliability of 0.05 and 0.1. Data were analyzed using multiple linear regression analysis and the Variying Parameter Model (VPM). The analysis of the VPM is used to find out more about each dependent variable socio-economic factors in maize farming. VPM is made when the coefficient of the dummy is significant, thus indicating the presence of the impact area of land use to see how influential the area of land to the income of maize farmers.

4. Results and Discussion

4.1. The Results of The Test Determination

The results of the analysis of multiple linear regression is the test of determination (R^2) of 0.738 or 73.8%. This suggests that the dependent variable (farmer income) value is 73,8% is explained by variants of the independent variable in the form of the price of seed, land area, price of the wages of labor, experience, number of family dependants, the price of urea fertilizer, price of fertilizer phomska, education, and age. While the rest of 26.2% is explained by other variables not included in the research model. Other variables such as cost of medicines, the season, the cost of depreciation of equipment, market demand, management capabilities of farmers, etc.

4.2. The Result of The Simultaneous Test (F Test)

Table 1. Simultaneous Test (F Test)

Model		*Sum of Squares*	Df	*Mean Square*	F	Sig.
1	*Regression*	17,499	10	1,750	16,637	.000***
	Residual	6,206	59	0,105		
	Total	23,704	69			

Source: Primary Data Analysis
Description: *** Significant at α 0.05

Based on the Table can be seen that the probability value is 0.000 that is smaller than α ($\alpha = 0.05$ and 0.1). This shows that H_0 is rejected and H_1 is accepted. It means that the variable price of seed (X_1), land area (X_2), the price of wage-labour outside (X_3), experience (X_4), number of family dependants (X_5), the price of urea fertilizer (X_6), the price of fertilizer phonska (X_7), education (X_8), Age (X_9) and land ownership (D_1) jointly affect the income of maize farmers in the District of Marga Tiga on the planting period December 2017 – February 2018.

4.3. The Result of Significance of Regression Coefficient Test (t Test)

Table 2. Significance of Regression Coefficient Test (t Test)

Model	*Unstandardized Coefficient* B	t	Sig
1 (Constant)	14,015	1,073	,288^ns
Price of seed (X1)	,397	,569	,571 ^ns
Land area (X_2)	1,036	7,027	,000***
The price of labour (X_3)	,771	1,755	,085**
experience (X_4)	,112	,944	,349 ^ns
Number of family dependants (X_5)	-,073	-,371	,712 ^ns
Price of urea (X_6)	,259	,371	,712 ^ns
Price of Phonska (X_7)	-1,600	-2,667	,010***
Education (X_8)	,054	,738	,464 ^ns
Age (X_9)	-,148	-,475	,637 ^ns
Land Ownership (D_1)	,242	2,917	,005***

Source: Primary Data Analysis

Description: ***Significant at α 0.05, **Significant at α 0.1, ns= non significant

Based on the results of the regression analysis the value of significance (Sig.) the price of seed by 0,571 greater than α (α = 0.05 and 0.1) so it is not significant. The results of regression analysis of land area (X_2) has a significance value of 0.000 is smaller than α (α = 0.05 and 0.1) so significant. Based on the regression analysis of the price of wage-labour outside (X_3) has a significance value at 0.085 smaller than α (α = 0.05 and 0.1) so significant. Based on the results of the regression analysis experience has a significance value of 0,349 greater than α (α = 0.05 and 0.1) it is not significant. Based on the results of regression analysis of the number of dependents of the family (X_5) has a value of signfkansi of 0,712 greater than α (α = 0.05 and 0.1) it is not significant. Based on the results of the regression analysis the price of urea fertilizer (X_6) has a significance value of 0,712 greater than α (α = 0.05 and 0.1) it is not significant. Based on the results of the regression analysis the price of fertilizer phonska (X_7) has a significance value of 0,010 smaller than α (α = 0.05 and 0.1) so significant. Based on the results of regression analysis of education (X_8) has a significance value of 0,464 greater than α (α = 0.05 and 0.1) it is not significant. Based on the results of regression analysis age (X_9) has a significance value of 0,637 greater than α (α = 0.05 and 0.1) it is not significant. The ownership of the land (D_1) has a significance value of 0.005 is smaller than α (α = 0.05 and 0.1) so significant. The coefficient of the dummy which are significant to the ownership of the land need to be analyzed further to determine the influence of socio-economic factors in greater detail and separate.

IOP Conf. Series: Materials Science and Engineering **633** (2019) 012054 doi:10.1088/1757-899X/633/1/012054

4.4. The Result of Varying Parameter Model (VPM)

Table 3. Varying Parameter Model Test (VPM)

Variable	Land Rent	Owned Land Own
Constant	19,601	19,601
	(0,000)	(0,000)
Price Of Seed (X_1)	3,247	– 3,222
	(0,728)	(0,456)
Land Area (X_2)	1,183	0,791
	(0,000***)	(0,000***)
The Price Of Labour (X_3)	0,001	1,326
	(0,999)	(0,027***)
Experience (X_4)	0,315	– 0,098
	(0,119)	(0,460)
Number of Family Dependant (X_5)	– 0.093	0.038
	(0,796)	(0,862)
Price of Urea (X_6)	1,111	– 0,580
	(0,230)	(0,498)
Price of Phonska (X_7)	– 1,443	– 1,675
	(0,087**)	(0,019***)
Education (X_8)	0,006	0,233
	(0,937)	(0,129)
Age (X_9)	– 0,369	– 0,099
	(0,500)	(0,787)

Source: Primary Data Analysis

Description: ***Significant at α 0.05 **Significant at α 0.1

Based on Table 3 it can be known that there are differences in the influence of socio-economic factors land area and the price of fertilizer phonska against the income of farmers with ownership of land rent, as well as there is the influence of socio-economic factors land area, the price of wage labor, and the price of fertilizer phonska against the income of farmers with ownership of their own land.

4.5. The Result of The Classical Assumption Test

Table 4. Multicollinearity Test

Model		Collinearity Statistics Tolerance	VIF
1	(Constant)		
	Price of seed (X1)	,818	1,222
	Land area (X_2)	,457	2,186
	The price of labour (X_3)	,432	2,313
	experience (X_4)	,421	2,378
	Number of family dependants (X_5)	,635	1,574
	Price of urea (X_6)	,862	1,161
	Price of Phonska (X_7)	,804	1,243
	Education (X_8)	,761	1,315
	Age (X_9)	,394	2,537
	Land Ownership (D_1)	,874	1,144

Source: Primary Data Analysis

Based on Table 4 it can be seen that there is no variable that has value tolerance less than 0.10 and no variable that has a value of VIF greater than 10, which means socio-economic factors that affect the income of maize farmers in the District of Marga Tiga free from multicollinearity. According to Priyatno [3], If the tolerance is more than 0.1 and VIF is less than 10 then it does not occur multicollinearity.

In the table model summary can be seen that the value of Durbin-Watson is 1,499. The value of du in the table of Durbin-Watson was 1,8375, so that the known value of 4 – du is 2,1625. This shows no problem of autocorrelation because the value of DW is greater than du and less than 4 – du.

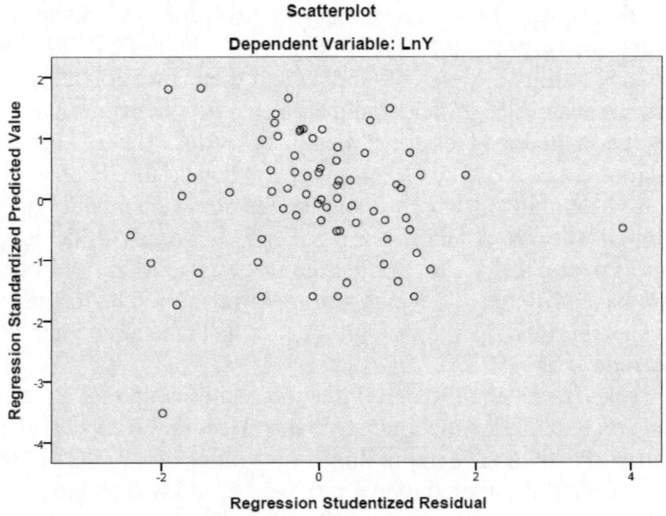

Figure 1. Heteroscedasticity Test

Based on figure 1, it is known that the pattern of dots do not form a specific pattern. The dots spread out with a pattern that is not clear above and below the number 0 on the axis Y. therefore, the socio-economic factors that affect the income of corn farmers in the District of Marga Tiga not affected by heteroscedasticity.

Normality test aims to determine whether the data distribution is normal or not. The terms in the regression analysis, namely data distribution should be normal. Testing using the Kolmogorov-Smirnov test [3].

Table 5. Normality Test

		Unstandarized Residual
N		70
Normal Parameters	Mean	.0000000
	Std. Deviation	.29989193
Most Extreme Differrences	Absolute	.075
	Positive	.073
	Negative	-.075
Kolmogorov-Smirnov Z		,624
Asymp. Sig. (2-tailed)		.831

Source: Primary Data Analysis

Based on Table 5 can be seen that sig. (2-tailed) is 0,831 where this value is greater than alpha (α = 0.05) so it is not significant. This shows that H_0 is accepted i.e. the residuals be normally distributed. According to Ghozali [4], the t test and F test assumes a value residuals follow normal distribution, if this assumption is violated then the test statistics are not valid for the number of samples is small.

5. Discussion and conclusions

Socio-economic factors basically is one of the factors that affect income. The use of such factors has an important role in the income of maize farmers in Marga Tiga District. Based on the regression analysis, the variables jointly price of seed (X_1), land area (X_2), the price of wage labor (X_3), experience (X_4), number of family dependants (X_5), the price of urea fertilizer (X_6), the price of fertilizer Phonska (X_7), education (X_8) Age (X_9), and dummy variable of land ownership (D_1) significantly affect the income of farmers in the District of Marga Three East Lampung Regency. This is in line with research conducted by Minai et., al [5].

But individually, each variable has different effects on the income of maize farmers. Area factor (X_2), labor (X_3), fertilizer prices phonska (X_7), and dummy variable of land ownership (D_1) individually significantly affect the income of corn farmers. Land area for individual influential real against the income of maize farmers due to the high area of land owned then it would be increasing the number of production and can improve revenue. Significant land area is in line with previous research [5], the price of labor to be significant due to the high labor paid then it will reduce the income of maize farmers, then the efficiency of use of labor is needed. The effect of wage labor on the income of farmers in line with the research conducted by Muttakin [6] which is one of the socio-economic factors that individually significant effect on income is the cost of labor. The price of fertilizer phonska individually significantly affect the income of corn farmers, this is due to the high price of fertilizer phonska will affect the cost of the means of production and will add to the expenditure of farmers. The influence of the significance of fertilizer prices phonska supported by research Igboji et.,al [7] who stated that the fertilizer used will affect the income of a farmer.

Varying parameters model is a method to determine the relationship of output-input condition if the use of the dummy variable on the model function regression effect real Rahmawati et., al, [8]. According to Ekananda [9] the result of the difference in the slope or the slope shows the differences in the average impact of all the observation. In this study, dummy variable of land ownership with land lease and land belonging to own which significantly shows the influence of the ownership of the land against the income of maize farmers in the District of Marga Tiga. Based on the analysis of advanced VPM (Variying Parameters of the Model) there are two variables that significantly affect the income of maize farmers with ownership of the land rental variables land area (X_2) and the variable price of fertilizer phonska (X_7). While on their own land, there are three significant variables or variables that influence the income of corn farmers is the variable land area (X_2), the price of wage labor (X_3), and the price of fertilizer phonska (X_7).

Based on the description above can be concluded that not all socio-economic factors affecting the income of farmers in the District of Marga Tiga. It is also supported by several previous studies [5,6,7].

References

[1] Suratiya K 2015. *The Science Of Farming*. Jakarta: Penebar Swadaya

[2] BPS 2016. *The District Of Marga Tiga dalam Angka 2016* (East Lampung:BPS-Statistics of East Lampung)

[3] Priyatno D 2009. *SPSS For Correlation Analysis, Regression, and Multivariate.* Yogyakarta: Gava Media

[4] Ghozali I 2014. *Econometrics: Theory, Concepts, and Applications With The IBM SPSS 22.* Semarang: Badan Penerbit UNDIP

[5] J M Minai., Nyairo and mbataru 2014. Analysis of Socio-Ekonomic Factors Affecting the Coffe Yields of Smallholder Farmers in Kirinyaga County Kenya. *Journal of Agricultural and Crop Research.* Vol. **2**(12)

[6] Muttakin D 2014. Socio-Ekonomic Factors That Affect The Income Of Oil Palm Farm, The Pattern Of Self-Help In The Vilage Kepau Jaya Kampar Regency. *Journal RAT.* Vol. **3**(1)

[7] Igboji Chidi 2015. Analysis of Socio-Ekonomic Factors and Profitability of Rice Production among Smallscale Farmers in Ebonyi State. *Journal of Agricultural and Veterinary Science.* Vol. **8**(2)

[8] Rahmawati A, Agustono, R Kunto 2018. Farming Of Organic Rice is White and Black: The Income Approach, The Function Of Cobb-Douglass and VPM. *Journal of Sustanable Agriculture.* Vol. **19**(1)

[9] Ekananda M 2015. *Ekonometrics Basic.* Jakarta: Mitra Wacana Media

IOP Conf. Series: Materials Science and Engineering **633** (2019) 012055 doi:10.1088/1757-899X/633/1/012055

Analysis of potential featured agriculture commodities and processed products in Sragen Regency

N Setyowati [1], D T Ardianto[2], I Khomah [1], and E Widiyanti[3]

[1] Department of Agribusiness, Faculty of Agriculture, Universitas Sebelas Maret,
[2] Indonesia Department of Visual Communication Design, Faculty of Art and Design, Universitas Sebelas Maret, Indonesia
[3] Department of Agricultural Extension and Communication, Faculty of Agriculture, Universitas Sebelas Maret, Indonesia

Email: noenk_setyo@yahoo.com

Abstract. This research aimed to identify the mapping of potential featured agricultural commodities and processed products in Sragen Regency. Data analysis used Exponential Comparative Method (ECM) and Borda Method. ECM Method showed that featured agriculture commodities at subdistrict level in Sragen Regency are Paddy, Pomelo Oranges, Corn, Banana Seedlings, Chili, Peanut, Melon, Watermelon, Sugarcane, Cassava, Arrowroot, Soybean, Fish, etc. Borda Method showed that the five featured agriculture commodities in Sragen Regency are Paddy with a value of 13,600.56; Corn 4,918.36; Sugar Cane 3,313.95; Pamelo Orange 1,466.64; and Peanut 904.64. Borda Method also showed that the featured processed products in the subdistrict level at Sragen Regency are Intip, Processed Fish, Rice, Marneng, Tempe Benguk, Ampyang, Emping Garut, Tofu, Keripik Tempe, Etawa Milk, Cornmeal, and others. Borda Method showed that the fifth rank featured processed products in Sragen Regency are Rice with the value of 7,530.2; Marneng 1,683; Emping Garut 1,241.99; Keripik Tempe 957.65; and Tempe Benguk 713.99. It means that the featured agriculture commodities in Sragen Regency feasible to economic development priority in order to support community economic welfare are Rice, Corn, Garut, Pamelo Orange, and Peanut. Meanwhile, the featured processed products in Sragen Regency feasible to develop are Rice, Marneng, Emping Garut, Keripik Tempe, and Tempe Benguk.

1. Introduction

Agricultural sector plays an important role in economy [1]. The economy is important to develop starting from regional economic development. Regional economic development is the local governments and their communities' process of managing existing resources and establishing cooperation between local governments, private institutions, and other sectors [2]. Resource utilization and management in regions is the community's opportunity of creating a job. This employment opportunity is related to accelerating the development of featured products in region. These featured products are expected to be market-oriented so that they can actually improve product competitiveness while providing added value to regional economic development [3]. Thus, regional economic development is expected to produce featured products that are able to compete domestically and abroad.

Agricultural sector has become an important sector in contributing to supporting national development [4], especially if the agricultural sector is supported by its processing industry.

IOP Conf. Series: Materials Science and Engineering **633** (2019) 012055 doi:10.1088/1757-899X/633/1/012055

Agroindustry is a major driver of agricultural development sector. Agroindustry can be described as industrial activities utilizing agricultural products as raw materials, designing, and providing equipment to reach the agricultural product processing industry [5]. The existence of agroindustry encourages the community to be more productive and advanced. In addition, agroindustry development can make local resources the featured product in the region [6]. Thus, the existence of local resources supported by the region's potencies will spur productive activities aiming at increasing income, expanding employment opportunities, diversifying food and economic activities, and increasing investment in the region [7].

2. Methods

2.1. Determining Featured Commodities/Products

The selection of commodity/featured agribusiness products starts with mapping the potential featured commodities/agribusiness products at subdistrict level. The mapping is done using some criteria: number of business units/farmers cultivating, market, raw material availability, and contribution to regional economy [8].

2.2. Determining Ranked Commodities/Products using Exponential Comparative Method

The ranking of featured products displayed in the business is done using Exponential Comparative Method (ECM), a method used to determine the priority order of decision alternatives with several criteria [9].

The selection of an alternative featured commodity product business was conducted based on the idea of a resource obtained through interviewing the staffs of the Ministries of Industry and Trade, Agriculture, and Food Security, and subdistrict staffs, and PPL (Agricultural Extension Officers). The analytical formulation of the Exponential Comparative Method (MPE) is written as follows:

$$\text{Total value (TNi)} = \sum_{j=1}^{m} (\text{RKij})^{\text{TKKj}}$$

Where:

TNi = Total alternative values to (i)

RK_{ij} = Relative importance degree of j-criteria for decision choice i

TKK_{ij} = Degree of interest to the decision criteria-j, TKK > 0; round

i = 1,2,3…n ; n= Number of decision choices

m = Number of decision choices

2.3. *Determining Featured Commodity/Product Ranking with Borda Method*

Based on the identification of commodities/featured agribusiness products at subdistrict level using Exponential Comparative Method (ECM), the selection of top 10 commodities / agribusiness products featured in Sragen regency level was carried out using Borda method. Considering the result of calculations with Borda method, it can be established maximally 10 commodities/featured agribusiness products at Sragen Regency level, so it will be identified as position/potential commodity/featured agribusiness products in Sragen. The formula for calculation using Borda method is written as follows:

$$Borda\ X\ Value = \sum \ (ECM\ X * N)$$

Where:

X = Commodity/Product X

ECM X = ECM Value of commodity/product
N = Ranking value of commodity/product X at subdistrict level

The mapping commodity/agribusiness product output is made in the form of distribution of featured commodities/ agribusiness products and commodity rank/featured agribusiness products in Sragen.

3. Results and Discussion

3.1. *Mapping the Featured Agricultural Commodities*

From the results of mapping using analysis of Exponential Comparison Method (MPE), it can be seen that Paddy is a featured agricultural commodity in 20 subdistricts of Sragen Regency. This is supported by the Sragen Regency's potency to cultivate Paddy. In fact, Sragen Regency is famous for its Organic Rice Centers.

Corn becomes the leading commodity in 15 subdistricts in Sragen. This is supported by agricultural land area in the subdistrict suitable to corn farming. Sugarcane also becomes a featured agricultural commodity in 10 subdistricts in Sragen. This is due to the existence of the Mojo Sugar Factory in Sragen Regency requiring sugarcane raw materials so that many farmers plant sugarcane.

Based on an analysis using Exponential Comparative Method (ECM), it can be found that some featured agricultural commodities are produced in addition to those aforementioned: Pomelo Orange, Peanut, Melon, Watermelon, sweet potato, Garut, Soybean, Pisang Mas seeds, Chili, Fish, Crops, etc.

3.2. *Determining the Potential Agricultural Commodities*

Borda method aims to identify potential featured agricultural commodities at Sragen Regency level. The next step is to sort ECM value from the largest value to the lowest value. The following are the results of the determination of the potential Leading Agricultural Commodities in Sragen Regency.

Table 1. Ranking of Potential Leading Agricultural Commodities in Sragen Regency.

Agricultural Commodities Featured	Borda Value	Ranking
Paddy	13,600.56	1
Corn	4,918.36	2
Sugarcane	3,313.95	3
Orange Pummelo	1,466.64	4
Peanut	904.64	5

Considering the result of analysis using Borda method, it can be seen that the potential featured Agricultural Commodities in Sragen have five rankings. On the first rank there is Paddy with Borda value of 13,600.56. Paddy in Sragen Regency is the leading agricultural commodity in each subdistrict. Each sub-district in Sragen Regency cultivates Paddy. The highest Paddy production can be found in Sidoharjo sub-district with Paddy production of 60,140 tons in 2016.

Corn is on the second rank in Sragen. Mondokan Subdistrict has greatest potency in cultivating corn plants. This is because the corn plant itself has simple criteria for planting and does not require too much water so that many farmers plant corn in dry season.

Sugarcane is on the third rank in the potential featured Agricultural Commodities in Sragen. This is supported by the existence of the Mojo Sugar Factory in Sragen so that many farmers still work in sugarcane farming. Pomelo Orange is on the fourth rank in Featured Agricultural Commodities in Sragen with Borda value of 1,466.64. Pomelo Orange is cultivated widely in Plupuh and Gemolong subdistricts. Plupuh Subdistrict has Pomelo Orange centers located in Jembangan, Pungsari, Gedongan, and Sambirejo Villages. Meanwhile, in Gemolong Subdistrict, Pomelo Orange center is

IOP Conf. Series: Materials Science and Engineering **633** (2019) 012055 doi:10.1088/1757-899X/633/1/012055

located in Jatibatur Village. Peanut occupies the fifth rank in the potential featured Agricultural Commodities in Sragen. Peanuts are cultivated widely in Sumberlawang, Tanon, Mondokan, Miri, and Plupuh sub-districts. This is because peanut cultivation is easier as it does not take too much water.

3.3. Mapping the Featured Processed Products Commodities

Considering the result of analysis using Exponential Comparative Method (ECM), it can be seen there are several featured processed product commodities: Intip, Mokaf, Tempe Benguk, Marneng, Cassava Chips, Soy Milk, Karak, Rice, Shredded Catfish, Banana Chips, Ampyang, Keripik Tempe, Emping Garut, Honey, Banana Sale, Peyek, Instant Jamu, and others. There is a diversity of Featured Processed Products in Sragen Regency. This is because of many processed food diversification and economic activities. Considering the results of featured agricultural commodities, it can be seen that Paddy is on the first position, so that it can also be processed into a diversity of products. Not only Paddy becomes a Featured Processed Product Commodity but there are also Intip and Karak processed from the rice. Sragen Regency becomes an Organic Rice center as well.

3.4. Determining the Potential Featured Processed Product Commodities

The next step is to sort MPE value from the largest value to the lowest one. The results of the determination of the potential featured Processed Product Commodities in Sragen Regency are presented below.

Table 2. Ranking of Potential Featured Product Commodities in Sragen Regency

Featured Processed Product Commodities	Borda Value	Ranking
Rice	7,530.02	1
Marneng	1,683.00	2
Emping Garut	1,241.99	3
Keripik Tempe	957.65	4
Tempe Benguk	713.99	5

Rice commodity is on the first rank of potential featured Processed Product Commodities in Sragen. The need for rice in Sragen has been highly sufficed and even there is surplus productivity. This also makes Sragen Regency called the Rice Center. In addition, the marketing of Sragen rice extends outside to surrounding areas.

The second featured processed product is Marneng, with Borda value of 1,683.00. There are several Marneng processed product centers including, among others, Plupuh, Tanon, and Sukadono Subdistricts. Marneng business center in Tanon Subdistrict has the largest number of SME units processing marneng innovatively with varying tastes.

Emping Garut is the third processed product commodity in Sragen Regency. Emping Garut is made of raw materials deriving from *Garut* tubers in produced widely in Tangen, jenar, Miri, and Mondhokan subdistricts. The production capacity of emping Garut is 30 kg/day, but this production is still seasonal in nature because Garut tuber raw material is still difficult to get continuously.

Keripik Tempe is on the fourth rank of the featured processed products in Sragen Regency. Keripik Tempe is produced widely in Masaran and Karang Malang Subdistricts. Keripik Tempe business is driven by home industry undertaken by Women Farmer Group (WFG). The production process still uses traditional technology.

The fifth rank of featured processed products is occupied by Tempe Benguk with Borda value of 713.99. Tempe Benguk is made of raw materials deriving from Benguk Beans. Tempe Benguk is produced widely in Mondokan and Gondang Subdistricts. This Tempe Benguk has great potency but the sustainability of raw materials in Sragen Regency is not adequate for production purpose.

4. Conclusion

The Featured Agricultural Commodities feasible to be prioritized for their development in Sragen include Paddy, Corn, Sugarcane, Pomelo Orange, and Peanut. Meanwhile, the type of featured processed product commodities potential to be used as a development priority as a community economic support in Sragen regency includes, among others, Rice, Marneng, Keripik Garut, Keripik Tempe, and Tempe Benguk.

References

[1] Pantow S, Palar S, and Wauran P 2015 *J. Berkala Ilmiah Efisiensi* **15** 100-12
[2] Susanto H 2014 *J. of Rural and Development* **5** 63-18
[3] Yolamalinda 2014 *J. of Economic and Economic Education* **3** 27-15
[4] Ayubi AA 2014 *J. Ekonomi Pembangunan* **12** 1-15
[5] Udaya I G B 2011 *Singhadwala* **44** 3-6
[6] Syakur U M and Hakim A 2013 *J. Ekonomi dan Perbankan Syariah* **1** 49-23
[7] Prawoto N 2010 *J. Ekonomi dan Studi Pembangunan* **11** 1-19
[8] Bank Indonesia 2010 *Pengembangan KPJu Unggulan UMKM Eks Karesidenan Madiun* (Kediri: Bank Indonesia)
[9] Marimin 2004 *Teknik dan Aplikasi Pengambilan keputusan Kriteria Majemuk* (Jakarta: Gramedia Widiasarana Indonesia)

IOP Conf. Series: Materials Science and Engineering **633** (2019) 012056 doi:10.1088/1757-899X/633/1/012056

Household consumption of broilers meat in Kudus Central Java

M B Zulkarnain, W Rahayu, E Antriyandarti and S W Ani

Department of Agribusiness, Faculty of Agriculture, Universitas Sebelas Maret, Surakarta, Indonesia 57126

E-mail: mahmudcancer@student.uns.ac.id

Abstract. Broilers meat is one of protein intake sources from meat in Indonesia, but the consumption still very low compared to the standard set by Food and Agriculture Organization (FAO). This study investigates the pattern of animal food consumption, examines the determinants of broilers meat consumption and analyzes the demand elasticity in the household. This study uses Panel Data Analysis with Random Effect Model to estimate the factors influencing broilers meat consumption. The determinants are the number of family members, household income, the price of broilers meat, egg, rice, beef, fish, and the residential area. The results show that the most influential factor is the price of broilers meat, with inelastic demand elasticity.

1. Introduction

The current consumption of animal protein in Indonesia is still very low compared to the standards set by Food and Agriculture Organization (FAO) [1]. One of animal protein sources at a relatively affordable price and easily obtained is broilers meat [2]. Indonesian people like to consume broilers meat for their daily consumption. Although the price of broilers meat is volatile, this commodity is still popular. Chicken meat prices are relatively affordable and influenced by price fluctuations of other agricultural commodities, which depend on the season. In addition, chicken meat is very easy to find in the market [3]. However, the rate of consumption growth for broilers meat is still fluctuating from year to year, particularly in Kudus District with the highest inflation rate among Central Java regions [4].Broilers meat is one of the commodities which contributes to inflation. Also, broilers meat has a big influence on urban poverty line value besides rice and other food commodities. Therefore, it is very important to investigate the household consumption of main food commodity in the areas where inflation is high.

Households in Kudus District have the highest average consumption of goods and services in Central Java [5]. Poultry is the most desirable food compared to beef, buffalo meat, and other meats [6]. Therefore, this study attempts to investigate the pattern of household meat consumption in Kudus District, focusing on broilers meat consumption and the sensitivity. This is the first study that examines the household consumption of broilers meat by using a monthly micro-data, thus provides a significant contribution on similar subject.

2. Theoretical Framework

In this study, the households are assumed to maximize their utility for broilers meat consumption. Household food consumption patterns are determined by the type and frequency of meals which reflect their eating habits. The criteria of frequency can be categorized as 'often' (4-6 times per week), 'quite often' (3 times per week), 'enough' (1-2 times per week), 'rare' (2-3 times per month), and 'very rare'once or less per month) [7].

The determinants of household consumption are divided into 2 factors, namely economic and non-economic factors [8]. The demand is defined on influencing factors consisting the price of its good and other goods, the number of population and income per capita. The relationship between demand and the influencing factors [9] is demonstrated in the equation (1).

$$Qd = f(P, Px, Py, Y) \quad (1)$$

Where Qd is the demand for good, P is the price of the item, Px is the price of substitute goods, Py is the price of complementary goods, and Y is the income per capita. Then, the degree of sensitivity was measured by using the elasticity. Elasticity is a measure of a variable's sensitivity to a change in another variable. In business and economics, elasticity refers the degree to which individuals, consumers or producers change their demand or the amount supplied in response to price or income changes [3]. There are some types of elasticities, namely Price Elasticity (Ep), Cross Elasticity (Ec), and Income Elasticity (Ei). If the value of price elasticity ($|\varepsilon|$) is > 1, < 1, and = 1 hence classified as elastic, inelastic and unitary, respectively. Furthermore, the goods would be categorized as superior goods (luxury goods), basic goods or normal goods, and inferior goods if the elasticity of income is $Ei>1$, $0<Ei<1$, and $Ei< 0$, respectively. Then, the classification according to cross elasticity is substitute goods ($\varepsilon_c> 0$), complementary goods ($\varepsilon_c <0$), and neutral goods ($\varepsilon_c = 0$) [10].

3. Methods

3.1. Empirical Model

In this study, 100 household samples were chosen, consisting of 50 respondents from each rural and urban areas, respectively. Based on the theoretical framework, the following empirical model was applied to estimate the determinants of household consumption [10] [11].

$$LnY_{it} = Ln\,\beta_0 + \beta_1 LnX_{1it} + \beta_2 LnX_{2it} + \beta_3 LnX_{3it} + \beta_4 LnX_{4it} + \beta_5 LnX_{5it} + \beta_6 LnX_{6it} + \beta_7 LnX_{7it} + d_1 D_{1it} + d_2 D_{2it} + d_3 D_{3it} + v_{it} \quad (2)$$

Where :

β_0 : Constant
β_i : Regression coefficient as variable of demand elasticity
LnY_{it} : Consumption of broilers meat (kg)
LnX_{1it} : Number of family members (people)
LnX_{2it} : Household income (IDR/month)
LnX_{3it} : Broilers meat price (IDR/kg)
LnX_{4it} : Chicken eggs price (IDR/kg)
LnX_{5it} : Rice price (IDR/kg)
LnX_{6it} : Beef price (IDR/kg)
LnX_{7it} : Fish price (IDR/kg)
D_1 : Dummy for household head occupation
If the occupation has permanent employees = 1, otherwise = 0
D_2 : Dummy of household members under 10 years old
If there are household members under 10 years old = 1, otherwise = 0
D_3 : Dummy for region residence
If the households live in urban area = 1, otherwise = 0
α : Unobserved effect
μ_{it} : Error
v_{it} : Composite error term
i : Respondent-i
t : Month-t

IOP Conf. Series: Materials Science and Engineering **633** (2019) 012056 doi:10.1088/1757-899X/633/1/012056

3.2. Analytical methods

The protein intake pattern of household consumption from meat was analyzed by using a table of specific criteria according to food materials and the consumption frequency of animal protein (chicken eggs, broilers meat, beef and fish) [7]. The Panel data analysis was taken into account to estimate the factors that influence the consumption of local broilers meat, also to find out the sensitivity degree (elasticity) of household demand for broilers meat in Kudus District. The Breusch-Pagan test was applied to investigate the existing types of effects (Random Effects or Pooled Models) [11]. Due to the rejection of the null hypothesis of individual effects, the Ordinary Least Square (OLS) method was not employed. The Hausman test was implemented to investigate further types of effects (Fixed or Random Effects) on the functions of household consumption. Finally, the Random Effect Model was conducted and indicated by the preliminary results [11] [12].

4. Results and Discussion

This study was conducted in Kudus District where the Consumer Price Index (CPI) is the highest in all Central Java province. Household consumption of broilers meat in this district is the largest compared to other meats [5], thus high inflation may had an effect on the consumption [13].

Table 1 shows the pattern of household meat consumption which is calculated from the types of meals' ingredients and frequency [8]. The type of food sources for protein intake from meat are seen from the food ingredients such as chicken eggs, broilers meat, beef, and fish. The most frequent consumed food stuffs is broilers meat. One household was categorized in 'often frequency', 15 households in quite frequent categories, 16 households in the fair category, 46 households in the rare category, and 22 households in the very rare category. Egg is the second top preferred commodity consumed, while the least consumed is beef (96% of households rarely consuming it) because it is more expensive than other foodstuffs. Furthermore, households consider beef as luxury goods. The affordable price is the main reason for the household to consume broilers meat in various circles.

Table 1. Patterns (type and frequency) of household meat consumption

Type	Often	Often enough	Enough	Rarely	Very rarely	*TOTAL*
Egg	2	8	18	50	22	100
Broilers meat	1	15	16	46	22	100
Beef	0	0	1	3	96	100
Fish	0	5	4	28	63	100

Table 2 shows the estimation of equation (2) and illustrates the determinants of broilers meat consumption. This study found that all the independent variables jointly affected the consumption of broilers meat. The estimated parameter of the number of family members, household income, the price of broilers meat, egg, beef, fish, as well as the residential areas are positively significant to the consumption of broilers meat ($p < 0.05$). While the estimated parameter of rice price is negatively significant to it. On the other hand, household head occupation and the presence of children under 10 years old do not affect the consumption ($p < 0.05$).

There are positive relationships between number of household members, household income, broilers meat price, beef price, fish price, and the consumption of broilers meat. This indicates that the increase values of those variables would lead to the increase of broilers consumption. In contrast, the increase of rice price causes broilers meat consumption to decrease. The most influential factor is broilers meat price, indicated by the biggest regression coefficient (0.671). The demand of broilers meat easily changes due to the price changes [14]. The dummy of region residence is positively effecting the broilers meat consumption. It is suggested that households who live in urban areas consume more broilers meat than those in rural areas. This finding is in line with a certain previous study [15].

Table 3 presents the sensitivity/elasticity of demand for consumption of broilers meat. It can be seen that the income elasticity is 0.035, indicating that an increase of household income by 1% will lead to demand increase by 0.035%. The income elasticity is positive, indicating broilers meat is categorized as basic goods, where the demand will increase when the income increases [16]. Furthermore, the cross elasticity of egg, beef, and fish prices are 0.013, 0.007, and 0.005, respectively. This means if the beef price rises by 1% leads to the increasing demand of broilers meat by 0.013% and vice versa. It also means if fish price rises by 1%, the demand for broilers meat will increase by 0.005% and vice versa. The cross elasticity of beef and fish prices are positive, indicating beef and fish are the substitutes of broilers meat ($Ec>0$). The cross elasticity of rice price is –0.062, means the increase in rice price by 1% leads to the decrease of demand for broilers meat by 0.062% and vice versa. The cross elasticity of rice price is negative, indicates that rice is a complementary for broilers meat ($Ec<0$).

Table 2. Determinants of broilers meat consumption in Kudus district

Variable	Regression Coefficient	Standard Error	P > |z|
(Constant)	-7.110^{***}	0.436	0.000
Number of Household Member(X_1)	0.099^{**}	0.045	0.025
Household Income (X_2)	0.038^{**}	0.017	0.022
Broilers Meat Price (X_3)	0.670^{***}	0.013	0.000
Egg Price(X_4)	0.013^{**}	0.005	0.003
Rice Price(X_5)	-0.061^{**}	0.031	0.049
Beef Price(X_6)	0.006^{**}	0.003	0.052
Fish Price(X_7)	0.005^{**}	0.002	0.018
Household Head Occupation (D_1)	0.038^{ns}	0.027	0.152
Children Under 10 Years Old (D_2)	-0.028^{ns}	0.031	0.364
Residential Area (D_3)	0.069^{**}	0.028	0.012
Prob > Chi^2 = 0.0000	R^2	= 0.8572	
Number of Observations = 1200	Hausman test	= 0.9999	
U = 100, V = 12	Breusch-pagan test	= 0.0000	

*Significant at 10% level; **Significant at 5% level; ***Significant at 1% level; ^{ns}Not significant

Table 3. Elasticity of broilers meat demand

Variable	Elasticity		
	Income	Cross	Price
Household Income (X_2)	0.035		
Egg Price (X_4)		0.013	
Rice Price (X_5)		– 0.062	
Beef Price (X_6)		0.007	
Fish Price (X_7)		0.005	
Broilers Meat Price (X_3)			0.671

The elasticity of broilers meat price is 0.671, means that if the price increases by 1% the demand in broilers meat increases by 0.671% and vice versa. The price of broilers meat has a linear proportional relationship with the demand. The demand is inelastic, the changes of broilers meat demand is smaller than that of its price[10]. This result is contrary to the law of demand [9]. This is due to the willingness to purchase broilers meat is higher than that of other sources. Broilers meat consumption is influenced by seasonal events and local cultures such as holiday celebration including the Maulidan (Shura) tradition, *Ruwah* tradition (before Ramadan), Ramadan, Shawwal, then Christmas and New Year [16] [17]. This finding suggests that the responses of households are different according to common consumer behaviour. Due to the strong influence of local cultures, households buy broilers meat at a

IOP Conf. Series: Materials Science and Engineering **633** (2019) 012056 doi:10.1088/1757-899X/633/1/012056

high price to meet their important seasonal needs. Thus, the demand will increase even though the prices rise.

5. Conclusions

The estimation showed that broilers meat is the most frequent consumed meats compared to others. The determinants of broilers meat consumption are the number of family members, household income, broilers meat price, egg price, rice, beef, fish, and residence area. The most influential factor is the price of broilers meat since it directly affected the household purchasing power. Broilers meat is classified as staple goods (positive income elasticity). The value of cross elasticity indicates beef, eggs and fish are substitute goods of broilers meat, while rice is a complementary goods. Positive price elasticity indicating that broilers meat price has a direct proportional relationship to demand. In addition, such elasticity is inelastic, therefore it is suggested that the changes of broilers meat demand have a smaller percentage than that of price.

References

[1] Syafrizal M, Ciptadi G, and Budiarto A 2017 *J. Ternak Tropika* **18** 1 pp 51-57
[2] Khomsan A 2010 *Food and Nutrition for Health* (Jakarta : Raja Grafindo Persada)
[3] Pracoyo T K and Antyo P 2006 *Basic Aspects of Microeconomics* (Jakarta : Raja Grafindo Persada)
[4] Ditjen PKH Kudus 2017 Food Security Database http://www.kuduskab.go.id/p/69/_database_ketahanan_pangan
[5] BPS Jawa Tengah [Central Bureau of Statistics Central Java] 2016 Official News Statistics: Development of the Consumer Price Index/Inflation in Central Java No. 81/12/33/th. XI
[6] BPS Kudus [Central Bureau of Statistics Kudus District] 2017 *Kudus District in Figures 2017* Official Gazette of Statistics (Statistical Data)
[7] Kusumawati T D, Marwanti S, Ani S W 2013 *Staple Food Availability Analysis and Rice Food Household Consumption Pattern in Nogosari District, Boyolali Distict* (Surakarta : Faculty of Agriculture UNS)
[8] Supartono K A and Mochamad A 2011 *J. Indonesian Applied Economics* **5** 1 pp 44-56
[9] Sukirno S 2005 *Introduction to Microeconomics Theory* (Jakarta : Raja GrafindoPersada)
[10] Antriyandarti E 2012 *Microeconomics for Agricultural Economics* (Yogyakarta : Nuha Litera)
[11] Baltagi B H 2005 *Economic Analysis of Panel Data3rd Edition* (New York: John Wiley & Sons Inc.)
[12] Park H M 2011 *Practical Guides to Panel Data Modeling: A Step-by-Step Analysis Using Stata* (Minami Uonuma: International University of Japan)
[13] Ermon M N 2012 *J. Kajian Ekonomi* **1** 1 p 4
[14] Yildirim I and Ceylan M 2008 Urban and Rural Households' Fresh Chicken Meat Consumption Behaviors in Turkey *Nutrition & Food Science* **38** 2 pp 154-163
[15] Sukirno S 2005 *Modern Microeconomics* (Jakarta : Raja Grafindo Persada)
[16] Ministry of Trade 2013 *Report of Dynamics Analysis of Indonesian Food Consumption* (Jakarta: Centre of Domestic Trade Policy-Ministry of Trade)
[17] Ani S W and Antriyandarti E 2018 *IOP Conf. Ser.: Earth Environ. Sci.* **142** 012057

Nutritional booklet and social media: Their effects on adolescents' fattening-food knowledge and consumption

Firmansyah[1], D G Tamtomo[2], R Cilmiaty[3]
[1]Master Program of Human Nutrition, Sebelas Maret University, Ir. Sutami Street 36A, Kentingan, Surakarta 57126, Indonesia,
[2]Medicine Faculty, Sebelas Maret University, Ir. Sutami Street 36A, Kentingan, Surakarta 57126, Indonesia
[3]Doctoral Program of Medicine, Sebelas Maret University, Ir. Sutami Street 36A, Kentingan, Surakarta 57126, Indonesia

E-mail: fsyah0202@gmail.com

Abstract. One cause of adolescents' fat-consumption imbalance is inadequate knowledge of nutrition and socialization about good eating habits. Adolescents like searching for food information through social media like Facebook and Instagram before consuming. This study was to analyse the effect of nutrition booklet and social media on the adolescents' fattening-food knowledge and consumptions. The sample of this quasi-experiment was 96 eleventh-grade Science students of SMAN 1, 3, and 5 Surakarta selected using probability proportional size sampling. This study was conducted in one month. Nutritional education intervention with social media and nutritional booklet was conducted six times in two weeks. The data were analysed using Wilcoxon, Kruskal Wallis, and Logistic Binary Regression with SPSS. Nutritional education with nutrition booklet influenced the fattening-food knowledge (p = 0.000) and consumption ($p < 0.05$). Nutritional education with social media influenced the fattening-food knowledge ($p = 0.001$) and consumption ($p < 0.05$). There were different scores of the fattening-food knowledge ($p = 0.001$) and consumption ($p = 0.005$) in the nutritional education group with nutritional booklet and social media. Nutritional booklet and social media influenced adolescents' fattening-food knowledge and consumption. There were different fattening-food knowledge and consumption in the adolescents using nutritional booklet and social media.

1. Introduction

Indonesia, one of the developing countries in the world, nowadays undergoes an epidemic transition whereby the prevalence of infectious diseases is still high and followed by the increase of non-communicable diseases (NCD) commenced from adolescents [1]. The phenomenon pertinent to the increase of NCD prevalence is supported by dietary changes to the consumption of highly saturated fat, high sugar, fast food, and low fibre [2]. The results of National Socio-Economic survey showed that the average of fat intake of Indonesian population reached 58.1 g/cup/day in 2002 and increased to 64.7 g/cup/day in 2009 [3]. In the meantime, in accordance with *Food and Agriculture Organization* [4] the proportion of saturated fat and trans fatty acids respectively reaches 8 and 1% from the aggregate energy. It means that with the sufficient energy 2000 kcal, adolescents need to limit their fat consumption at 56 g/day and the consumption of saturated fat at 18 g/day [5].

One of the causes of fat consumption imbalance is the inadequacy of nutrition-related knowledge and the lack of socialization in association with good eating habits [6]. About 65% of adolescents in

IOP Conf. Series: Materials Science and Engineering **633** (2019) 012057 doi:10.1088/1757-899X/633/1/012057

America have more nutritional status as a manifestation of unbalanced fat consumption [7]. The data of basic health research (*Riskesdas*, an abbreviated term in Indonesian language) in 2007, 2010, and 2013 manifested an increase in the prevalence of obesity (BMI > 25 kg/m) in all age groups, including adolescents [8]. The increasing prevalence of obesity in Indonesia is followed by an increase in obesity in various regions. According to the health profile report of Surakarta City in 2014, the incidence of obesity reached 27.55% in all age groups [9]. The nutrition-related knowledge is still believed to be an important factor influencing the nutritional behaviour of individuals, families and society [10].

In the digital era, adolescents incline to easily access information through social media and the internet. However, this is not optimally used for the efforts to increase knowledge-related nutrition and health by Indonesian youths [11]. The research in Sweden reveals that 46.5% of adolescents have the habit of finding out information about healthy food through social media such as Facebook and Instagram before consuming food [12]. Facebook and Instagram are types of social media that are capable of describing in real terms the form of information to be conveyed [13]. Adolescents nowadays tend to move from texts to the media displaying lots of images to search for information [14]. The research in neuroscience also shows that food images can affect brain activities associated with appetite [15]. According to Li et al, [16] social media can affect health through a couple of ways such as emotional, instrumental and information supports [17]. Furthermore, nutritional education media in the form of booklets can significantly improve the adolescents' knowledge of nutrition [18]. The level of knowledge of a person's nutrition affects the attitudes and behaviour in food selection which will ultimately affect the nutritional condition concerned [19].

The studies about the use of nutritional booklets and social media in improving adolescents' nutrition knowledge are still limited although people are getting worried about the importance of limiting the consumption of fatty foods [20]. Therefore, this study was conducted due to the high prevalence of obesity, lack of physical activities, and high levels of consumption of fatty foods in Indonesia. It aimed to explore more information about adolescents' knowledge and the perception of fattening food.

2. Method

2.1. Study design and subjects

This study used a quasi-experimental research with non-equivalent control group design. This study was conducted at three senior high schools in Surakarta City from August to December 2017. The research subjects were selected by using cluster random sampling and met the inclusion criteria: adolescents (students) aged 16-18 years old who studied at the eleventh grade. The result of anthropometric screening indicated that those students had overweighed nutritional status and had smartphones with them. This research protocol had been approved by the ethical committee of Sebelas Maret University Surakarta with number 354/IV/HREC/2017.

2.2. Data collection

The data with respect to *fattening foods* knowledge were measured based on the answers to questions given about fat which subsumed the subjects' knowledge regarding fat consumption which was comprised of the definition of fat, fat function, the types of fat source food, the amount of fat needed, and the effects of excessive fat consumption. The assessment of nutritional knowledge was carried out by giving scores whereby if the subjects answered correctly, they were subsequently given a score of 1. For those giving a wrong answer, they were then given a score of 0. The evaluation of the subjects' answers was categorized into three types according to Khomsan [21]. The categories extended to "low" if the answer was < 60%, "moderate" if the answer was in the range from 60-80%, and "high" if the correct answer was > 80%.

The data on the consumption level of *fattening foods* of the samples were obtained by conducting direct interview using 24-hour recall form and SQ-FFQ (Semi-Quantitative Food Frequency Questionnaire) form. The parameters used were composed of: the level of total fat consumption: Less

(< 15% of total energy); Enough (15-30% of total energy); and Excessive (> 30% of total energy). The consumption level of saturated fatty acids: Less (< 8% of total energy); Enough (8-10% of total energy); and Excessive (> 10% of total energy). The consumption level of monounsaturated fatty acids: Less (< 15% of total energy); and Enough (≥ 15% of total energy). The consumption level of polyunsaturated fatty acids: Less (< 10% of total energy); and Enough (≥ 10% of total energy) [22].

2.3. Statistical analysis

The data analysis used Statistical Package of Social Sciences (SPSS) software 16.0 version. The data were determined using Fisher's exact test. The logistic regression test was used to multivariate analysis with degree of significance.

3. Results

A total of 96 adolescents were included in this *Quasi Experimental Design* study with female proportion higher than male proportion. Table 1 showed the general characteristics of the research subjects based on three research treatments. The nutritional status in this study was based on BMI/A. The results showed that all subjects of this study had excessive nutritional status (overweight), 23.6 ± 0.62. The range of the subjects' ages in the three study groups was 15-17 years old, and they were mostly 16 years old. Most adolescents have no family history of overweight.

Table 1. The characteristic of 96 research subjects from 3 senior high schools in Surakarta City.

General characteristic	Nutritional booklet group		Social media group		Control group		Mean ± SD
	n	%	n	%	n	%	
Sex							
Male	8	23.5	15	45.5	9	31.0	
Female	26	76.5	18	54.5	20	69.0	
Height (cm)							162.5±5.54
Weight (kg)							62.8±4.54
Age (Years)							
15	7	20.6	6	18.2	2	6.9	
16	25	73.5	25	75.8	23	79.3	
17	2	5.9	2	6.1	4	8.3	
BMI							
Normal	0	0.0	0	0.0	0	0.0	
Overweight	34	100	33	100	29	100	23.6±0.62
Obesity	0	0.0	0	0.0	0	0.0	
Family History of Overweight							
Yes	13	38.2	12	36.3	14	48.2	
No	21	61.8	21	63.7	15	51.8	

The results of statistical test using Wilcoxon (Table 2) indicated that the average score of nutritional knowledge before intervention in the nutrition booklet group was 15.79; that in the social media group was 16.55; and that in the control group was 17.36. After being given an intervention, there was an increase in the average score in all research groups. The results also showed that the nutrition booklet group obtained the highest average score of nutritional knowledge after the intervention by 4.36. Based on the test, it can be discerned that nutritional education provided with the intervention of nutritional booklets and social media ($p = 0.01$) influences the adolescents' knowledge.

Table 2. The effect of nutritional booklet and social mediaon the adolescents' knowledge of fattening food.

Nutritional knowledge	The average score of pre-test and post-test	The average difference score	The minimal score	The maximal score	*p*
Nutritional booklet					
Before	15.79	4.36	15	22	0.00
After	20.14		19	24	
Social Media					
Before	16.55	1.52	16	23	0.00
After	18.07		20	24	
Control Group					
Before	17.36	0.26	15	22	0.266
After	17.62		15	23	

These results indicated a positive influence of nutritional booklet and social media on the adolescents' knowledge appertaining to fattening foods (Table 3).

Table 3. Multivariate analysis of the effect of nutritional booklet and social media on the adolescents' knowledge of fattening foods.

Predictor	B	*p*	Exp (B)	CI 95%		Nagelkerke R Square
				Lower limit	Upper limit	
Nutritional food	0.009	0.829	1.909	0.386	2.140	0.002
Social media	0.095	0.827	1.100	0.467	2.589	

Table 4 showed that the level of fat consumption of the subjects had a significant difference before and after given the treatment, with both nutritional booklets and social media. The level of subjects' fat consumption was obtained from the results of 24-hour recall and SQ-FFQ form to find out the trend of fattening types consumed by adolescents in Surakarta.

Table 4. The effect of nutritional booklet and social media on the adolescents' consumption level of fattening foods.

Consumption level	Nutritional booklet group		Social media group		Control group	
	Before	After	Before	After	Before	After
	n	n	n	n	n	n
Total fat						
Less	2	2	6	3	9	8
Enough	13	22	11	20	9	10
Excessive	19	10	16	10	11	11
PUFA						
Less	6	3	8	3	10	9
Enough	12	21	10	18	10	10
Excessive	16	10	15	12	9	10
MUFA						

Less	15	12	18	11	12	13
Enough	19	22	15	22	17	16
SAFA						
Less	12	10	10	4	16	16
Enough	22	24	23	29	13	13

4. Discussion

The study results using the Wilcoxon test indicated that there was an influence of nutritional education through nutritional booklet on the adolescents' nutritional knowledge ($p = 0.001$). The nutritional education using nutritional booklet could improve the adolescents' nutritional knowledge 1.909 times compared to the knowledge of those in the control group. This result is in line with the result of study conducted by Sukiyasa and Sukoco [23] on vocational school students whereby it showed that students' learning outcomes in the learning of automotive electrical system material taught using nutritional booklet media were higher than those taught using PowerPoint media. The difference in the average post-test scores in both groups showed that the nutritional booklet media was more effective in improving students' learning outcomes. The use of nutritional booklet media in learning made it easy for students to receive abstract lessons. The study result conducted by Noviyanto et al [24] in junior high school students showed that the nutritional booklet media of the human respiratory system could improve biology-learning outcomes. Nutritional booklet media could make it easier for students to understand the respiratory processes that occur in the human body so that they can understand the learning material of the respiratory system as a whole.

The other media used in the process of nutritional education is social media. The nutritional education through social media could improve adolescents' nutrition knowledge 1.100 times compared to the knowledge of those in the control group. The result of this study is in line with the study conducted by Utami et al [25] that there is an influence of Instagram-assisted project-based learning model on creative thinking skills of students on the X grade of SMA Negeri 8 Surakarta. It is supported by the data of students' average score in the experimental class (82.72) higher than that in the control class (77.12). The result of Lukitasari et al [26] study on high school students also shows that the use of Facebook as a means of increasing students' motivation and achievement is able to provide good changes to the Natural Sciences (IPA) subject, especially biology subject.

Social media is an online communication media that allows individuals and groups to create content, various kinds of information connected with one another. Adolescents today prefer to use technological sophistication to search for various kinds of information that they need and are considered not boring because various interesting contents in it can also be accessed at the same time. Accordingly, social media can be used as an educational medium which can reach a large number of targets that are not limited to space and time [27].

5. Conclusion

There is effect of nutritional booklet and social media on the knowledge and consumption level of fattening foods in adolescents. The given intervention is proven able to increase the adolescents' knowledge of fattening foods 1.909 and 1.100 times and is able to change the consumption level of fattening foods for adolescents in Surakarta.

Acknowledgment

We would like to express our gratitude to the headmasters of three senior high schools who had allowed us to conduct this study and to research subjects who had participated in our study. A huge gratitude is also expressed to Prof. Bhisma Murti, dr., MPH, M.SC, Ph.D who had guided us in analysing the statistics and to our enumerators and friends who had helped the accomplishment of this study.

IOP Conf. Series: Materials Science and Engineering **633** (2019) 012057 doi:10.1088/1757-899X/633/1/012057

References

[1] Kemenkes. 2013. Laporan Nasional Riset Kesehatan Dasar 2013. Badan Penelitian dan Pengembangan Kesehatan, Kementerian Kesehatan, Republik Indonesia

[2] Popkin BM. 2002. An Overview on Thenutrition and Its Health Implications: the Bellagio meeting. *Public Health Nutr*, vol. **5**, p 93-103

[3] Hardinsyah. 2011. Analisis konsumsi lemak, gula, dan garam penduduk Indonesia. *Jurnal Gizi Indonesia,* vol: **34**, p 92-100

[4] FAO. 2010. *Human Vitamin and Mineral Requirements.* Report of joint FAO/WHO Expert Consultation. Roma (IT): FAO

[5] Benyamin, Caballero, Lindsay Allen, Andrew Prentice. 2016. *Encyclopedia of Human Nutrition Third Edition*. Instituto de Salud Carlos III, Universidad de Navarra, Pamplona, Spain

[6] Eliana, Desy dan Solikhah. 2012. Pengaruh Buku Saku Gizi Terhadap Tingkat Pengetahuan Gizi Pada Anak Kelas 5 Muhammadiyah Dadapan Desa Wonokerto Kecamatan Turi Kabupaten Sleman Yogyakarta. *KES MAS*, vol. **6**, p 162-232

[7] Louise A. Berner, PhD; Debra R. Keast, PhD; Regan L. Bailey, PhD, RD; Johanna T. Dwyer, DSc, RD. 2013. Fortified Foods Are Major Contributors to Nutrient Intakes in Diets of US Children and Adolescents. *Journal of the Academy of Nutrition and Dietetics*, vol. **114**, p 1009-1022

[8] Kemenkes. 2013. Laporan Nasional Riset Kesehatan Dasar 2013. Badan Penelitian dan Pengembangan Kesehatan, Kementerian Kesehatan, Republik Indonesia

[9] Dinkes Kota Surakarta. 2014. Profil Kesehatan Kota Surakarta Tahun 2014. Surakarta. Dinas Kesehatan Kota Surakarta

[10] Demirozu, B., Pehlivan, A., Camliguney, A. 2012. Nutrition Knowledge and Behaviours of Children Aged 8-12 WHO Attend Sport Schools. *Procedia Social and Behavioral Sciences*, vol. **46**, p 4713-4717

[11] Primada. 2015. Hubungan Antropometri, Aktivitas Fisik, dan Pengetahuan Gizi dengan Asupan Energi dan Komposisi Makronutrien pada Remaja. *eJKI,* vol. **1**, p 90-99

[12] Christoper, Holmberg, Jhon E. Chaplin. Thomas Hillman, Christiana Berg. 2016. Adolescents' presentation of food in social media: An explorative study. *Appetite*, vol. **99**, p 121-129

[13] Findahl, O. 2014. Svenskarna och Internet 2014. Retrieved from https://www.iis.se/docs/SOI2014.pdf

[14] Fritz, G.K. 2014. Social media use and adolescents: a guide for parents. *The Brown University Child and Adolescent Behavior Letter*.

[15] Beaver, J. D., Lawrence, A. D., van Ditzhuijzen, J., Davis, M. H., Woods, A., & Calder, A. J. 2009. Individual differences in reward drive predict neural responses to images of food. *Journal of Neuroscience*, vol. **26**, p 1-9

[16] Li, J., Barnett, T., Goodman, E., Wasserman, R., Kemper, A. 2012. Approaches to Prevention and Management on Childhood Obesity: The Role of Social Networks and the Use of Social Media and Related Electronic Technologies. *American Heart Association Journal*, vol. **127**, p 1-8

[17] Lagiono. 2012. Pola Implementasi Jejaring Sosial Facebook sebagai Media dalam Pembelajaran. *Jurnal Ilmiah Kependidikan*, vol. **7**, p 37-43

[18] Agustin, Maria. 2014. Efektifitas Pendidikan Kesehatan Media Booklet Dibandingkan Audiovisual Terhadap Pengetahuan Orang Tua Tentang Karies Gigi Pada Anak Usia 5-9 Tahun Di Desa Makamhaji. *Jurnal Kesmas*, vol.**7**, p 1-16

[19] Hausmann, Jonathan S., Currie Touloumtzis, Matthew T. White, James A. Colbert, and Holly C. Gooding. 2017. Adolescent and Young Adult Use of Social Media for Health and Its Implications. *Journal of Adolescent Health*, vol. **30**, p 1-9

[20] Shepherd J, Harden A, Rees R, *et al.* 2016. Young people and healthy eating: A systematic review of research on barriers and facilitators. *Health Education Research*, vol. **21**, p 239–257

[21] Khomsan, A. 2000. *Teknik Pengukuran Pengetahuan Gizi*. Bogor (ID): Fakultas Pertanian, Institut Pertanian Bogor.

IOP Conf. Series: Materials Science and Engineering **633** (2019) 012057 doi:10.1088/1757-899X/633/1/012057

[22] EFSA. 2010. 2010. Scientific opinion on dietary reference values for fats, including saturated fatty acids, polyunsaturated fatty acid, monounsaturated fatty acids, trans fatty acids, and cholesterol. *EFSA Journal,* vol. **8**, p 1461-1508

[23] Sukiyasa, K and Sukoco. 2013. Pengaruh Media Pembelajaran Terhadap Hasil Belajar dan Motivasi Belajar Siswa Materi Sistem Kelistrikan Otomotif. Jurnal Pendidikan Vokasi, vol. 3, no. **1**, p 126-137

[24] Noviyanto, T., Juaeningsih, N., Rosyodatun, E. 2015. Penggunaan Media Pembelajaran Sistem Pernapasan Manusia untuk Peningkatan Hasil Belajar Biologi. Edusains, vol**. 7**, no. 1, p 57-36

[25] Utami, R., Probosari, R., Fatmawati, U., 2015. Pengaruh Model Pembelajaran Project Based Learning Berbantu Instagram Terhadap Kemampuan Berfikir Kreatif Siswa kelas X SMA Negeri 8 Surakarta. Bio-Pedagogi, vol. **4**, no. 1, p 46-52

[26] Lukitasari, M., Pujiati, Kristianto, I. 2015. Penggunaan Jejaring Sosial (facebook) Sebagai Sarana Meningkatkan Motivasi dan Prestasi Belajar Biologi. Jurnal Pendidikan Informatika dan Sains, vo. **4**, no. 1, p 91-98

[27] Shah, V and Kotsenas, M. 2017. Sosial Media Tips to Enhance Medical Education. Academic Radiology, vo; **24**, no. 6, p 747-752

IOP Conf. Series: Materials Science and Engineering **633** (2019) 012058 doi:10.1088/1757-899X/633/1/012058
***** Formatting Issue - Best Available Paper/Graphic *****

Optimization of pyrolysis condition for bioactive compounds of wood vinegar from oil palm empty bunches using response surface methodology (RSM)

H A Oramahi[1], Elvi Rusmiyanto P. Wardoyo[2,3] and Kustiati[2]

[1] Faculty of Forestry, Universitas Tanjungpura, Jalan Prof. Dr, Hadari Nawawi, Pontianak, West Kalimantan 78124, Indonesia
[2] Faculty of Mathematics and Natural Sciences, Universitas Tanjungpura, Jalan Prof. Dr, Hadari Nawawi, Pontianak, West Kalimantan 78124, Indonesia

E-mail: elvirusm1971@gmail.com

Abstract. Optimized of the phenol and acid total of wood vinegar from oil palm empty bunches using response surface methodology was carried out. Response surface methodology (RSM) with central composite design (CCD) was employed to optimize the parameters of pyrolysis process, namely pyrolysis temperature (X_1) and pyrolysis times (X_2). Two variables, namely pyrolysis temperature (°C) and pyrolysis time (min) with 5 levels (-1.414, -1, 0, 1 and 1.414) were used to design the optimization model. The two variables, a temperature of 420 °C and a pyrolysis temperature of 122.40 minutes were the optimized conditions for the response variables of phenol total content. Meanwhile, the temperature of 422 °C and the pyrolysis temperature of 118.5 minutes were the optimized conditions for the response variables of the total acid content. The regression equation obtained for phenol and acid total content were $Y = 2.97 + 0.12 X_1 + 0.03 X_2 + 0.23 X_1^2 - 0.24 X_2^2 – 0.24 X_1.X_2$, and $Y = 9.89 + 0.63 X_1 + 0.70 X_2 - 0.56 X_1^2 + 0.12 X_2^2 – 1.70 X_1.X_2$, respectively. The optimal conditions for the highest yield of phenol and acid total were 3.13 and 10.04 %, respectively.

1. Introduction

Wood vinegar prepared from many different sources of wood have been use as safe natural inhibitors with various applications, such as antifungal, antimicrobial, and antioxidant activities [1, 2, 3]. Wood vinegar obtained from pineapple plant waste biomass has been reported possessing antioxidant activity, and rich of phenolic compounds especially guaiacol, catechol, methoxy catechol and syringol [4]. Wood vinegar, also called liquid smoked, a yellowish or dark brown condensate made from wood carbonization process at a high temperature in the absence of oxygen [5].

Various studies about a different kind of wood, pyrolysis temperature, and temperature time for the wood vinegar yield and bioactive compound have been published. The temperature of pyrolysis and types of wood are dominant factor for yield of wood vinegar and bioactive component of wood vinegar including phenol and phenolic compounds [6, 7]. For example, the highest yield of phenol and phenolic compound in wood vinegar made from palm kernel shells was about 70% area with increasing temperature up to 475 °C [7].

IOP Conf. Series: Materials Science and Engineering **633** (2019) 012058 doi:10.1088/1757-899X/633/1/012058

Response surface methodology (RSM) approach succeeded in optimizing the extraction conditions of the phenolic for antioxidants from the fruits of *Euterpe oleracea* [8] and *Artemisia absinthium* [9]. Optimization of operating parameter extraction of antioxidant from *Coriandrum sativum* seeds was studied [10] and it was found that optimum condition of subcritical water extraction was obtained at temperature of 200 °C, pressure of 30 bar and extraction time of 28.3 min.

However, it has not previously been used to optimization of bioactive compound including phenol and acid total in wood vinegars made from oil palm empty bunches. The aim of the study is to develop a maximum yield of phenol and acid total of wood vinegar from oil palm empty bunches.

2. Materials and Methods

2.1. Pyrolysis of wood vinegar

Wood vinegar was produced from burning wood particles of oil palm empty bunches. The material was collected from Bumi Pratama Khatulistiwa (BPK) Ltd. and was converted into particle with a disk mill, and air dried of moisture content (12.0%). The pyrolysis process was conducted at the Engineering Laboratory, Faculty of Agriculture Engineering, Universitas Gadjah Mada, Yogyakarta, Indonesia.

2.2. Analysis of phenol total

Phenol total of wood vinegar from oil palm empty bunches was calculated by using the method of spectrochopy [11]. Wood vinegar (1mL) was diluted 1.000 times. One mL of the diluted sample was put into 5 mL 2% Na_2CO_3 solution. After 10 minutes, 0.5 mL of Folin Ciocalteau suspension was added to the solution. Absorbance of the solution at 750 nm was measured by a UV-Vis spectrophotometer (Shimadzu 1601, Shimadzu Manufacturing Co. Ltd, Japan). The phenol total content in each wood vinegar was measured by comparing with that of pure phenol suspension.

2.3. Analysis of Acid Total

The acidity of the wood vinegar was calculated using the method of AOAC [12]. Wood vinegar (1 mL) was diluted to 100 mL with deionised (DI) water. The 3 drops PP (Phenolphthalein) indicator was added into it, and the solution was titrated by 0.1 N NaOH.

2.4. Experimental Design

CCD was applied to optimize phenol and acid total content made from oil palm empty bunches. Operation condition process including pyrolysis temperature (X_1) and pyrolysis time (X_2) were the selected independent variables (table 1).

Table 1. The level of variable used for the central composite design

Independent Variable	Symbol	Coded variable level				
		- α (-1.44)	-1	0	1	+ α (1.44)
Temperature of pyrolysis (°C)	X_1	329.30	350	400	450	470.70
Times of pyrolysis (min)	X_2	98.79	105	120	135	141.21

These variables were low and high factor settings are coded as -1 and 1, the midpoint coded as 0. The factor settings of trails that run along axes drawn from the middle of the cube through the centers of each face of the tube are coded as -1.414 or 1.414, and total of 13 experimental runs.

For optimum point prediction, second-order polynomial equation is following:

$$Y = \beta_0 + \sum_{i=1}^{k} \beta_i x_i + \sum_{i=1}^{k} \beta_{ii} x_i^2 + \sum_{i} \sum_{j} \beta_{ij} x_i x_j + \varepsilon \quad (1)$$

IOP Conf. Series: Materials Science and Engineering **633** (2019) 012058 doi:10.1088/1757-899X/633/1/012058

where β_0 β_i, β_{ii} and β_{ij} are the regression coefficients for intercept, linear, quadratic and interaction terms, respectively, and x_i, x_j are the independent variables [14].

2.5. *Statistical analysis*

SAS software (version 8.2, SAS Institute Inc., NC. USA) were used for the analysis of the central composite design.

3. Results and Discussion

3.1. *Maximizing the phenol total content of wood vinegar*

For determining the temperature pyrolysis (X1), and times of pyrolysis (X2) within the range of 329.30-470.70 °C, and 98.79-141.21 minutes, respectively, the trials were designed based on a CCD. The phenol total compounds of wood vinegar from oil palm empty bunches was further optimized using RSM [13]. Two factor levels in coded and un-coded values are shown in table 1. The experimental design and the observed data for phenol total content is shown in table 2.

Table 2. Real value of variables and response variable of *central composite design*

Run	Temperature (°C; X_1)	Times (min; X_2)	Responses	
			Phenol total (Y_1)	Acid total (Y_2)
1	350 [a)] (-1)[b)]	105[a)] (-1)[b)]	1.72	6.02
2	350 (-1)	135 (1)	2.64	11.01
3	450 (1)	105 (-1)	3.16	10.83
4	450 (1)	135 (1)	3.10	9.04
5	400 (0)	120 (0)	3.11	10.93
6	400 (0)	120 (0)	3.03	8.97
7	400 (0)	120 (0)	2.72	9.33
8	400 (0)	120 (0)	2.64	9.55
9	400 (0)	120 (0)	3.28	10.69
10	470,70 (1.414)	120 (0)	3.51	9.74
11	329,30 (-1.414)	120 (0)	4.06	8.20
12	400 (0)	141,21(1.414)	2.88	11.19
13	450 (0)	98,78 (-1.414)	3.33	9.50

Where: a) Condition of treatment and b) Code

Some factors contribute to maximizing the phenol and phenolic compound of wood vinegar were temperature, feed size and feed rate [7, 15]. Similar results were obtained indicating that high coefficient of determination showed that the model can be efficiently be applied to prediction biogas productivity [16].

Figures 1 show the three dimensional (3D) response surface curve of graphical illustration on the effect of pyrolysis temperature and pyrolysis time on the phenol total compound of wood vinegar.

IOP Conf. Series: Materials Science and Engineering **633** (2019) 012058 doi:10.1088/1757-899X/633/1/012058
***** Formatting Issue - Best Available Paper/Graphic *****

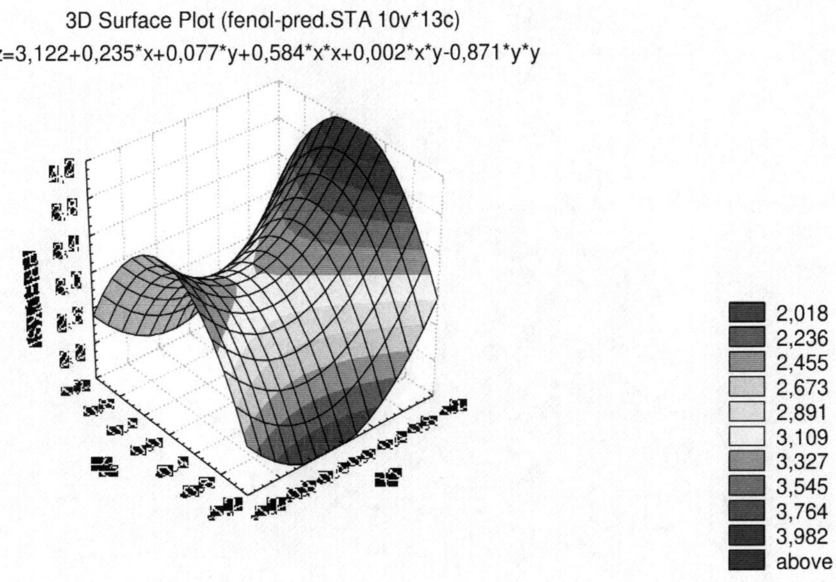

Figure 1. Response surface curve for phenol total compound of wood vinegar showing the interaction between pyrolysis temperature (X_1; °C) and times pyrolysis (X_2; min)

The maximizing of phenol total compound of wood vinegar was selected as the responses for the combination of independent variables (table 2). The stationer point of phenol total content of wood vinegar, pyrolysis temperature, and pyrolysis times were found to be 420 °C, 122.40 minutes while the maximum of phenol total content of wood vinegar was 3.13 %.

The responses of the CCD fitted with a second-order polynomial equation, illustrating the phenol total content, is given by following:

$$Y = 2.97 + 0.12\ X_1 + 0.03\ X_2 + 0.23\ X_1^2 - 0.24\ X_2^2 - 0.24\ X_1.X_2$$

3.2. *Maximizing the acid total content of wood vinegar*

The acid total compounds of wood vinegar from oil palm empty bunches was further optimized using RSM. Two factor levels in coded and un-coded values are shown in table 1. The experimental design and the observed data for acid total content is shown in table 2. The ANOVA was statistically significant ($p<0.05$) and suggested that at least one of the parameters of the models can explain the experimental variation for acid total content.

Figures 2 shows the three dimensional (3D) response surface curve of graphical illustration on the effect of pyrolysis temperature and pyrolysis time on the acid total compound of wood vinegar.

IOP Conf. Series: Materials Science and Engineering **633** (2019) 012058 doi:10.1088/1757-899X/633/1/012058

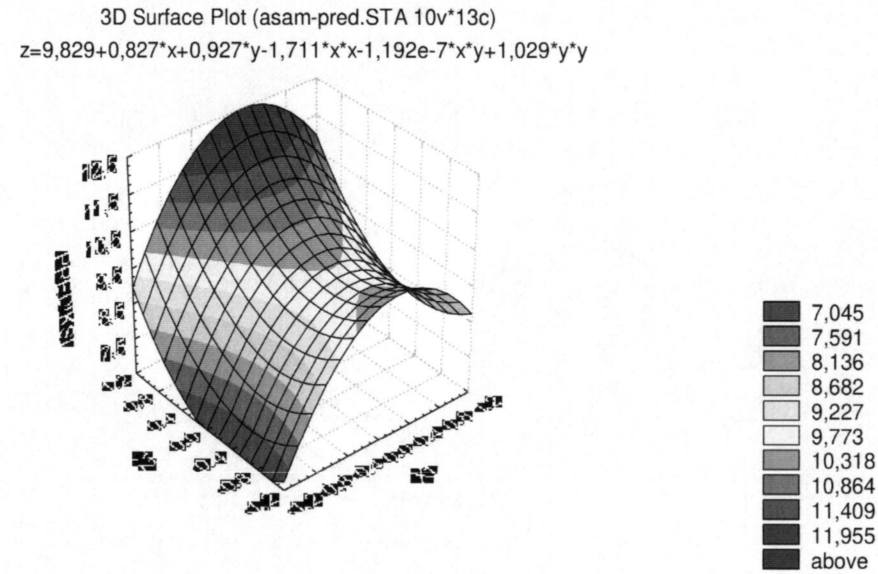

Figure 2. Response surface curve for acid total compound of wood vinegar showing the interaction between pyrolysis temperature (X_1; °C) and times pyrolysis (X_2; min)

The maximizing of acid total content of wood vinegar was selected as the responses for the combination of independent variables (table 2). The stationary point acid total content of wood vinegar was pyrolysis temperature and pyrolysis times were found to be 422 °C, 118.50 minutes while the maximum acid total content of wood vinegar was 10.04 %.

The responses of the CCD fitted with a second-order polynomial equation, illustrating the acid total content, is given by following:

$$Y = 9.89 + 0.63 X_1 + 0.70 X_2 - 0.56 X_1^2 x + 0.12 X_2^2 - 1.70 X_1.X_2$$

4. Conclusions

Optimization by RSM-CCD suggested the pyrolysis temperature and pyrolysis time were 420 °C and 122.40 minutes and the maximum phenol total content of wood vinegar was 2.98 %. Meanwhile, the maximum acid total content of wood vinegar was 10.04 % and the pyrolysis temperature and pyrolysis times were 422 °C dan 118.50 minutes.

Acknowledgments
The authors would like to acknowledge the financial support by Universitas Tanjungpura and Ministry of Research, Technology and Directorate General of Higher Education of Indonesia in fiscal year 2018, through a research grant for International research collaboration and scientific publication.

References

[1] Kiarie-Makara MW, Yoon H-E and Lee D-K 2010 *Entomol. Res.* **40** 97–103

[2] Ma C, Song K, Yu J, Yang L, Zhao C, Wang W, Zu G and Zu, Y 2013 *J. Anal Apl Pyrol* **104** 38-47

[3] Kim KH, Jeong HS, Kim JY, Han GS, Choi IG and Choi JW 2012 *Chemosphere* **89** 688-93

[4] Mathew S, Zakaria, ZA and Musa, NF 2015 *Process Biochem.* **50** 1985-92

IOP Conf. Series: Materials Science and Engineering **633** (2019) 012058 doi:10.1088/1757-899X/633/1/012058
*** Formatting Issue - Best Available Paper/Graphic ***

[5] Lee SH, H'ng PS, Lee AN, Sajap AS, Tey BT and Salmiah 2011 *J. Appl. Sci.* **10** 2440-6

[6] Akhtar J and Amin NS 2012 *Renew. Sust. Energy Rev.* **16** 5101-9

[7] Kim S J, Jung S H and Kim, J S 2010 *Bioresour. Technol.* **101** 9294-300

[8] Pompeu DR, Silva EM and Rogez H 2009 *Bioresour. Technol.* **100** 6076-82

[9] Şahin, S, Aybastıer Ö, and Işık E 2013 *Food Chem.* **141** 1361-8

[10] Zeković Z, Vidović S, Vladić J, Radosavljević R, Cvejin A, Elgndi MA and Pavlić B 2014 *The J. Supercrit. Fluids* **95**,560-6

[11] Senter SD, Robertson JA and Meredith FI 1989 *J. Food Sci.* **54** 1259–68

[12] AOAC, association of official analytical chemists 1990 *Official Methods of Analysis.* AOAC, Arlington

[13] Montgomery DC 2008 *Design and Analysis of Experiments* (New York: John Wiley & Sons)

[14] Sodeifian G, Ardestani, NS and Sajadian SA 2018 *J. For. Res.* 1-11

[15] Demiral I and Ayan EA 2011 *J. Bioresour. Technol.* **102** 3946–51

[16] Safari M, Abdi R, Adl, M and Kafashan J 2018 Renew. Energy. **118** 368-75

IOP Conf. Series: Materials Science and Engineering **633** (2019) 012059 doi:10.1088/1757-899X/633/1/012059

Economic glocalization: A sustainable local food production system in Wonosobo Indonesia

A Wibowo[1,2], R Karsidi[2], B Sudardi[3] and M Wijaya[4]

[1] Department of Agriculture Extension and Communication, Sebelas Maret University, Jl. Ir. Sutami 36 A, Kentingan, Surakarta,57126, Indonesia
[2] Doctoral Program in Development Extension, Graduate School, Sebelas Maret University, Jl. Ir. Sutami 36 A, Kentingan, Surakarta, 57126, Indonesia
[3]Department of Sociology, Sebelas Maret University, Jl. Ir. Sutami 36 A, Kentingan, Surakarta, 57126, Indonesia
[4]Department of Indonesia Language, Sebelas Maret University, Jl. Ir. Sutami 36 A, Kentingan, Surakarta, 57126, Indonesia

Email: agungwibowo@staff.uns.ac.id

Abstract. Economic glocalization has been developed in the local food production system in Wonosobo. Glocalization is indicated by the existence of sustainability in the food production system and local empowerment. The aim of this study was to describe economic glocalization in the local food production system (carica and purwaceng) in Wonosobo. The qualitative method used in this study was based on the paradigm of constructivism with a phenomenological approach including knowledge, attitude and behavior. The result of this study showed that local products of carica (*Carica candamsrcensis)* and purwaceng (*Pimpinella pruatjan*) were able to compete with global product capitalism. The impact of the glocalization of carica and purwaceng was to develop local economic institutions such as productive economic enterprises and cooperatives. In addition, economic glocalization in the local food production system in Wonosobo had also the effect of the increase production capacity, product development, market expansion, local economy and eco-tourism.

1. Introduction

Dieng, Indonesia is known as a producer of vegetables and fruits because it has fertile soil with a very high organic carbon content, a high cation exchange capacity, Ca and K content, and an adequate N, P, and Na content [1]. The main commodities produced there such as potatoes, cabbages, chilies, and tomatoes have been exported to other towns and even to other islands, and they were main source of livelihood for the local residents. Fortunately, Dieng also has a special and unique local product, carica (*Carica candamsrcensis)* and purwoceng (*Pimpinella pruatjan*), which can only grow well in that area. This fact is also supported by evidences showing the products are unable to grow well or produce fruit in other regions with almost the same topography. Attempts have been made to cultivate carica

IOP Conf. Series: Materials Science and Engineering **633** (2019) 012059 doi:10.1088/1757-899X/633/1/012059

plants in other areas such as on the slopes of Lawu Mountain in Karanganyar District, but it failed to thrive and its fruit had a different taste and aroma. For this reason, the plants can be included in the category of Geographical Indications. Along with eco-tourism development in the region of Dieng have led to increase demand of carica and purwaceng both local and foreign tourists. At the present time, the production of two processed products is developing rapidly, and their market distribution is continuing wider. The increase these products led to the formation of a local food production system in Dieng, Wonosobo. Thus, this system was used as a tool for the development of economic glocalization.

Glocalization has been described as the process of "creating a product or service that is directed at a global market but adapted to a local culture" [2] [3]. Glocalization, which involves the interaction of local and global cultures, creates heterogeneity in life. Based on the idea of postmodernists, glocalization has two sides of a reciprocal influence between global and local culture. On one side, the strength of identity of the local (traditional) culture means that the global culture is unable to eliminate the local culture, but on the other side, the global culture absorbs elements of the local culture. The phenomenon of glocalization in society has started a contetention between local cultures and global (capitalist) cultures [4].. Connecting global with local has subsequently posed a huge challenge for community development. Glocalization forces a community to evaluate the consequences of developing its region and repositioning the community in a global context [3] [5]. Hence, glocalization is a good illustration of the combination and connection of local and global contexts while continuing to maintain a significant contribution of different cultural communities and contexts [5]. Glocalization has been successfully used in various fields such as marketing, education, theology, and so on, as an effective and prudent way of utilizing global ability to fulfil local demand. This is one method for meeting these needs pragmatically. Glocalization is implemented to internationalize the crisis of communication which profits practitioners and researchers by evading the application of Western frameworks and interpretations in non-Western crisis situations [6] [7] This has been done by demonstrating the approach of a case that involves the multinational McDonald's Corporation and its foreign subsidiary, McDonald's Japan [8].

Community empowerment is an endeavour to improve the dignity and pride of communities that are unable to break free from the poverty trap and the shackles of underdevelopment. The success of community development is determined by the participation of the community and the role of the relevant government. One great innovation of the concept of glocalization is that local communities are not portrayed as passive "actors" in the process. This can be interpreted as meaning that local communities have the power to create their own vision about the interfusion of hybrid cultures, creole, and cosmopolitan cultures.This research describes a local community that has local potential and local traditions capable of developing glocalization to deal with the rapid infiltration of capitalis.

2. Research Methods

This research focused on the community living in and around the highland of Dieng, Wonosobo which is an area producing geographical indication plants namely carica and purwaceng. In this area, there is a great deal of local wisdom, which is densely packed with local cultural values and has managed to synergize with the potential of the beautiful natural surroundings to attract large numbers of tourists to this area. This research used a qualitative method which centered on an investigation of the ways people interpret a social life that continues to be guided by the synergy of culture and local wisdom in developing the existing local potential. The study follows the method that connect the subjective characteristics of a social reality, a method which is highly capable of generating an understanding of the perspective of stakeholders, and also enables the researcher to view various things as they are viewed by their subjects [9]. This research was a case study. According [10], the way of distinguishing the type of a research study is by identifying the type of research question that is posed from the start with respect to operational connections that require separate time tracking and do not simply depend on frequency or appearance. This research also refers to the opinion of [11] in which the researcher explores a bounded system (a case) or multiple bounded systems (cases) over time, through detailed in-depth data collection involving several sources of information, and reports the description of a case or case-based theme.

The data collection in this research refered to [12] in that the researcher collected and then organized and also analyzed the data about certain cases related to the problems addressed by the researcher so that the data can subsequently be compared or connected with other data (in multiple cases) while adhering to holistic and contextual principles. In this study, the researcher understands the potential of social behaviour occurring in an inter-relationship activity between individuals in a community, specifically concerning the extent to which the macroscopic structural factor and institutional factor influence the relationship between the individuals. The social behaviour that was studied to be understood in depth includes the aspect of local cultural values that have developed and local wisdom of farmers in line with the economic glocalization that was taking place to develop the potential of local food.

The primary data source for qualitative data is words and actions, while supplementary data includes documents and other sources [13]. The techniques for data collection used in this research were in-depth interviews with important community figures, participant observation of the daily lives of the farmers, focus group discussions involving stakeholders in order to seek an understanding of the data and information developing in the community, and a documentary study. In each case, the analysis is processed using an interactive model of analysis [14]. In this model of analysis, the three components of the analysis are data reduction, data presentation, and verification.

3. Result and Discussion

Efforts to prevent the emergence of local food products dominated by global food products are by building a sustainable local food production system. The development of a local food production system had been carried out in Wonosobo by integrating various potentials in the region including local institution, local wisdom, local government and society so as to create economic glocalization of local food products (Figure 1.).

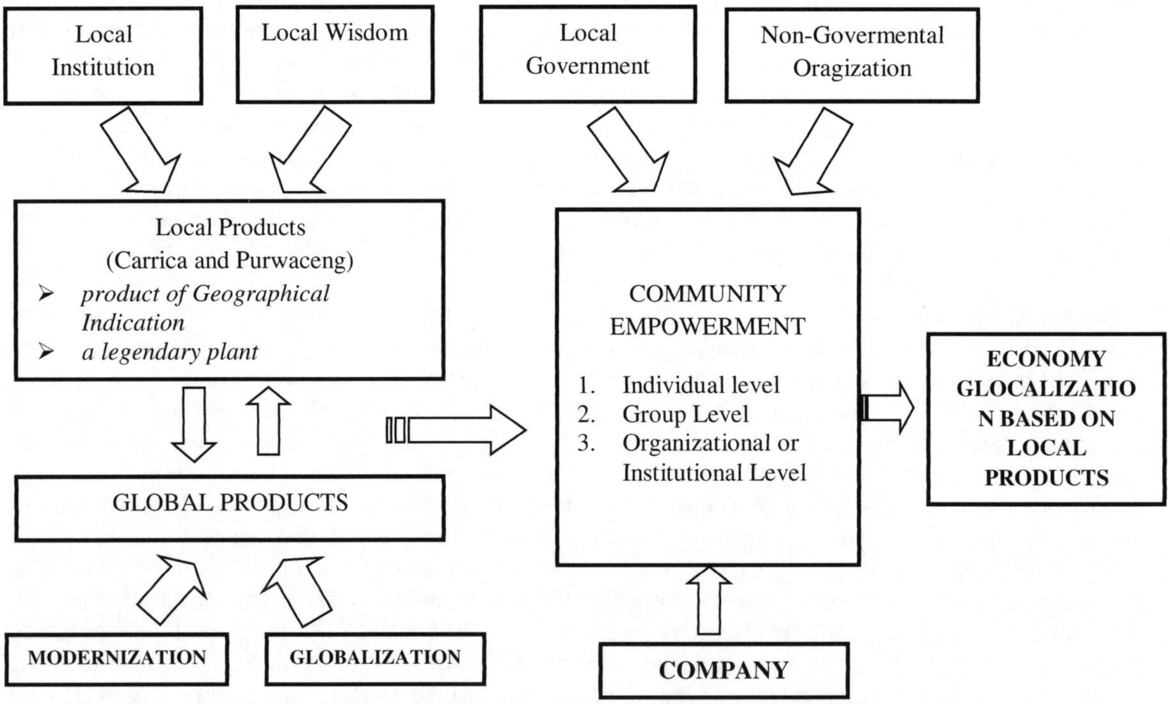

Figure 1. Model of the economic glocalization as a sustainable local food production system

3.1. Carica and purwaceng plants as a product of Geographical Indication.

Local products in Wonosobo Regency that are able to compete with global products are Carrica and Purwaceng. Both of these commodities are able to compete because they are products of geographical indications and are legendary plants. Carica is a small tree or shrub which is non-woody and similar to a normal papaya (*Carica papaya L.*), but it has more branches and the dimension of all the parts of the plant are smaller. Its average height is 1-2 meters, its stamen has a long stem, growing to a length of 15 cm, and its pistil is larger, with a short, hard stem. The Dieng papaya is oval shaped and measures 6-10 cm in length and 3-4 cm in width. A fruit that is ripe is shaped like an upside-down egg measuring 6–15 cm x 3–8 cm, with hard, yellowish-orange flesh and a slightly sour but fragrant taste, and a lot of seeds that are wrapped in a white, watery sarcotesta. The unripe fruit has a dark green skin which turns yellow as the fruit ripens. The seeds are black, copious, and densely packed. The fruit contains sap which gradually becomes less as the fruit ripens. This sap contains proteolytic papain [15].

The purwaceng plant, meanwhile, also grows only at an altitude of over 1,400 meters above sea level, and more specifically only in certain parts of the Kejajar Regency in Dieng, such as the village of Sikunang, Kejajar. The physical appearance of the purwaceng plant is like a small bush which exhibits a vine-like growth over the surface of the soil, similar to the gotu kola plant and mountain clover. It has small reddish green leaves with a diameter of 3.5 cm. This plant is usually processed into the form of a health drink for increasing stamina, mainly in the form of purwaceng tea, purwaceng coffee, and purwaceng milk. According to local farmers, purwaceng is believed to have aphrodisiac properties that increase male stamina. As a result of the development of various tourist objects in the Dieng region and the increasing number of visitors in this areas, purwaceng has become more popular and in greater demand in other towns and cities and even abroad.

3.2. Purwaceng as a legendary plant that has been mythicized and supports glocalization

It is described as a legendary plant used as a tonic by kings and other members of the Javanese courts. In Indonesia, the use of medicinal plants or other plants with aphrodisiac properties is generally based on myth, belief, or experience. The purwaceng plant has been mythicized as a plant that has the ability to give men a stronger erection. Originally, this myth was only known in the mountain region where the purwaceng plant grew, but nowadays the myth has become more widespread as a result of the increasing number of tourists visiting the Dieng area.

In terms of people's beliefs, many people believe that a wild plant that can only grow in a particular place must have medicinal properties. This belief is still upheld by rural communities in Indonesia. In terms of experience, people who have visited the various tourist objects in Dieng return home with stories to share with their friends and families about the purwaceng drinks. These stories are passed on by word of mouth by the purwaceng drink sellers in Dieng, and this has evidently had an impact on the demand for the purwaceng plant packaged in the form of a drink of tea or coffee.

3.3. The role of Local institutions, local wisdom, local government and local community

Local institutions and local wisdom that have existed in the community contribute to the survival of local products in line with many global products in the face of modernization and globalization. Local wisdom originating from local communities and combined with local potential can be used as an example in facing today's global life. In terms of environmentally sound development, the Dieng community is well aware of the importance of nature for the sustainability of future generations. This awareness creates a great respect for the people of Dieng for the existence of nature by not exploiting nature excessively for economic benefits. Meanwhile in terms of glocalization, the Dieng community can utilize globalization through the tradition of dreadlocks as a tool to make local wisdom and local traditions owned as a global center and development model. Although it is undeniable that local wisdom possessed by the Dieng community is no longer pure and tends to be seen as a result of a combination of foreign influences.

The survival of local products is inseparable from the synergy of the role of local governments, non-governmental organizations and companies. The role of the three entities is seen in empowering both at the individual, group and organizational and institutional levels. On the basis of the role of the three entities, local communities have a collective awareness and critical awareness that moves massively and sustainably and has independence which leads to economic indegenous based vocalization. This economic globalization will create sustainability if the three entities each have a balanced role. The basic thing as a condition of economic glocalization will be, if a region has local products.

3.4. Carica and purwaceng plants as tools for developing productive economic enterprises.
The carica fruit is a plant that is unique to the Dieng area and can be processed to become a variety of products, such as candied fruit in syrup and carica porridge. The purwaceng plant, on the other hand, is processed in the form of purwaceng tea, purwaceng coffee, and purwaceng milk. All these products are now well known due to the developments in the tourist sector in Dieng. The current situation is that farmers are continuing to cultivate carica and purwaceng plants because of the good prospects and the ability to advance local productive economic enterprises.

There are a number of indicators which prove that the carica and purwaceng plants have supported the development of productive economic enterprises. *First*, carica and purwaceng plants can be processed in various ways to produce a variety of products. *Second,* carica and purwaceng have encouraged the development of new enterprises, both in the Dieng region itself and in other nearby areas. Carica and purwaceng have become a trade mark of the region and designated as the traditional food of Dieng Plateau in Wonosobo, Central Java. *Third,* numerous business groups and cooperatives have developed which are able to accommodate carica and purwaceng producers both in Dieng itself and in other regions, thus acting as a motor to stimulate the local economy. *Fourth*, the existence of carica and purwacengenterprises has helped the government to reduce unemployment. The existence of these two plants which are location specific and have unique characteristics ismaking them increasingly popular as gifts for tourists and visitors to take home, and this has indirectly led to a rise in levels of community welfare, specifically for communities in and around the Dieng region.

4. Conclusion
The impact of economic glocalization of carica and purwaceng plants is the development of local economic institutions such as productive economic enterprises and cooperatives. In addition, economic glocalization in the system of local food production in Wonosobo also has the effect of increasing production capacity, product development, market expansion, local economy, and ecotourism.

5. Acknowledgment
The authors gratefully acknowledge to Sebelas Maret University for funding the research through Research Grant for Doctoral Dissertation (No. 1076/UN27.21/PM/2017).

6. References
[1] Santoso D J S Adiningsih and Heryadi 1989 N, S, P and K status of soils in islands outside Java. Sulphur fertilizer policy for lowland and upland rice cropping systems in Indonesia, Aciar Proceedings p 77-82
[2] Maynard Michael L 2003 From Global to Glocal: How Gillete's Sensor Excel Accommodates to Japan. *Keio Communication Review* **5** 57-75
[3] Suutarinen Tuomas 2015 Glocalisation of Global Market Forces and The Repositioning of a Peripheral Russian Mining Community. *Barents Studies: Peoples, Economics and Politics*. **13** 55-81

[4] Robertson R 2013 Situating glocalization: A relatively authobiographical intervention. In G S Drori M A Hollerer & P Walgenbach (Eds) Global themes and local variations in organization and management: Perspectives on glocalization pp 25-36 New York: Rountledge

[5]. Griborescu Adriana and Alexandra Saif 2017 The concept of glocalization and its incorporation in global brands' marketing strategies *International Journal of Business and Management Invention Romania*: National University of Political Studies and Public Administration **6** 1 70-74

[6] Patel Fay and Hayley Lynch 2013 Glocalization as an alternative to Internationalization in Higher Education: Embedding Positive Glocal Learning Perspectives. *International Journal of Teaching and Learning in Higher Education* **25** 2 223-230

[7] Lehmberg D & Hicks J 2018 A 'glocalization' approach to the internationalizing of crisis communication. *Business Horizons 61* 3 357-366 doi:10.1016/j.bushor.2018.01.002

[8] Suutarinen Thomas Glocalizaion of global market forces and the repositioning of a peripheral Russan mining community Barents Studies: People, economic and politics 1 3 55-81

[9] Daymon Christine dan Holloway Immy 2008 *Metode Riset Kualitatif dalam Public Relations dan Marketing Communications* Yogyakarta: Penerbit Bentang

[10] Yin Robert K 1987 *Case Study Research: Design and Methods*. Beverly Hills California : Sage Publication

[11] Creswell John W 2006 *Qualitative Inquiry and Research Design: Choosing among Five Approaches* Thousand Oaks CA: Sage

[12] Pawito 2007 *Penelitian Komunikasi Kualitatif* Yogyakarta: LKIS

[13] Moleong Lexy J 2009 *Metodologi Penelitian Kualitatif* Bandung: PT Remaja Rosdakarya

[14] Miles Matthew B dan Huberman A Michael 1992 *Analisis Data Kualitatif* (Terj. Tjetjep Rohendi Rohidi) Jakarta : Universitas Indonesia

[15] Hendro S 2005 Seri Agribisnis: Berkebun 21 Jenis Tanaman Buah Jakarta: Penebar Swadaya

IOP Conf. Series: Materials Science and Engineering **633** (2019) 012060 doi:10.1088/1757-899X/633/1/012060

Farming occupation in the views of farmer families in rural area

E Widiyanti[1,2], R Karsidi[3], M Wijaya[4] and P Utari[5]

[1]Communication of Post-Graduate School, Universitas Sebelas Maret
[2]Extension and Agricultural Communication, Faculty of Agriculture, Universitas Sebelas Maret
[3]Faculty of Teacher's Training and Education, Universitas Sebelas Maret
[4,5] Faculty of Political and Social Science, Universitas Sebelas Maret

E-mail: emi_widiyanti@student.uns.ac.id

Abstract. As main actors in food production system, farmers receive much of the world's attention due to the food crisis phenomenon happening all over the world. However, the attention to small-scale farmers tends to be neglected, especially regarding to the crisis of farmer regeneration. Thus it is interesting to understand how small-scale farmers and their families think of farming occupations. To comprehend the meanings underlying the farmer families' views on farming occupations, the researchers applied Hebert Blumer's symbolic interactionism in descriptive qualitative study using paradigm construction. The data were collected by interviews with a number of farmers and their families in Tegalmade Village, Mojolaban District, Sukoharjo Regency. The results showed that there are different views among older generation (farmers) and younger generation (farmers' children). The identified views were positive, neutral and negative. The differences in their views are caused by different interaction processes, leading to different action orientations. The young generation has positive views when interacting with their families and youth community engaged in agriculture especially in agriculture entrepreneurship. These positive views result in action orientations to work on the farming occupations. The older generation has positive views when interacting with fellow farmers, views that they want to continue to make farming as their main livelihood.

1. Background

Farmers, as the main actors in providing food for the community, have received great attention from the world, especially by the United Nations amid the food crisis. The UN acknowledgment could be seen at *The World Food Day* on October 16, 2012 on the topic "*Small-Scale Farmers as A Key to Feeding the World*", as well as on the publication of the United Nations' paper entitled "*Small Farmers Feed the World*". The United Nations has also drafted a Declaration on the Rights of Farmers in *Human Rights of Peasants and Other People Working in Rural Areas* as a constructive effort to address the problems of food crisis, poverty and rural marginalization [1].

The world's attention to farmer existence does not stop at the UN declaration. Many countries in the world have been conducting researches on the interests in the agricultural sector to formulate strategies to attract their people, especially the younger generations, to pursue agriculture in the context of farmer regeneration. Indications of the abandonment of agricultural sector by the younger generation have become common phenomena experienced by many countries in the world. Ben White tried to trace the indications that occurred in several countries such as Mali, Ethiopia, Egypt, India and Africa, found that there are indications of young people avoiding agriculture-based occupations and even sometimes choosing to be unemployed or underemployed rather than returning to rural areas [2]. Another research in Nigeria found that although Agricultural Science was one of the core subjects in the secondary school curriculum in the state of Kogi in Nigeria, his study on 600 youths from 12 secondary schools found that the youths generally have negative attitudes toward agriculture as a future occupation [3]. Other phenomenon on the youth's disinterest to stay in the rural areas and pursue farming occupations is caused by the view that farming occupations are not prestigious and farmers are not considered as successful people in their community [4].

Several studies that have been conducted in several countries show that the main problem in farmer regeneration is the negative view of the youths on the farming occupation. Thus it is very interesting to study on how the farmer families, in this case the farmers themselves and their children as the next generation, perceive and construct meanings regarding to farming occupations.

This study seeks to apply Herbert Blumer's symbolic interactionism approach to identify the views and meanings of the farmers and their families towards farming occupations. In Herbert Blumer's symbolic interactionism, individuals construct meaning through the communication process because meaning is not an intrinsic thing or idea; it takes other people to construct meanings. There are three assumptions in Hebert Blumer's symbolic interactions, namely: humans act on others on the basis of the meaning that others have, meaning is constructed in the interactions among people, and that meaning is modified through interpretation process. Blumer explained that there are three ways to look at the origin of meanings. The first approach is to consider meaning as an intrinsic thing, the second approach is to see meaning as something brought to an object by people who have the meaning of the object, and the third approach is considering meaning as a "social product" or "a creation constructed in and through people's defining activities when they interact "[5].

In the view of symbolic interactionism, when interacting, communicators not only interact with other people and with social objects, but also communicate with themselves when making decisions about how to act on the social object. Kuhn refers to this as an action plan guided by verbal attitudes or statements that show values towards actions [6]. Broadly speaking, it can be concluded that Blumer's main idea is that people act on something based on the object's meaning which is obtained through interaction processes.

This study examines on how meanings of the farmers and their families regarding to farming occupations are obtained through their interaction with their social systems and how their actions are oriented towards farming occupations in relation to the meaning construction process.

2. Research Method

This study applied qualitative method designed in the framework of a phenomenological approach. The determination of the research location was conducted purposively in the Tegalmade Village, Mojolaban Subdistrict, Sukoharjo Regency, with the following considerations: Sukoharjo Regency is one of the areas that experiences farmer crisis [7], Tegalmade Village of Mojolaban District is one of the agricultural areas that are starting to lose its agricultural land due to the construction of housing areas and industrialization, the existence of industrialization in this rural areas provides references to other jobs besides farming to the younger generation.

The data were collected through in-depth interviews with 6 informants consisting of two farmers, one farm worker and three young people who are farmers' children. To meet the data validity, two techniques from Lincoln and Guba [8]were used namely triangulation between in-depth interviews and observation data, as well as member check by peer-checking the data with other researchers to obtain the correct understanding or to find deficiencies to be corrected. The data analysis method in this phenomenology research was conducted in several stages [9]namely: the initial phase, the researchers described fully the experienced phenomena into written language; Horizontalization stage, the researchers inventorized statements - important statements relevant to the topic; meaning cluster phase, the researchers classified statements into themes or units of meaning; description of essence phase, the researchers constructed a comprehensive description of the meaning and experience essence of the subject and Reporting of the research results, the researchers provided more understanding regarding to the way someone construct meaning on a social object, which in this study was farming occupation.

3. Research Results and Discussion

The farmers in this study are rural farmers. Rural farmers, according to Wolf [10] are villagers who grow crops, in which, they grow crops and raise livestock in rural areas, not in enclosed spaces (greenhouses) in the city center or in aspidistra boxes. They are also not agricultural entrepreneurs. The views and meaning on farming occupations were focused on four things, namely income, social status, work comfort and old age insurance. These four factors were derived from Maslow's theory of needs and the two factors theory proposed by Frederick Herzberg [11].

The research results on the meaning of farming occupations identified three views namely negative, neutral, and positive as presented in the following table:

Table 1. Views of Farming Occupation by the Young Generation

View	Meaning	Process (interaction)	Result (action)
Negative	▪ Low and uncertain income ▪ Identical with low social status ▪ Uncomfortable occupation due to the physical environment and the characteristics of farming occupations	▪ Farmer-mate and school-mate Peer interaction ▪ Interaction with older farmer generation	▪ The expectation from the older generation that the young generation do not become farmers like themselves ▪ The expectation of the young generation to have an occupation with higher income and social status. ▪ The intention to work in other sectors due to work reference from other parties such as friends, acquaintances or teachers. ▪ Low opportunity to work as farmers
Neutral	▪ Farming is one of available occupation. ▪ Low social status can be changed by improving agricultural technology and land management	▪ Interacting with family ▪ Interacting with play-mate peer at home	▪ Continuing the parents work as farmer and managing the land inherited from the family ▪ Applying agricultural technology to improve income and social status ▪ The opportunity to work farmers
Positive	▪ The independence factor and authority to time management in farming occupations	▪ Interacting with family ▪ Interacting with friends who are engaged in farming and agropreneur.	▪ Making farming occupation as an alternative work choice because the job opportunity and requirement in the formal sector are becoming more and more difficult ▪ The high opportunity to work as farmers

Source: Primary Data Analysis, 2018

The young generation's views on farming occupation were mostly negative. These negative views were the result of their interaction with their peers or with schoolmates. The negative views resulted in action orientation to work in other sectors according to the references they got from the school environment or from their teacher. Whereas the negative views obtained from interaction with the older generation were simply because the older generation did not want the younger generation to become farmers like themselves. However, there were positive views of the younger generation when they interacted with the youth community who engaged in the world of agriculture and those who do farming and run business in agricultural commodity as well as entrepreneur farmers (agropreneur), for example, some of the young people joined a Whatsapp group of agropreneur, in which they obtain the positive views. One of the meanings of the positive views is that farming is a form of entrepreneurship and independence.

Unlike the younger generation, the older generation gave more positive views to the work of farming. They, in these views, gave meaning to farming as a comfortable occupation because it is their culture and they get to have the freedom to manage their time and resources. These views were obtained in their interaction with friends in their neighborhood who also depend on farming. The followings review the farming occupations by the older generation.

Table2. Views of Farming Occupation by the Older Generation

View	Meaning	Process (interaction)	Result (action)
Negative	▪ An occupation with many risks such as crop failure due to weather/climate changes	▪ Interacting with their children or other acquaintances	▪ Not orientating on income level but only on the fulfillment of the basic needs (food) ▪ Opportunity to abandon farming occupations when there are other occupations with higher income and due to increasing daily needs
Neutral	▪ Appropriate level of income to meet the daily needs	▪ Interacting with fellow farmers who live in rural area	▪ Limitations of education, skill, and age to access other occupations ▪ Opportunity to stay as farmers
Positive	▪ Comfortable occupation due to familiar socialization of work system since early childhood. ▪ Freedom in time and resource management. ▪ Reliability to meet the basic need (of food) up to old age	▪ Interacting with fellow farmers and fellow villagers, or other acquaintances and their families	▪ Optimizing available resources (land) ▪ Utilizing available spare time to work on other sectors such as construction worker, carpenter, etc. ▪ Keep farming occupation as the main livelihood

Source: Primary data analysis, 2018

There are various views in the farmers' families toward farming occupations, in which both the young generation (farmers' children) and the older generation (farmers) have various views on this work when viewed from the level of income, inherent social status, working comfort and old age insurance. The views that were identified were positive, neutral and negative. Each kind gives different meaning on the views. The meanings and views that occurred were resulted from different interaction processes which also produced different action orientations. This is in accordance with Herbert Blumer's statement that people act toward others on the basis of the meanings that other people have and meanings that are created in the interactions among them.

The younger generation has positive meanings and views when they interact with their own families and when they are in the youth community who are involved in agriculture business. These positive views will result in the young generation's orientation to work on the farming occupations. Likewise with the older generation or the farmers themselves, they will give positive meanings when they interact with fellow farmers in their environment, or when they have the sense of similar way of life and fate. In other words, when interacting with people who share the same fate, the farmers will continue to make farming occupation as their main job.

4. Conclusion

There are various views among the older generation (farmers) and the younger generation (farmers' children). The identified views are positive, neutral and negative. The differences in the views are the results of different interaction processes that lead to different action orientations. The younger generation has positive views and meanings when they interact with their families and in the community of young people who are engaged in agriculture especially agricultural entrepreneurship (agropreneur). These positive meanings and views will result in an action orientation to work on the farming occupations. In which the older generation or the farmers themselves have positive views and meanings when they interact with fellow farmers which means that they want to continue to make farming occupations as their main livelihood.

5. Suggestion

In order to encourage the young generation to be interested in farming occupations, a social process that will result in positive meaning among the young generation is needed. One of the social processes that can be pursued is strengthening youth activist groups or communities that engage in the field of agriculture especially agropreneur. More interactions that occur among young people in agricultural groups will further encourage them to be interested in agriculture.

Acknowledgment

The authors thank the participating farmers and all those who helped the research. This research was supported by the Communication of Post-Graduate School, Universitas Sebelas Maret.

References

[1] Syahyuti 2013 Pemahaman Terhadap Petani Kecil sebagai Landasan Kebijakan Pembangunan Pertanian Understanding the Small Farmers as the Basis for Agriculture Development Policy *Forum Penelitian Agro Ekonomi* **31** (1) 15-29

[2] White B 2012 Agriculture and the Generation Problem: Rural Youth, Employment and the Future of Farming *IDS Bulletin* **43** (6)

[3] Ibitoye S J 2011 Attitude of Youths towards Career in Agriculture in Kogi State, Nigeria. *Int. J. Appl. Eng.Res.* **6** (14) 1683-1693

[4] Paisley C 2014 Involving Young People in Agricultural Development: why it's critical for the sustainability of the sector *Palawija Newsletter* **31** (1)

[5] West R and Lynn H T 2017 *Pengantar Teori Komunikasi Pertanian Analisis dan Aplikasi Introducing Communication Theory Analysis and Application* (Jakarta: Salemba Humanika. Jakarta)

[6] Littejohn S W and Karen A F 2011 *Theory of Human Communication* (Long Grove: Wafeland Press)

[7] LIPI (2015) *Minat Bertani Generasi Muda Menurun, Indonesia Terancam Krisis Petani* (http://lipi.go.id/berita/minat-bertani-generasi-muda-menurun-indonesia-terancam-krisis-petani/10836)

[8] Lincoln Y S and Guba E B 1985 *Naturalistic Inquiry* (Beverly Hills: Sage Publications)

[9] Hasbiansyah, O 2008 Pendekatan Fenomenologi: Pengantar Praktik Penelitian dalam Ilmu Sosial dan Komunikasi *Jurnal MEDIATOR* **9** (1)

[10] Wolf E R 1985 *Petani Suatu Tinjauan Antropologis* (Jakarta: Rajawali)

[11] Purwanto D 2006 *Komunikasi Bisnis* (Jakarta: Erlangga)